Basic Electronics

for Engineers and Scientists

/Basic Electronics

for Engineers and Scientists

Russell E. Lueg

PROFESSOR OF ELECTRICAL ENGINEERING
UNIVERSITY OF ALABAMA

Erwin A. Reinhard

ASSOCIATE PROFESSOR OF ELECTRICAL ENGINEERING
UNIVERSITY OF ALABAMA

INTEXT EDUCATIONAL PUBLISHERS
College Division of **Intext**

Scranton   San Francisco   Toronto   London

The Intext Series in Electrical Engineering

Robert F. Lambert —— consulting editor

PROFESSOR OF ELECTRICAL ENGINEERING
UNIVERSITY OF MINNESOTA

Library of Congress Catalog Card Number 77-177299
ISBN 0-7002-2296-0

Dedicated to: Maurita and Irene

Preface

This electronics text has been written to provide instructors with a high degree of flexibility in its utilization. The first two chapters should be studied by students who have no background in electric circuits. Chapter 3 can be considered the beginning chapter for students who are familiar with such topics as loop-current analysis, node voltage methods, network reduction techniques; i.e., application of Ohm's Law and Kirchhoff's Laws. To understand the material in the first eleven chapters it is necessary that the reader be familiar with the solution of linear algebraic equations. These chapters easily contain sufficient material for a one-semester course in basic electronics.

Beginning with Chapter 12 it is necessary that the student be familiar with the solution of linear differential equations. If one is familiar with steady state a-c circuit analysis methods, Chapter 12 can be omitted; if one understands Laplace transform methods, Chapter 19 can be omitted. Thus this text can be used in a first course in Electrical Engineering either where electric circuits and electronics are both studied, or in an electronics course preceded by a course in electric circuits.

The authors suggest that instructors consider these three study options depending on the needs or desires of the student.

OPTION I. As a first course in electrical engineering:

   Chapters 1–11 in sequence

OPTION II. As a first course in electronics following a circuits course:

   Chapters 3–11 in sequence plus other chapters as desired

OPTION III. As a first course in electric circuits:

   Chapters 1, 2, 12, 19

The authors decided to make Chapters 12 and 19 more than just a quick survey of a-c circuit analysis and Laplace transforms. It is for this reason that the chapters listed in Option III provide a firm basis for an introductory course in electric circuits. Therefore, this text could be used as a two-semester sequence—the first course would be in electric circuits followed by a second course which would emphasize electronics.

The material in this text emphasizes the practical or macroscopic use of electronic devices and minimizes the microscopic or internal physics description of diodes, transistors, SCRs, vacuum tubes, etc.

The overall format is divided into three reasonably distinct areas. The first area (Chapters 1–11) covers such topics as electric circuit analysis methods, diodes, transistors (the BJT, FET, and MOSFET), vacuum tubes, amplifiers, digital logic circuits, and the design of practical electronic circuits. In the second area (Chapters 12–18) the intent is to acquaint the reader with a variety of topics that will provide him with some concept of such phenomena, devices, and systems as power supplies, filters, audio and frequency modulation, radio propagation, lasers, radio astronomy, the photoelectric effect, gas diodes, silicon controlled rectifiers, etc. Chapter 12 A-C Circuit Analysis Methods, Chapter 13 Frequency Response, and Chapter 14 Nonlinear Graphical Analysis are considered a part of this second area. The authors generally assign a portion of this material either during the semester or at the end of the semester to broaden the vision of the student in regard to the wide spectrum of electronic devices that are currently in use and to some of the analytical methods that are used to solve the varied problems encountered in electronic circuits and systems. The third area Chapters 19–21 covers analog computers, feedback, and servomechanisms, with Laplace transforms introduced as a working tool. Although the analog computer is introduced as a tool for solving differential equations, emphasis is also placed on the use of the computer or the use of operational amplifiers for synthesizing servo transfer functions.

This text is used in a four-credit-hour course at the University of Alabama (44 one-hour lecture and 15 three-hour laboratory periods). The first ten chapters are covered in detail (the design examples of Chapter 6 are discussed in the laboratory) during the semester. In addition Chapter 11 along with portions of two or three of the survey chapters (Chapters 15, 16, 17, or 18) are discussed briefly at the option of the instructor and as time permits. This introductory course in electronics and resistive electric circuits is required of all students in the college of engineering (an occasional nonengineering student takes the course) and about half of the students take the course at the beginning of their sophomore year.

This text has been used in multilith form for several semesters at the University of Alabama. The authors are thus indebted to the many students who graciously accepted and endured the difficulties of studying from such copy. The authors are further indebted to the other instructors who used this manuscript, and to Dr. O. P. McDuff, Head of the Department of Electrical Engineering, and Dr. W. E. Lear, Dean of the College of Engineering, for their active encouragement and support during the writing of this text.

Russell E. Lueg
Erwin A. Reinhard

Tuscaloosa, Alabama
October, 1971

Contents

xiv   /   *Contents*

1 / *Basic Concepts*

In 1883 Thomas A. Edison noted during his experiments on incandescent lamps that a current flowed between a heated filament and a positively charged metal plate. However, it wasn't until 1897 that use of the "Edison Effect" was made by J. A. Fleming at England. Fleming used his new two-element device, or diode, to rectify received radio signals. The rectification aided greatly in the reception of information from radio waves. Since Edison and Fleming, the field of electronics has undergone and is undergoing a spectacular development. From the triode vacuum tube of 1907 of Lee DeForest, which opened the door to the electronic age by making electronic amplification of electrical signals possible, to the micro and molecular circuits of today, which allow complicated electronic circuits to be fabricated in their entirety on a chip the size of your fingernail, one can only speculate as to what the future holds in store for the imaginative designer of electronic circuits. Today one finds electronics playing an integral and highly important role in our way of life, through systems such as: 1) data processing equipment, digital and analog computers, compact electronic desk calculators; 2) communication equipment, a very sophisticated telephone network, radio and television receivers and transmitters, weather satellite and space vehicle telemetry equipment; 3) control and guidance equipment, autopilots, guided missiles and vehicles, automated industrial processes; and 4) home and personal equipment, hi-fidelity stereo sets, medical electronic devices such as the pacemaker, automobile ignition systems, speed control of appliance motors, light illumination control. A detailed listing of how and where electronics is being used would be a text in itself.

The reasons for the popularity of electronic devices are reliability, flexibility, reproducibility, small size, low cost, and low energy consumption. Because of the reliability and reproducibility of electronic components and devices such as transistors, diodes, and integrated circuits, the designer of electronic systems (such as amplifiers, computers, communication systems, etc.) can accurately predict the performance of an actual system by using well known analytical circuit methods. Needless to say, confidence in predicting the performance of hardware through theoretical calculations alone is of utmost importance to the engineer since he can design a system to meet a specific application before building a prototype.

It is also interesting to note that recent advances in microelectronics have served to improve reliability, reproducibility, and performance criteria and to reduce energy consumption, cost, and size. Usually improvement in one direction tends to be offset by disadvantages in another when a product is changed, but in the case of electronics almost every improvement seems to be unqualifiedly good.

To understand how an electronic device or system operates, it is necessary for one to be familiar with electric circuit analysis techniques. Chapter 2 is devoted to analysis techniques applied to linear resistive networks while this chapter is concerned with the basic electrical quantities of voltage, current, energy, power, and a description of the resistance circuit parameter.

1-1. The RMKS System of Units

There are four fundamental units in any system of measurements; that is, length, mass, time, and charge. In the rmks (rationalized meter, kilogram, second) system of units, length is measured in meters, mass in kilograms, time in seconds, and charge in coulombs. Compared to the English system of units, the meter is equivalent to 39.37 in., the kilogram to 2.22 lb avoirdupois, and the coulomb to the accumulated charge of 6.24×10^{18} electrons.

1-2. Circuit Variables

Only steady state, linear circuit analysis methods as applied to resistive circuits will be considered in this chapter. In electric circuits (the concept of a circuit is discussed in the next chapter) three variables are of prime importance: current, voltage and power. Perhaps the most elegant way of describing or defining these quantities would be to do so in terms of electric field theory but a simpler approach will be used here. Man is constantly trying to understand more about the nature of physical phenomena, and to describe his observations of the various phenomena in precise terms. Models, diagrams, descriptions and mathematics, which are the engineer's and scientist's technical language, are all used to aid man in describing his physical world. Once man understands the laws of nature and then writes these laws in the technical language of mathematics he is able to utilize these laws and thus build many practical and useful machines and devices. For the engineer to be able to predict the behavior of natural phenomena or a man-made machine from theoretical considerations is invaluable. If man were not able to describe his observations in such a way that he could use the underlying laws of nature in a logical and intelligent fashion the startling technological revolution of this century would not be possible.

Man must be precise in describing the laws of nature and frequently he finds it necessary to define certain basic terms, which then represent the starting point of engineering analysis and design. To emphasize the man-made nature of certain technical concepts consider the definition of the basic unit of electrical current (the ampere) as defined by an act of Congress in 1894: "The unit of current shall be what is known as the international ampere and is the practical equivalent of the unvarying current (direct current or d-c), which when passed through a solution of nitrate of silver in water in accordance with standard specifications, deposits silver at the rate of 0.001118 of a gram per second."

By way of further definition the coulomb is the amount of electric charge conveyed by one ampere in one second. Since it is recognized that electrons are the "fluid" in motion in our electrical conductor, it turns out that it takes 6.24×10^{18} electrons per second to cause the 0.001118 gram of silver to be deposited as stated in the preceding paragraph.

The electromotive force which energizes the electrons and causes them to move is usually defined in terms of voltage. Voltage is always defined as a difference in potential energy between two points. Specifically, this potential energy difference is defined as the work or energy in joules that is expended in moving a unit positive charge, one coulomb, from a low electrical potential energy point to a higher electrical potential energy point. Voltage is analogous to the gravitational potential energy difference, which is defined as the work or energy change in joules in moving a unit mass, one kilogram, between two spatial points of different heights (with respect to some common reference or datum point). An electrical charge placed between two such points (assuming that an electrical potential energy difference exists between the two points) will be urged to move toward one point or the other depending on the sign of the electrical charge. More discussion on the direction of motion will be given later. The force urging these charges to move is termed voltage. The international volt is defined as 1/1.01830 of the voltage of a normal Weston cell.

The watt is the unit of electrical power and is defined as the conversion of electrical energy (conversion to some other form, e.g., thermal as in a toaster or heater) at the rate of one joule per second. Thus if one coulomb per second (one ampere) moves through a potential difference of one volt (recall that *one* joule of energy is involved in the transfer of *one* coulomb of charge through a one *volt* electrical potential difference) the resulting unit is termed power and has the dimensions of one joule per second or one watt.

SUMMARY OF ELECTRICAL CIRCUIT TERMINOLOGY

Ampere.   The ampere is the basic unit of electrical current and is the rate of flow of electrical charges expressed in coulombs per second,

$$I = \frac{\Delta Q}{\Delta t} \quad \text{coulombs/sec or amps} \qquad (1\text{-}1)$$

where ΔQ = total flow of charge in coulombs in a time interval of Δt seconds.

EXAMPLE 1-1.   If in a uniform stream of electrons, it is observed that 2×10^{20} electrons flow past a given point every 3 seconds, what is the current in amperes past this point?

Solution: Using Eq. 1-1 and the fact that there are 6.24×10^{18} electrons in a coulomb of charge,

$$I = \frac{\Delta Q}{\Delta t} = \frac{\dfrac{2 \times 10^{20} \text{ electrons}}{6.24 \times 10^{18} \dfrac{\text{electrons}}{\text{coulomb}}}}{3 \text{ sec}} = 10.7 \frac{\text{coulombs}}{\text{sec}} = 10.7 \text{ amps}$$

Volt.   The volt is the basic unit of electromotive force and is the work or energy involved in moving electrical charges from one point to another expressed in joules per coulomb.

$$V = \frac{\Delta W}{\Delta Q} \text{ joules/coulomb or volts} \qquad (1\text{-}2)$$

where ΔW = total energy in joules involved in moving ΔQ coulombs of charge.

EXAMPLE 1-2.   It is known that it takes 60 joules of energy to move 5 coulombs of charge between two points. What is the potential difference in volts between these two points?

Solution: Using Eq. 1-2,

$$V = \frac{\Delta W}{\Delta Q} = \frac{60 \text{ joules}}{5 \text{ coulombs}} = 12 \frac{\text{joules}}{\text{coulomb}} = 12 \text{ volts}$$

Watt.   The watt is the basic unit of electrical power and is the rate of energy conversion involved in moving electrical charges from one point to another expressed in joules per second,

$$P = \frac{\Delta W}{\Delta t} \text{ joules/sec or watts} \qquad (1\text{-}3)$$

where ΔW = total energy in joules converted in a time interval of Δt seconds. Note that if Eqs. 1-1 and 1-2 are substituted into Eq. 1-3,

$$P = \frac{\Delta W}{\Delta t} = \frac{V \cdot \Delta Q}{\Delta t} = V \left(\frac{\Delta Q}{\Delta t} \right) = V \cdot I \qquad (1\text{-}4)$$

which gives the power in terms of the easily measured quantities of voltage and current.

EXAMPLE 1-3.   Find the power converted in moving the 5 coulombs of charge in Example 1-2 if the time required to move the charges between the two points is a) 2 seconds b) 0.02 seconds.

Solution: Using Eq. 1-3,

a) $P = \dfrac{\Delta W}{\Delta t} = \dfrac{60 \text{ joules}}{2 \text{ sec}} = \dfrac{30 \text{ joules}}{\text{sec}} = 30 \text{ watts}$

b) $P = \dfrac{\Delta W}{\Delta t} = \dfrac{60 \text{ joules}}{0.02 \text{ sec}} = \dfrac{3000 \text{ joules}}{\text{sec}} = 3000 \text{ watts} = 3 \text{ kilowatts}$

As expected a larger power is obtained when energy is converted in a shorter time, that is, at a higher rate. Another way of obtaining the solution to this problem is to use Eq. 1-4. Note that $V = 12$ volts.

a) $P = V \cdot I$

$= 12 \text{ volts} \left(\dfrac{5 \text{ coulombs}}{2 \text{ sec}} \right) = (12 \text{ volts})(2.5 \text{ amps})$

$= 30 \text{ volt-amps} = 30 \text{ watts} \qquad (check)$

b) $P = V \cdot I$

$= 12 \text{ volts} \left(\dfrac{5 \text{ coulombs}}{0.02 \text{ sec}} \right) = (12 \text{ volts})(250 \text{ amps})$

$= 3000 \text{ volt-amps} = 3000 \text{ watts} \qquad (check)$

Note that *one* volt-ampere is equivalent to *one* watt. This is independent of the voltage and current magnitudes as long as their product is one, that is, one volt × one ampere, 100 volts × 0.01 ampere, 0.5 volt × 2 amperes all result in one watt of power.

1-3. Symbolic Notation

Even though the discussion up to this point has been quite correct it has not been complete. As mentioned earlier man must evolve precise models and mathematical formulations for observed physical phenomena to avoid unnecessary and undesirable mistakes and confusions. Electrical charges come in two varieties—those with positive and those with negative signs. In addition electrical energy can be considered as gained or lost; and finally, currents, voltages, and power can be considered as time varying as well as time invariant quantities.

Because current flows from one point to another and because voltages are defined *between* two points and never at just one point, it is desirable that double subscript notation be used to emphasize which two points are under consideration. For now the letters *a* and *b* arbitrarily refer to any two points in space; that is, I_{ab} means the current from point *a* to point *b* via a specified path.

Current. Before the advent of the atomic theory of matter scientists (such as Benjamin Franklin) decided that electric current was caused by a flow of positive charges and even though it is now known that electrons are the "fluid" in a conducting circuit, convention still dictates that *conventional* current flow is due to positive charges. The symbol for an average or direct current (non-time-varying) is I and for an instantaneous current (time varying) the symbol is i. Using double subscript notation,

$$\left.\begin{array}{l} I_{ab} = - I_{ba} \\ I_{ab} = - I_{ab} \text{ (electron)} \\ I_{ab} = I_{ba} \text{ (electron)} \end{array}\right\} \tag{1-5}$$

where I_{ab} (electron) means a flow of negatively charged electrons instead of a flow of positive charges. A reversal of all subscripts as well as the substitution of lower case i for upper case I results in equally valid identities.

Voltage. Since the terms "voltage drop" and "voltage rise," meaning potential energy drop and rise, respectively, appear often in this text, much confusion will be eliminated by choosing symbols that differentiate between the quantities. Voltage drops will hereafter be designated by the symbol V or v and voltage rise by E or e. Using double subscript notation,

$$\left.\begin{array}{l} V_{ab} = - V_{ba} = E_{ba} = - E_{ab} \\ E_{ab} = - E_{ba} = V_{ba} = - V_{ab} \end{array}\right\} \tag{1-6}$$

The symbol V_{ab} refers to the voltage drop *from* point a *to* point b while E_{ab} refers to the voltage rise from point a to point b. One might refer back to the gravitational potential energy analog. If the spatial point a was two meters above spatial point b, the potential energy drop per unit mass from point a to point b, say PED_{ab}, would be 2 meter-kilogram/kilogram (or meter). If one insisted in talking about the potential energy rise from point a to point b, say PER_{ab}, then PER_{ab} would be -2 meters. In a similar manner, PED_{ba} would be -2 meters and PER_{ba} would be $+2$ meters. So either a drop or a rise can be used as the unit of measure from one point to another point independent of whether an actual rise or drop, that is, which one is positive, exists from the first point to the second point. If equations for such systems are written properly, the mathematics takes care of the signs and the sign of the result can be properly interpreted from whatever viewpoint (rise or drop) taken. Many different schemes are used to indicate whether a voltage drop or a voltage rise is being indicated and Fig. 1-1 shows several common graphical notations which are all equivalent. It is noted that in addition to double subscripts, polarity markings and arrows (in conjunction with a voltage drop or voltage rise symbol) can convey the same information. Lower case v's and e's can be substituted for the upper case V's and E's respectively, without loss of validity in Eq. 1-6.

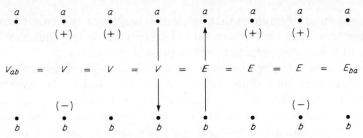

FIG. 1-1. Graphical notation used for voltage from one spatial point
to a second point.

Power. Power in the conventional sense is defined as the product of
the voltage drop V and the conventional current I or

$$P_{ab} = V_{ab} \cdot I_{ab} \qquad (1\text{-}7)$$

which is identical to Eq. 1-4 except double subscripts have been added.
Equation 1-7 represents the electrical power being *lost* (electrical energy
converted to some other form) by the device or element between points
a and b. If Eq. 1-7 results in a negative quantity, then electrical power is
being *gained* (energy converted to electrical form) by the element or device
between points a and b. Note that in view of Eqs. 1-5 and 1-6,

$$P_{ab} = V_{ab}I_{ab} = (-V_{ba})(-I_{ba}) = V_{ba}I_{ba} = P_{ba}$$

which is expected since the electrical power lost or gained between two
points does not have an associated spatial direction.

Instantaneous power can be written as

$$p_{ab} = v_{ab} \cdot i_{ab} \qquad (1\text{-}8)$$

Additional examples on how to view and use correctly the defined
concepts of current, voltage and power will be presented in the next section which is primarily concerned with a discussion of Ohm's law and the
resistance parameter.

1-4. Resistance Circuit Parameter

In 1826, Georg Ohm published a paper showing that the direct (non-
time-varying) current through a conductor varied in direct proportion to
the voltage applied across the terminals of the conductor. The constant
of proportionality between the voltage and the current is a constant which
is now called the resistance of a piece of material.

Since the technical language of science or engineering is mathematical, it is necessary to describe an actual or practical component or device
in terms of a model which, in turn, can be couched in mathematical terminology. In electronic circuits it is also necessary to model the diodes,
transistors, and vacuum tubes so that they, in turn, can be included in the
mathematical analysis process. The modeling of these active circuit elements will be presented in later chapters.

Although it is assumed that the reader is already somewhat familiar with the term resistance, a qualitative, descriptive review concerning the nature of this circuit parameter will be presented next.

Resistance. The resistance parameter is simply a constant of proportionality between voltage and current (Ohm's Law). To understand adequately the nature of resistance, one must delve into the microscopic behavior of matter; however, the concept of resistance, as observed by Ohm, is predominantly a macroscopic phenomenon. Although the gaseous forms of matter, including ionized and nonionized gases, are important in electrical investigations, only matter in solidified form is usually considered in discussing the circuit quality called resistance. A brief, simplified discussion of solid-state phenomena will aid in explaining the nature of resistance.

Solid matter consists of a structure which is either crystalline or amorphous in nature. Amorphous solids, such as glass, do not have an easily understood atomic or molecular array and will not be discussed here. Crystalline solids have been studied in detail, and the basic structure is known to be a latticelike array of atoms. Metals such as copper, aluminum, nickel, and so on, as well as other materials (such as carbon) used in fabricating electrical resistive components are good examples of matter in crystalline form. If one visualizes a three-dimensional lattice structure with imaginary lines forming the lattice array, the individual atoms which make up the material are situated at the intersections of the lines (called lattice points). Most materials have a rather complicated lattice structure, but common table salt (NaCl) has an easy-to-visualize cubelike array of alternative sodium and chlorine ions. The lattice array for NaCl is called a face-centered cubic (FCC). It is sketched in Fig. 1-2.

An important fact to understand is that the ions remain essentially fixed or immobile at the lattice points, although at temperatures above absolute zero ($0°K$) there is a small thermal vibration about the equilibrium point. If too much energy is imparted to the material, the ions may

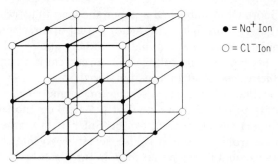

$\bullet = Na^+$ Ion

$\circ = Cl^-$ Ion

FIG. 1-2. The face-centered cubic lattice structure of NaCl.

move about so violently as to cause the material to liquefy or even to vaporize into the gaseous state. If Fig. 1-2 depicted the crystalline structure of a good conductor (dry NaCl is a good insulator), an important addition to the sketch would have to be made; that is, a "cloud" of free electrons would have to be shown drifting through the lattice structure. This does not mean that the piece of material is electrically charged, for the electron charge is exactly counterbalanced by the positive charges on the lattice ions. As might be expected, the free-electron cloud is the primary reason current can flow in a conductor.

A reasonably accurate qualitative description of resistance can be obtained by a consideration of the free-electron theory for good conductors. The free-electron model assumes that a large number of valence electrons are free to move unimpeded. From a classical standpoint, the electrons collide with the atoms or ions in the lattice array, and this, in turn, impedes the electrons' movement and causes a loss of energy. This atom-electron interaction causes a friction or resistance to electron flow to develop. From the more modern wave-mechanical standpoint, the electrons (sometimes called an electron wave) are reflected by the potential-energy barriers of the atomic array, with the electrons, in turn, losing energy to the recoiling atoms.

The atomic theory of matter was unknown in the days of Ohm. He simply observed that the voltage across and the current through a conductor were related by a constant. Ohm's law is usually expressed as

$$v_{ab} = i_{ab} \cdot R \qquad (1\text{-}9)$$

or

$$i_{ab} = \frac{1}{R} v_{ab} = G \cdot v_{ab} \qquad (1\text{-}10)$$

where R, called the resistance of the material, has the dimensions of ohms, $G = 1/R$ has the dimensions of mhos (mho is ohm spelled backward), and G is called the conductance of a material. Equation 1-9 is general, in that it relates a time varying voltage and time varying current. In the case of direct current (d-c), that is, non-time-varying voltages and currents, the lower case letters are replaced with upper case letters as

$$V_{ab} = I_{ab} \cdot R \qquad (1\text{-}11)$$

The physical element which has the quality called resistance is termed a resistor. The usual symbol for a resistor is -\/\/\-, and this is shown in Fig. 1-3. Figure 1-3 also illustrates the proper polarity for v and the direction for i in order for R (in Eq. 1-9) to be a positive constant. The conventional current i must be chosen in the direction of the voltage drop v across the resistor. If i is reversed, then Eq. 1-9 becomes

$$v_{ab} = -i_{ba} \cdot R$$

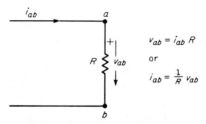

FIG. 1-3. The descriptive symbol for the resistance parameter R, and the usual voltage-current relationships.

Resistance Formula. The electrical resistance of a conductor of length l with uniform cross-sectional area A is usually expressed in one of two ways

$$R = \rho \frac{l}{A} \text{ ohms} \tag{1-12}$$

or

$$R = \frac{1}{\sigma} \frac{l}{A} \text{ ohms} \tag{1-13}$$

where ρ is the resistivity of the material, in ohm-meters, and σ is the conductivity of the material, in mhos per meter. Comparison of Eqs. 1-12 and 1-13 shows that

$$\rho = \frac{1}{\sigma} \tag{1-14}$$

Equation 1-12 or 1-13 should be intuitively obvious in that the resistance of a resistor should be directly proportional to its length and inversely proportional to its area. (The flow of electrons in a wire is analogous to the flow of water in a pipe.)

For annealed copper at room temperature (20°C), macroscopic measurements show that $\rho = 1.72 \times 10^{-8}$ ohm-meter or $\sigma = 5.82 \times 10^7$ mhos/meter. In calculating the resistance of a resistor, one usually makes use of either Eq. 1-12 or Eq. 1-13 and obtains the proper value for ρ or σ from data such as those given in Table 1-1.

TABLE 1-1

Metal	Resistivity at 20°C Ohm-Meters
Copper, soft-drawn	1.76×10^{-8}
Copper, standard annealed	1.72×10^{-8}
Aluminum, commercial hard-drawn	2.83×10^{-8}
Gold	2.44×10^{-8}
Platinum	10.1×10^{-8}
Silver	1.63×10^{-8}
Steel, carbon	$35 \quad \times 10^{-8}$ (approx.)

EXAMPLE 1-4. Determine the resistance and conductance of a solid, soft-drawn copper wire 0.1 in. in diameter and 50 ft long. Assume that the wire is at room temperature of 20°C.

Solution: From Table 1-1, the resistivity ρ is 1.76×10^{-8} ohm-meter at 20°C. Using Eq. 1-12,

$$R = \rho \frac{l}{A} = \frac{1.76 \times 10^{-8} \times 50 \times 12 \times 2.54 \times 10^{-2}}{\left[\dfrac{\pi \times (0.1 \times 2.54 \times 10^{-2})^2}{4}\right]}$$

$$= \frac{1.76 \times 10^{-8} \times 6 \times 2.54 \times 4}{\pi \times 10^{-6} \times (2.54)^2}$$

$$R = 0.0529 \text{ ohm } (\Omega)$$

The conductance is given by $G = 1/R = 1/0.0529 = 18.9$ mhos ($\mho$).

Quite obviously, the product of the resistivity and conductivity of a conductor is unity, and one term is the reciprocal of the other. Perhaps not so obvious is the fact that the resistance or conductance of a metal is greatly affected by temperature, by the metallurgical condition of the metal, and by impurities in the metal.

Needless to say, one could spend a great deal more time studying the resistive properties of metallic materials, but this is not necessary in order to understand the remaining material in this text, which is concerned with the analysis of electronic circuits composed of discrete components.

1-5. Energy and Power

In this section, the energy and power relationships in the resistor will be discussed. Energy is, perhaps, the most basic concept of man's knowledge of the physical world. Heat is a form of energy, as is mechanical motion, but electric energy is of fundamental importance in the study of electric circuits.

Power, as defined by Eq. 1-4, is the rate of doing work, and it is repeated here as

$$P = \frac{\Delta W}{\Delta t}$$

In terms of the instantaneous voltage and current, power is written as (see Eq. 1-8)

$$p_{ab} = v_{ab} \cdot i_{ab} \text{ joules/sec, or watts} \tag{1-15}$$

The power and energy relationships of the resistor circuit parameter will now be developed, using the R parameter voltage-current relationships of the preceding section.

The power in a resistor is obtained by using Ohm's law ($v_{ab} = i_{ab} \cdot R$)

and Eq. 1-15 or

$$p_{ab} = (i_{ab} \cdot R)\, i_{ab}$$
$$= (i_{ab})^2 \cdot R \tag{1-16}$$

or

$$p_{ab} = v_{ab} \cdot \left(\frac{v_{ab}}{R}\right)$$

$$= \frac{(v_{ab})^2}{R} \tag{1-17}$$

Some observations from Eqs. 1-16 or 1-17 can be made. Power is always positive (electrical energy converted to some other form) since the square of the electrical quantities (current or voltage) is involved. This energy loss in a resistor is frequently termed the "i squared R loss" ($i^2 \cdot R$) without regard to current direction for power calculations. Again, the relationships are true for arbitrary time varying voltages and currents. For the d-c case, upper case letters are used.

The electric energy given to a resistor is lost in the form of heat, and the conversion process is irreversible (as demonstrated above mathematically) because the heat cannot be converted into electric energy by the resistor.

1-6. Power Sources

A power source is a source of electric energy. The most common power source is the voltage generator. An ideal voltage source or generator maintains a constant voltage across its terminals, regardless of the current demanded from the source. The ideal voltage source does not have the qualities of resistance and, therefore, has zero internal voltage and power losses. Although, in practice, there is no such device as an ideal voltage source, it is convenient, in theoretical analyses, to use ideal sources. This may be justified by considering Fig. 1-4.

A practical voltage source is generally designed to deliver any value

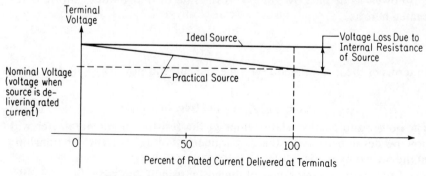

FIG. 1-4. Behavior of a practical voltage source.

of current up to some maximum determined by many design factors, such as size, efficiency, heating limitations, and so on. Over this operating range, the terminal voltage will differ from that of an ideal voltage source by the amount of voltage loss in the internal resistance of the source, as shown in Fig. 1-4. In a well-designed source of general application, this loss will amount to less than 2 or 3 percent of the nominal voltage when the source is delivering its maximum rated current, and it will be still smaller when lesser amounts of current are being delivered. Correspondingly, little error is introduced by neglecting this voltage loss and considering the source to be ideal. If this loss is to be taken into account, it is still common practice to consider the voltage source as being ideal but as having an internal resistance in series with the ideal voltage source ahead of the available external terminals, as depicted in Fig. 1-5. The volt-

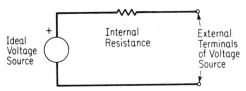

FIG. 1-5. Practical voltage source representation.

age appearing at the external terminals will then be the voltage of the ideal source (which is constant) minus the internal voltage loss. This loss is a function of the current being delivered through the internal series resistance. For a well-designed voltage source, the internal series resistance will be small (to keep voltage loss low) so, quite frequently, it will be negligible when compared to the resistances comprising the network utilizing the energy of the voltage source. Often this is sufficient justification for neglecting the internal resistance thus considering it to be zero and replacing it with zero resistance, that is, a short circuit.

An ideal current source is a power source which maintains a constant output current, regardless of the voltage across its terminals. In practice, the current source is not nearly so commonly used as is the voltage source, but it must be considered as an important concept, from both a practical and a theoretical standpoint. The ideal current source has no resistive quality and, therefore, has zero internal current and power losses. An examination of a practical current source, similar to that for the voltage source, can be made by studying Fig. 1-6.

A practical current source is generally designed to operate in the neighborhood of some rated voltage in which the current loss will be some few percent of the nominal current. Little error is introduced by neglecting this current loss and assuming the current source to be ideal over this operating range. If this loss is to be taken into account, the common

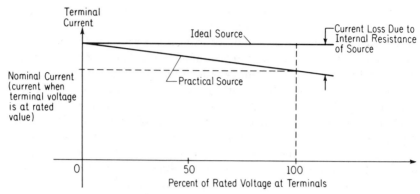

FIG. 1-6. Behavior of a practical current source.

practice is still to consider the current source as being ideal but as having an internal resistance in parallel with the ideal current source to account for the current loss ahead of the available external terminals, as depicted in Fig. 1-7. The current appearing at the external terminals will then be

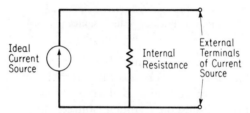

FIG. 1-7. Practical current-source representation.

the current of the ideal source minus the current loss through the internal parallel (shunt) resistance. A well-designed current source will have a large internal shunt resistance, so, quite frequently, it will be orders of magnitude greater than the resistances comprising the network utilizing the energy of the current source. In such cases, it may justifiably be considered to be infinity (∞) and can be replaced by an infinite resistance; that is, delete the internal parallel resistance from the circuit, leaving the parallel or shunt path open.

Power sources applied to resistor combinations constitute a circuit and it now remains to develop techniques to find the resultant voltages and currents. The next chapter presents several alternative methods of solving circuits.

Problems

Current, Voltage, and Power

1-1. A microammeter reads a current of 12 microamperes (μa) in a circuit. How many electrons per second is this?

1-2. If a current flow of 800 amps exists for $\frac{1}{4}$ sec in a series circuit, how many electrons pass through any point in the circuit?

1-3. A cubic meter of copper contains about 8.5×10^{28} free or valence electrons. If a No. 10 copper wire carries a rated current of 40 amps, what is the average velocity of the electrons as they drift through the wire? (Number 10 copper wire has a diameter of 0.102 in.)

1-4. A lead-acid 12 volt storage battery (such as used in an automobile) has a 120 ampere-hour rating. As practically all batteries have a nominal rating based on an 8 hour rate of discharge, such a battery can deliver 15 amperes continuously for 8 hours before being exhausted or almost fully discharged. a) What power does the battery deliver if it delivers 15 amperes? b) What energy does the battery deliver over the full 8 hour discharge period. c) What energy does the battery deliver if it delivers 220 amperes for 3 minutes (this is a typical current delivered by such a battery when one starts a car)? You may assume that the battery terminal voltage remains constant for this problem.

1-5. Repeat Prob. 1-4 if two such batteries are connected in series.

1-6. Repeat Prob. 1-4 if two such batteries are connected in parallel.

1-7. If electric energy costs 3 cents per kilowatt-hour how much does it cost to burn a 120 volt, 100 watt light bulb for 24 hours?

1-8. A kilowatt-hour is a standard way of measuring electric energy. How many joules are there in a kilowatt-hour?

1-9. A typical rate schedule for electric power delivered to a residence is as follows:

$1.25 for the first 20 Kw-hrs
3¢ per Kw-hr for the next 40 Kw-hrs
2¢ per Kw-hr for the next 240 Kw-hrs
1.2¢ per Kw-hr for all over 300 Kw-hrs

Using this rate schedule how much would it cost to burn a 120 volt, 100 watt light bulb continuously for 30 days?

1-10. Using the rate schedule in Prob. 1-9 how much would it cost to run a 120 volt, one horsepower pump motor continuously for 30 days? One horsepower = 746 watts.

1-11. How much current does it take to run a $\frac{3}{4}$ horsepower, 120 volt motor? (*Hint:* see Prob. 1-10.)

Resistance

1-12. What is the resistance of a bus bar of standard annealed copper at 20°C which is 5 in. wide by $\frac{1}{2}$ in. thick and 40 ft long? What is the conductance of this bar?

1-13. What must be the width of an aluminum bus bar that is $\frac{3}{4}$ in. thick and 40 ft long if the bar must have the same resistance as the copper bus bar of Prob 1-12?

1-14. A certain 1 in. diameter cable has a resistance of 0.0175 ohm per 1000 ft of length. Of what is the cable made?

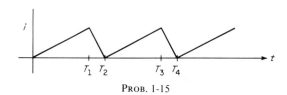

PROB. 1-15

1-15. For the waveform in the accompanying figure, i represents the current through a resistor. Sketch the waveform of the voltage across the resistor.

1-16. What is the power in a resistor of 1000 ohms carrying a current of 0.25 amp?

1-17. How much energy is dissipated by the resistor in Prob. 1-16 if the current flows for 1 day? If energy costs 3 cents per kilowatt-hour, how much does the electricity cost?

Sources

1-18. If the lead acid battery of Prob. 1-4 consists of 6 individual cells connected in series with each cell having an internal resistance of 0.01 ohm, draw the circuit model that describes this battery as a practical voltage source.

1-19. If the lead acid battery of Prob. 1-4 has an internal resistance of 0.06 ohm, a) how much energy is lost in the battery over the 8 hour nominal discharge period? b) what is the terminal voltage of the battery? c) how much energy is delivered to the load?

2 / Linear Resistive Circuit-Analysis Techniques

The analysis and design of electronic systems is generally a three step operation. First an equivalent circuit representation of the electronic device is sought. The equivalent circuit parameters will usually be a function of the region over which the device is operated and are *not* valid under all conditions of voltage, current, power, frequency, temperature, etc. However, the equivalent circuit representation permits a large body of knowledge (circuit theory) to be brought to bear on the analysis and design problem. The second step has to do with finding an analytical solution for the problem using these various circuit analysis techniques. The third step is to construct the system according to the analytical solutions and to verify experimentally the analytical predictions by direct measurement.

This chapter is concerned with the second step. The objective is to present the two most general methods of hand analysis, namely, the loop current and node voltage methods of circuit analysis. It should be stated that even though only resistive networks which are energized by direct currents (d-c) are considered, the laws and methods are easily extended to more general networks. This will be done in later chapters. It is also recognized that many important (from a calculation time-saving viewpoint) network theorems could be presented but time and objectives preclude such presentations. Thevenin's and Norton's theorems will be included because of their wide applicability in simplifying a network for certain applications.

The techniques of circuit analysis are many and varied, but they all rest firmly on the foundation of Ohm's and Kirchhoff's laws. Ohm's law

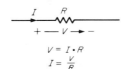

$$V = I \cdot R$$
$$I = \frac{V}{R}$$

FIG. 2-1. Ohm's law giving the voltage-current relationships for a resistor.

defines the voltage-current relationship (developed in Chapter 1 and given in Fig. 2-1) governed by the properties of a single element, such as resistance, whereas Kirchhoff's laws enable one to write the equations for various combinations of the elements comprising a network, which is the first and most important step in solving for unknown circuit voltages and/or currents.

2-1. Geometrical Representation

Unfortunately, before one can proceed with the more interesting applications of an element's voltage-current relationship in actual circuit applications, it is necessary to study some basic circuit terminology, definitions, and concepts. Although it is sometimes tedious—perhaps even dull and boring—to study and memorize terminology and definitions, the authors earnestly hope that the reader will attempt diligently to master the material in this section. A few hours extra study on the geometry and terminology of networks should assist greatly in the understanding, appreciation and utilization of the circuit-analysis methods and applications which follow.

The first question is: What is meant by an electric circuit? An electric circuit (often called an electric network) consists of a simple or complex combination of electric components such as resistors, capacitors, inductors, motors, generators, transistors, vacuum tubes, batteries, transformers and so on. Usually, the components are physically or conductively tied together by means of a conductor such as a copper wire. Although circuits can be inductively coupled (as by a transformer) or capacitively coupled (as by a capacitor), it is generally assumed, in circuit theory, that electric energy remains within the network and is not radiated away. (It is permissible to convert electric energy into heat energy which can then be radiated from the circuit.)

The actual or physical circuit is of little use to the theoretical analyst unless the physical circuit entities can be replaced by equivalent circuit concepts which, in turn, can be couched in descriptive mathematical language. In Chapter 1 it was shown how a physical element, called a resistor, could be replaced by a quality called resistance. Similar equivalent circuits or concepts can be employed to replace other actual or physical electric elements or components. Although it is quite important for the engineer to be able to replace an actual circuit by the proper descriptive equivalent circuit, the material in this chapter is concerned with the analysis of the equivalent circuit rather than with its determination. Later chapters will deal with obtaining the equivalent circuit for various electronic systems.

If the solution of an electric network is to be obtained by either the loop-current or the node-voltage method of analysis, it is obviously most

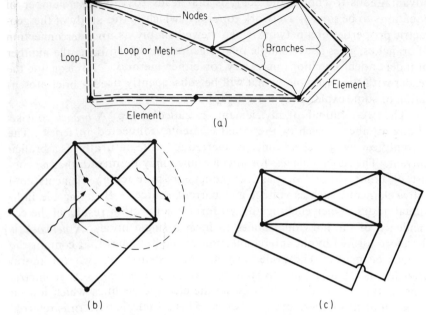

FIG. 2-2. Geometric patterns of the connection of elements forming a), b) planar networks, and c) a nonplanar network.

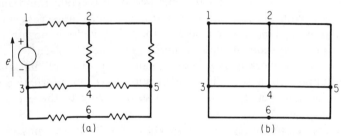

FIG. 2-3. a) A typical electric circuit or network. b) The graph of part a.

tracing through the network without passing through a node or an element more than once, and returning to the original node, paths 12431, 125631, and 125431 are seen to be loops originating and terminating at node 1. A mesh is defined as a basic or simple loop which does not contain any subloops; therefore, of the loops just mentioned, only loop 12431 is considered a mesh. Other meshes that can be identified are 2542 and 34563.

With this terminology in hand, the first order of business is to review the transition from branch currents to loop currents—a topic usually first

advantageous to choose the analysis that leads to the least numbe
equations to be solved. Network topology, which is the study of the g
metric properties of a network when viewed simply as an interconnecti
of branches, gives rise to simple relationships for determining the numb
of independent equations necessary for either method. To acquaint th
reader with the terminology that will be subsequently used a brief presentation of some explanatory material follows.

The two terminals of any element are called *nodes*. A *branch* consists
of any number of such two-terminal elements, connected in series. The
current flowing in such a branch is referred to, quite naturally, as a branch
current. The series *elements* forming a branch may be mixed; that is, consist of combinations of passive (nonenergy-producing) elements and
active elements, such as voltage and current sources. Bringing the individual nodes of elements together to form a connection results in the co-
incidence of the individual nodes to form a single node. A *network* is
therefore formed by the interconnection of the branches that contain the
various elements. The node forming the junction of two (or more)
branches is quite frequently referred to as a *junction*. A geometric
representation is obtained by viewing the elements as lines which join at
nodes to form a pattern of the network. This geometric pattern is referred
to as a *graph* of the network. If all the lines between various nodes can be
drawn on a plane without any crossing another, the network is said to be
planar. A *loop* is any closed path obtained by traversing the branches of
the network with the restriction that the loop may not cross any node
more than once. A *mesh* is defined as a loop that encloses no other loops.
A pictorial illustration of these terms is given in Fig. 2-2a. It is to be
noted that the network in Fig. 2-2b is planar, although at first glance, it
does not appear to be so. The dashed lines show an alternative way of
drawing the same network which is not so deceiving to the eye as is the
solid line sketch.

The application of the above terms to a specific circuit may be understood by examining the circuit and its graph (Fig. 2-3). Figure 2-3a shows
a circuit with 8 elements (1 voltage source and 7 resistors) connected in an
arbitrary manner. There are 6 numbered nodes appearing in Fig. 2-3a.
From a topological standpoint, it is convenient to redraw Fig. 2-3a in the
form of a graph. The graph of the circuit, showing all the nodes but with
the elements replaced by lines, is given in Fig. 2-3b.

The specification of the nodes by number enables Fig. 2-3b to be used
to illustrate three additional, important circuit terms—the branch, the
loop, and the mesh. For example, 312, 365, and 24 constitute branches
but path 245 does not, since it involves parts of 2 different branches (2
and 45); nor does path 31, since it is only a part of branch 312. Since
loop is defined as any closed path that is traversed by beginning at a nod

encountered in a physics course in electricity. Once this transition is understood, the direct approach of using loop currents in circuit analysis should not appear puzzling nor create any difficulties owing to their "fictitious" nature.

2-2. Branch Currents

The use of branch currents, in solving a network, begins by identifying the currents in each of the branches of the network. Consider Fig. 2-4 which shows a two-mesh circuit containing two d-c voltage sources and three resistors. If one is asked to solve this circuit problem, several questions come immediately to mind: 1) What does solving a circuit mean? 2) What is the meaning of the battery symbol ─┤ı│├─ for E_1 and E_2? 3) Why have the branch currents I_1, I_2, and I_3 been chosen, as shown in Fig. 2-4? The answers to these questions, in respective order, are as follows:

1. A circuit is in effect solved if either all the node voltages or all the branch currents are known. If other quantities, such as charge, power, or energy, are desired, these can be obtained from the already-determined voltages or current.

2. The longer lines indicate the positive terminal of a d-c voltage source, or battery, and the short lines indicate the negative terminal. A positive charge gains potential energy in moving from the negative to the positive terminal (a voltage rise). Referring to Fig. 2-4, the battery presents a voltage rise from terminal d to terminal a ($E_{da} = E_1$) or a voltage drop from terminal a to terminal d ($V_{ad} = E_1$), where E_1 is considered to be the magnitude of the voltage of the battery. Note that $E_{da} = E_1 = V_{ad}$.

3. The directions of the branch currents in Fig. 2-4 were chosen *completely at random*. In circuit analysis, one can assume initially that currents flow in any direction, but, once the choice has been made, one must adhere rigidly to the assumed direction for the remainder of the analysis. The mathematical solution will give a positive or negative value for each current. If positive, the d-c current will be flowing in the direction as-

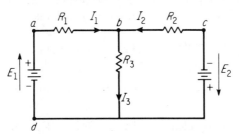

FIG. 2-4. 2-mesh circuit with designated branch currents.

sumed; if negative, the d-c current will be flowing in the direction opposite to that assumed.

The circuit of Fig. 2-4 is solved by using both Ohm's and Kirchhoff's laws. For convenience, the statement of Ohm's laws is repeated here: Ohm's law states that the ratio of voltage to current is a constant (in linear-circuit theory) called resistance. Kirchhoff's laws are based on the laws of conservation of matter and energy; namely, the fact that "the summation of currents entering (or leaving) a junction, at any instant of time, is zero," which is a restatement of the law of the conservation of matter, while the law of conservation of energy supports Kirchhoff's voltage law (potential energy), which stipulates that "the summation of voltage drops (or rises) around a closed loop, at any instant of time, equals zero." Using Kirchhoff's current law, one can write, for the currents entering node b,

$$I_1 + I_2 + (-I_3) = 0 \qquad (2\text{-}1)$$

Note that for the purposes of writing the equation, *all* currents were assumed *entering* node b, although current I_3 in Fig. 2-4 is shown directed *away* from node b. Hence the negative sign $(-I_3)$ in Eq. 2-1 is to give agreement between the expression of the law and the directions assumed for each of the currents. Note if one quoted the law as "the sum of all currents leaving the node must be equal to zero," the resulting equation would be

$$-I_1 - I_2 + I_3 = 0$$

which is Eq. 2-1 multiplied by a scalar (-1) which does not affect the mathematical properties of an equation. Similarly, quoting the law as "the sum of the currents entering the node must equal the sum of the currents leaving the node" results in

$$I_1 + I_2 = I_3$$

which is mathematically equivalent to Eq. 2-1. In whatever manner the law is quoted or visualized by the analyst, the important point is to write the correct equation in terms of the desired visualization.

The application of Kirchhoff's second law (in conjunction with Ohm's law) around loops *abda* and *cbdc* (using "the sum of the voltage drops around a closed loop must equal zero") yields

Loop *abda* $\qquad V_{ab} + V_{bd} + V_{da} = 0$

$$R_1 I_1 + R_3 I_3 - E_1 = 0 \qquad (2\text{-}2)$$

or

Loop *cbdc* $\qquad V_{cb} + V_{bd} + V_{dc} = 0$

$$R_2 I_2 + R_3 I_3 + E_2 = 0 \qquad (2\text{-}3)$$

Voltage rises could have been used rather than voltage drops. Setting "the sum of voltage rises around loop *abda* equal to zero" gives

$$E_{ab} + E_{bd} + E_{da} = 0$$
$$\downarrow \qquad \downarrow \qquad \downarrow$$
$$-R_1 I_1 - R_3 I_3 + E_1 = 0$$

which is Eq. 2-2 multiplied by a scalar (-1). For loop *cbdc*, the sum of voltage rises set equal to zero gives

$$E_{cb} + E_{bd} + E_{dc} = 0$$
$$\downarrow \qquad \downarrow \qquad \downarrow$$
$$-R_2 I_2 - R_3 I_3 - E_2 = 0$$

which is Eq. 2-3 multiplied by a scalar (-1). Again, the important point is to write the equation consistent with the chosen viewpoint of Kirchhoff's voltage law (drops or rises). The relationship

$$V_{12} = -E_{12} = E_{21}$$

can be used to change from drops to rises or vice-versa whenever convenient to do so. Many times, one will use a voltage rise symbol to indicate a voltage source as in Fig. 2-4. For loop *abda*

$$V_{ab} + V_{bd} + V_{da} = 0$$
$$\downarrow \qquad \downarrow \qquad \downarrow$$
$$V_{ab} + V_{bd} - E_{da} = 0$$
$$\downarrow \qquad \downarrow \qquad \downarrow$$
$$R_1 I_1 + R_3 I_3 - E_1 = 0$$

where in the intermediate step the voltage drop symbol (V_{da}) across a source was converted to an equivalent voltage rise symbol $(-E_{da})$ since the circuit diagram symbol chosen for the source voltage was a rise.

Equations 2-1, 2-2, and 2-3 give a set of three simultaneous equations with three unknowns as follows,

$$I_1 + \quad I_2 - \quad I_3 = 0$$
$$R_1 I_1 + \quad 0 I_2 + R_3 I_3 = E_1$$
$$0 I_1 + R_2 I_2 + R_3 I_3 = -E_2$$

The unknowns I_1, I_2 and I_3 in these three simultaneous equations can be solved for either by the method of substitution or by the method of determinants. A numerical example will be given shortly.

The reader may, at this point, have several questions to ask about the sudden appearance of Eqs. 2-1, 2-2, and 2-3. For example: Why is node *b* the chosen node for summing currents: Furthermore, why aren't currents summed at nodes *a*, *c*, and/or *d*? Also, what governed the choice of the pair of loops *abda* and *cbdc* instead of the loops *bcdab* and *bcdb* or perhaps *cbdc* and *dabd* (and so on)? All these questions can be rephrased into one: How many equations must one write to solve a circuit? The general answer is that one must write a complete set of independent equations. It goes without saying that the set must contain as many equations as there are unknown quantities to be determined. This leads to the question of how many unknown quantities must be initially specified prior to writing the equations. Equations 2-1, 2-2, and 2-3 represent the necessary number of branch-current and element-voltage equations which must be written to solve the circuit of Fig. 2-4. Note that these equations include 1) all the branch currents and 2) all the element voltages (thereby assuring that all the elements have been taken into account) shown in Fig. 2-4. Thus, it can be inferred that the necessary number of equations must include all the branch currents and all the element voltages. A more definitive and elegant statement, based on topological considerations regarding the necessary and sufficient equations required to solve a circuit, will be presented in a later section.

Equations 2-1, 2-2, and 2-3 are useful in explaining Kirchhoff's laws, but an analyst rarely uses branch currents in solving a circuit. Instead, the method of loop (or mesh) currents is almost universally used in writing the necessary number of network equations. This method presents an orderly way to write the circuit equations, requires fewer equations than does the branch-current method and with a few restrictions, leads to writing the final form of the equations by inspection.

Consider the same circuit as before, but with loop currents flowing in the network, rather than branch currents (see Fig. 2-5). The loop-current method assumes that a current *flows around an entire loop.* (For simplicity and orderliness, meshes are usually chosen as the loops.)

Enough loop or mesh currents must be chosen so that a current flows through each branch, ensuring that each element will be accounted for in the describing equations. (More on how the choice can be made will be discussed later.) The concept of the loop or mesh current is sometimes confusing, when first encountered because it seems to be a "fictitious" or "artificial" current. The branch currents can be considered the "real" currents in a circuit, because they are the currents measured when an

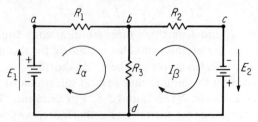

Fig. 2-5. 2-mesh circuit with designated loop currents.

ammeter is placed in a branch in the circuit. As will be seen later, the branch currents are always expressible as an algebraic sum of certain loop currents; thus, no information is lost by choosing loop currents instead of branch currents in solving a network.

The loop-current method of writing equations uses Kirchhoff's summation-of-voltages law. Using loop *abda* (which is traversed by loop current I_α) and loop *bcdb* (which is traversed by loop current I_β) and using voltage drops, the following equations are obtained:

$$\left. \begin{array}{l} V_{ab} + V_{bd} + V_{da} = 0 \\ V_{bc} + V_{cd} + V_{db} = 0 \end{array} \right\} \tag{2-4}$$

or

$$\left. \begin{array}{l} \underbrace{(R_1 I_\alpha)}_{V_{ab}} + \underbrace{(R_3 I_\alpha - R_3 I_\beta)}_{V_{bd}} + \underbrace{(-E_1)}_{V_{da}} = 0 \\ \underbrace{(R_2 I_\beta)}_{V_{bc}} + \underbrace{(-E_2)}_{V_{cd}} + \underbrace{(R_3 I_\beta - R_3 I_\alpha)}_{V_{db}} = 0 \end{array} \right\} \tag{2-5}$$

Upon reorganizing these equations

$$(R_1 + R_3) I_\alpha - R_3 I_\beta = E_1 \tag{2-6}$$

and

$$-R_3 I_\alpha + (R_2 + R_3) I_\beta = E_2 \tag{2-7}$$

Now, by comparing the branch currents in Fig. 2-4 with the loop currents in Fig. 2-5, the following identities must hold if these networks are indeed identical:

$$I_1 = I_\alpha \tag{2-8}$$

since I_α is the only loop current traversing the left-hand branch. The relationship is positive, since both currents traverse the branch in the same direction. Now

$$I_2 = -I_\beta \tag{2-9}$$

since I_β is the only loop current traversing the right-hand branch. The relationship is negative, since the assumed positive direction of I_β is counter to that of the I_2 in the right-hand branch. Now

$$I_3 = I_\alpha - I_\beta \tag{2-10}$$

since the net loop current flowing downward (the same direction as that assumed for I_3) is the difference between the two loop currents which are flowing in opposite directions in the center branch.

Equations 2-8, 2-9, and 2-10 can be looked upon as transforming the three branch currents into two loop currents. Rewriting Eq. 2-1, that is,

Kirchhoff's current law

$$I_1 + I_2 + (-I_3) = 0 \qquad [2\text{-}1]$$

and solving for I_3 results in

$$I_3 = I_1 + I_2 \qquad (2\text{-}11)$$

Upon substituting the redefined currents I_1 and I_2 given by Eqs. 2-8 and 2-9, respectively, into Eq. 2-11 gives

$$I_3 = I_\alpha - I_\beta \qquad (2\text{-}12)$$

which is identical to Eq. 2-10. Equation 2-10 then forms the basis of the transition from the 3 branch currents to the two loop currents. Consider what happens if the redefinitions of the branch currents, given by Eqs. 2-8, 2-9, and 2-10, are substituted into the original branch-current equations, which, for convenience, are repeated here, that is,

$$R_1 I_1 + R_3 I_3 - E_1 = 0 \qquad [2\text{-}2]$$

and

$$R_2 I_2 + R_3 I_3 + E_2 = 0 \qquad [2\text{-}3]$$

Upon making the substitution, these equations become

$$R_1(I_\alpha) + R_3(I_\alpha - I_\beta) - E_1 = 0 \qquad (2\text{-}13)$$

and

$$R_2(-I_\beta) + R_3(I_\alpha - I_\beta) + E_2 = 0 \qquad (2\text{-}14)$$

Rearranging Eqs. 2-13 and 2-14 gives

$$(R_1 + R_3)I_\alpha - R_3 I_\beta = E_1 \qquad (2\text{-}15)$$

and

$$-R_3 I_\alpha + (R_2 + R_3)I_\beta = E_2 \qquad (2\text{-}16)$$

Note that Eqs. 2-15 and 2-16 are identical to Eqs. 2-6 and 2-7, respectively; thus, the equivalence of the two sets of equations has been demonstrated on a purely mathematical basis as well as reasoned from the current equivalences that must be met.

The above development is the only proof that will be offered to show how loop currents can replace branch currents. Note that, by using loop currents, the number of equations necessary to solve a circuit is reduced (in this example from three to two), as compared to using branch currents. In a multiloop circuit, a reduction in the number of equations necessary to obtain a solution is welcome indeed. It should be apparent that this reduction comes about because of the implicit inclusion of Kirchhoff's current law in the loop currents. Wherever two or more loop currents traverse the same branch, the branch current is eliminated from direct consideration and hence removed as an unknown current variable.

A numerical example will now be given to demonstrate further the equivalence of branch and loop currents in solving a network.

EXAMPLE 2-1. Consider the circuit shown in Fig. 2-6. Solve for all the branch currents, and determine the voltage drop V_{ce} across R_4 as well as the power supplied by the 80 volt source.

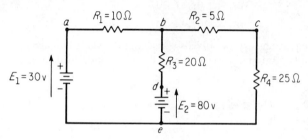

FIG. 2-6. Circuit for Example 2-1.

Solution: First solve for the branch currents, using element-voltage and branch-current equations. Figure 2-7 is a repeat of Fig. 2-6, but with all the nodes identified and with the branch currents chosen in an arbitrary manner.

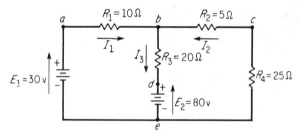

FIG. 2-7. Circuit of Fig. 2-6 redrawn with branch currents identified.

Using Kirchhoff's voltage law of summing voltage drops gives, for loop *abdea,*

$$V_{ab} + V_{bd} + V_{de} + V_{ea} = 0 \longleftrightarrow R_1 I_1 + R_3 I_3 + E_2 - E_1 = 0 \qquad (2\text{-}17)$$

and, for loop *bcedb,*

$$V_{bc} + V_{ce} + V_{ed} + V_{db} = 0 \longleftrightarrow -R_2 I_2 - R_4 I_2 - E_2 - R_3 I_3 = 0 \quad (2\text{-}18)$$

Using Kirchhoff's current law gives, for currents entering node *b,*

$$I_1 + I_2 - I_3 = 0 \qquad (2\text{-}19)$$

Rewriting Eqs. 2-17 and 2-18 in terms of the branch currents, the known voltage sources, and the known resistances, one obtains

$$10\,I_1 + 20\,I_3 + 80 - 30 = 0 \qquad (2\text{-}20)$$

and

$$-5\,I_2 - 25\,I_2 - 80 - 20\,I_3 = 0 \tag{2-21}$$

Equations 2-19, 2-20, and 2-21 represent three independent equations in three unknowns, which can be rearranged for convenience as

$$1I_1 + 1I_2 - 1I_3 = 0 \tag{2-22}$$
$$10I_1 + 0I_2 + 20I_3 = -50 \tag{2-23}$$
$$0I_1 - 30I_2 - 20I_3 = 80 \tag{2-24}$$

This set of equations can be solved in a straightforward manner by either of two methods—substitution or determinants. Using solution by determinants

$$I_1 = \frac{\begin{vmatrix} 0 & 1 & -1 \\ -50 & 0 & 20 \\ 80 & -30 & -20 \end{vmatrix}}{\Delta} \quad I_2 = \frac{\begin{vmatrix} 1 & 0 & -1 \\ 10 & -50 & 20 \\ 0 & 80 & -20 \end{vmatrix}}{\Delta} \quad I_3 = \frac{\begin{vmatrix} 1 & 1 & 0 \\ 10 & 0 & -50 \\ 0 & -30 & 80 \end{vmatrix}}{\Delta}$$

where

$$\Delta = \begin{vmatrix} 1 & 1 & -1 \\ 10 & 0 & 20 \\ 0 & -30 & -20 \end{vmatrix}$$

Evaluation of the determinants results in

$$I_1 = \frac{1600 - 1500 - 1000}{300 + 600 + 200} = -\frac{9}{11} \text{ amp}$$

$$I_2 = \frac{1000 - 800 - 1600}{1100} = -\frac{14}{11} \text{ amps}$$

and

$$I_3 = \frac{-1500 - 800}{1100} = -\frac{23}{11} \text{ amps}$$

The voltage drop V_{ce} across R_4 is given by

$$V_{ce} = -I_2 R_4 = -\left(-\frac{14}{11}\right)(25) = \frac{350}{11} \text{ volts}$$

Note that the negative sign $(-I_2 R_4)$ results from the fact that the voltage drop was desired in a direction counter to the assumed positive direction for I_2.

The power lost between nodes d and e is given by

$$P_{\text{LOST}} = V_{de}I_{de} = E_2 I_3 = (80 \text{ volts})\left(-\frac{23}{11} \text{ amps}\right) = -\frac{1840}{11} \text{ watts}$$

This *negative power loss* in E_2 represents a *positive power gain* to the network. Indeed the 80 volt source is furnishing (1840/11) watts to the network. Had I_3 been positive in this problem then the 80-volt source would be receiving (charging or storing electrical energy) instead of delivering power.

The bulk of the electrical energy used today is created in two general ways: 1) by an electromechanical process such as a steam turbine driving an electric generator or alternator; 2) by an electrochemical process such as found in the lead-acid storage batteries used in automobiles. The thermoelectric (the principle of the thermocouple) and the photoelectric (the principle of the solar cell) processes are two other ways of obtaining electric energy. Note that for a source the voltage-current product expression *must* be used to determine the source power since there is no resistance associated with the ideal source symbol. Any source power loss would be accommodated through a resistance in series with the ideal source symbol and would have to be charged against the source by making a separate calculation.

The use of loop-current methods can reduce the number of equations necessary to solve the circuit, since the branch currents are automatically taken care of in writing the loop equations. Figure 2-8 shows the circuit diagram properly

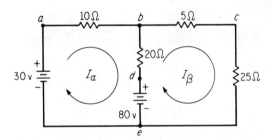

FIG. 2-8. Circuit of Fig. 2-6 redrawn for loop-current solution.

labeled for using loop currents. The loop-or-mesh-current equations can be written by inspection as

$$V_{ab} + V_{bd} + V_{de} + V_{ea} = 0 \leftrightarrow 10I_\alpha + 20(I_\alpha - I_\beta) + 80 - 30 = 0$$

or

$$30I_\alpha - 20I_\beta = -50 \tag{2-25}$$

$$V_{bc} + V_{ce} + V_{ed} + V_{db} = 0 \leftrightarrow 5I_\beta + 25I_\beta - 80 + 20(I_\beta - I_\alpha) = 0$$

or

$$-20I_\alpha + 50I_\beta = 80 \tag{2-26}$$

Equations 2-25 and 2-26 can be solved either by substitution or determinants. Again using determinants

$$I_\alpha = \frac{\begin{vmatrix} -50 & -20 \\ 80 & 50 \end{vmatrix}}{\Delta} \qquad I_\beta = \frac{\begin{vmatrix} 30 & -50 \\ -20 & 80 \end{vmatrix}}{\Delta}$$

where

$$\Delta = \begin{vmatrix} 30 & -20 \\ -20 & 50 \end{vmatrix} = (30)(50) - (-20)(-20) = 1100$$

Thus,

$$I_\alpha = \frac{(-50)(50) - (-20)(80)}{1100} = -\frac{9}{11} \text{ amp}$$

and

$$I_\beta = \frac{(30)(80) - (-50)(-20)}{1100} = \frac{14}{11} \text{ amps}$$

Upon identifying the branch currents of Fig. 2-7 with the loop currents of Fig. 2-8, one notes that

$$I_1 = I_\alpha = -\frac{9}{11} \text{ amp}$$

$$I_2 = -I_\beta = -\frac{14}{11} \text{ amps}$$

and

$$I_3 = I_\alpha - I_\beta = -\frac{9}{11} - \frac{14}{11} = -\frac{23}{11} \text{ amps}$$

which checks with the previously calculated values of the branch currents I_1, I_2, and I_3. The voltage drop V_{ce}, in terms of the loop currents, is given by

$$V_{ce} = I_\beta R_4 = \left(\frac{14}{11}\right)(25) = \frac{350}{11} \text{ volts}$$

which checks with the previously calculated value. The power lost between nodes *d* and *e* is

$$P_{\text{LOST}} = V_{de} I_{de} = V_{de}(I_\alpha - I_\beta) = (80 \text{ volts})\left(-\frac{23}{11} \text{ amps}\right) = \frac{-1840}{11} \text{ watts}$$

or a gain from the 80 volt source of $(1840/11)$ watts as previously calculated. This simple example demonstrates that the loop- or mesh-current technique is generally superior to the branch-current method in that fewer equations need to be solved.

Attention will now be concentrated exclusively on loop currents. The techniques of choosing loop currents will be presented. Utilization of the topological information will lead to answering some of the questions concerning the necessary and sufficient equations that must be written in order to obtain a solution for any network.

2-3. Method of Loop Currents

It has been demonstrated that the utilization of loop currents is based on Kirchhoff's voltage law, which states that the summation of all voltage drops (or rises) around a closed loop, at any instant of time, must be zero. Thus, it is necessary to choose the loops, treat these loops as if they were directed loop currents, write the equation expressing the voltage drop (or rise) for each element in the loop in terms of a parameter and certain loop currents, and then equate the sum of these drops (or rises) to zero.

Choosing the Loop Currents. The fundamental question, previously posed, asked how many loops would be necessary to get an independent set of equations that would result in the complete solution of a network. This is answered by network topology, with the aid of the geometric pattern of the network. No attempt will be made to derive formally or to verify this answer, other than to comment that its form will be argued as being intuitively correct by making certain observations as it is applied to circuits in several examples. Thus, if, in a network, e is the number of elements, n is the number of nodes, and N_{cs} is the number of current sources, then

$$N_i = e + 1 - (n + N_{cs}) \tag{2-27}$$

where N_i is the number of unknown loop currents necessary to obtain a proper set of independent equations. An independent group of equations equal in number to that given by Eq. 2-27, obtained by utilizing the unknown loop-current approach to a network, is called a proper set. It is to be noted that more than one proper set will generally exist for a given network. Now, with Eq. 2-27, one knows the necessary number of unknown loop currents required to solve the network. This is precisely the required number. A set of equations smaller in number than that given by Eq. 2-27 may result in more unknowns than equations, which means that the set is not solvable. This condition can be recognized easily, since such a set is not solvable, thus forcing the analyst to correct the condition. However, it is also possible that a set of equations smaller in number than that given by Eq. 2-27, derived from a set of loop currents whereby each element in the network is traversed by at least one loop current and hence all elements are seemingly accounted for, may be an independent set of equations and therefore may be solvable. This leads to erroneous results which may not be recognized easily by the analyst. Such a set usually leads to placing some artificial constraint (physically invalid) on the branch currents. An example illustrating such a constraint will be presented shortly.

A set of equations larger in number than that given by Eq. 2-27 will result in some of the equations being redundant. If a proper set is included in this larger set, then each equation in excess of the proper set is expressible as a linear combination of the proper-set equations and hence is redundant. The next problem, therefore, is how to choose the minimum number of equations to ensure that they form a proper set. It is possible, by delving deeper into topology, to set up a criterion whereby one is assured of choosing loop currents in such a manner that the resultant equations are independent and form a proper set. This criterion gives rise to a simple rule of thumb that will suffice in almost all cases; furthermore, as will be seen, it will be most desirable to choose the loops in a very systematic manner, thus assuring the independence of the equations, except for some very special cases. *The rule is that each successive choice of a loop should contain at least one element that has not been traversed by the loops previously chosen.* As mentioned before, if all the elements are traversed by a number of loops which is less than the number given by Eq. 2-27, it is probable that some constraint is being imposed on the circuit, and this must be removed by a suitable choice of the remaining loops. The following discussion illustrates some of these points.

Consider the geometric representation of a network such as that given in Fig. 2-9a. One needs to determine the number of loop currents necessary for the complete solution of the network and also determine a suitable choice of loop currents leading to a proper set of equations. It is to be assumed that no current sources are present. Examination of Fig. 2-9a shows the number of elements to be 9 and the number of nodes to be 6. Thus, by Eq. 2-27, the necessary number of unknown loop currents is given by

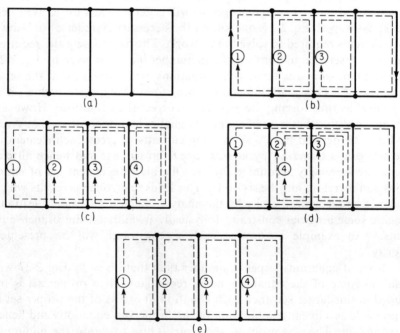

FIG. 2-9. a) Geometric representation of a network. b) Choice of 3 of the necessary 4 loop currents. c) Choice of a fourth loop current resulting in the removal of the constraint and leading to an independent set of equations. d) Choice of a fourth loop current which does not remove the constraint. e) Choice of a set of loop currents obeying the rule of including an additional element with each successive choice of a loop current.

$$N_i = (e + 1) - (n + N_{cs}) = (9 + 1) - (6 + 0) = 4 \qquad (2\text{-}28)$$

The choice of 3 loop currents obeying the rule of thumb is depicted in Fig. 2-9b. However, it is noted that all elements have been traversed by these 3 loop currents, and therefore the fourth loop current cannot possibly satisfy the rule. Upon further examination of Fig. 2-9a, it is observed that the rightmost branch current, being equal to loop current 1, is therefore equal to the leftmost branch current, since it is also equal to loop current 1. Thus, a constraint has been put on the two branch currents (namely, that they be equal to each other), and this must be removed by a suitable choice of the remaining loop current. The choice of the fourth loop current in Fig. 2-9c accomplishes this objective, but that in Fig. 2-9d does not. Only the former choice gives rise to a proper set of equations. Finally, a set of loop currents obeying the rule of thumb—that each successive loop current must contain an element not traversed by the previously chosen loop currents—is depicted in Fig. 2-9e. It should be noted that each of these currents is also a so-called mesh current. These mesh currents are often referred to descriptively as the "window pane" currents.

Writing the Loop-Current Equations. The examples will be concerned with networks that have only voltage sources as the active elements. Although this may seem restrictive, it will eventually be demonstrated how a practical current source can be converted to an equivalent practical voltage source, so any general resistive network could be manipulated into this class.

Now consider the circuit of Fig. 2-10. It will be used to illustrate how to obtain equations by the method of loop currents.

Symbolic Form of Equations. The first step in the solution is the choice of the unknown loop currents. Examining Fig. 2-10 shows that there are 9 elements, 8 nodes, and no current sources, so the number of equations necessary for a proper set (which is the necessary number of unknown loop currents as given by Eq. 2-27) is

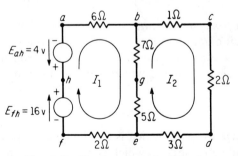

FIG. 2-10. Circuit to be solved by the method of loop currents.

$$N_i = (e + 1) - (n + N_{cs}) = (9 + 1) - (8 + 0) = 2 \qquad (2\text{-}29)$$

At this point, it is desirable to set up a technique for choosing the unknown loop currents of planar circuits with no current sources (such as this example) that will always lead to a proper set of equations and—perhaps of almost equal importance—will enable the analyst to write the equations, in a very compact form, by visual inspection of the circuit. (The efficiency of the latter point will be fully appreciated as more complex circuits are analyzed.)

To begin, the particular closed loops that are chosen are called meshes. Recall that a mesh is defined as a loop that encloses no other loops. The current associated with a mesh is quite often referred to as a mesh current rather than a loop current. From topology, it can be shown that the number of meshes in a planar circuit will be equal to the necessary number of independent loop equations plus the number of current sources present in the circuit. This is so, since the branch containing a current source places a constraint on the mesh currents traversing this branch and results in an equation describing their relationship in this branch. (Additional comments on this type of network will be given later.) The two mesh currents for the example are shown in Fig. 2-10. It is advantageous to choose them all in the same sense, that is, either all in a clockwise sense or all in a counterclockwise sense. This gives a *symmetry* to the resultant equations which will reduce error and facilitate writing the reduced form of the equations by inspection. It is now necessary to apply Kirchhoff's voltage law to the two meshes in order to obtain the two independent equations. It is desirable to apply this law by writing the summation of voltage drops around the mesh in the direction of the mesh current. This sum, by Kirchhoff's law, must be equal to zero. Hence, symbolically, for mesh *abgefha*,

$$V_{ab} + V_{bg} + V_{ge} + V_{ef} + V_{fh} + V_{ha} = 0 \qquad (2\text{-}30)$$

and for mesh *bcdegb*,

$$V_{bc} + V_{cd} + V_{de} + V_{eg} + V_{gb} = 0 \qquad (2\text{-}31)$$

Substitution from the Circuit. Examining Eq. 2-30, term by term, in view of the circuit diagram in Fig. 2-10, gives the following:

1. The voltage drop V_{ab} which is the voltage drop across the 6-ohm resistor from node *a* to node *b* which is in the direction of the mesh current I_1 is

$$V_{ab} = +6 \cdot I_1 \qquad (2\text{-}32)$$

2. The voltage drop V_{bg} which is the voltage drop across the 7-ohm resistor from node *b* to node *g*, which is in the direction of the mesh current I_1 but counter to the direction of the mesh current I_2, with the net

current from node b to node g being $(I_1 - I_2)$ is

$$V_{bg} = +7 \cdot (I_1 - I_2) \tag{2-33}$$

3. The voltage drop V_{ge} which is the voltage drop across the 5-ohm resistor from node g to node e with the net current from node g to node e being $(I_1 - I_2)$ is

$$V_{ge} = +5 \cdot (I_1 - I_2) \tag{2-34}$$

4. The voltage drop V_{ef} which is the voltage drop across the 2-ohm resistor from node e to node f which is in the direction of the mesh current I_1 is

$$V_{ef} = +2 \cdot I_1 \tag{2-35}$$

5. The voltage drop V_{fh} which is equal to the negative of the voltage rise from node f to node h is

$$V_{fh} = -(E_{fh}) = -(+16) = -16 \tag{2-36}$$

6. The voltage drop V_{ha} which is equal to the voltage rise from node a to node h is

$$V_{ha} = E_{ah} = +4 \tag{2-37}$$

Substitution of these expressions (Eqs. 2-32 through 2-37) into Eq. 2-30 gives

$$6 \cdot I_1 + 7 \cdot (I_1 - I_2) + 5 \cdot (I_1 - I_2) + 2 \cdot I_1 - 16 + 4 = 0 \tag{2-38}$$

With a similar line of reasoning,

$$V_{bc} = +1 \cdot I_2, \qquad V_{cd} = +2 \cdot I_2, \qquad V_{de} = +3 \cdot I_2 \tag{2-39}$$

The voltage drop V_{eg} is the voltage drop across the 5-ohm resistor from node e to node g, and is in the direction of the mesh current I_2 but counter to the direction of the mesh current I_1. Thus the net current from node e to node g is $(I_2 - I_1)$ which obtains

$$V_{eg} = +5 \cdot (I_2 - I_1) \tag{2-40}$$

In a like manner, the voltage drop from node g to node b is

$$V_{gb} = +7 \cdot (I_2 - I_1) \tag{2-41}$$

The substitution of Eqs. 2-39, 2-40, and 2-41 into Eq. 2-31 gives

$$1 \cdot I_2 + 2 \cdot I_2 + 3 \cdot I_2 + 5 \cdot (I_2 - I_1) + 7 \cdot (I_2 - I_1) = 0 \tag{2-42}$$

Gathering the coefficients of each of the two unknown loop currents and rearranging Eqs. 2-38 and 2-42 gives

$$(6 + 7 + 5 + 2) I_1 - (7 + 5) I_2 = 16 - 4 \tag{2-43}$$

and

$$-(5 + 7) I_1 + (1 + 2 + 3 + 5 + 7) I_2 = 0 \tag{2-44}$$

Intermediate Observations. A number of conclusions can be drawn from the forms of Eqs. 2-43 and 2-44 which apply to loop-current analysis based on 1) no current sources present, 2) meshes being chosen as the loops, 3) all mesh currents being chosen in the same sense, and 4) Kirchhoff's voltage law being used in the form of the summation of the voltage drops around the mesh in the direction of the mesh current with the sum being set equal to zero.

Note that in Eq. 2-43, written for mesh 1, the coefficient of the unknown current I_1 is the positive sum of all the resistance values in mesh 1. In Eq. 2-44, written for mesh 2, the coefficient of the unknown current I_2 is the positive sum of all the resistance values in mesh 2. In general, in equation j, written for mesh j, the coefficient of the unknown current I_j is the *positive sum* of all the resistance values in mesh j.

Note also that in Eq. 2-43, written for mesh 1, the coefficient of the unknown current I_2 is the negative sum of all the resistance values in the branch common to mesh 1 and mesh 2. In Eq. 2-44, written for mesh 2, the coefficient of the unknown current I_1 is the negative sum of all the resistance values in the branch common to mesh 2 and mesh 1. Since the branch common to two meshes is unique, the coefficient of the unknown current I_k in mesh-equation j is equal to the coefficient of the unknown current I_j in mesh-equation k and is the *negative sum* of all the resistance values in the branch common to meshes j and k.

One further observation can be made with respect to the voltage sources encountered in traversing a mesh in the direction of the mesh current. Since voltage drops are being expressed, a source presenting a voltage rise in the direction of the mesh current has a negative sign affixed to its absolute value, but a source presenting a voltage drop in the direction of the mesh current has a positive sign affixed to its absolute value, as illustrated in Eq. 2-38. Now take the algebraic sum of all the voltage rises (recall that a positive voltage drop in a given direction is considered as a negative voltage rise in the same direction) of all the voltage sources encountered around the mesh in the direction of the mesh current, and add this sum to both sides of the mesh equation. When added to the left side of the mesh equation, this sum exactly cancels the sum of the factors representing the voltage sources, since they are expressed as voltage drops. The right side of the mesh equation, which was initially zero, becomes the net sum of the voltage rises of all the sources encountered in traversing the mesh in the direction of the mesh current. This is essentially the physical interpretation of the rearrangement that took place in the step going from Eqs. 2-38 to 2-43. Mathematically, it is equivalent to adding the same quantity to both sides of an equality. In terms of the final forms of the describing equations, it can be stated that the sum of the voltage drops of the passive elements is set equal to the net sum of the voltage rises of the voltage sources as the mesh is traversed in the direction of the mesh current. The final step is the summing of the coefficients of the un-

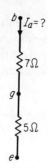

FIG. 2-11. Branch *bge* of the circuit of Fig. 2-10.

known loop currents in Eqs. 2-43 and 2-44, giving

$$20I_1 - 12I_2 = 12$$

and

$$-12I_1 + 18I_2 = 0$$

(2-45)

Mathematically, Eq. set 2-45 is a set of linear algebraic equations. These equations can be solved by using determinants. The solution of Eq. set 2-45 gives

$$I_1 = \frac{\begin{vmatrix} 12 & -12 \\ 0 & 18 \end{vmatrix}}{\begin{vmatrix} 20 & -12 \\ -12 & 18 \end{vmatrix}} = \frac{(12)(18)}{(20)(18) - (-12)(-12)} = 1.0 \text{ amp}$$

and

$$I_2 = \frac{\begin{vmatrix} 20 & 12 \\ -12 & 0 \end{vmatrix}}{\begin{vmatrix} 20 & -12 \\ -12 & 18 \end{vmatrix}} = \frac{-(-12)(12)}{(20)(18) - (-12)(-12)} = \frac{2}{3} \text{ amp}$$

If any particular branch current is desired, it can be obtained in terms of the mesh currents. Suppose one wants to know the branch current flowing from node *b* to node *e* through the series combination of the 7-ohm and 5-ohm resistor, that is, the current indicated in Fig. 2-11. By examining Fig. 2-10, this branch current is seen to be

$$I_a = I_1 - I_2$$

or

$$I_a = 1 - \tfrac{2}{3} = \tfrac{1}{3} \text{ amp}$$

Any other branch currents can be found similarly in terms of the mesh currents and thus the circuit problem is considered to be solved.

Using the rules that have been outlined, it is now possible to write directly the equations in the final form, by visual inspection of a network. Consider the following example.

EXAMPLE 2-2.   Using the technique of employing mesh currents as the unknown loop currents, write the set of linear algebraic equations for the network given in Fig. 2-12.

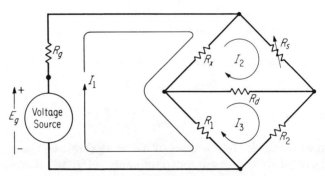

FIG. 2-12. Circuit for Example 2-2.

Solution: Inspection of Fig. 2-12 indicates that there are 3 meshes. As a check, the number of elements is 7, the number of nodes is 5, and, by Eq. 2-27, the necessary number of unknown loop currents needed is

$$n_i = (e + 1) - (n + N_{cs}) = (7 + 1) - (5 + 0) = 3 \qquad (2\text{-}46)$$

which confirms the visual observation. To utilize the systematic manner of writing the equations, choose the 3 mesh currents in a clockwise sense, as indicated in the figure. Thus, by inspection, the describing equations become

$$(R_g + R_x + R_1)I_1 \qquad -(R_x)I_2 \qquad -(R_1)I_3 = E_g$$
$$-(R_x)I_1 + (R_x + R_s + R_d)I_2 \qquad -(R_d)I_3 = 0$$
$$-(R_1)I_1 \qquad -(R_d)I_2 + (R_1 + R_d + R_2)I_3 = 0$$

$$(2\text{-}47)$$

This particular circuit is a common type of Wheatstone bridge network that is used for measuring an unknown resistance R_x by tuning (varying) a known variable resistance R_s until the voltage across the resistance R_d (representing a detector such as a voltmeter, ammeter, galvanometer, and so on) is zero and hence the current through R_d is also zero. Under balanced conditions, a relationship can be deduced giving R_x as a function of R_s and the fixed known resistances R_1 and R_2. From the network, it can be seen that the condition for balance is for the mesh current I_2 to be equal to the mesh current I_3. Under this condition, the net current in the detector branch is zero. The determinant solution for I_2 is

$$I_2 = \frac{\begin{vmatrix} R_g + R_x + R_1 & E_g & -R_1 \\ -R_x & 0 & -R_d \\ -R_1 & 0 & R_1 + R_d + R_2 \end{vmatrix}}{\Delta} \qquad (2\text{-}48)$$

and for I_3 is

$$I_3 = \frac{\begin{vmatrix} R_g + R_x + R_1 & -R_x & E_g \\ -R_x & R_x + R_s + R_d & 0 \\ -R_1 & -R_d & 0 \end{vmatrix}}{\Delta} \qquad (2\text{-}49)$$

For the balance condition,

$$I_2 - I_3 = 0$$

or, from Eqs. 2-48 and 2-49 (using determinant evaluation),

$$(-E_g)[(-R_x)(R_1 + R_d + R_2) - (-R_1)(-R_d)]$$
$$-(E_g)[(-R_x)(-R_d) + R_1(R_x + R_s + R_d)] = 0$$

which, when evaluated, gives

$$-R_x R_2 + R_1 R_s = 0$$

or

$$R_x = \frac{R_1}{R_2} \cdot R_s \qquad (2\text{-}50)$$

Thus, the unknown resistance R_x is given in terms of the known fixed-resistance values R_1 and R_2 and the value of the known variable resistance R_s necessary to balance the bridge.

This example serves as a practical illustration of the use of mesh currents in solving a specific circuit problem.

Summary of the Method of Loop Currents. A basic tool for the solution of any resistive network has been developed in the form of the method of loop currents. The necessary number of loop currents or, analogously, the minimum number of independent equations that must be utilized to form a proper set is given by Eq. 2-27, in terms of the numbers of elements, nodes, and current sources comprising the network to be analyzed. A technique of using so-called mesh currents has been demonstrated which, with some practice, enables one to write immediately the coefficients of the unknown currents, by inspection of the network. In all cases, a set of linear algebraic equations results. This set of equations can be solved by any number of techniques, for example, through the use of determinants or by the method of substitution.

The circuits that have been analyzed were energized by voltage sources alone. In the more general case, both voltage sources and current sources would be present in the network. The inclusion of current sources in a network necessitates certain modifications to the procedures used in the loop-current analysis. An alternative procedure is to convert any current sources that appear in the network to equivalent voltage sources, thereby forcing the network into the form just analyzed. This conversion or demonstration of equivalence of voltage and current sources will now be discussed.

2-4. Equivalence of Sources

It has previously been mentioned that practical voltage sources can be converted to practical current sources, and vice versa. Consider the practical voltage source shown in Fig. 2-13a. It is desired to construct a practical current source, as shown in Fig. 2-13b, that will be equivalent to the practical voltage source with respect to the network to the right of terminals a and b. In other words, no distinction can be discerned be-

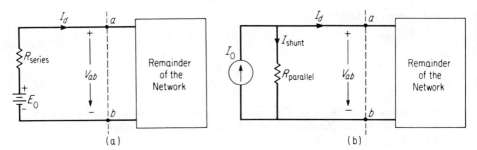

FIG. 2-13. a) Practical voltage source. b) Practical current source.

tween the two sources, as far as any network configuration to the right of the dashed line is concerned. If the two sources are to be equivalent then, for the same terminal condition on the sources, each must deliver the same current I_d and each must present the same terminal voltage drop V_{ab}. Consider the situation where the terminals a and b are open-circuited (oc) so that $I_{d_{oc}} = 0$, as shown in Fig. 2-14. If the two sources are to be equivalent, the voltage drop $V_{ab_{oc}}$ must be the same for each circuit, that is,

$$V_{ab_{oc}} = E_0 = I_0 \cdot R_{\text{parallel}} \qquad (2\text{-}51)$$

Now consider the situation where the terminals *a* and *b* are short-circuited (sc) so that $V_{ab_{sc}} = 0$, as shown in Fig. 2-15. Again, if the two sources are to be equivalent, the current $I_{d_{sc}}$ must be the same for each circuit, that is,

$$I_{d_{sc}} = \frac{E_0}{R_{\text{series}}} = I_0 \qquad (2\text{-}52)$$

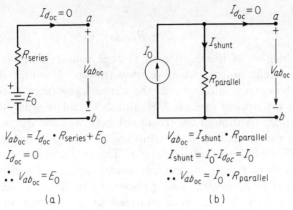

$$V_{ab_{oc}} = I_{d_{oc}} \cdot R_{series} + E_0$$
$$I_{d_{oc}} = 0$$
$$\therefore V_{ab_{oc}} = E_0$$

$$V_{ab_{oc}} = I_{shunt} \cdot R_{parallel}$$
$$I_{shunt} = I_0 - I_{d_{oc}} = I_0$$
$$\therefore V_{ab_{oc}} = I_0 \cdot R_{parallel}$$

(a)   (b)

FIG. 2-14. Open-circuit conditions for a) voltage source, and b) current source.

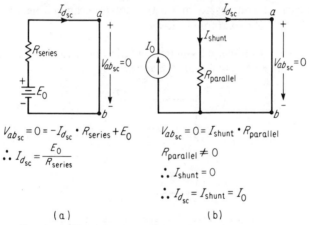

$$V_{ab_{sc}} = 0 = -I_{d_{sc}} \cdot R_{series} + E_0$$
$$\therefore I_{d_{sc}} = \frac{E_0}{R_{series}}$$

$$V_{ab_{sc}} = 0 = I_{shunt} \cdot R_{parallel}$$
$$R_{parallel} \neq 0$$
$$\therefore I_{shunt} = 0$$
$$\therefore I_{d_{sc}} = I_{shunt} = I_0$$

(a)   (b)

FIG. 2-15. Short-circuit conditions for a) voltage source, and b) current source.

Equations 2-51 and 2-52 can be used to obtain the two parameters for converting from one source to another; that is, if E_0 and R_{series} are known for a voltage source, then

$$I_0 = \frac{E_0}{R_{series}}$$

and

$$R_{parallel} = R_{series}$$

give the parameters of the equivalent current source in terms of E_0 and R_{series}. In like manner,

$$E_0 = I_0 \cdot R_{parallel}$$

and

$$R_{\text{series}} = R_{\text{parallel}}$$

give the parameters of the equivalent voltage source in terms of I_0 and R_{parallel} of a current source. A note of warning: If either the battery polarity or the current direction of the current generator is reversed, a negative sign will appear in Eqs. 2-51 and 2-52 and it must be included in the derivation. Although the treatment here was for d-c sources and resistances, it can be extended to arbitrary time-varying sources and resistances. (This will be discussed later.) The transformation was introduced here to show that if any current sources appear in a network they can be changed to equivalent voltage sources, so that the network will then fit the category that has been analyzed by the method of loop currents. In the same vein, the next circuit-analysis technique will treat networks excited by current sources alone, and it should now be obvious that any voltage sources that might be present in a network can be replaced by equivalent current sources to make the network amenable to this technique of solution. An example follows.

EXAMPLE 2-3.   Consider the voltage source in Fig. 2-16. Find the equivalent current source with respect to terminals *a* and *b*.

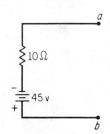

FIG. 2-16. Voltage source of Example 2-3.

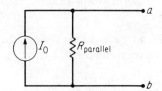

FIG. 2-17. Schematic of equivalent current source of Example 2-3.

Solution: The problem amounts to finding I_0 and R_{parallel} in Fig. 2-17. The expression for the short circuit current in Fig. 2-16 is

$$I_{ab_{\text{sc}}} = I_0 = \frac{-E_0}{R_{\text{series}}} = \frac{-45 \text{ volts}}{10 \text{ ohms}} = -4.5 \text{ amps}$$

Note the negative sign, since the polarity of E_0 is reversed from that in the derivation while the direction of I_0 is not. If the direction of I_0 is also reversed, then its value would be $+4.5$ amps since the reversed I_0 would be the negative of the current shown on the diagram. The parallel resistance is

$$R_{parallel} = R_{series} = 10 \text{ ohms}$$

The equivalent current source is shown in Fig. 2-18. It should be noted that Figs.

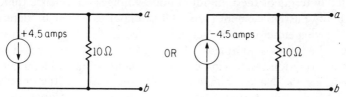

FIG. 2-18. Equivalent current source of Example 2-3.

2-16 and 2-17 are *not* equivalent to the *left* of terminals a and b. This is most easily demonstrated by considering a given condition, say the open circuit condition across terminals a and b. In the case of the voltage source, no power is expended since the current through the series resistor is zero while all the source current from the current generator passes through the parallel resistor and gives a power of $(I_0)^2 \cdot R$. These conditions are certainly *not* the same and points out that equivalent sources are equivalent only with respect to the circuitry external to the terminal pair. This is important if questions are asked pertaining to the source and its resistance. The original source configuration *must* be used to answer these questions.

A voltage source and its series resistance is often referred to as a Thévenin generator; a current source and its parallel resistance is often called a Norton generator. Additional material concerning Thévenin's and Norton's theorems, which encompass the Thévenin and Norton generators, will be found in a later section.

Although almost all problems in electrical circuits could be analyzed by using the loop-current method, other techniques often result in a simpler set of equations to be solved. The node-voltage method, which complements the loop-current method, is presented in the next section.

2-5. Method of Node Voltages

Another technique that is quite useful in the analysis of electrical networks is the method of node voltages. This is based on Kirchhoff's current law, which is restated: The algebraic sum of all currents directed away from (or into) a node, at any instant of time, must be equal to zero. An analogous statement is that, at any instant of time, the sum of the currents leaving a node must be equal to the sum of the currents entering the node. Recall that in the loop-current method the circuit was completely analyzed by finding all the loop currents, since any branch current could then be evaluated in terms of the loop currents, and hence allowing the potential difference across any passive element (using Ohm's law) in any branch to be determined. The node-voltage method involves the reverse order; that is, the potentials or voltages at all the nodes are found and knowing these, the potential difference between any two nodes can be evaluated in terms of the individual node voltages, and hence the current through any passive element (using Ohm's law) can then be determined. Thus, knowing either a proper set of loop currents or a proper set of node voltages constitutes a complete solution to the network, since any other desired information can be explicitly or implicitly obtained from either set.

Choosing the Node Voltages. As in the case of the loop-current method, the first question that arises is how to obtain a proper set of node voltages. Again, network topology is called upon to give an expression for the number of unknown node-pair voltages that must be considered in order to obtain a proper set of equations. It can be demonstrated by topological methods that if, in a network, n is the number of nodes and N_{vs} is the number of voltage sources, then

$$N_v = (n - 1) - N_{vs} \tag{2-53}$$

where N_v is the number of unknown node-pair voltages necessary to obtain a proper set of equations. It is now in order to explain what is meant by node-pair voltages. If the currents flowing through each element in a network are to be solved for by using the method of node voltages, what is of significance is the difference of potential across an element, that is, the potential difference between the terminals or nodes of the element. Now arbitrarily raising or lowering the potential at the two nodes of a linear element by the same value has no effect on the difference of potential between the nodes and hence has no effect on the current flowing through the element. Extending this use of the difference of potential to a network allows any node to be chosen as a reference node, with the potentials of all other nodes to be determined with respect to the potential of this reference node. Consider the diagram in Fig. 2-19 where points a, b, and c are nodes in a network and point 0 is an external point of zero potential or zero volts, that is, a water pipe or a copper rod driven into moist earth. Now V_{a0} represents the voltage of node a relative to point 0 or is simply the voltage drop from node a to point 0. If point a is at a higher potential than ground (zero volts), then V_{a0} is positive since a drop in voltage is evidenced in going from node a to point 0 (ground). If point a is at a lower potential than ground then V_{a0} is negative since a rise in voltage is encountered in going from node a to point 0 (ground). Now from Kirchhoff's law, the voltage from one point to another is independent of the path taken so that

$$V_{ab} = V_{a0} + V_{0b} = V_{a0} - V_{b0} \tag{2-54}$$

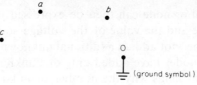

FIG. 2-19. Arbitrary node configuration.

which implies that the voltage drop from node *a* to node *b* is the difference between the two node voltages, each taken with respect to ground. Now in like manner

$$V_{ac} = V_{a0} - V_{c0} \quad \text{and} \quad V_{bc} = V_{b0} - V_{c0} \qquad (2\text{-}55)$$

Consider that

$$V_{ab} = V_{ac} + V_{cb} = V_{ac} - V_{bc} = (V_{a0} - V_{c0}) - (V_{b0} - V_{c0}) = V_{a0} - V_{b0}$$
$$(2\text{-}56)$$

which is identical to Eq. 2-54. This shows that the potential difference is independent of the choice of a reference point. Hence any one node in the network can be chosen as the reference node and all other node voltages are determined with respect to this reference node. Since only voltage differences are used in circuit analysis and differences are unaffected by raising or lowering *all* voltages, it is most convenient to consider the reference node to be at zero volts (ground). All node voltages are specified with respect to the potential of the reference node, and are often termed node-pair voltages. With this in mind, the second subscript on all node voltages is understood to refer to the reference node, which for simplicity is omitted in writing the describing equations. The reference node explains the significance of the number of unknown node voltages (pair implied) as given by Eq. 2-53, being, first of all, less by one than the number of nodes in the network, since the voltage of the reference node is arbitrarily assumed to be zero.

Consider the case when a voltage source appears in the network. Let one of its terminals or nodes be considered as an unknown node voltage.

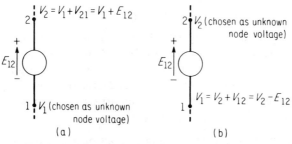

FIG. 2-20. Effect of the voltage source on the choice of unknown node voltages.

The second terminal or node can then be expressed in terms of this un-known node voltage and the value of the voltage source (see Fig. 2-20). This second node does not add an additional unknown node voltage to the network so the two nodes have added only one unknown node voltage to the network. This explains why the number of unknown node voltages, as given in Eq. 2-53, is further reduced by the number of voltage sources appearing in the network.

Writing the Node-Voltage Equations. As with the method of loop currents, the goal to be achieved here is to develop a systematic approach that leads to the ability to write the equations, in a compact form, by visual inspection of the network. With this in mind, the networks will first be limited to combinations of resistive elements excited by current sources. It has already been demonstrated how voltage sources can be converted to equivalent current sources, so that insisting on one type of source is not really a restriction. Furthermore, for convenience in writing the equations, conductances (reciprocal of resistance) will be used. The procedure will again be explained in terms of a specific case.

Consider the network shown in Fig. 2-21. It will be used to illustrate the method of node voltages in the solution of a circuit.

General Observation. The elements comprising this circuit are con-ductances (measured in mhos) and d-c current sources. Again, note that the solution offers techniques which will be extended, with little difficulty, to more general networks. Conductances are used, rather than resist-ances, to simplify the form of the describing equations, where

$$G_i(\text{mhos}) = \frac{1}{R_i(\text{ohms})} \tag{2-57}$$

Symbolic Form of Equations. From the network shown in Fig. 2-21a, there are three nodes and no voltage sources, so the number of unknown node voltages, given by Eq. 2-53, is

$$N_v = (n - 1) - N_{vs} = (3 - 1) - 0 = 2 \tag{2-58}$$

Let the reference node be c (indicated by the ground symbol in Fig. 2-21b). Hence

$$V_c = 0 \tag{2-59}$$

Let the unknown node voltages be V_a and V_b, as indicated in Fig. 2-20b. It is now necessary to write an expression of Kirchhoff's current law at node a and at node b. Consider node a first. Assume that all branch cur-rents are leaving the node as shown in Fig. 2-21c, and therefore, by Kirch-hoff's current law, their sum must be equal to zero or, symbolically,

$$I_1 + I_2 + I_3 + I_4 + I_5 = 0 \tag{2-60}$$

Next consider node b. Again, assume that all branch currents are leaving the node as shown in Fig. 2-21d, and therefore, by Kirchhoff's current

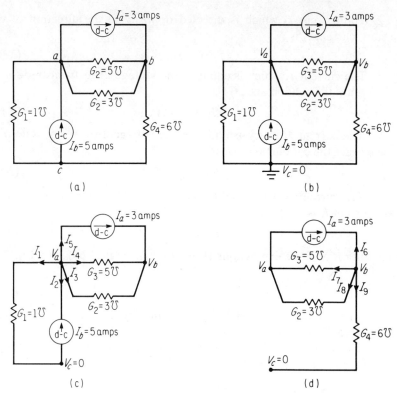

FIG. 2-21. a) Circuit to be solved by method of node voltages. b) Choice of the node voltages. c) Choice of currents for node *a*. d) Choice of currents for node *b*.

law, their sum must be equal to zero or, symbolically,

$$I_6 + I_7 + I_8 + I_9 = 0 \tag{2-61}$$

Note that the choice of the directions of the branch currents at node *b* is independent of the choice of the directions of the branch currents at node *a*, even though branches common to the two nodes are involved. It is necessary only that the nodal equations be expressed properly for whatever choice is made at each of the nodes.

Substitution from the Circuit. For these nodal equations, it is now necessary to express each of the currents in terms of the circuit parameters. Examining Eq. 2-60, term by term, in view of the circuit diagram in Fig. 2-21c, gives the following:

1. The current I_1 which is equal to the voltage drop from node *a* to node *c* times the conductance G_1 is

$$I_1 = G_1 \cdot (V_a - V_c) = G_1 \cdot (V_a - 0) = G_1 \cdot V_a \tag{2-62}$$

2. The current I_2 which is directed opposite to the direction of the constant current I_b is

$$I_2 = -I_b \qquad (2\text{-}63)$$

3. The current I_3 which is equal to the voltage drop from node a to node b times the conductance G_2 is

$$I_3 = G_2 \cdot (V_a - V_b) \qquad (2\text{-}64)$$

4. The current I_4 which is equal to the voltage drop from node a to node b times the conductance G_3 is

$$I_4 = G_3 \cdot (V_a - V_b) \qquad (2\text{-}65)$$

5. The current I_5 which is directed in the same direction as the constant current I_a is

$$I_5 = +I_a \qquad (2\text{-}66)$$

Substitution of these expressions (Eqs. 2-62 through 2-66) into Eq. 2-60 gives

$$G_1 \cdot V_a - I_b + G_2 \cdot (V_a - V_b) + G_3 \cdot (V_a - V_b) + I_a = 0 \qquad (2\text{-}67)$$

In a similar manner, for nodal Eq. 2-61, in view of the circuit diagram given in Fig. 2-21d, the terms are given by:

1. The current I_6 which is directed opposite to the direction of the constant current I_a is

$$I_6 = -I_a \qquad (2\text{-}68)$$

2. The current I_7 which is equal to the voltage drop from node b to node a times the conductance G_3 is

$$I_7 = G_3 \cdot (V_b - V_a) \qquad (2\text{-}69)$$

3. The current I_8 which is equal to the voltage drop from node b to node a times the conductance G_2 is

$$I_8 = G_2 \cdot (V_b - V_a) \qquad (2\text{-}70)$$

4. The current I_9 which is equal to the voltage drop from node b to node c times the conductance G_4 is

$$I_9 = G_4 \cdot (V_b - V_c) = G_4 \cdot (V_b - 0) = G_4 \cdot V_b \qquad (2\text{-}71)$$

Substituting these expressions (Eqs. 2-68 through 2-71) into Eq. 2-61 gives

$$-I_a + G_3 \cdot (V_b - V_a) + G_2 \cdot (V_b - V_a) + G_4 \cdot V_b = 0 \qquad (2\text{-}72)$$

The nodal equations (Eqs. 2-67 and 2-72) can be rearranged to give the set of linear algebraic equations

$$\left. \begin{array}{l} (G_1 + G_2 + G_3) \cdot V_a \quad\quad - (G_2 + G_3) \cdot V_b = I_b - I_a \\ -(G_2 + G_3) \cdot V_a + (G_2 + G_3 + G_4) \cdot V_b = I_a \end{array} \right\} \qquad (2\text{-}73)$$

or, with the values of the parameters given in Fig. 2-21a,

$$\left. \begin{array}{l} (1 + 3 + 5)V_a \qquad -(3 + 5)V_b = 5 - 3 \\ -(3 + 5)V_a + (3 + 5 + 6)V_b = 3 \end{array} \right\} \qquad (2\text{-}74)$$

Intermediate Observations. Several conclusions can be drawn from the forms of Eq. set 2-73 or 2-74 which apply to the node-voltage method of analysis based on 1) no voltage sources present, 2) for each nodal equation, all currents being assumed to flow away from the node, and 3) Kirchhoff's current law being expressed as the sum of all the branch currents leaving the node with the sum being set equal to zero.

The following observations can be made, concerning Eq. set 2-73 or 2-74 that lead to the general form of the coefficients of the unknown node voltages. Consider the first nodal equation, in either set, written for node a. The coefficient of the unknown node voltage V_a is the *positive sum* of all the conductances having one terminal connected to node a. The coefficient of the unknown node voltage V_b is the *negative sum* of all the conductances having one terminal connected to node b and the other terminal connected to node a. Similar relationships are found in the second nodal equation, in either set, written for node b. In the second equation, the coefficient of the unknown node voltage V_b is the positive sum of all the conductances having one terminal connected to node b. The coefficient of the unknown node voltage V_a is the negative sum of all the conductances having one terminal connected to node a and the other terminal connected to node b. In general, in nodal equation j, written for node j, the coefficient of the unknown node voltage V_j is the positive sum of all the conductances that have one terminal connected to node j while the coefficient of the unknown node voltage V_k is the negative sum of all the conductances having one terminal connected to node k and the other terminal connected to node j. Since the sum of the conductances having one terminal connected to node k and the other terminal connected to node j is unique, the coefficient of the unknown node voltage V_k in nodal equation j is equal to the coefficient of the unknown node voltage V_j in nodal equation k.

One further observation can be made. Since the left-hand sides of the nodal equations represent currents leaving the node, then the right-hand sides represent currents entering the node. Thus, as can be seen in Eq. set 2-73 or 2-74, the right-hand side of a nodal equation is the algebraic sum of all currents entering the node from constant-current sources having one of their terminals connected to the node. Care must be exercised in determining the sign of each right-hand term, as a current from a constant-current source which is in a direction leaving the node is negative with respect to this sum. This is illustrated by the second term on the right-hand side in the first equation.

The final step is performing the arithmetic in Eq. set 2-74 to give

$$\left.\begin{array}{r} 9V_a - 8V_b = 2 \\ \\ -8V_a + 14V_b = 3 \end{array}\right\} \qquad (2\text{-}75)$$

and

Solving the equations by the use of determinants gives

$$V_a = \frac{\begin{vmatrix} 2 & -8 \\ 3 & 14 \end{vmatrix}}{\begin{vmatrix} 9 & -8 \\ -8 & 14 \end{vmatrix}} = \frac{(2)(14) - (-8)(3)}{(9)(14) - (-8)(-8)} = \frac{52}{62} = 0.839 \text{ volt}$$

and

$$V_b = \frac{\begin{vmatrix} 9 & 2 \\ -8 & 3 \end{vmatrix}}{62} = \frac{(9)(3) - (2)(-8)}{62} = \frac{43}{62} = 0.694 \text{ volt}$$

Suppose that the current flowing from node a to node b through the conductance G_3 is desired. This current is given by

$$G_3 \cdot (V_a - V_b) = (5)(0.839 - 0.694) = 0.725 \text{ amp}$$

Using the technique outlined here, it is now possible to write the equations, in the form of Eq. set 2-75, by visual inspection of the circuit. Consider the following example.

EXAMPLE 2-4. Using the method of node voltages, write the set of linear algebraic equations for the network given in Fig. 2-22.

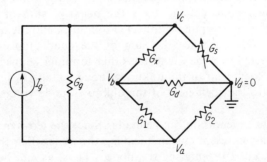

FIG. 2-22. Circuit for Example 2-4.

Solution: In a comparison with Example 2-2, it is evident that the same type of bridge network is indicated here, with the bridge and the detector elements taken as conductances rather than resistors and the bridge being driven by a current source rather than a voltage source. Inspection of Fig. 2-22 shows 4 nodes in the network and no voltage sources. The number of unknown node voltages (from Eq. 2-53) is

$$N_v = (n - 1) - N_{vs} = (4 - 1) - 0 = 3 \qquad (2\text{-}76)$$

This is equal to the unknown loop currents, given by Eq. 2-46, so this network represents a case where either method (loop currents or node voltages) results in the same number of equations to be solved. Now, utilizing the technique developed in the previous section, the equations can be written, at each node, by visual inspection of the network. With node d being the reference node as indicated, the equations are

$$
\begin{aligned}
(G_1 + G_2 + G_g) \cdot V_a & & -(G_1) \cdot V_b & & -(G_g) \cdot V_c &= -I_g \\
-(G_1) \cdot V_a &+ (G_1 + G_d + G_x) \cdot V_b & & -(G_x) \cdot V_c &= 0 \\
-(G_g) \cdot V_a & & -(G_x) \cdot V_b &+ (G_s + G_x + G_g) \cdot V_c &= I_g
\end{aligned}
$$
$$(2\text{-}77)$$

In this analysis of the bridge network, the condition for balance is that the node voltage V_b be identically zero. Under this balanced condition, the voltage difference across the detector is zero and hence the current through the detector is also zero. Note that a judicious choice was made for the reference node. If either node a or node c had been chosen as the reference node, then the potential difference across the detector $(V_b - V_d)$ would have to be zero for a balanced condition. This would involve two variables rather than one, and illustrates the point that if only partial information rather than a complete solution is desired for a network, the choice of the reference node can influence the amount of computation necessary to obtain this information. This same statement applies to the choice of loop currents when only partial information is desired. The determinant form of the solution for V_b is

$$
V_b = \cfrac{
\begin{vmatrix}
G_1 + G_2 + G_g & -I_g & -G_g \\
-G_1 & 0 & -G_x \\
-G_g & I_g & G_s + G_x + G_g
\end{vmatrix}
}{
\begin{vmatrix}
G_1 + G_2 + G_g & -G_1 & -G_g \\
-G_1 & G_1 + G_d + G_x & -G_x \\
-G_g & -G_x & G_s + G_x + G_g
\end{vmatrix}
}
$$

and, for $V_b = 0$ and expanding on column 2, one obtains for $\Delta \neq 0$

$$(-1)(-I_g)[(-G_1)(G_s + G_x + G_g) - (-G_x)(-G_g)]$$
$$+ (-1)(I_g)[(G_1 + G_2 + G_g)(-G_x) - (-G_g)(-G_1)] = 0$$

from which

$$G_x = \frac{G_1 G_s}{G_2} \qquad (2\text{-}78)$$

which is the same relation obtained in Eq. 2-50. This demonstrates that both methods of analysis of this bridge network yield the same relationship as, indeed, they must, since they are simply different analytical approaches to the same problem.

Consider a somewhat more involved network.

EXAMPLE 2-5. Using the method of node voltages, write the set of linear algebraic equations for the network given in Fig. 2-23.

Solution: Networks of the type depicted in Fig. 2-23 are often encountered as the result of cascading several basic sections to produce a desired relationship for the overall configuration. The network having 11 elements, 5 nodes, and 1 current source would require 6 unknown loop currents (by Eq. 2-27) if solved by the method of loop currents while 4 unknown node voltages are required (by Eq. 2-53) if solved by the method of node voltages. This is a case where one method, namely that of node voltages, has an advantage over the method of loop currents, in that two less unknowns are involved, resulting in a smaller set of algebraic equations to be solved. By inspection of the network, the equations are

$$
\begin{aligned}
(1 + 2)V_a \quad &-2V_b \quad &+0V_c \quad &+0V_d = 2 \\
-2V_a + (2 + 4 + 3 + 2)V_b \quad &-2V_c \quad &+0V_d = 0 \\
+0V_a \quad -2V_b + (2 + 4 + 3 + 2)V_c \quad &-2V_d = 0 \\
+0V_a \quad +0V_b \quad -2V_c + (2 + 4 + 3)V_d = 0
\end{aligned}
\tag{2-79}
$$

The equations may be further simplified by performing the sums indicated in the expressions for the coefficients, giving

$$
\left.
\begin{aligned}
3V_a - 2\ V_b + 0\ V_c + 0V_d &= 2 \\
-2V_a + 11V_b - 2\ V_c + 0V_d &= 0 \\
0V_a - 2\ V_b + 11V_c - 2V_d &= 0 \\
0V_a + 0\ V_b - 2\ V_c + 9V_d &= 0
\end{aligned}
\right\}
\tag{2-80}
$$

The solution for V_a is

$$
V_a = \frac{
\begin{vmatrix}
2 & -2 & 0 & 0 \\
0 & 11 & -2 & 0 \\
0 & -2 & 11 & -2 \\
0 & 0 & -2 & 9
\end{vmatrix}
}{
\begin{vmatrix}
3 & -2 & 0 & 0 \\
-2 & 11 & -2 & 0 \\
0 & -2 & 11 & -2 \\
0 & 0 & -2 & 9
\end{vmatrix}
} = \frac{2018}{2647} = 0.762 \text{ volt}
$$

In a similar manner, the solutions for the other nodal voltages are

$$
V_b = \frac{380}{2647} = 0.144 \text{ volt}
$$

$$
V_c = \frac{72}{2647} = 0.027 \text{ volt}
$$

and

$$
V_d = \frac{16}{2647} = 0.006 \text{ volt}
$$

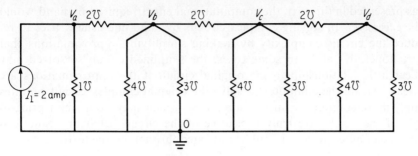

FIG. 2-23. Circuit for Example 2-5.

Summary of the Method of Node Voltages. The technique of utilizing node voltages, as a method of circuit analysis, provides a complementary analytical tool to the method of loop currents. For a given network, it may result in a proper set of equations that is smaller in number than would be required by the method of loop currents. It also lends itself to situations where only partial information is desired, such as the potential of one point in a network with respect to another point. As in the case with the method of loop currents, a proper set of node-voltage equations, for a resistive network with d-c current sources, results in a set of simultaneous linear algebraic equations.

Quite frequently, circuits having both voltage sources and current sources must be analyzed. It has been demonstrated that the practical current sources (outside the source terminals) could be converted to equivalent practical voltage sources and the network analyzed by the method of loop currents. Alternatively, the practical voltage sources (outside the source terminals) could be converted to equivalent practical current sources and the method of node voltages could be employed in the analysis of the network. In either case, it has been demonstrated how the equations can be immediately written, in the final numerical form for solution, by visual inspection of the network. If some additional rules are imposed on the choice of variables and the manner of writing the equations, equations for circuits containing both voltage and current sources can be written directly without too much trouble.

One further note, many computer programs now exist for obtaining the solutions to electrical circuits. The authors heartily endorse their use after Kirchhoff's laws are understood and have been applied to circuits for a sufficient number of times to enable one to gain an intuitive judgement as to the behavior of circuit variables. One must be able to verify any computer solution which is questionable and situations exist where a computer is not available or uneconomical from a cost or time viewpoint. The computer solutions become much more important when the complexity of the circuits is such that even though the analysis is always the same as presented in this text, the computational effort required by hand would be excessive. In solving complex circuits by hand one usually tries to reduce the circuit complexity by making simplifying approximations that are hopefully valid. In some cases, the simplified circuit so solved may bear little relationship to the original circuit if the approximations are weak. Using the computer to solve the original complex circuit requires only more computer computational time which may be a small price to pay if the results are truly indicative of the circuit behavior. Solutions from computer programs will be indicated throughout the text.

2-6. Dependent Sources

The voltage and/or current sources discussed to this point have been independent in the sense that they do not depend on any other voltage or current in the network. Many electrical and electronic devices have the characteristic that a source generating energy in one portion of the device *depends* on a voltage or current in some other portion of the device. Such voltage and/or current sources are termed *dependent* or *controlled* energy sources. This dependency must be considered in the equivalent circuit representation of the device. Analyses of circuits containing dependent sources require that such sources be properly taken into account when writing the equations. To illustrate the technique of handling circuits containing dependent sources, several examples follow.

EXAMPLE 2-6. Solve the circuit given in Fig. 2-24 by the method of loop currents.

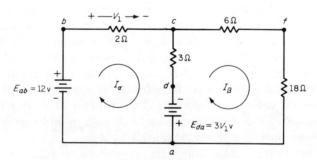

FIG. 2-24. Circuit for Example 2-6.

Solution: Note the voltage source in the center branch is a dependent source since its value *depends* on the controlling voltage V_1 developed across the 2-ohm resistor. The two loop current equations are

$$\text{Loop } abcda \qquad V_{ab} + V_{bc} + V_{cd} + V_{da} = 0$$

or

$$-12 + 2I_\alpha + 3(I_\alpha - I_\beta) - 3V_1 = 0$$

Loop *cfadc* $\qquad V_{cf} + V_{fa} + V_{ad} + V_{dc} = 0$

or

$$6 I_\beta + 18 I_\beta + 3 V_1 + 3 (I_\beta - I_\alpha) = 0$$

Note that the dependent source introduced the unknown controlling voltage V_1 into the set of loop current equations. An additional equation is necessary to relate this unknown to the loop currents and through use of Ohm's law is given by

$$V_1 = 2 I_\alpha$$

With this relationship, the loop current equations become

$$-I_\alpha - 3I_\beta = 12$$
$$3I_\alpha + 27I_\beta = 0 \qquad\qquad (2\text{-}81)$$

The solution gives $I_\alpha = -18$ amperes and $I_\beta = 2$ amperes. The important point in this solution is that once the equation for the dependent source is inserted into the original equation set, only the loop currents remain as unknowns as in previous sections. Note however, that writing the equations by inspection is difficult since the dependent source will have an effect on one or more of the coefficients. If not sure, one should always invoke Kirchhoff's laws as the starting point for writing the equations.

Consider the same circuit with the dependent voltage source dependent on a current rather than a voltage as shown in Fig. 2-25.

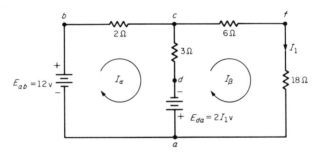

FIG. 2-25. Modified circuit for Example 2-6.

The loop-current equations become

Loop *abcda* $\qquad V_{ab} + V_{bc} + V_{cd} + V_{da} = 0$

or

$$-12 + 2I_\alpha + 3(I_\alpha - I_\beta) - 2I_1 = 0$$

Loop *cfadc* $\qquad V_{cf} + V_{fa} + V_{ad} + V_{dc} = 0$

or

$$6I_\beta + 18I_\beta + 2I_1 + 3(I_\beta - I_\alpha) = 0$$

In this instance, the controlling branch current I_1 has been introduced into the set of loop current equations and must be expressed in terms of the loop currents.

Inspection of the circuit shows

$$I_1 = I_\beta$$

with this relationship, the loop current equations become

$$5I_\alpha - 5I_\beta = 12$$
$$-3I_\alpha + 29I_\beta = 0$$

The solution gives $I_\alpha = \dfrac{174}{65}$ amperes and $I_\beta = \dfrac{18}{65}$ ampere which is totally un-related to the first solution. Naturally with a change in a single element a re-distribution of all voltages and currents is to be expected. This example points out that with dependent sources present, usually, but not always, an additional equation per each such source will be required to relate the controlling variable to the proper set of voltages or currents being used to effect the solution.

A circuit containing a dependent current source follows in the next example.

EXAMPLE 2-7. Solve the circuit given in Fig. 2-26. Show that an energy balance exists between energy sources and energy dissipators.

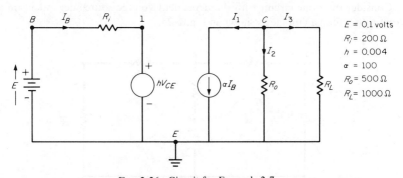

FIG. 2-26. Circuit for Example 2-7.

Solution: Note that the circuit contains a dependent voltage source (depends on the circuit node voltage V_{CE}) as well as a dependent current source (depends on the branch or mesh current I_B). Later it will be demonstrated that this circuit is an accurate representation of a transistor amplifier in a common emitter con-figuration with the values of several parameters dependent on the particular transistor used. For now however, the emphasis is on obtaining the solution for the network. Using the method of node voltages and writing the current equation at node C (current leaving the node) gives

$$I_1 + I_2 + I_3 = 0$$
$$\downarrow \quad \downarrow \quad \downarrow$$
$$\alpha I_B + \frac{V_{CE}}{R_o} + \frac{V_{CE}}{R_L} = 0$$

This equation contains an unknown current I_B, resulting from the dependent current source, which is expressible as

$$I_B = \frac{V_{BE} - V_{1E}}{R_i} = \frac{E - hV_{CE}}{R_i} \tag{2-82}$$

With this relationship the node voltage equation becomes

$$V_{CE}\left(\frac{h \cdot \alpha}{R_i} - \frac{1}{R_o} - \frac{1}{R_L}\right) = \frac{\alpha E}{R_i} \tag{2-83}$$

Substituting the values of the parameters into Eq. 2-83 gives

$$V_{CE} = -50 \text{ volts}$$

From Eq. 2-82,

$$I_B = 1.5 \times 10^{-3} \text{ amps or } 1.5 \text{ ma}$$

Energy balance check; power *losses* are tabulated:

E---$P_E = V_{EB}I_{EB} = (-E)I_B = (-0.1)(1.5 \times 10^{-3})$ $\quad = -0.15 \text{ mw}$

which is electrical energy given to circuit;

R_i---$P_{R_i} = (I_B)^2 \cdot R_i = (1.5 \times 10^{-3})^2(200)$ $\quad = 0.45 \text{ mw}$

which is electrical energy transformed to heat and radiated from circuit;

hV_{CE}---$P_{hV_{CE}} = V_{1E}I_{1E} = (hV_{CE})I_B = (.004)(-50)(1.5 \times 10^{-3}) = -0.3 \text{ mw}$

which is electrical energy given to circuit;

αI_β---$P_{\alpha I_\beta} = V_{CE}(\alpha I_\beta) = (-50)(100)(1.5 \times 10^{-3})$ $\quad = -7.5 \text{ watts}$

which is electrical energy given to circuit;

R_o---$P_{R_o} = (V_{CE})^2/R_o = (-50)^2/500$ $\quad = 5 \text{ watts}$

which is electrical energy transformed to heat and radiated from circuit;

R_L---$P_{R_L} = (V_{CE})^2/R_L = (-50)^2/1000$ $\quad = 2.5 \text{ watts}$

which is electrical energy transformed to heat and radiated from circuit;

$$\text{Sum} = 0$$

The energy balance checks out. In later chapters, the ultimate source of the energy provided to the network by the dependent sources will be discussed.

It has been demonstrated that the analysis of networks containing dependent voltage and/or current sources is not very different from circuits containing only independent sources. The only real difference is that additional equations relating the dependent sources to the chosen circuit variables are required before the solution can be effected.

2-7. Further Application of the Methods of Loop Currents and Node Voltages

Loop-Current Analysis of a Network with Current Sources. It was observed in Eq. 2-27 that the number of current sources in a network has an effect on the number of unknown loop currents needed for the complete solution of a network by the loop-current method. This section will show what that effect is and will demonstrate how properly to take it into account in order to maintain a systematic manner of writing the loop equations by inspection of the network. Recall that a current source delivers a constant current independent of the voltage across its terminals. This therefore implies that the branch current in a branch containing a current source is known and, indeed, is exactly equal to the value of the current source. An elementary example will illustrate the procedure for developing the proper equation-writing technique.

EXAMPLE 2-8.   Discuss the solution to the network given in Fig. 2-27.

Solution: Figure 2-27a shows that the network is composed of 5 elements, 4 nodes, and 1 current source, so the necessary number of *unknown* loop currents, as given by Eq. 2-27, is

$$N_i = e + 1 - (n + N_{cs})$$
$$= (5 + 1) - (4 + 1) = 1 \qquad [2\text{-}27]$$

The first observation is that this number is less by 1 (the number of current sources in the network) than the number of meshes (2) in the network. The choice of the single loop current in Fig. 2-27b gives rise to the equation

$$R_1 i_1 + R_3 i_1 + R_4 i_1 + R_2 i_1 = 0$$

which results in

$$i_1 = 0$$

This certainly is not a valid solution, since the single current source (assumed to be nonzero) will cause a current to flow in each of the two resistive branches. The choice of the single loop current in Fig. 2-27c is inadequate, for two reasons: It is quite obvious that the branch containing the series combination of the resistors R_3 and R_4 is not accounted for, and, furthermore, the current i_1 is the only loop current through the branch containing the current source and is therefore equal to the current I_B, which makes it a known rather than an unknown loop current. This last condition means that the leftmost branch current, which is equal to the loop current i_1, is constrained to be equal to I_B.

It is quite apparent that a similar situation exists for the choice of the single loop current in Fig. 2-27d.

It is reasonable to ponder what would happen if one chose the set of mesh currents, as the unknown loop currents, that always assured a proper set of equations when no current sources were present. Consider such a choice, as shown in Fig. 2-27e. Symbolically, the equations for mesh 1 and mesh 2 are,

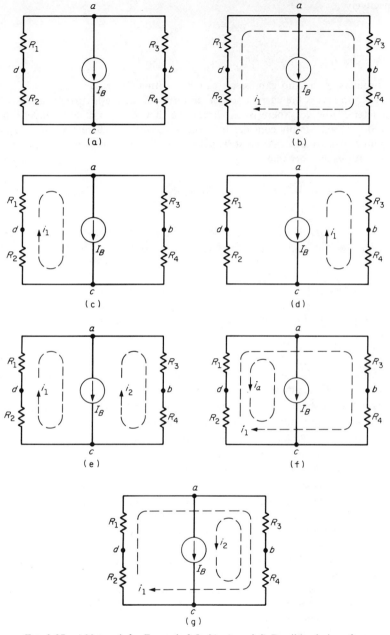

FIG. 2-27. a) Network for Example 2-8. b), c), and d) Possible choice of a single loop current. e) Choice of mesh currents. f), and g) Possible choice of a single unknown loop current and a single known loop current.

respectively,

$$v_{cd} + v_{da} + v_{ac} = 0$$

and

$$v_{ab} + v_{bc} + v_{ca} = 0$$

The problem is that no expression can be written for v_{ac} or v_{ca}, which is the voltage drop across the current source, in terms of the mesh currents and a fixed parameter. However, there is an alternative that leads to a set of independent equations. First, simply consider the voltage across the current source, say v_{ac}, as an unknown voltage that must be solved for, in addition to the mesh currents. The equations then become

$$R_2 i_1 + R_1 i_1 + v_{ac} = (R_1 + R_2)i_1 + 0i_2 + v_{ac} = 0 \qquad (2\text{-}84)$$

and

$$R_3 i_2 + R_4 i_2 + v_{ca} = 0i_1 + (R_3 + R_4)i_2 - v_{ac} = 0 \qquad (2\text{-}85)$$

since

$$v_{ca} = -v_{ac}$$

Equations 2-84 and 2-85 are two equations with three unknowns, namely, i_1, i_2, and v_{ac}. Another equation is needed, and this can be obtained by observing that, in the branch containing the current source, the 2 loop currents must satisfy the relation

$$i_1 - i_2 = I_B \qquad (2\text{-}86)$$

Now Eq. 2-86 along with Eqs. 2-84 and 2-85 form an independent set of three equations with three unknowns. This technique can be used for any network containing any number of current sources. The steps are as follows: 1) choose mesh currents as the unknown loop currents; 2) write the equations for the mesh currents, treating the voltage drops across the current sources as additional unknowns; and 3) for each branch containing a current source, write the relationship that must be satisfied for the mesh currents traversing that branch.

The foregoing steps have outlined a procedure that will lead to the solution of any network containing current sources. However, upon closer observation, it can be seen to have some features that make it rather unattractive as a means of solving such networks. First of all, unknown voltages as well as mesh currents appear in the equations, and therefore the equations are no longer simply loop-current equations. With this procedure, the number of equations will always be equal to the number of meshes *plus* the number of current sources. But, from Eq. 2-27, the necessary number of unknown loop currents is equal to the number of meshes *minus* the number of current sources. Thus, this procedure has involved more equations than are absolutely necessary to solve the network.

In the example, if only 1 unknown loop current need be chosen, then that indicated in Fig. 2-27b seems to be the proper one of the three possible choices. The other two, by virtue of passing through the branch containing the current source, become known loop currents. Suppose, now, that a *known* loop current is chosen in addition to the unknown loop current. This can readily be accomplished by making it the *only* loop current traversing the branch containing the current source. Then the choice of this second known loop current does not add an unknown, since it is of known value. Such a combination is indicated in Fig. 2-27f, where

$$i_a = (-)I_B \tag{2-87}$$

and is therefore a known loop current. The equation for loop (*abcda*), which is the path of the unknown loop current, is given symbolically by

$$v_{ab} + v_{bc} + v_{cd} + v_{da} = 0 \tag{2-88}$$

or

$$i_1 R_3 + i_1 R_4 + (i_1 - i_a)R_2 + (i_1 - i_a)R_1 = 0 \tag{2-89}$$

from whence

$$(R_1 + R_2 + R_3 + R_4)i_1 = (R_1 + R_2)i_a \tag{2-90}$$

Since the value of the current source I_B is known, substituting Eq. 2-87 into Eq. 2-90 gives

$$(R_1 + R_2 + R_3 + R_4)i_1 = (R_1 + R_2)(-I_B)$$

which is the single equation with 1 unknown loop current that can be solved to give

$$i_1 = \frac{(R_1 + R_2)(-I_B)}{R_1 + R_2 + R_3 + R_4} \tag{2-91}$$

The choice of the unknown and known loop currents in Fig. 2-27g will also lead to a single equation with one unknown.

A systematic manner of choosing the unknown and known loop currents which will lead to the ability of writing the set of linear algebraic equations by inspection is most desirable. In order to capitalize on the procedure outlined in Sec. 2-3, the following steps can be used to achieve this goal:

1. To choose the unknown loop currents, assume that the branches containing current sources are removed from the circuit, and choose the remaining meshes as the loops.

2. Choose all unknown loop currents in the same sense, either clockwise or counterclockwise.

3. Choose a known loop current by having the loop traverse one, and only one, branch containing a current source. This means that as many such loops must be chosen as there are current sources.

4. Choose all known loop currents in the same sense. (A good procedure is to choose them in a sense opposite to that of the unknown loop currents, so they may be distinguished easily.)

5. Write Kirchhoff's voltage law in the form of the summation of the voltage drops around the loop in the direction of the unknown loop currents. This must be equal to zero.

This procedure is best illustrated by an example.

EXAMPLE 2-9. Following the preceding steps, obtain and solve the set of linear algebraic equations for the network shown in Fig. 2-28.

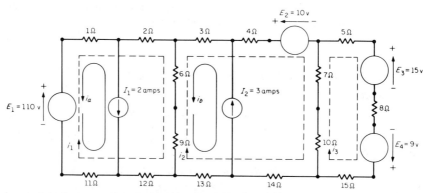

FIG. 2-28. Circuit for Example 2-9.

Solution: In the network of Fig. 2-28, the number of elements is 21, the number of nodes is 17, and the number of current sources is 2, and therefore, by Eq. 2-27, the necessary number of unknown loop currents is given by

$$N_i = e + 1 - (n + N_{cs}) = (21 + 1) - (17 + 2) = 3$$

The unknown loop currents (i_1, i_2, and i_3) are chosen by visualizing the deletion of the branches containing the current sources and choosing the unknown loop currents as the mesh currents of the remaining network, all directed in a clockwise sense and indicated by the dashed lines. The known loop currents (i_a, i_b) are then chosen, with each traversing a branch containing a current source and any other convenient branches to complete the loop, as long as the other branches do not contain current sources. The 2 known currents, directed in a counterclockwise sense, are shown by the solid lines. It now remains to write Kirchhoff's voltage law for each of the 3 loops, equating the sum of the voltage drops around a closed loop to zero. Thus, for

Loop 1:

$$1(i_1 - i_a) + 2i_1 + 6(i_1 - i_2 + i_b) + 9(i_1 - i_2 + i_b)$$
$$+ 12i_1 + 11(i_1 - i_a) - E_1 = 0$$

Loop 2:

$$3(i_2 - i_b) + 4i_2 + E_2 + 7(i_2 - i_3) + 10(i_2 - i_3) + 14i_2$$
$$+ 13(i_2 - i_b) + 9(i_2 - i_1 - i_b) + 6(i_2 - i_1 - i_b) = 0$$

Loop 3:

$$5i_3 + E_3 + 8i_3 - E_4 + 15i_3 + 10(i_3 - i_2) + 7(i_3 - i_2) = 0$$

$$(2\text{-}92)$$

Rearranging Eq. set 2-92 gives

$$\left.\begin{array}{l}(1 + 2 + 6 + 9 + 12 + 11)i_1 - (6 + 9)i_2 - (0)i_3 \\ \qquad = E_1 + (1 + 11)i_a - (6 + 9)i_b \\ -(6 + 9)i_1 + (3 + 4 + 7 + 10 + 14 + 13 + 9 + 6)i_2 - (7 + 10)i_3 \\ \qquad = -E_2 + (0)i_a + (3 + 13 + 9 + 6)i_b \\ -(0)i_1 - (7 + 10)i_2 + (5 + 8 + 15 + 10 + 7)i_3 \\ \qquad = E_4 - E_3 + (0)i_a + (0)i_b \end{array}\right\} (2\text{-}93)$$

The first observation is that the coefficients of the unknown loop currents have the same relationships as they would have had if the circuit had been treated by the method of mesh currents, with the branches containing the current sources removed. Thus, the left-hand side of the equation set can be readily obtained by inspection of the network. The right-hand side is also the same with respect to the voltage sources, but there are additional terms due to the current sources. Recall that the right-hand side of the equation can be thought to represent the summation of the fixed voltage rises encountered around the loop. Now a *known* current flowing through a resistor produces a *known* voltage. Consider the case when the known loop-current sense is opposite to the unknown loop-current sense in any branch common to both loop currents. Then the known loop current produces a voltage rise in any resistor in the common branch *in* the direction of the unknown loop-current sense. This, then, represents a positive term on the right-hand side of the equation. Alternatively, consider the case when the known loop-current sense is in the same direction as the unknown loop-current sense in any branch common to both loops. Then the known loop current produces a voltage drop in any resistor in the common branch *in* the direction of the unknown loop-current sense. This, then, represents a negative term on the right-hand side of the equation. This means that, by checking the senses of the known loop currents with respect to the unknown loop current for each loop equation, the sign of the terms involving the known loop currents can be ascertained. The coefficient of each known loop current in each loop equation will simply be the summation of the resistors that are common to the respective known loop currents and the particular unknown loop current. Finally, each known loop current is replaced by its value in terms of its relation to the constant-current source that it traverses. Adding the coefficients in Eq. set 2-93 gives

$$\left.\begin{array}{l}(41)i_1 - (15)i_2 - (0)i_3 = (12)i_a - (15)i_b + E_1 \\ -(15)i_1 + (66)i_2 - (17)i_3 = (0)i_a + (31)i_b - E_2 \\ -(0)i_1 - (17)i_2 + (45)i_3 = (0)i_a + (0)i_b + E_4 - E_3 \end{array}\right\} (2\text{-}94)$$

Now, from the network in Fig. 2-28,

$$E_1 = 110, E_2 = 10, E_3 = 15, E_4 = 9, i_a = (-)I_1 = -2, i_b = (+)I_2 = 3$$

giving

$$\left.\begin{array}{l}41i_1 - 15i_2 - 0i_3 = +41 \\ -15i_1 + 66i_2 - 17i_3 = +83 \\ -0i_1 - 17i_2 + 45i_3 = -6 \end{array}\right\} (2\text{-}95)$$

which is the desired set of linear algebraic equations.

Solving the set of equations by using determinants gives

$$i_1 = \frac{\begin{vmatrix} 41 & -15 & -0 \\ 83 & 66 & -17 \\ -6 & -17 & +45 \end{vmatrix}}{\begin{vmatrix} 41 & -15 & -0 \\ -15 & 66 & -17 \\ -0 & -17 & +45 \end{vmatrix}} = \frac{(+1)(41)\begin{vmatrix} 66 & -17 \\ -17 & +45 \end{vmatrix} + (-1)(-15)\begin{vmatrix} 83 & -17 \\ -6 & 45 \end{vmatrix}}{(+1)(41)\begin{vmatrix} 66 & -17 \\ -17 & +45 \end{vmatrix} + (-1)(-15)\begin{vmatrix} -15 & -17 \\ 0 & 45 \end{vmatrix}}$$

$$= \frac{164,416}{99,796} = 1.65 \text{ amps}$$

$$i_2 = \frac{\begin{vmatrix} 41 & 41 & -0 \\ -15 & 83 & -17 \\ -0 & -6 & 45 \end{vmatrix}}{99,796} = \frac{176,628}{99,796} = 1.77 \text{ amps}$$

$$i_3 = \frac{\begin{vmatrix} 41 & -15 & 41 \\ -15 & 66 & 83 \\ -0 & -17 & -6 \end{vmatrix}}{99,796} = \frac{53,420}{99,796} = 0.537 \text{ amp}$$

Node-Voltage Analysis of a Network with Voltage Sources. The number of unknown node voltages is affected by the number of voltage sources that appear in the circuit, as indicated by Eq. 2-53. However, the treatment, by the method of node voltages, of a network containing a voltage source differs only slightly from the networks previously analyzed. It was shown that, with one node of the voltage source considered to be the unknown voltage, the voltage of the other node is then known relative to the first node. The treatment of such a network is best explained by way of examples.

EXAMPLE 2-10. Discuss the solution to the network given in Fig. 2-29.

Solution: There are 3 nodes and 1 voltage source in the network shown in Fig. 2-29a. Therefore, from Eq. 2-53

$$N_v = (n - 1) - N_{vs}$$
$$= (3 - 1) - 1 = 1 \qquad [2\text{-}53]$$

The choosing of the reference node is still arbitrary and the choice of the unknown node voltage follows in a straightforward manner. In Fig. 2-29b, node *a* is chosen as the reference node; hence v_a is zero. Now the voltage at node *b* is known, since it must be E_{ba} volts lower in potential than node *a*. Therefore, node *c* represents the unknown node voltage. In a like manner, if node *b* is chosen as the reference node so that v_b is zero, then the voltage at node *a* must be E_{ba} volts higher in potential than node *b*, and node *c* again represents the unknown voltage. If node *c* is chosen as the reference node, then either node *a* or node *b* may represent the unknown voltage, but not both, since knowing one fixes the other. This

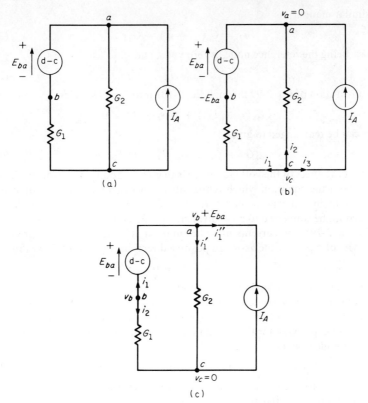

FIG. 2-29. a) Circuit for Example 2-10. b) Node *a* chosen as reference node, with node *c* representing the unknown node voltage.
e) Node *c* chosen as reference node, with node *b* representing
the unknown node voltage.

choice of the reference node is given in Fig. 2-29c, with node *b* representing the
unknown node voltage.

Some care must be exercised in writing the equations, and this is best illustrated by considering, in turn, the two cases represented in Figs. 2-29b and 2-29c.

In Fig. 2-29b, Kirchhoff's current law must be applied at node *c*. Considering
the assumed positive directions of each of the currents as shown in the figure

$$i_1 + i_2 + i_3 = 0 \qquad (2\text{-}96)$$

For the current i_1, one would normally write

$$i_1 = G_1(v_c - v_b) \qquad (2\text{-}97)$$

but v_b is known to be E_{ba} volts lower in potential than the reference node, or

$$v_b = -E_{ba} \qquad (2\text{-}98)$$

Therefore, Eq. 2-97 becomes

$$i_1 = G_1[v_c - (-E_{ba})] = G_1(v_c + E_{ba}) \qquad (2\text{-}99)$$

In a similar manner,

$$i_2 = G_2(v_c - v_a) = G_2 v_c \qquad (2\text{-}100)$$

since v_a, being the reference node, is zero volts, and

$$i_3 = I_A \qquad (2\text{-}101)$$

The substitution of Eq. 2-99 through Eq. 2-101 into Eq. 2-96 gives

$$G_1(v_c + E_{ba}) + G_2 v_c + I_A = 0 \qquad (2\text{-}102)$$

which can be rearranged to give

$$(G_1 + G_2)v_c = -I_A - G_1 E_{ba} \qquad (2\text{-}103)$$

Note that the effect of the voltage source is to give an additional term on the right-hand side of the equation which represents known currents. This should not be unexpected, since it was demonstrated, in Chapter 1, that a practical voltage source could be converted to an equivalent current source.

In Fig. 2-29c, the currents must be summed at node b, since it represents the unknown voltage. For the assumed positive directions of current flow, this sum is

$$i_1 + i_2 = 0 \qquad (2\text{-}104)$$

Now the current i_1 is directed from node b to node a through a fixed voltage potential which is independent of the current; hence no relationship exists between the current i_1 and the difference of potential between node b and node a. In order to obtain an expression for the current i_1, Kirchhoff's current law must be further applied at node a, where

$$-i_1 + i_1' + i_1'' = 0 \qquad \text{or} \qquad i_1 = i_1' + i_1'' \qquad (2\text{-}105)$$

Subjectively, one views i_1 splitting into the 2 currents $i_1' + i_1''$ which, when substituted into Eq. 2-104 gives

$$i_1' + i_1'' + i_2 = 0 \qquad (2\text{-}106)$$

Now expressions can be obtained for each of the currents, namely,

$$i_1' = G_2(v_a - v_c) = G_2[(v_b + E_{ba}) - 0] \qquad (2\text{-}107)$$

$$i_1'' = -I_A \qquad (2\text{-}108)$$

and

$$i_2 = G_1(v_b - v_c) = G_1 v_b \qquad (2\text{-}109)$$

Substituting Eqs. 2-107 through 2-109 into Eq. 2-106 gives

$$G_2(v_b + E_{ba}) - I_A + G_1 v_b = 0 \qquad (2\text{-}110)$$

or

$$(G_1 + G_2)v_b = I_A - G_2 E_{ba} \qquad (2\text{-}111)$$

As shown in Example 2-10, the effect of the voltage source is to give an additional term to the right-hand side of the equation which represents known current sources. The choice of the reference node and the node representing the unknown voltage will affect the sign of this term. These two cases show how to account properly for a voltage source when using the method of node voltages, and the technique remains the same no matter how complex the network may become. It is simply a matter of practice to facilitate the writing of equations when voltage sources appear in the network.

Consider the circuit of Example 2-1 which was solved by the method of loop currents. It is desired to solve the same circuit by using the method of node voltages. The circuit diagram is repeated in Fig. 2-30.

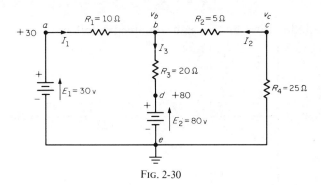

FIG. 2-30

The solution is accomplished by using node e as the reference node and noting that the unknown node voltages are v_b and v_c, since $v_a = E_1$ and $v_d = E_2$ as indicated in Fig. 2-30. Applying Kirchhoff's current law to node b, one obtains

$$\frac{v_b - v_a}{R_1} + \frac{v_b - v_c}{R_2} + \frac{v_b - v_d}{R_3} = 0 \tag{2-112}$$

and in a similar manner to node c, one obtains

$$\frac{v_c - v_b}{R_2} + \frac{v_c}{R_4} = 0 \tag{2-113}$$

Substituting the known circuit values into Eqs. 2-112 and 2-113 gives

$$\frac{v_b - 30}{10} + \frac{v_b - v_c}{5} + \frac{v_b - 80}{20} = 0$$

and

$$\frac{v_c - v_b}{5} + \frac{v_c}{25} = 0$$

Upon rearranging,

$$\left(\frac{1}{10} + \frac{1}{5} + \frac{1}{20}\right)v_b - \frac{1}{5}v_c = \frac{30}{10} + \frac{80}{20}$$

and

$$-\frac{1}{5}v_b + \left(\frac{1}{5} + \frac{1}{25}\right)v_c = 0$$

The equation set to be solved is

$$\frac{7}{20}v_b - \frac{1}{5}v_c = 7$$

$$-\frac{1}{5}v_b + \frac{6}{25}v_c = 0$$

The solution for v_b and v_c can be obtained as follows:

$$v_b = \frac{\begin{vmatrix} 7 & -\dfrac{1}{5} \\ 0 & \dfrac{6}{25} \end{vmatrix}}{\begin{vmatrix} \dfrac{7}{20} & -\dfrac{1}{5} \\ -\dfrac{1}{5} & \dfrac{6}{25} \end{vmatrix}} = \frac{42/25}{22/500} = \frac{42 \times 20}{22} = \frac{420}{11} \text{ volts}$$

and

$$v_c = \frac{\begin{vmatrix} \dfrac{7}{20} & 7 \\ -\dfrac{1}{5} & 0 \end{vmatrix}}{\begin{vmatrix} \dfrac{7}{20} & -\dfrac{1}{5} \\ -\dfrac{1}{5} & \dfrac{6}{25} \end{vmatrix}} = \frac{7/5}{22/500} = \frac{350}{11} \text{ volts}$$

These results can be compared with those obtained by using loop-current methods by now computing the branch currents, that is,

$$I_1 = \frac{v_a - v_b}{R_1} = \frac{30 - 420/11}{10} = -\frac{9}{11} \text{ amp}$$

$$I_2 = \frac{v_c - v_b}{R_2} = \frac{350/11 - 420/11}{5} = -\frac{14}{11} \text{ amps}$$

and

$$I_3 = \frac{v_b - v_d}{R_3} = \frac{420/11 - 80}{20} = -\frac{23}{11} \text{ amps}$$

They are identical, as they should be.

A final example will treat a more complex network containing voltage sources.

EXAMPLE 2-11.   By the method of node voltages, determine the complete solution for the network given in Fig. 2-31.

Solution: There are 6 nodes and 2 voltage sources, so the number of unknown node voltages, given by Eq. 2-53, is

$$N_v = (n - 1) - N_{vs} = (6 - 1) - 2 = 3$$

Note that 3 loop currents are necessary for a complete solution, so either method will result in the same number of equations making up a proper set. The network in Fig. 2-31b shows node a as the reference node. Node b represents an unknown node voltage; thus, the voltage at node e is known, since it is E_{eb} volts lower in potential than the voltage at node b. Node f also represents an unknown node voltage; therefore, the voltage at node c is known, since it is E_{fc} volts higher in potential than the voltage at node f. Node d represents the third unknown node voltage. Kirchhoff's current law now needs to be applied at each of the nodes representing an unknown node voltage. The assumed positive directions of the currents at each node are shown in Fig. 2-31b.

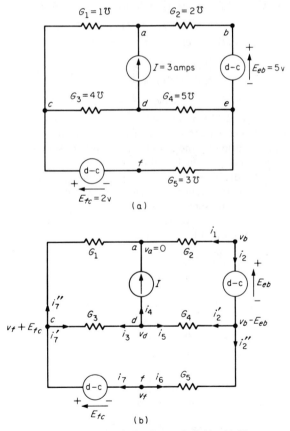

FIG. 2-31. a) Circuit for Example 2-11. b) Nomenclature for node voltage solution.

At node b the describing equation is

$$i_1 + i_2 = i_1 + i_2' + i_2'' = 0$$

or

$$G_2(v_b - 0) + G_4[(v_b - E_{eb}) - v_d] + G_5[(v_b - E_{eb}) - v_f] = 0$$

At node d the describing equation is

$$i_3 + i_4 + i_5 = 0$$

or

$$G_3[v_d - (v_f + E_{fc})] + I + G_4[v_d - (v_b - E_{eb})] = 0$$

At node f the describing equation is

$$i_6 + i_7 = i_6 + i_7' + i_7'' = 0$$

or

$$G_5[v_f - (v_b - E_{eb})] + G_3[(v_f + E_{fc}) - v_d] + G_1[(v_f + E_{fc}) - 0] = 0$$

The proper set of equations becomes

$$\left.\begin{aligned}
(G_2 + G_4 + G_5)v_b - G_4 v_d - G_5 v_f &= (G_4 + G_5)E_{eb} \\
-G_4 v_b + (G_3 + G_4)v_d - G_3 v_f &= G_3 E_{fc} - G_4 E_{eb} - I \\
-G_5 v_b - G_3 v_d + (G_1 + G_3 + G_5)v_f &= -G_1 E_{fc} - G_3 E_{fc} - G_5 E_{eb}
\end{aligned}\right\} \quad (2\text{-}114)$$

and, upon substitution of the given numerical values,

$$\left.\begin{aligned}
10v_b - 5v_d - 3v_f &= 40 \\
-5v_b + 9v_d - 4v_f &= -20 \\
-3v_b - 4v_d + 8v_f &= -25
\end{aligned}\right\} \quad (2\text{-}115)$$

Examination of Eq. set 2-114 or Eq. set 2-115 again shows that the coefficients of the unknown voltages are the same as if the voltage sources were replaced by short circuits, that is, bringing the 2 nodes of a voltage source together to form a single node representing a single unknown node voltage. Thus, with a little practice, the left sides of the equations can be written by visual inspection of the network. The form of the right-hand sides depends upon which nodes of the equations are picked to represent unknown node voltages and which node voltages are then known as a result of the presence of voltage sources in the network.

A determinant solution of Eq. set 2-115 gives

$$v_b = \frac{\begin{vmatrix} 40 & -5 & -3 \\ -20 & 9 & -4 \\ -25 & -4 & 8 \end{vmatrix}}{\begin{vmatrix} 10 & -5 & -3 \\ -5 & 9 & -4 \\ -3 & -4 & 8 \end{vmatrix}} = \frac{25}{159} = +0.16 \text{ volt}$$

$$v_d = \frac{\begin{vmatrix} 10 & 40 & -3 \\ -5 & -20 & -4 \\ -3 & -25 & 8 \end{vmatrix}}{159} = \frac{-715}{159} = -4.50 \text{ volts}$$

and

$$v_f = \frac{\begin{vmatrix} 10 & -5 & 40 \\ -5 & 9 & -20 \\ -3 & -4 & -25 \end{vmatrix}}{159} = \frac{-845}{159} = -5.31 \text{ volts}$$

2-8. Network Relationships

Several special formulations and techniques require mention because of their use, although the techniques already discussed could always be used to obtain any information obtained by the methods which follow. However, in many instances, these methods may offer a simpler approach to the solution by not requiring as much computational work as with the preceding techniques. Among the topics to be treated are the voltage divider formula, current splitting formula, network reduction, driving point resistance and transfer relationships.

Voltage Divider Formula. Consider two resistors in series as shown in Fig. 2-32 where the voltage V across the series combination is presumed

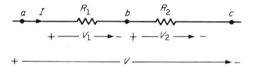

FIG. 2-32. Voltage divider network.

known. Since the same current must flow through both resistors (series connected) Kirchhoff's voltage law gives

$$V_{ac} = V_{ab} + V_{bc}$$
$$\downarrow \qquad \downarrow \qquad \downarrow$$
$$V = V_1 + V_2$$

or

$$V = I \cdot R_1 + I \cdot R_2 \rightarrow I = \frac{V}{R_1 + R_2} \qquad (2\text{-}116)$$

Ohm's law gives

$$V_1 = I \cdot R_1 \qquad \text{and} \qquad V_2 = I \cdot R_2$$

and substituting for I from Eq. 2-116 gives

$$V_1 = \frac{R_1}{R_1 + R_2} \cdot V \qquad \text{and} \qquad V_2 = \frac{R_2}{R_1 + R_2} \cdot V \qquad (2\text{-}117)$$

Equation 2-117 gives the individual resistor voltages in terms of the voltage across the series combination and the resistance parameter values. The relative values of the resistors determine how the total voltage divides across the two resistors and this ratio is given by

$$\frac{V_1}{V_2} = \frac{R_1}{R_2}$$

Such divider networks can be used to scale down a given voltage that may be inaccessable to direct control or measurement.

Current Splitting Formula. Consider two resistors in parallel as shown in Fig. 2-33 where the current I into the parallel combination is

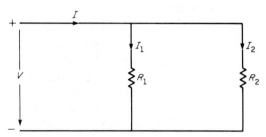

FIG. 2-33. Current splitting network.

presumed known. Kirchhoff's current law gives

$$I = I_1 + I_2$$

or

$$I = \frac{V}{R_1} + \frac{V}{R_2} \rightarrow V = \frac{R_1 R_2}{R_1 + R_2} \cdot I \qquad (2\text{-}118)$$

Ohm's law gives

$$I_1 = \frac{V}{R_1} \qquad \text{and} \qquad I_2 = \frac{V}{R_2}$$

and substituting for V from Eq. 2-118 gives

$$I_1 = \frac{R_2}{R_1 + R_2} \cdot I \qquad \text{and} \qquad I_2 = \frac{R_1}{R_1 + R_2} \cdot I \qquad (2\text{-}119)$$

Equation 2-119 gives the individual branch currents in terms of the total current into the two branches and the resistance parameter values. The ratio of the two currents is given by

$$\frac{I_1}{I_2} = \frac{R_2}{R_1}$$

Analogous to the voltage divider network, this parallel combination could be used to scale down the total current in any desired proportion between the two branches.

Network Reduction. Networks driven by a single source can be solved by systematically reducing the network. The reduction is accomplished by utilizing the relationships shown in Fig. 2-34.

This technique has the disadvantage that the identity of portions of

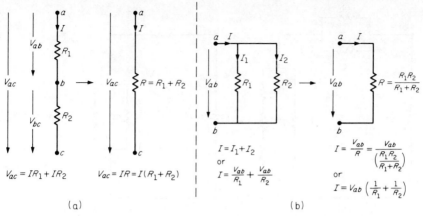

FIG. 2-34. Combining resistors. a) Resistors in series. b) Resistors in parallel.

the actual physical circuit is lost when equivalences are made. This is acceptable in cases where only partial information about a network is desired rather than a total solution.

The equivalences in Fig. 2-34 are shown to be valid for a series or parallel combination of two resistor elements. It can be stated that:

1. The equivalent resistance of N resistors in series is the sum of the N resistors, that is,

$$R_{eq} = R_1 + R_2 + \cdots + R_N \qquad (2\text{-}120)$$

2. The *reciprocal* of the equivalent resistance of N resistors in parallel is the sum of the *reciprocals* of the N resistors, that is,

$$\frac{1}{R_{eq}} = \frac{1}{R_1} + \frac{1}{R_2} + \cdots \frac{1}{R_N} \qquad (2\text{-}121)$$

The proof of these two equations is left as an exercise in applying Kirchhoff's and Ohm's law to such network configurations.

An example will show the application of these relationships.

EXAMPLE 2-12. In the circuit given in Fig. 2-35a, it is desired to find the current and power being furnished by the 20 volt source.

Solution: The reduction is begun in portions of the network most removed from the source and successive applications of Eqs. 2-120 and 2-121 are used to reduce the network to a single equivalent resistance as shown in Fig. 2-35d. It is obvious that the physical identity of the original circuit is lost with the exception of the 20 volt source and the source current I_{12}. This current is the only variable required to obtain the information called for in the problem statement. From Fig. 2-35d, using Ohm's law, the current is

$$I_{12} = \frac{V_{21}}{5} = \frac{20}{5} = 4 \text{ amps}$$

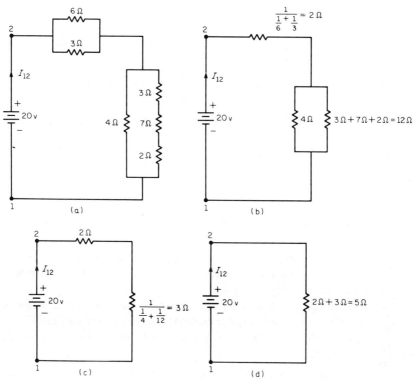

FIG. 2-35. a) Circuit for Example 2-12. b), c), and d) Network reduced by combining series and parallel combinations of resistors.

and the power *furnished* to the network by the 20 volt source is

$$P_{\text{furnished}} = -P_{\text{lost}} = -(V_{12}I_{12}) = V_{21}I_{12} = (20)(4) = 80 \text{ watts}$$

With the simplified network in Fig. 2-35d, it would be a relatively easy task to find the current provided to the network for any value of the source voltage. The 5-ohm resistance is the equivalent network resistance as viewed from the source. It determines the current delivered by the source and as such is termed the *driving point* resistance or input resistance. There are arrangements of resistors in special networks that are neither in parallel nor in series and require special transformations if they are to be combined into single equivalent resistances. Due to their special nature and limited use, they will not be treated here. The interested reader is referred to Δ to Y and Y to Δ transformations treated in most circuit texts.[1]

Driving Point Resistance. The resistance of a network as viewed from the terminals of the driving energy source is termed the driving point

[1] R. E. Lueg and E. A. Reinhard, *Basic Electric Circuits for Engineers*, International Textbook Company, 1967.

resistance. Network reduction can be used to determine the driving point or input resistance presented to the source terminals. The driving point resistance for a given source can also be obtained by determining the ratio of the source voltage to the source current delivered to the network. Consider the following example.

EXAMPLE 2-13. For the circuit of Fig. 2-35a, find the driving point resistance using the method of loop currents. The circuit is redrawn in Fig. 2-36 with the loop currents indicated.

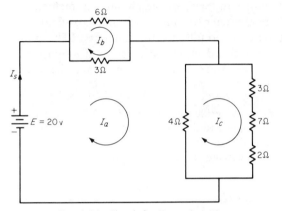

FIG. 2-36. Circuit for Example 2-13.

Solution: The equations are given by

$$7I_a - 3I_b - 4I_c = E$$
$$-3I_a + 9I_b + 0I_c = 0$$
$$-4I_a + 0I_b + 16I_c = 0$$

from which

$$I_a = \frac{\begin{vmatrix} E & -3 & -4 \\ 0 & 9 & 0 \\ 0 & 0 & 16 \end{vmatrix}}{\begin{vmatrix} 7 & -3 & -4 \\ -3 & 9 & 0 \\ -4 & 0 & 16 \end{vmatrix}} = \frac{144E}{720} = \frac{E}{5}$$

From Fig. 2-36 and Eq. 2-122, the current supplied by the source is

$$I_s = I_a = \frac{E}{5}$$

and the driving point resistance is

$$R_{\text{driving point}} = \frac{E}{I_s} = \frac{E}{E/5} = 5 \text{ ohms}$$

Note the source magnitude cancels out since the driving point resistance is a function of the parameters of the network external to the source itself. This value of 5 ohms is the same as the value obtained by the method of network reduction.

Transfer Relationship. In many passive (no energy sources applied) networks it is useful to know what relationship exists between two different circuit variables in terms of the circuit parameters. Networks having an input where the energy source is applied, and an output where the energy is ultimately used fall into this category. The transfer relationship between the output and input variables is most important in many applications. An example should demonstrate the concept of transfer relationships.

EXAMPLE 2-14. Consider the circuit in Fig. 2-37a. The resistor R is variable. A source is to be applied to the terminal pair 1–0 and the value of the resistor R should be such that the output voltage measured at terminal pair 3–0 is one-tenth the source magnitude as indicated in Fig. 2-37b.

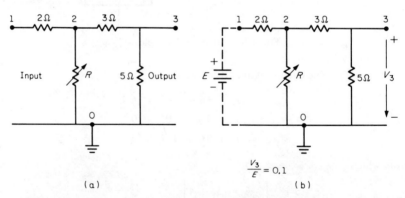

FIG. 2-37. a) Circuit for Example 2-14. b) Source and output variables indicated.

Solution: The node voltage equations for the network in Fig. 2-37 (currents leaving the nodes) are

$$\text{Node 2} \quad \left(\frac{1}{2} + \frac{1}{3} + \frac{1}{R}\right) \cdot V_2 \quad - \frac{1}{3} V_3 = E/2$$

$$\text{Node 3} \quad -\frac{1}{3} V_2 + \left(\frac{1}{3} + \frac{1}{5}\right) \cdot V_3 = 0$$

from which

$$V_3 = \frac{\begin{vmatrix} \dfrac{5}{6} + \dfrac{1}{R} & E/2 \\[2mm] -\dfrac{1}{3} & 0 \end{vmatrix}}{\begin{vmatrix} \dfrac{5}{6} + \dfrac{1}{R} & -\dfrac{1}{3} \\[2mm] -\dfrac{1}{3} & \dfrac{8}{15} \end{vmatrix}} = \frac{E/6}{\dfrac{1}{3} + \dfrac{8}{15R}} \qquad (2\text{-}123)$$

Now the transfer ratio between the output variable V_3 and the source E can be formed from Eq. 2-123 and is given by

$$\frac{V_3}{E} = \frac{5}{10 + \dfrac{16}{R}} \qquad (2\text{-}124)$$

Eq. 2-124 is in turn set equal to 0.1 and gives

$$R = 0.4 \text{ ohm}$$

as the value of R which will give the desired transfer relationship.

PROBLEMS

Geometrical Representation

2-1. Determine the number of nodes, branches, loops, and meshes in the accompanying circuits. Identify your choices by the letter symbols shown on the diagram. Comment as to whether any redundancies or unnecessary lettering are shown on the diagram.

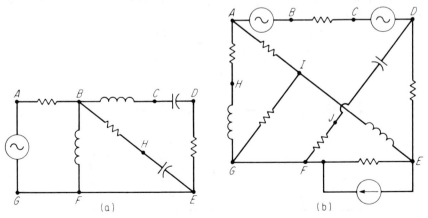

(a)   (b)

PROB. 2-1

2-2. Using the figures of Prob. 2-1, determine the number of active and passive elements in each circuit.

2-3. Draw neat planar graphs for the circuits shown in Prob. 2-1. Are the networks planar or nonplanar?

2-4. Determine I in each of the accompanying graphs.

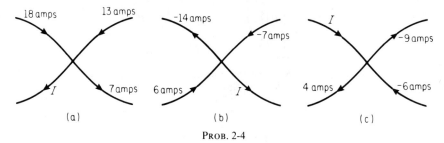

(a)   (b)   (c)

PROB. 2-4

2-5. In many systems of practical interest, the currents i_1, i_2, and i_3 vary sinusoidally: $i_1 = I_0 \sin \omega t$, $i_2 = I_0 \sin (\omega t + 120°)$, and $i_3 = I_0 \sin (\omega t - 120°)$. Using the accompanying figure, determine i_4.

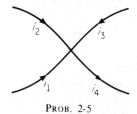

PROB. 2-5

Method of Branch Currents

2-6. A branch current solution for the circuit shown in the accompanying figure yields

$$
\begin{aligned}
I_{ab} &= -1.0 \text{ amp} & I_{cd} &= 1.0 \text{ amp} \\
I_{bf} &= 1.0 \text{ amp} & I_{df} &= 0.67 \text{ amp} \\
I_{bc} &= -2.0 \text{ amp} & I_{de} &= 0.33 \text{ amp} \\
I_{cf} &= -3.0 \text{ amp}
\end{aligned}
$$

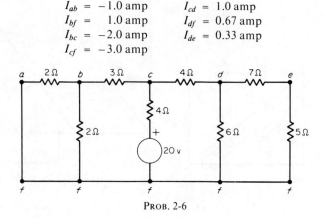

PROB. 2-6

a) Use Kirchhoff's current law at nodes b,c, and d and verify the correctness of the solution. The branch voltages are given as

$$
\begin{array}{ll}
V_{ab} = -2 \text{ volts} & V_{cd} = 4 \text{ volts} \\
V_{bf} = 2 \text{ volts} & V_{df} = 4 \text{ volts} \\
V_{bc} = -6 \text{ volts} & V_{de} = 4 \text{ volts} \\
V_{cf} = 8 \text{ volts} &
\end{array}
$$

b) Verify that the branch voltages are correct using the given branch currents and Ohm's law. c) Does $V_{dc} + V_{cb} + V_{fd} = 0$? Should it? d) Does $V_{ed} + V_{dc} + V_{cd} + V_{da} = 0$? Should it? What branch voltage is missing to complete the equation?

2-7. Using the method of branch currents, solve for the branch currents in the accompanying circuits. Indicate the branch currents on the circuit diagrams.

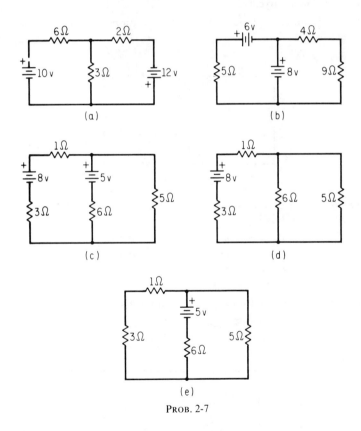

PROB. 2-7

2-8. Compare the networks and results of parts c, d, and e, in Prob. 2-7, and try to deduce an important property of networks composed of linear elements. (*Hint:* These circuits demonstrate the principle of *superposition*.)

2-9. Using branch currents, determine the voltage v_{ac} in the accompanying figure.

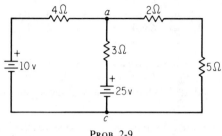

PROB. 2-9

Method of Loop Currents

2-10. Find the currents I_B and I_C in the accompanying figure, using loop-current analysis.

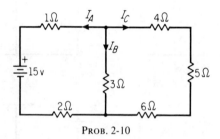

PROB. 2-10

2-11. A branch current solution for the circuit given in the accompanying figure results in the following:

$$
\begin{aligned}
I_{ab} &= 5.75 \text{ amps} & I_{cd} &= 0.5 \text{ amp} \\
I_{bf} &= 4.25 \text{ amps} & I_{df} &= 0.333 \text{ amp} \\
I_{bc} &= 1.5 \text{ amps} & I_{de} &= 0.167 \text{ amp} \\
I_{cf} &= 1.0 \text{ amp}
\end{aligned}
$$

and a loop current solution for the same circuit results in

$$
\begin{aligned}
I_1 &= 5.75 \text{ amps} & I_3 &= 0.5 \text{ amp} \\
I_2 &= 1.5 \text{ amps} & I_4 &= 0.167 \text{ amp}
\end{aligned}
$$

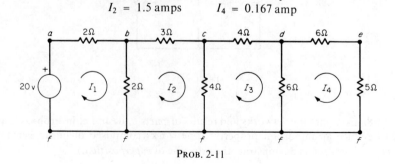

PROB. 2-11

a) Verify the correctness of the solution by cross checking the loop current and branch current values using Kirchhoff's current law. b) You proudly verify that the above currents are correct. Your boss takes your results and makes a calculation or two using Kirchhoff's voltage law. He brings the work back to you and states that one of the 6 ohm resistors is apparently not labeled properly. He's right—find which of the 6 ohm resistors is incorrectly labeled and label it correctly. Note: the error was made in transcribing the circuit from a rough diagram to the finished diagram; thus, you can assume that the above branch and loop currents are correct and the diagram as shown is partially wrong.

2-12. If the excitation voltage in Prob. 2-11 is increased from 20 to 40 volts, will all of the branch or loop currents be doubled? Prove your conclusion by setting up loop current equations for the circuit.

2-13. A branch current solution for the circuit shown in accompanying figure results in

$$
\begin{aligned}
I_{ba} &= 0.167 \text{ amp} & I_{dc} &= 0.667 \text{ amp}\\
I_{bf} &= 0.167 \text{ amp} & I_{df} &= 0.667 \text{ amp}\\
I_{cb} &= 0.333 \text{ amp} & I_{ed} &= 1.33 \text{ amps}\\
I_{cf} &= 0.333 \text{ amp}
\end{aligned}
$$

and a node voltage solution for the circuit yields

$$
\begin{aligned}
V_{bf} &= 0.33 \text{ volt}\\
V_{cf} &= 1.33 \text{ volts}\\
V_{df} &= 4.0 \text{ volts}\\
V_{ef} &= 13.33 \text{ volts}
\end{aligned}
$$

a) Check to see if the given node voltages are correct by using the given branch currents and Ohm's law. b) Does $V_{dc} + V_{cf} + V_{fd} = 0$? Should it? c) Does $V_{ed} + V_{dc} + V_{bf} + V_{fe} = 0$? Should it? d) Does $V_{cb} + V_{ba} + V_{fb} = 0$? Should it? e) Notice that I_{ba} in Prob. 2-13 equals I_{de} in Prob. 2-11. This is an example of the Theorem of Reciprocity which states "If a voltage source in a given branch in a circuit produces a current I in a second branch, and then if that voltage source is transferred to the second branch it will produce the same current I in the first branch." The theorem has been demonstrated here by example. The alert reader will note that an error in the figure to Prob. 2-11 has been corrected in the figure accompanying this problem.

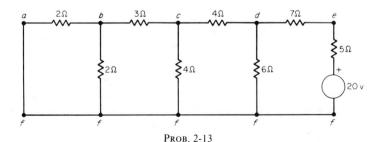

PROB. 2-13

2-14. Find the current I_A in the accompanying figure.

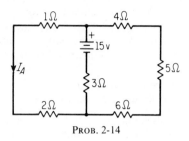

PROB. 2-14

2-15. Find the current I_A in the accompanying figure.

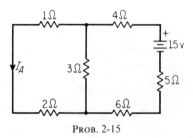

PROB. 2-15

2-16. Compare the circuits and results of Probs. 2-10, 2-14, and 2-15, and try to deduce an important property of networks composed of linear elements and a *single* independent voltage source. (*Hint:* These circuits demonstrate the property of reciprocity. See Prob. 2-13e.)

2-17. a) Find the power being dissipated in the 5-ohm resistor in the accompanying figure. b) Find the power being furnished by the 36-volt battery.

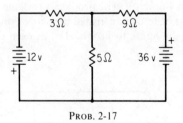

PROB. 2-17

2-18. a) Find the power being furnished to the network by the 2-volt battery in the accompanying figure. b) For the circuit in part a, find the power dissipated in each resistor and the power furnished by each battery. c) Does the sum of the powers being dissipated in the several resistors equal the powers being furnished by the batteries?

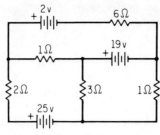

PROB. 2-18

2-19. Using the accompanying figure, determine a) the current through R_3, b) the power dissipated in R_3, and c) the power supplied by E_1 and E_2.

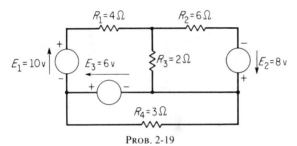

PROB. 2-19

2-20. Using the accompanying figure, a) determine the power supplied by each of the three sources, b) determine the power dissipated in each of the three resistors, and c) compare the results of parts a and b.

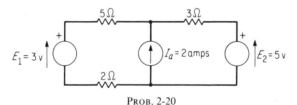

PROB. 2-20

Method of Node Voltages

2-21. Using the accompanying figure, a) determine how many loop equations are necessary for the solution of the network shown, b) indicate which loops could be used, and c) write the node-voltage equations necessary for solution of the network, but *do not solve.*

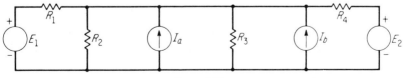

PROB. 2-21

2-22. Use node voltages to determine the voltage difference between nodes *a* and *b* in the accompanying figure.

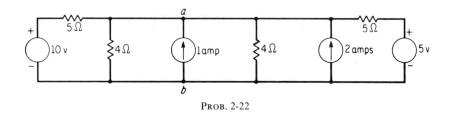

PROB. 2-22

2-23. Find the unknown node voltage in the accompanying figure.

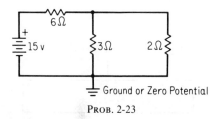

PROB. 2-23

2-24. Find all the unknown node voltages in the accompanying figure.

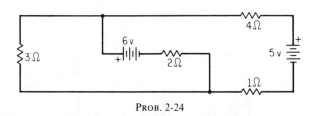

PROB. 2-24

2-25. Find all the unknown node voltages in the accompanying figure.

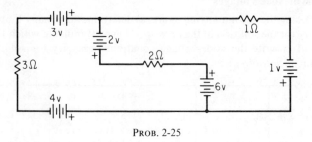

PROB. 2-25

2-26. a) Find the voltage difference across the 4-ohm resistor in the accompanying figure. b) Find the power dissipated in the 2-ohm resistor.

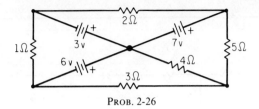

PROB. 2-26

General

2-27. The circuit shown in the accompanying figure has been set up in the laboratory and by use of voltmeters and ammeters the following readings are taken:

V_{10} = 3.87 volts	I_1 = 2.07 amps	I_4 = 2.13 amps
V_{20} = 0.77 volts	I_2 = 1.29 amps	I_5 = −1.35 amps
V_{30} = 7.53 volts	I_3 = 0.78 amp	

a) A friend shows you the circuit and points out that although 10 watt resistors have been used in every case one of the resistors gets quite warm while another gets hot and starts smoking if the circuit is left on for over 2 or 3 minutes. He can't remember which resistors are giving trouble; can you help him find the trouble by using the schematic and the given voltage and current values? b) As your friend leaves he states that he is not sure that all of the resistor values are correct as indicated on the diagram. He is right as one of the 5-ohm resistors is incorrectly marked. Find the incorrectly marked resistor and determine its correct value.

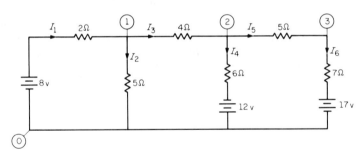

PROB. 2-27

2-28. You and your co-worker are measuring voltages between the various nodes in the accompanying circuit. You read the following voltages and currents.

V_{10} = 100 volts	I_1 = 5.19 amps	I_5 = 0.40 amp
V_{20} = 70.4 volts	I_2 = 3.20 amps	I_6 = 1.94 amps
V_{30} = 67.8 volts	I_3 = 1.97 amps	I_7 = 1.28 amps
V_{40} = 42.2 volts	I_4 = 1.57 amps	I_8 = 1.69 amps

a) Assuming that the above values are correct, your co-worker reads V_{13} as 32.2 volts, V_{23} as 2.6 volts, and V_{14} as −58 volts. Has he read all values correctly? b) Identify the following loop currents with the indicated loops or meshes

I_a with loop 1201 I_c with loop 3403
I_b with loop 2402 I_d with loop 13421

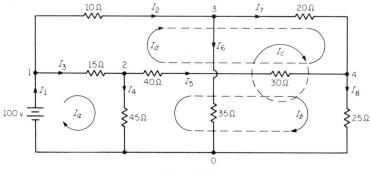

PROB. 2-28

Next write the four loop current equations that are necessary in order to obtain the solution to the circuit variables. c) Identify the loop currents of part b) with the given branch currents above. Substitute the proper loop current values into the equations of part b) and see if the equations are in proper balance.

2-29.

$V_{10} = 41.4$ volts	$I_1 = 1.78$ amps	$I_6 = 0.25$ amp
$V_{20} = 33.7$ volts	$I_2 = 0.59$ amp	$I_7 = 0.68$ amp
$V_{30} = 39.1$ volts	$I_3 = 0.77$ amp	$I_8 = 1.10$ amps
$V_{40} = 33$ volts	$I_4 = 0.42$ amp	$I_9 = 1.78$ amps
$V_{50} = 24.9$ volts	$I_5 = -0.33$ amp	

Your instructor has asked you to obtain the solution to the circuit shown in the accompanying figure by any means. Your roommate had the course last semester and gives you the accompanying answers a) verify that the answers given are indeed correct; b) your roommate recalls that one of the 6-ohm resistors is incorrectly labeled. Is he right and if so what ohmic value should be attached to the resistor in question?

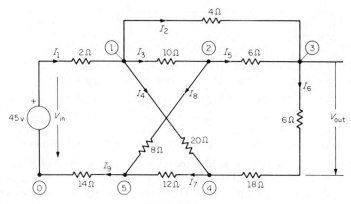

PROB. 2-29

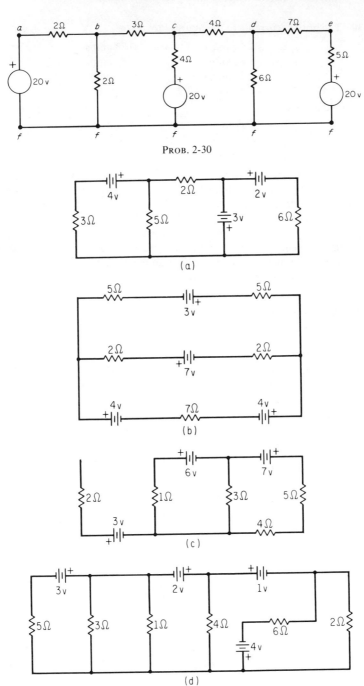

PROB. 2-30

(a)

(b)

(c)

(d)

PROB. 2-31

2-30. The branch currents to the figure shown are

$$I_{ab} = \quad 4.58 \text{ amps} \qquad I_{cd} = \quad 0.83 \text{ amps}$$
$$I_{bf} = \quad 5.42 \text{ amps} \qquad I_{df} = \quad 1.67 \text{ amps}$$
$$I_{bc} = -0.83 \text{ amps} \qquad I_{de} = -0.83 \text{ amp}$$
$$I_{cf} = -1.67 \text{ amps}$$

The node voltages are

$$V_{af} = 20 \text{ volts} \qquad V_{df} = 10 \text{ volts}$$
$$V_{bf} = 10.8 \text{ volts} \qquad V_{ef} = 15.85 \text{ volts}$$
$$V_{cf} = 13.3 \text{ volts}$$

a) Is Kirchhoff's current law verified at nodes b, c, and d? b) Is Kirchhoff's voltage law verified around loops $abcdfa$ and $dfed$? c) Compare the branch current I_{df} in this problem with the sum of the three branch currents I_{df} from Probs. 2-6, 2-11, and 2-13. Does $I_{df}(2\text{-}30) = I_{df}(2\text{-}6) + I_{df}(2\text{-}11) + I_{df}(2\text{-}12)$? Do the same with I_{cf}. You are observing the principle of superposition for linear circuits. The superposition theorem states that "in a linear circuit the algebraic sum of the currents in any given branch due to the voltage sources taken one at a time is equal to the current in the branch when all voltage sources are considered simultaneously."

2-31. Use the easiest and quickest method to solve the accompanying networks for all currents, or, alternatively all voltage differences.

Series and Parallel Addition of Resistors

Use a systematic network reduction method to obtain the solution to the following problems. Check your answer using loop or node equations, if you wish.

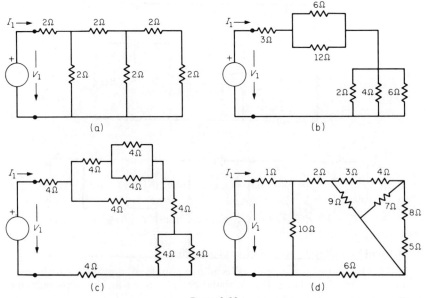

PROB. 2-32

2-32. Determine the input or driving point resistance V_1/I_1 for the circuits shown in the accompanying figures.

2-33. Determine the output voltage V_o in the accompanying figures.

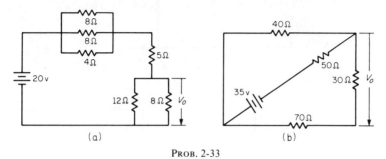

(a)

(b)

PROB. 2-33

Current Splitting Formula

2-34. Determine the current in each branch of the following circuits using the current splitting formula.

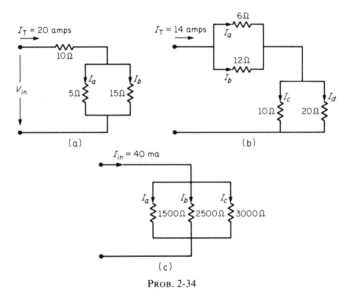

(a)

(b)

(c)

PROB. 2-34

2-35. Determine the output voltage V_o in Prob. 2-33b using the current splitting formula.

2-36. If $V_1 = 80$ volts in Prob. 2-32b determine the power loss a) in the 2Ω resistor; b) in the 12Ω resistor.

Driving Point Resistance and Dependent Sources

2-37. Using any circuit analysis technique you deem appropriate, determine the driving point impedance V_1/I_1 for the circuits shown in the accompanying figures. (Recall that "K" means kilo or thousand.)

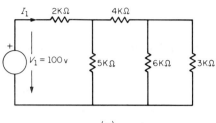

(a)

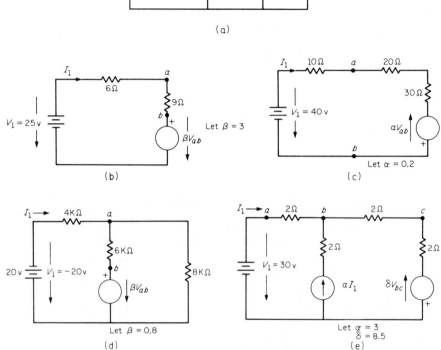

(b)   (c)

(d)   (e)

PROB. 2-37

2-38. Work Prob. 2-37c with a) $\alpha = 9$; b) $\alpha = -0.2$; c) $\alpha = -2$; d) $\alpha = -1$. Comment on the results.

General

2-39. By means of the current-splitting theorem and/or other network reduction methods, determine I_1.

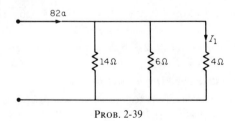

PROB. 2-39

2-40. Using the accompanying circuit determine a) V_o, and b) the power dissipated in the 4Ω resistor.

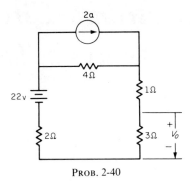

PROB. 2-40

2-41. In the accompanying figure all resistors are 5 ohms. Determine the power supplied to the circuit by the 50-volt battery.

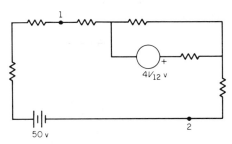

PROB. 2-41

2-42. Using the accompanying circuit determine all node voltages (with node 0 used as the reference node) by the node voltage method and find the power dissipated in the 10-ohm resistor.

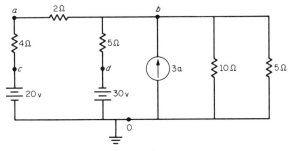

PROB. 2-42

3 / Electronic Circuit-
Analysis Techniques

Circuit-analysis techniques have become a highly developed subject and are used extensively in the study of electrical engineering. Almost any piece of electrical apparatus or hardware, whether it is a motor, generator, transformer, vacuum tube, or solid-state device, when operated in its linear region is describable in terms of conventional circuit parameters. These parameters include the passive elements such as the resistance R, the inductance L, and the capacitance C, and such active elements as voltage and current sources. It is assumed that the reader is now familiar with the loop current and node voltage method of analysis. Only a brief review of these two analysis techniques will be given here. Circuit variables, electrical source types, network reduction through series and parallel combination of resistors, dependent source types with analytical treatment of same, source equivalences and transfer relationships should also be familiar to the reader. If you are not familiar with these topics or wish a treatment in depth, the authors suggest that you study Chapters 1 and 2 before proceeding through this chapter.

This chapter describes the more popular circuit-analysis methods that are used in analyzing electronic circuits, but the reader should recognize that these methods apply to conventional circuits as well. The entire foundation of circuit analysis rest on Kirchhoff's two laws and Ohm's law. Briefly, Kirchhoff's two laws state 1) that the algebraic summation of voltage drops around a closed loop equals zero, and 2) that the algebraic summation of currents entering a junction equals zero. Ohm's law states that the voltage and current in a linear circuit under steady-state conditions are related by a constant. Because Ohm developed this concept for d-c circuits, the constant has often been regarded as a resistance only; however, the development of a-c circuit analysis techniques has caused us to call this voltage-to-current ratio an *impedance*, where the voltages, currents, and impedances are in phasor form. The study of a-c circuits is taken up in Chapter 12.

The topics to be considered in this chapter are loop currents, node voltages, Thévenin's and Norton's theorems, graphical-analysis methods, two port network analysis, and equivalent circuits as applied to resistive circuits.

3-1. Loop Currents and Node Voltages

Loop-current and node-voltage methods are two of the more basic and general means of analyzing electric circuits. The loop-current method sums voltage drops, which are due to the loop currents, around a closed loop; and the node-voltage method sums currents, which are due to the node voltages, leaving a junction. Both methods are explained in detail in Chapters 1 and 2 and only a brief review will be given here.

Before solving any problem, one must decide what is required in the solution. To solve for a particular current in an element or branch, loop currents are normally used in the solution. If a particular voltage is desired, node voltages would probably be the better method of solution. In the following examples, the loop currents and then the node voltages will be used to solve for the circuit voltages and currents. The method to be used depends on several factors, among them being the user's familiarity and preference, network information being sought and relative amount of computational work required for each method.

Loop-Current Solution.   In using loop currents to effect a network solution, one must first determine the number of independent loops in the circuit. In the circuit of Fig. 3-1a there are only two independent loops,

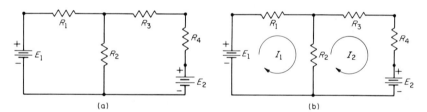

FIG. 3-1. a) Network for loop-current solution. b) One possible choice of loop currents.

and thus only two loop currents are necessary to obtain enough simultaneous equations to solve for the circuit voltages, currents, or powers. Many paths may be assigned for the loop currents, and one possible set of paths (so-called meshes or window panes) is shown in Fig. 3-1b. Once the current paths have been assumed, Kirchhoff's first law, concerning voltage drops (or rises) around a closed loop, must be strictly obeyed. The loop-current equations for the circuit of Fig. 3-1b are

$$(R_1 + R_2)I_1 - R_2I_2 = E_1$$
$$-R_2I_1 + (R_2 + R_3 + R_4)I_2 = -E_2 \qquad (3\text{-}1)$$

To solve for the current through R_1, solve for I_1 which, in determinant form, is

$$I_1 = \frac{\begin{vmatrix} E_1 & -R_2 \\ -E_2 & R_2 + R_3 + R_4 \end{vmatrix}}{\begin{vmatrix} R_1 + R_2 & -R_2 \\ -R_2 & R_2 + R_3 + R_4 \end{vmatrix}} \quad (3\text{-}2)$$

To solve for the current through R_3 or R_4, solve for I_2. To determine the current through R_2, the solutions for both I_1 and I_2 must be obtained, as the current through R_2 is equal to $I_1 - I_2$.

EXAMPLE 3-1. If $V_2 = 0.5\,V_{bf}$ and $V_3 = 2\,V_{dc}$ in Fig. 3-2, solve for I_1 and I_2.

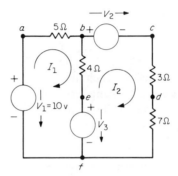

FIG. 3-2. Circuit for Example 3-1.

Solution: First write the equations in symbolic form

$$\left. \begin{array}{ll} \text{Loop } abefa & V_{ab} + V_{be} + V_{ef} + V_{fa} = 0 \\ \text{Loop } bcdfeb & V_{bc} + V_{cd} + V_{df} + V_{fe} + V_{eb} = 0 \end{array} \right\} \quad (3\text{-}3)$$

Next, apply Ohm's law and knowledge of the voltage sources to each symbolic voltage

$$\left. \begin{array}{ll} \text{Loop } abefa & 5I_1 + 4(I_1 - I_2) + V_3 - 10 = 0 \\ \text{Loop } bcdfeb & V_2 + 3I_2 + 7I_2 - V_3 + 4(I_2 - I_1) = 0 \end{array} \right\} \quad (3\text{-}4)$$

Before substituting the indicated or given relations for the dependent voltage sources V_2 and V_3 in Eq. set 3-4, clearly identify these voltages as follows:

$$V_2 = 0.5\,V_{bf}$$

where

$$V_{bf} = -5I_1 + 10 \quad (3\text{-}5)$$

or

$$V_{bf} = 4(I_1 - I_2) + V_3 \quad (3\text{-}6)$$

or

$$V_{bf} = V_2 + 3I_2 + 7I_2 \tag{3-7}$$

Observe that although all three of these expressions (Eqs. 3-5, 3-6, or 3-7) are quite correct, Eq. 3-5 is much the simpler expression of the three and so will be used to complete the problem solution.

$$V_3 = 2V_{dc}$$

where

$$V_{dc} = -V_{cd} = -(3I_2) \tag{3-8}$$

Now substitute Eqs. 3-5 and 3-8 into Eq. set 3-4.

$$5I_1 + 4(I_1 - I_2) + 2(-3I_2) - 10 = 0$$
$$0.5(-5I_1 + 10) + 3I_2 + 7I_2 - 2(-3I_2) + 4(I_2 - I_1) = 0$$

Collecting terms results in

$$\left.\begin{array}{r} 9I_1 - 10I_2 = 10 \\ -6.5I_1 + 20I_2 = -5 \end{array}\right\} \tag{3-9}$$

The reader should confirm that the solution to Eq. set 3-9 is

$$I_1 = \frac{150}{115} \text{ amps}$$

$$I_2 = \frac{20}{115} \text{ amp}$$

Node Voltage Solution. A node is defined as a junction of any two or more branches in a circuit. Figure 3-3 repeats the circuit of Fig. 3-1a but

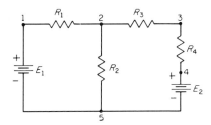

FIG. 3-3. Numbering of the nodes of part a of Fig. 3-1.

with the five nodes numbered. After the nodes have been identified, a reference node must be chosen. It is usually obvious as to which node should best be labeled the reference, and, once the reference has been chosen, it is imperative to use it throughout the problem solution. Node 5 is as good a choice as any for the reference in Fig. 3-3, since node 5 appears to be the ground terminal and also has three branches tied to it. The node voltage solution is based on Kirchhoff's second law concerning the fact that the sum of currents leaving (or entering) a junction must equal zero, or

Node

1   Since $V_{15} = E_1$, a known voltage, no equation need be written

2    $\dfrac{V_{25} - V_{15}}{R_1} + \dfrac{V_{25} - V_{35}}{R_3} + \dfrac{V_{25}}{R_2} = 0$

3    $\dfrac{V_{35} - V_{25}}{R_3} + \dfrac{V_{35} - V_{45}}{R_4} = 0$

4   Since $V_{45} = E_2$, a known voltage, no equation need be written

Combining the known node voltages with the unknown node voltages at nodes 2 and 3 and dropping the reference symbol in the double-subscript notation,

$$\left. \begin{aligned} \frac{V_2 - E_1}{R_1} + \frac{V_2 - V_3}{R_3} + \frac{V_2}{R_2} = 0 \\ \frac{V_3 - V_2}{R_3} + \frac{V_3 - E_2}{R_4} = 0 \end{aligned} \right\} \qquad (3\text{-}10)$$

Upon reorganizing the equations in a form suitable for determinant solution,

$$\left. \begin{aligned} \left(\frac{1}{R_1} + \frac{1}{R_2} + \frac{1}{R_3}\right)V_2 - \frac{1}{R_3}V_3 = \frac{E_1}{R_1} \\ -\frac{1}{R_3}V_2 + \left(\frac{1}{R_3} + \frac{1}{R_4}\right)V_3 = \frac{E_2}{R_4} \end{aligned} \right\} \qquad (3\text{-}11)$$

Using Eq. set 3-11, V_2 or V_3 can be found by the usual determinant or simultaneous-equation methods. Once V_2 and V_3 are known, all currents and powers can be calculated. For instance, the current through R_3 directed from node 2 toward node 3 is $(V_2 - V_3)/R_3$.

EXAMPLE 3-2.   Using Fig. 3-4 find the voltage drop across the 4-ohm resistor.

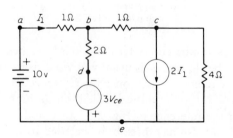

FIG. 3-4. Circuit for Example 3-2.

Solution: Using node voltages with node e as the reference node and first writing the equations in symbolic form results in

Node

$$a \qquad V_{ae} = 10$$

$$b \qquad \frac{V_{be} - V_{ae}}{1} + \frac{V_{be} - V_{de}}{2} + \frac{V_{be} - V_{ce}}{1} = 0$$

$$\left.\rule{0cm}{3cm}\right\} \qquad (3\text{-}12)$$

$$c \qquad \frac{V_{ce} - V_{be}}{1} + 2I_1 + \frac{V_{ce}}{4} = 0$$

$$d \qquad V_{de} = -3V_{ce}$$

Next identify the dependent current source as

$$2I_1 = 2\left(\frac{V_{ae} - V_{be}}{1}\right)$$

$$= 2(10 - V_{be})$$

Now make the necessary substitutions into the middle two equations of Eq. set 3-12.

Node

$$b \qquad \frac{V_{be} - 10}{1} + \frac{V_{be} - (-3V_{ce})}{2} + \frac{V_{be} - V_{ce}}{1} = 0$$

$$c \qquad \frac{V_{ce} - V_{be}}{1} + 2(10 - V_{be}) + \frac{V_{ce}}{4} = 0$$

Collecting terms results in

$$(1 + \tfrac{1}{2} + 1)V_{be} + (\tfrac{3}{2} - 1)V_{ce} = 10$$
$$(-1 - 2)V_{be} + (1 + \tfrac{1}{4})V_{ce} = -20$$

Solving for V_{ce} which is the voltage across the 4-ohm resistor

$$V_{ce} = \frac{\begin{vmatrix} \dfrac{5}{2} & 10 \\[2mm] -3 & -20 \end{vmatrix}}{\begin{vmatrix} \dfrac{5}{2} & \dfrac{1}{2} \\[2mm] -3 & \dfrac{5}{4} \end{vmatrix}} = \frac{-50 + 30}{\dfrac{25}{8} + \dfrac{3}{2}} = \frac{-20}{\dfrac{37}{8}}$$

$$= -4.3 \text{ volts}$$

Actually since it was not specified whether V_{ce} or V_{ec} is to be identified as the voltage across the 4-ohm resistor

$$V_{ec} = -V_{ce} = +4.3 \text{ volts}$$

is also an acceptable answer.

3-2. Thévenin's and Norton's Theorems

Thévenin's theorem is one of the most popular and useful tools ever developed for circuit analysis. It states that any two-terminal linear resistive network can be replaced by a voltage source E_o and a series resistance R_o. Thévenin's theorem is much broader in scope than this since it is also applicable to nonresistive or mixed element networks as long as they are not magnetically coupled to an external network. It is stated here in context of the resistive networks being examined. The voltage source E_o is the voltage appearing at the open-circuited pair of terminals, and the series resistance R_o is the driving-point resistance "looking into" the open-circuited pair of terminals. Figure 3-5a shows two general linear resistive

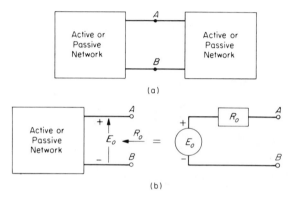

Fig. 3-5. a) Two linear networks connected by only two terminals. b) Replacement of a network with an equivalent Thévenin circuit.

networks connected by only two paths. Thévenin proved that either or both of these two networks can be replaced by a voltage source and a series resistance. Figure 3-5b illustrates how the left-hand network is reduced to a Thévenin circuit by first opening the circuit at terminals AB and then finding both the open circuit voltage appearing at AB, or E_o, and the driving-point resistance at terminals AB or R_o. The driving-point resistance R_o can be obtained by 1) applying an arbitrary voltage source E_1 to terminals AB as shown in Fig. 3-6, 2) removing all independent

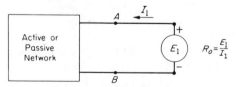

Fig. 3-6. Circuit used to solve for the driving point or input resistance R_o.

voltage and current sources but leaving all dependent sources (a dependent source is one that (in any way) depends on or is proportional to the input voltage E_1 or input current I_1), and 3) calculating E_1/I_1, which is defined as R_o. An example will help to clarify this calculation.

EXAMPLE 3-3.   Find Thévenin's equivalent network to the right of terminals AB for the circuit shown in Fig. 3-7a.

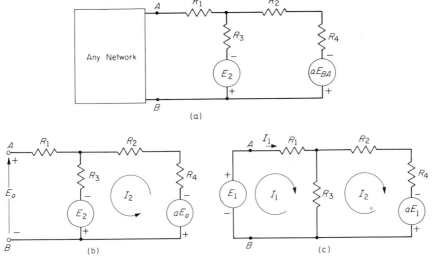

FIG. 3-7. Circuit for Example 3-3. b) Circuit configuration used to find the Thévenin generator voltage E_o. c) Circuit configuration used to solve for the Thévenin equivalent resistance R_o.

Solution: To find E_o, the open circuit terminal voltage, we first remove the network to the left of terminals AB. This causes a new circuit to appear as shown in Fig. 3-7b with the open-circuit voltage between terminals A and B indicated by E_o.   From Fig. 3-7b, the open-circuit terminal voltage is calculated as follows:

Solve for I_2 (which is an assumed loop current) by applying K.V.L. around the loop:

$$I_2 = \frac{E_2 - aE_o}{R_2 + R_3 + R_4}$$

Since there is no voltage drop across R_1 (terminals A and B are open-circuited, hence no current flow),

$$E_o = I_2 R_3 - E_2 = \frac{E_2 - aE_o}{R_2 + R_3 + R_4} R_3 - E_2$$
$$= \frac{E_2 R_3 - aR_3 E_o - E_2(R_2 + R_3 + R_4)}{R_2 + R_3 + R_4}$$

or

$$E_o = -\frac{(R_2 + R_4)E_2}{(a + 1)R_3 + R_2 + R_4} \tag{3-13}$$

To find R_o, the driving-point resistance at terminals AB, apply a voltage source E_1 at terminals AB and remove all independent sources. To remove a voltage source, replace the generator or voltage source with a short circuit, but remove a current source by leaving the circuit open after the source is removed. Note the change in nomenclature on the dependent voltage source to agree with the applied voltage E_1. Figure 3-7c shows the resulting circuit that can be used to calculate the driving-point resistance. Loop currents offer a convenient means of solution for the ratio E_1/I_1, as follows:

$$(R_1 + R_3)I_1 - R_3I_2 = E_1$$
$$-R_3I_1 + (R_2 + R_3 + R_4)I_2 = aE_1$$

$$I_1 = \frac{\begin{vmatrix} E_1 & -R_3 \\ aE_1 & R_2 + R_3 + R_4 \end{vmatrix}}{\begin{vmatrix} R_1 + R_3 & -R_3 \\ -R_3 & R_2 + R_3 + R_4 \end{vmatrix}} = \frac{E_1(R_2 + R_3 + R_4) + aE_1R_3}{(R_1 + R_3)(R_2 + R_3 + R_4) - R_3^2}$$

$$I_1 = \frac{E_1[R_2 + R_4 + (a + 1)R_3]}{R_1(R_2 + R_3 + R_4) + R_3(R_2 + R_4)}$$

thus

$$R_o = \frac{E_1}{I_1} = \frac{R_1(R_2 + R_3 + R_4) + R_3(R_1 + R_4)}{R_2 + R_4 + (a + 1)R_3} \tag{3-14}$$

It should be noted that if a is negative it would be possible for R_o to become negative if $aR_3 > (R_2 + R_3 + R_4)$. The phenomenon of a negative driving point resistance rarely occurs in most practical circuits, and if it does occur it usually is considered undesirable and must be corrected.

Norton's theorem can be considered a corollary to Thévenin's theorem or can be derived independently. Since Norton's equivalent network is derivable from Thévenin's equivalent network, all circuit assumptions that were made to derive Thévenin's network are also necessary and valid for Norton's network. Figure 3-8 shows the comparison of the two net-

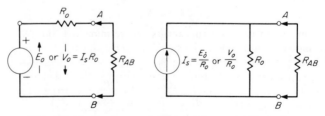

Fig. 3-8. Comparison of Thévenin's and Norton's equivalent circuits.

works. Norton's equivalent network consists of a constant-current generator and a shunt resistance, as compared to the constant voltage generator and series resistance of Thévenin's network.

To derive Norton's equivalent network from Thévenin's, simply place a constant-current generator, whose value is V_o/R_o in parallel with R_o. To illustrate that these two networks produce identical results at terminals AB, note that, when terminals AB are open, the voltage appearing across the terminals is V_o for the Thévenin network and $I_s R_o = V_o$ for the Norton network. When terminals AB are shorted, a current of $V_o/R_o = I_s$ flows through the shorted terminals for both Thévenin's and Norton's equivalent circuits. The reader should be able to verify that the current through any resistance placed across the terminals AB (such as, R_{AB} in Fig. 3-8) is $I_{AB} = E_o/(R_o + R_{AB})$ for both the Thévenin and Norton circuits. Keep in mind that the two equivalent circuits behave in an identical fashion only at the external terminals AB. Internally, the two equivalent circuits are not identical; for instance, when the terminals AB are open, energy is dissipated in the Norton circuit but not in the Thévenin circuit. A numerical example of a circuit reduced to its Thévenin and Norton equivalents follows.

EXAMPLE 3-4. In Fig. 3-9a, find the Thévenin equivalent circuit for the network to the left of terminals AB.

Solution: Figure 3-9b shows the independent voltage sources removed (by replacing the sources with a short circuit) and terminals AB opened (the load resistance R_1 removed). From this new circuit the driving-point resistance R_o can be calculated:

$$R_o = 4 + \frac{2 \times 3}{2 + 3} = 4 + 1.2 = 5.2 \text{ ohms}$$

The open-circuit terminal voltage V_o can be calculated with the aid of Fig. 3-9c. Perhaps the easiest way of calculating this voltage is to 1) assume a loop current I, as shown; 2) notice that no current flows through the 4-ohm resistance, thus $V_{AB} = V_{CB} = V_o$; and 3) solve for I and then V_{CB}.

$$I = \frac{5 + 15}{2 + 3} = 4 \text{ amps}$$
$$V_{CB} = 3I - 15 = 3 \times 4 - 15$$
$$= -3 \text{ volts}$$

Thévenin's equivalent circuit can now be inserted in place of the previous circuit, as shown in Fig. 3-9d. The only similarity in the two circuits of Figs. 3-9a and 3-9d is that the same current flows through R_1 in each case (hence the same voltage appears across R_1 and the same energy is dissipated in R_1).

The reader should verify that Fig. 3-9e is the correct Norton equivalent circuit for this example.

There is another popular way of determining the Thévenin equivalent

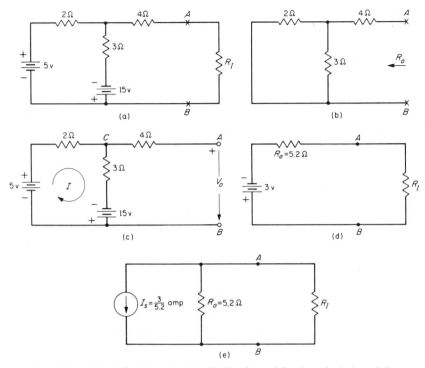

FIG. 3-9. a) Circuit for Example 3-4. b) Circuit used for the calculation of R_o. c) Circuit used for the calculation of V_o. d) Thévenin's equivalent circuit for part a. e) Norton's equivalent circuit for part a.

circuit using what is commonly termed the open- and short-circuit calculations. Observe the two Thévenin circuits shown in Fig. 3-10. The

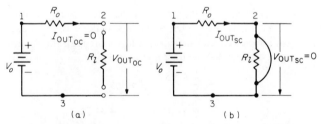

FIG. 3-10. a) Open-circuit configuration ($R_l = \infty$ or is physically *removed* from the circuit). b) Short-circuit configuration ($R_l = 0$ or is physically *replaced* by a short circuit).

external load resistor R_l is removed in Fig. 3-10a while the load resistor is shorted in Fig. 3-10b. In the open circuit case ($R_l = \infty$)

$$I_{\text{out}_{oc}} = \left. \frac{V_o}{R_o + R_l} \right|_{R_l = \infty} = 0$$

Summing voltage drops around loop 1231 gives

$$V_{12} + V_{23} + V_{31} = 0$$

or

$$I_{out_{oc}} R_o + V_{out_{oc}} - V_o = 0$$

or

$$V_{out_{oc}} = V_o - I_{out_{oc}} R_o = V_o - (0) R_o$$
$$= V_o \qquad (3\text{-}15)$$

Note that the Thévenin voltage V_o is the same as the open-circuit output voltage $V_{out_{oc}}$ when R_l is removed ($R_l = \infty$) from the circuit.

For the short-circuit case ($R_l = 0$) and again summing voltages around loop 1231

$$V_{12} + V_{23} + V_{31} = 0$$
$$I_{out_{sc}} R_o + 0 + (-V_o) = 0$$

therefore,

$$I_{out_{sc}} = \frac{V_o}{R_o} \qquad (3\text{-}16)$$

By dividing Eq. 3-16 into Eq. 3-15 one obtains

$$\frac{V_{out_{oc}}}{I_{out_{sc}}} = \frac{V_o}{V_o/R_o} = R_o \qquad (3\text{-}17)$$

The results shown in Eq. 3-17 emphasize that the open-circuit voltage $V_{out_{oc}}$ divided by the short-circuit current $I_{out_{sc}}$ gives the Thévenin circuit resistance R_o. This technique is useful not only in hand calculations and computer solutions but is almost universally used in the experimental determination of the Thévenin equivalent circuit of an actual circuit.

EXAMPLE 3-5.   Determine the Thévenin equivalent circuit of Example 3-4 using open- and short-circuit calculations.

Solution: First redraw the circuit of Fig. 3-9a with R_l removed for the open-circuit calculation. This will allow one to calculate $V_{out_{oc}}$ which is equal to V_o, the Thévenin generator voltage. This is shown in Fig. 3-11a.

Using node voltages as the solution technique, sum currents at node b since this is the only unknown node voltage. In symbolic form

Node

$$b \qquad \frac{V_{ba}}{2} + \frac{V_{bd}}{3} + \frac{V_{bc}}{4} = 0 \qquad (3\text{-}18)$$

or

$$\frac{V_{be} - V_{ae}}{2} + \frac{V_{be} - V_{de}}{3} + \frac{V_{be} - V_{out_{oc}}}{4} = 0$$

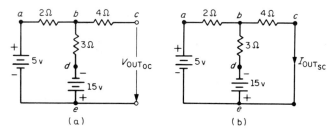

FIG. 3-11. a) Circuit of Fig. 3-9a with R_l removed. b) Circuit of Fig. 3-9a with R_l replaced by a short circuit.

Next substitute specific circuit values into Eq. 3-18

$$\frac{V_{be} - 5}{2} + \frac{V_{be} + 15}{3} + 0 = 0 \qquad (3\text{-}19)$$

where it should be evident that $V_{be} = V_{out_{oc}}$ since there is no current in the 4-ohm resistor. Solving for V_{be} from Eq. 3-19 results in

$$\left(\frac{1}{2} + \frac{1}{3}\right) V_{be} = -5 + 2.5$$

$$V_{be} = \frac{-2.5}{5/6} = -3 \text{ volts}$$

Thus $V_{out_{oc}} = V_{be} = -3$ volts which is the desired Thévenin voltage.

Next solve for $I_{out_{sc}}$ with R_l replaced by a short circuit as shown in Fig. 3-11b. Again using node voltage analysis by summing currents at node b results in

$$\frac{V_{be} - 5}{2} + \frac{V_{be} + 15}{3} + \frac{V_{be}}{4} = 0$$

and solving for V_{be}

$$\left(\frac{1}{2} + \frac{1}{3} + \frac{1}{4}\right) V_{be} = -5 + 2.5$$

$$V_{be} = -\frac{2.5}{13/12} = -\frac{30}{13} \text{ volts}$$

Now notice that $I_{out_{sc}} = V_{bc}/4 = V_{be}/4$, since nodes c and e are at the same electrical potential.

$$I_{out_{sc}} = \frac{-30}{4 \times 13} = \frac{-7.5}{13} \text{ amp}$$

The Thévenin resistance R_o is given by Eq. 3-17 as

$$R_o = \frac{V_{out_{oc}}}{I_{out_{sc}}} = \frac{-3}{-7.5/13} = \frac{13}{2.5} = 5.2 \text{ ohms}$$

The Thévenin equivalent circuit is shown in Figs. 3-12a or b. (Compare to Fig. 3-9d.)

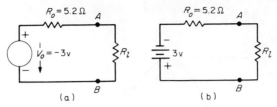

FIG. 3-12. Two ways of drawing the Thévenin equivalent circuit for Fig. 3-9a to the left of terminals AB.

3-3. Maximum Power Transfer Theorem

It is often desirable to determine the value of a load resistance R_l which will allow a generator of internal resistance R_g to deliver maximum power to R_l. Figure 3-13 shows a voltage generator E_g in series with a

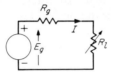

FIG. 3-13. Circuit used to illustrate the maximum power transfer theorem.

resistance R_g. It will be assumed that R_g is fixed but that R_l can be varied in value. If the circuit seems too elementary, recall from the previous section that, by Thévenin's theorem, a quite complex linear circuit can be reduced to a simple series circuit containing a resistance and a voltage source. By inspection the current I is

$$I = \frac{E_g}{R_g + R_l}$$

and the power P_l transferred from E_g to R_l is

$$P_l = I^2 R_l$$

$$= \left(\frac{E_g}{R_g + R_l}\right)^2 R_l = \frac{R_l}{(R_g + R_l)^2} E_g^2$$

To determine the value of R_l that will cause P_l to be a maximum take the derivative of P_l with respect to R_l and set this equal to zero.

$$\frac{dP_l}{dR_l} = \frac{(R_g + R_l)^2 - R_l 2 (R_g + R_l)}{(R_g + R_l)^4} E_g^2 = 0$$

or

$$R_g^2 + R_l^2 - 2R_l^2 = 0$$

and thus $R_l = R_g$ for maximum power transfer to R_l.

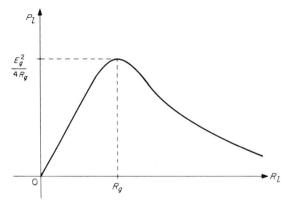

FIG. 3-14. Power P_l developed in R_l versus R_l.

If the reader objects to the use of differential calculus this result could also be deduced by graphing the power developed in R_l versus values of R_l in terms of R_g as depicted in Fig. 3-14. This maximum power condition is very important when the signal level is low and as much energy as possible must be extracted or when the signal level is high and efficiency is important.

3-4. Graphical Analysis Methods

Almost all analytical circuit-analysis methods are based on the solution of linear equations with constant coefficients. Unfortunately, a great many practical problems involve nonlinearities of some type (such as the power-delivery limitation of a dependent voltage or current source). If only one nonlinear element or device is situated in an otherwise linear circuit, graphical-analysis techniques are not only useful but quick and simple to apply.

As stated in the preceding section, almost any linear circuit can be reduced by Thévenin's theorem to a voltage source and a series resistance. Consider the simple case of a d-c source and a resistance in series with a nonlinear element, as shown in Fig. 3-15a. Assume that experi-

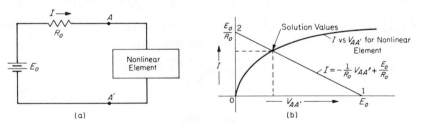

FIG. 3-15. a) Circuit with a nonlinear element included. b) Graphical solution for current in part a.

mental tests on the nonlinear element yield the I versus $V_{AA'}$ curve shown in Fig. 3-15b. Since the resistance of the nonlinear element varies with the current, as evidenced by the changing slope, it is expedient to resort to a graphical method to determine the current in the circuit of Fig. 3-15a. The rationale of the analysis is to match the I versus $V_{AA'}$ characteristic of the linear circuit to the left of terminals AA' with the I versus $V_{AA'}$ characteristics of the nonlinear circuit to the right of terminals AA'. In other words the current I and the voltage $V_{AA'}$ must simultaneously satisfy both characteristics.

To continue with the analysis, temporarily remove the nonlinear element from the circuit of Fig. 3-15a. With terminals AA' open, the current I goes to zero, and $V_{AA'} = E_o$. Plot this point 1 on Fig. 3-15b. Now place a short circuit across terminals AA'. The current I goes to E_o/R_o, and $V_{AA'} = 0$. Also plot this point 2 on Fig. 3-15b. For any other value of resistance placed across terminals AA' (other than infinite or zero ohms), the current I and voltage $V_{AA'}$ must lie between the two points previously plotted. Since E_o and R_o were assumed to be linear elements, the only possible path between these two points must be a straight line. A moment's reflection and use of Ohm's law should indicate that the plot between these two points is the voltage-current curve of the equation

$$V_{AA'} = E_o - IR_o$$

or

$$I = -\frac{1}{R_o} V_{AA'} + \frac{E_o}{R_o}$$

which in terms of I and $V_{AA'}$ is the equation of a straight line with a slope of $-(1/R_o)$, an ordinate intercept of E_o/R_o and an abscissa intercept of E_o. Furthermore, this straight line will always be characteristic for a Thévenin equivalent circuit. Thus, a straight line is drawn between points 1 and 2 on Fig. 3-15b, and the intersection of this line and the current-voltage characteristic of the nonlinear device is the only possible equilibrium solution of I and $V_{AA'}$ that simultaneously satisfies both characteristics. This straight line will be termed the *d-c load line* in the chapters that follow.

EXAMPLE 3-6. Figure 3-16a is a circuit showing a nonlinear device in series with a 2500-ohm resistance and a 10-volt battery. If Fig. 3-16b is a plot of I versus V_{12} for the nonlinear device, find the current I flowing in the series circuit under steady-state conditions.

Solution: Construct the straight line characteristic of the voltage source and series resistance given by

$$I = -\frac{1}{2500} V_{12} + \frac{10}{2500}$$

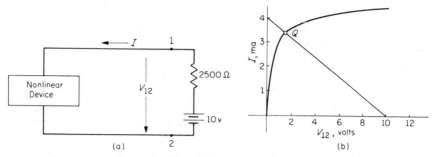

FIG. 3-16. a) Electric circuit with the linear and nonlinear portions clearly identified and separated. b) Graphical solution for I (and V_{12}) for the circuit shown in part a.

for which

$$I = 0 \rightarrow \underbrace{V_{12} = 10}_{\substack{\text{abscissa} \\ \text{intercept}}} \text{ volts} \qquad \text{and} \qquad V_{12} = 0 \rightarrow \underbrace{I = 4}_{\substack{\text{ordinate} \\ \text{intercept}}} \text{ma (milliamperes)}$$

The point marked Q is the solution point, and $I = 3.4$ ma satisfies both characteristics.

3-5. Two-Port Network Analysis

All of the previously mentioned circuit-analysis methods were used to study networks having only one pair of external terminals, that is, one-port[1] networks. Figure 3-17a illustrates such a one-port circuit, while parts b and c depict a two-and a three-port circuit, respectively. As will be seen in later chapters, almost all transistor or vacuum-tube circuits have a pair of input and a pair of output terminals and thus are justifiably called two-port circuits. This section deals only with linear networks, and, although many new terms are introduced, Kirchhoff's laws, Ohm's law, loop currents, and node voltages form an integral part of the overall analytic techniques.

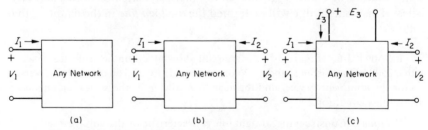

FIG. 3-17. a) One-port circuit. b) Two-port circuit. c) Three-port circuit.

[1] The term one-port, two-terminal, and one-terminal pair are used synonymously in the literature. In general, n-port, $2n$-terminal, and n-terminal pair can be indicated.

It can be shown that it is possible to describe the behavior of a two-port network in terms of its external characteristics; that is, the determination of the variables V_1, I_1, V_2, I_2 will specify the behavior of any linear network in the block of Fig. 3-17b. Conversely, if the network is completely specified, any two of the external voltages and/or currents can be described in terms of the remaining two. Since the combinations of four quantities taken two at a time is six, there are six possible sets of parameters that will describe the behavior of the network. These six sets are usually termed the z, y, h, a, b, and g parameters.

EXAMPLE 3-7. An example of a two-port network is shown in Fig. 3-18. The four variables V_1, I_1, V_2, and I_2 are unknown quantities. The two port network of interest is shown isolated from the rest of the circuit in Fig. 3-18b. The

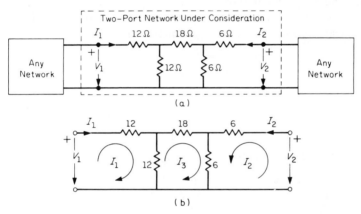

(a)

(b)

FIG. 3-18. a) Two-port network situated between two other networks. b) Two-port network of part a isolated for further study.

purpose of this example is to show how two circuit equations can be written solely in terms of the external variables V_1, I_1, V_2, and I_2. In other words all internal variables, such as I_3, can be eliminated from the two desired simultaneous equations. Using the loop currents as indicated (to agree with the standard nomenclature for the two-port current variables) the loop-current equations by inspection are

1. $24I_1 + 0I_2 - 12I_3 = V_1$
2. $0I_1 + 12I_2 + 6I_3 = V_2$
3. $-12I_1 + 6I_2 + 36I_3 = 0$

Solution: Solving for I_3 in the third equation

$$I_3 = \frac{I_1}{3} - \frac{I_2}{6}$$

and substituting I_3 into the first two equations obtain

$$24I_1 - 12\left(\frac{I_1}{3} - \frac{I_2}{6}\right) = V_1$$

$$12I_2 + 6\left(\frac{I_1}{3} - \frac{I_2}{6}\right) = V_2$$

Combining and rearranging terms results in

$$V_1 = 20I_1 + 2I_2$$
$$V_2 = 2I_1 + 11I_2$$

If desired I_1 and I_2 could be written in terms of V_1 and V_2 or other combinations of variables could be obtained. The importance of this sort of manipulation will become evident as one studies the subsequent sections.

This Example 3-7 will be reviewed in more detail after three of the possible parameter sets have been introduced. In the study of electronic circuits the z, y, and h parameters are the most useful and will be the only ones discussed in this text. These parameters are defined as follows:

z parameters (or impedance parameters or open-circuit parameters):

$$\left.\begin{array}{c} V_1 = z_{11}I_1 + z_{12}I_2 \\ V_2 = z_{21}I_1 + z_{22}I_2 \end{array}\right\} \tag{3-20}$$

y parameters (or admittance parameters or short-circuit parameters):

$$\left.\begin{array}{c} I_1 = y_{11}V_1 + y_{12}V_2 \\ I_2 = y_{21}V_1 + y_{22}V_2 \end{array}\right\} \tag{3-21}$$

h parameters (or hybrid parameters):

$$\left.\begin{array}{c} V_1 = h_{11}I_1 + h_{12}V_2 \\ I_2 = h_{21}I_1 + h_{22}V_2 \end{array}\right\} \tag{3-22}$$

where the voltages and currents are expressed in r.m.s.[2] terms if the sources are sinusoidal in nature.

The z's, y's, and h's are determined solely by the characteristics of the network, and it can be seen that the dimensions of the z parameters are ohms, the y parameters are mhos, and the h parameters are a mixture of ohms, mhos, and dimensionless quantities. Thus, a common name given to the z parameters is "impedance"[3] or "resistance," to the y "admittance"[4] or "conductance," and to the h "hybrid."

Since it is necessary to find some way to evaluate these parameters, it is desirable to find them in the easiest way possible. The following expressions represent very convenient ways to find these parameters either

---

[2] The term r.m.s. stands for root-mean-square. Voltmeters and ammeters that are used to read sinusoidally time-varying voltages and currents are almost always calibrated in terms of the r.m.s. value. The 120-240 volt electrical circuits used in the home represent r.m.s. values. A detailed discussion of how the r.m.s. units are defined and evaluated is given in Chapter 12 on A-C Circuit Analysis.

[3] Impedance is a more general term which includes resistance.

[4] Admittance is a more general term which includes conductance.

analytically or experimentally.   The expressions 3-23, 3-24, and 3-25 follow directly from Eqs. 3-20, 3-21, and 3-22, respectively.

z parameters:

$$z_{11} = \left. \frac{V_1}{I_1} \right|_{I_2=0}$$
driving-point or input impedance (resistance) when the output is open circuited

$$z_{12} = \left. \frac{V_1}{I_2} \right|_{I_1=0}$$
reverse transfer impedance (resistance) with the input open circuited

$$z_{21} = \left. \frac{V_2}{I_1} \right|_{I_2=0}$$
forward transfer impedance (resistance) with the output open circuited

$$z_{22} = \left. \frac{V_2}{I_2} \right|_{I_1=0}$$
driving-point or output impedance (resistance) when the input is open circuited

(3-23)

y parameters:

$$y_{11} = \left. \frac{I_1}{V_1} \right|_{V_2=0}$$
input admittance (conductance) when the output is short circuited

$$y_{12} = \left. \frac{I_1}{V_2} \right|_{V_1=0}$$
reverse transfer admittance (conductance) when the input is short circuited

$$y_{21} = \left. \frac{I_2}{V_1} \right|_{V_2=0}$$
forward transfer admittance (conductance) when the output is short circuited

$$y_{22} = \left. \frac{I_2}{V_2} \right|_{V_1=0}$$
output admittance (conductance) when the input is short circuited

(3-24)

h parameters:

$$h_{11} = \left. \frac{V_1}{I_1} \right|_{V_2=0}$$
input impedance (resistance) with the output short circuited

$$h_{12} = \left. \frac{V_1}{V_2} \right|_{I_1=0}$$
reverse voltage gain with the input open circuited

$$h_{21} = \left. \frac{I_2}{I_1} \right|_{V_2=0}$$
forward current gain with the output short circuited

$$h_{22} = \left. \frac{I_2}{V_2} \right|_{I_1=0}$$
output admittance (conductance) with the input open circuited

(3-25)

It should be evident that Eq. set 3-20 is in the proper format for determining the z parameters from the solutions in Example 3-7. The z parameters are

$$R_{11} \triangleq z_{11} = 20 \text{ ohms} \qquad R_{12} = z_{12} \triangleq 2 \text{ ohms}$$
$$R_{21} \triangleq z_{21} = 2 \text{ ohms} \qquad R_{22} = z_{22} \triangleq 11 \text{ ohms}$$

If the network is composed of resistive elements only, then the impedance parameters are resistive. Furthermore, the definitions suggest the test conditions to be imposed on the network in order to determine the values of the parameters in the laboratory. As an example, R_{11} and R_{21} are obtained when the output is open circuited. This condition is shown in Fig. 3-19 for the network of Example 3-7. Note that since the network is

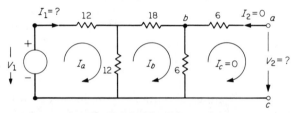

FIG. 3-19. Circuit for determining R_{11} and R_{21} for network of Example 3-7.

linear, any voltage V_1 can be applied to the network since the parameters are defined in terms of *ratios* of circuit variables. Increasing or decreasing the applied source level will not affect the ratio. This can also be demonstrated by leaving the source in symbolic form and noting its eventual cancellation. Using loop currents, the equations are

$$24I_a - 12I_b = V_1$$
$$-12I_a + 36I_b = 0$$

so that

$$I_a = \frac{V_1}{20}$$

$$I_b = \frac{V_1}{60}$$

Since $I_2 = 0$ and $I_c = 0$ and $I_1 = I_a = V_1/20$

$$R_{11} = \left. \frac{V_1}{I_1} \right|_{I_2 = 0} = \frac{V_1}{V_1/20} = 20\Omega$$

Also observe that

$$V_2 = V_{ab} + V_{bc} = 6I_2 + 6(I_b - I_c)$$
$$= 0 + 6\left(\frac{V_1}{60}\right) = \frac{V_1}{10} \text{ volts}$$

Thus

$$R_{21} = \frac{V_2}{I_1}\bigg|_{I_2=0} = \frac{V_2/10}{V_2/20} = 2\Omega$$

It is left to the reader to set up the circuit for determining R_{12} and R_{22}. Of course any technique of circuit analysis may be used to determine the parameters. The parameter R_{11} is the driving point resistance, i.e., the resistance looking into the input terminals with the output terminals open circuited. Network reduction as shown in the sequence in Fig. 3-20 could also be used to obtain R_{11}.

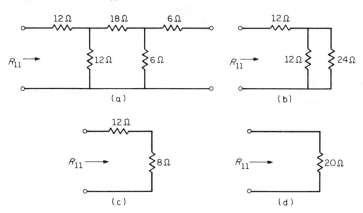

FIG. 3-20. Sequence of network reduction steps to obtain R_{11}.

Similar procedures are utilized to determine any of the other parameters. For example, y_{12} and y_{22} are obtained when the input is

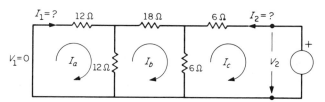

FIG. 3-21. Circuit for determining y_{12} and y_{22} for network of Example 3-7.

short circuited. This condition is shown in Fig. 3-21 for the network of Example 3-7. Applying an excitation V_2 the loop current equations are

$$24I_a - 12I_b + 0I_c = 0$$
$$-12I_a + 36I_b - 6I_c = 0$$
$$0I_a - 6I_b + 12I_c = -V_2$$

Solving for I_a and I_b

$$I_a = \frac{-V_2}{108}$$

$$I_c = \frac{-5V_2}{54}$$

By inspection of Fig. 3-21 note that

$$I_1 = I_a = \frac{-V_2}{108}$$

$$I_2 = -I_c = \frac{5V_2}{54}$$

Therefore,

$$y_{12} = \frac{I_1}{V_2}\bigg|_{V_1=0} = \frac{-V_2/108}{V_2} = -\frac{1}{108} \, \mho$$

$$y_{22} = \frac{I_2}{V_2}\bigg|_{V_1=0} = \frac{5V_2/54}{V_2} = \frac{5}{54} \, \mho$$

Another simpler example follows to illustrate the determination of the three sets of parameters most applicable to electronic circuits.

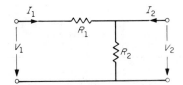

FIG. 3-22. Circuit for Example 3-8.

EXAMPLE 3-8. Find the z, y, and h parameters for the network shown in Fig. 3-22.

Solution:

z parameters:

$$z_{11} = \frac{V_1}{I_1}\bigg|_{I_2=0} = R_1 + R_2$$

$$z_{12} = \frac{V_1}{I_2}\bigg|_{I_1=0} = R_2$$

$$z_{21} = \frac{V_2}{I_1}\bigg|_{I_2=0} = R_2$$

$$z_{22} = \frac{V_2}{I_2}\bigg|_{I_1=0} = R_2$$

y parameters:

$$y_{11} = \frac{I_1}{V_1}\bigg|_{V_2=0} = \frac{1}{R_1}$$

$$y_{12} = \left.\frac{I_1}{V_2}\right|_{V_1=0} = -\frac{1}{R_1}$$

$$y_{21} = \left.\frac{I_2}{V_1}\right|_{V_2=0} = -\frac{1}{R_1}$$

$$y_{22} = \left.\frac{I_2}{V_2}\right|_{V_1=0} = \frac{1}{R_1} + \frac{1}{R_2} = \frac{R_1 + R_2}{R_1 R_2}$$

h parameters:

$$h_{11} = \left.\frac{V_1}{I_1}\right|_{V_2=0} = R_1$$

$$h_{12} = \left.\frac{V_1}{V_2}\right|_{I_1=0} = 1$$

$$h_{21} = \left.\frac{I_2}{I_1}\right|_{V_2=0} = -1$$

$$h_{22} = \left.\frac{I_2}{V_2}\right|_{I_1=0} = \frac{1}{R_2}$$

Although these parameters can be written down by inspection, it is well for the reader to study and verify the results.

3-6. Equivalent Circuits

At this point the reader may well ask how the z, y, and h parameters are useful in two-port network analysis. In this section it will be shown that these parameters allow the development of equivalent-circuit techniques that aid greatly in reducing a complex linear (active or passive) network to a more simple configuration. Since an exhaustive presentation of these equivalent circuits is beyond the scope of this text, only one equivalent circuit, involving the hybrid or h parameters, will be developed.

Rewriting Eq. set 3-22 for convenience,

$$V_1 = h_{11}I_1 + h_{12}V_2 \tag{3-26a}$$

$$I_2 = h_{21}I_1 + h_{22}V_2 \tag{3-26b}$$

it is now possible to develop a circuit that is equivalent to the original circuit as far as the two sets of *external* terminals are concerned. Keep in mind that the equivalent circuit may bear little physical resemblance to the actual circuit but that it does describe adequately the behavior of the circuit at the external terminals, and it is generally this external circuit behavior that is important. To calculate the *internal* currents, voltages, and powers, the *original* circuit must be analyzed.

The development of one equivalent circuit from Eq. set 3-26 is as follows: 1) Each term in Eq. 3-26a has the dimensions of voltage and thus

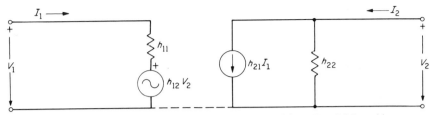

FIG. 3-23. Two-generator equivalent circuit derived from Eqs. 3-26a and b.

suggests a series circuit. 2) The terms in Eq. 3-26b, on the other hand, have the dimensions of current, thus suggesting a shunt or parallel configuration. Figure 3-23 is the result of this two-step analysis and is perhaps the most obvious equivalent circuit that can be developed from Eqs. 3-26a and 3-26b. Note that the two circuits of Fig. 3-23 are not conductively coupled, although many times an author will show an electrical connection between the two, as indicated by the dashed line. (This is done because in most practical circuits there is a common ground connection to all of the individual networks comprising the electronic device or system.) This addition does not change the equivalent circuit as there is no return path between the two circuits shown in Fig. 3-23.

Although it will not be shown here, it is possible to derive either a two-generator (voltage and/or current) or a one-generator equivalent circuit for two-port networks. If the two-generator equivalent circuit is used, a choice of two voltage generators, two current generators, or a voltage and a current generator is possible. The one-generator equivalent circuit is a bit more involved to derive than is the two-generator circuit.

EXAMPLE 3-9. a) Using the circuit of Example 3-8, show an equivalent circuit using h parameters. b) If $V_1 = 100$ volts and $R_1 = 5$ KΩ, $R_2 = 10$ KΩ (kilohms), find V_2 if a load resistance of 1 K is placed across terminals 22' as shown in Fig. 3-24.

Solution: a) First insert the h parameters obtained in Example 3-8 into Eq. set 3-26.

$$V_1 = R_1 I_1 + 1 V_2$$

$$I_2 = -1 I_1 + \frac{1}{R_2} V_2$$

Next, insert the proper values for the h parameters into the equivalent circuit of Fig. 3-23. The result of this analysis is shown in Fig. 3-24. Parameter h_{22} is equal to $1/R_2$ mho, but may be represented in the equivalent circuit as a shunt resistor of R_2 ohms.

b) We choose node voltage equations as a solution technique. Note that node c is the unknown node voltage with node d used as the obvious reference node.

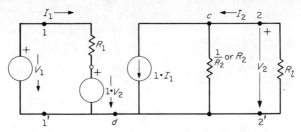

FIG. 3-24. Equivalent circuit for network of Example
3-9.

Using node c

$$\left(\frac{1}{R_2} + \frac{1}{R_l}\right) V_{cd} + (-I_1) = 0$$

or

$$\left(\frac{1}{R_2} + \frac{1}{R_l}\right) V_2 - I_1 = 0$$

The dependent current source has introduced an unknown current into the node voltage equation. This current must then be expressed in terms of node voltages. Applying Ohm's law to R_1 gives

$$I_1 = \frac{V_1 - V_2}{R_1} \rightarrow \frac{V_2 - V_1}{R_1} + I_1 = 0$$

Substituting the given values for V_1, R_1, R_2, and R_l gives

$$\frac{V_2 - 100}{5000} + I_1 = 0$$

$$\left(\frac{1}{10,000} + \frac{1}{1000}\right) V_2 - I_1 = 0$$

and rearranging the equations

$$\left. \begin{array}{r} V_2 + 5000I_1 = 100 \\ 11 V_2 - 10,000I_1 = 0 \end{array} \right\} \tag{3-27}$$

Solving for V_2 in Eq. set 3-27 by multiplying the top equation by a factor of 2 and adding

$$13 V_2 = 200$$

$$V_2 = \frac{200}{13} = 15.4 \text{ volts}$$

If one isn't familiar with the node voltage method he could consider two other solutions to this example. Using the current splitting formula

$$I_2 = (-I_1) \frac{R_2}{R_2 + R_l}$$

$$= (-I_1) \frac{10^4}{10^4 + 10^3} = -\frac{1}{1.1} I_1$$

Next note that

and

$$V_2 = -I_2 R_l = -10^3 I_2 = -10^3 \left(-\frac{I_1}{1.1} \right) = \frac{10^3}{1.1} I_1$$

$$I_1 = \frac{V_1 - V_2}{R_1} = \frac{100 - V_2}{5 \times 10^3}$$

thus

$$V_2 = \frac{10^3}{1.1} \left(\frac{100 - V_2}{5 \times 10^3} \right) = \frac{100}{5.5} - \frac{V_2}{5.5}$$

$$6.5 V_2 = 100$$

$$V_2 = 15.4 \text{ volts} \qquad \text{checks with previous result}$$

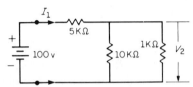

FIG. 3-25. Figure 3-22 repeated with V_1 = 100 v, R_1 = 5 KΩ, R_2 = 10 KΩ and with a 1 KΩ resistor placed across the output terminals.

As a further check return to the original circuit which is shown in Fig. 3-25.

$$R_{eq} = 5 \times 10^3 + \frac{10 \times 1}{10 + 1} \times 10^3 = \frac{6.5}{1.1} 10^3 \Omega$$

$$V_2 = 100 - 5 \times 10^3 I_1$$

where

$$I_1 = \frac{100}{R_{eq}}$$

$$V_2 = 100 - 5 \times 10^3 \frac{100 \times 1.1}{6.5 \times 10^3}$$

$$= 100 - \frac{5.5}{6.5} \times 100 = \frac{100}{6.5} = 15.4 \text{ volts} \qquad \text{checks again}$$

Problems

3-1. If, in Fig. 3-1b, $R_1 = R_2 = 10$ ohms, $R_3 = R_4 = 20$ ohms, $E_1 = 15$ volts, and $E_2 = 35$ volts, a) solve for the loop currents I_1 and I_2, and b) find the current flowing through R_2.

3-2. Using the values given in Prob. 3-1, solve for V_{25} in Fig. 3-3, using the method of node voltages.

3-3. Using the values given in Prob. 3-1, solve for the current flowing through R_2, using Thévenin's theorem.

3-4. For the circuit to the left of terminals AA' in the accompanying figure, find a) Thévenin's equivalent circuit, and b) Norton's equivalent circuit.

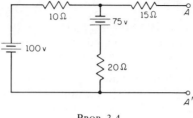

PROB. 3-4

3-5. Find both Thévenin's and Norton's equivalent circuits for the network to the right of terminals AB in Fig. 3-7c. Let $R_1 = R_2 = R_3 = R_4 = 15$ ohms, $E_1 = 60$ volts, and $a = 25$.

3-6. The nonlinear element used in the circuit of part a of the accompanying figure has the ampere-voltage characteristic given in part b of the figure. Find a) the current flowing through the nonlinear element, and b) the power in the nonlinear element.

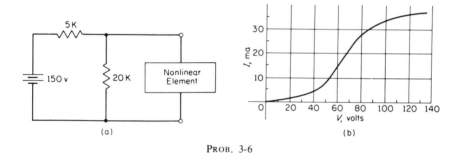

PROB. 3-6

3-7. In the accompanying figure, find a) the z parameters, b) the y parameters, and c) the h parameters for the circuit.

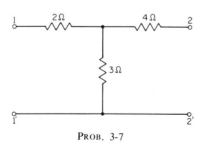

PROB. 3-7

3-8. In Prob. 3-7, use the two-port network relations and find an equivalent circuit using a) the z parameters, b) the y parameters, and c) the h parameters.

3-9. If a 1-ohm resistance is placed across terminals 22′ in the figure of Prob. 3-7 and $V_{11'} = 30$ volts, find $V_{22'}$ by a) using node voltages, b) using loop currents, c) using the *h*-parameter equivalent circuit, and d) using the *y*-parameter equivalent circuit.

3-10. Using the figure of Prob. 3-7, find the ratio $V_{22'}/V_{11'}$, using node-voltage analysis. (The ratio of an output to an input quantity is frequently called a *transfer function*.)

3-11. In the accompanying circuit all resistance values are in ohms. It is desired to solve the circuit for many different values of R_L so a Thévenin equivalent circuit is to be obtained with respect to the terminal pair, *A-B*. Derive and sketch the Thévenin equivalent circuit.

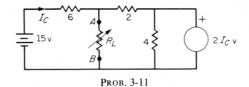

PROB. 3-11

3-12. For the accompanying circuit, find h_{11} and h_{21} for the *h*-parameter equivalent circuit of the network within the dashed lines. All resistance values are in ohms.

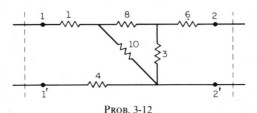

PROB. 3-12

3-13. Using the circuit of part a of the accompanying figure and the ampere-voltage characteristic of the nonlinear device of part b of the figure, graphically determine the value of R that will produce a maximum current flow in the circuit: a) for $E = 120$, b) for $E = 160$, and c) for $E = 200$ volts. (The graphical char-

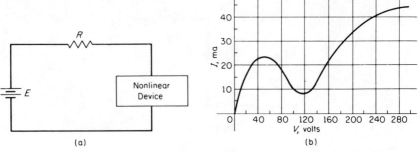

PROB. 3-13

acteristic shown in part b of the figure is typical of the tetrode vacuum tube which will be studied in Chapter 9.)

3-14. Repeat Prob. 3-10 if a 6-ohm resistance is placed across terminals 22′.

3-15. Using the circuit that accompanies Prob. 3-11 plot the power loss in R_L versus R_L. Identify graphically the value of R_L that causes maximum power loss in R_L. Use the maximum power transfer theorem to determine the value of R_L that produces maximum power loss in R_L. Do the graphical and analytical results compare favorably?

3-16. Determine $V_{22'}$ in Prob. 3–12 if a 5-ohm resistor is placed across terminals 22′ and a 10-volt source is placed across terminals 11′.

3-17. What resistance value would have to be placed across terminals 22′ in Prob. 3-12 in order to transfer maximum power into this resistor from an ideal voltage source placed across terminals 11′.

4 / *Electronic Fundamentals*

The purpose of this chapter is to introduce some of the elementary, basic, physical phenomena that are useful if one wishes to understand how transistors and vacuum tubes behave. The reader may find it expedient only to scan the material in this chapter and proceed quickly to Chapter 5.

4-1. Electron Ballistics

Electron ballistics concerns the behavior of an electron while under the influence of electric, magnetic, or electromagnetic fields. The study of electron ballistics is essential if one is to understand the operation of particle accelerators such as the cyclotron, the focusing and operation of cathode-ray tubes such as those used in oscilloscopes and TV sets, or the operation of high-frequency vacuum-tube devices.

In this chapter only the simplest field configurations will be studied. The first will be a uniform electrostatic field and the second a uniform magnetostatic field. It should also be noted that most relationships will involve vector quantities which have spatial direction as well as magnitude. Only simple examples where all motion is confined along a single spatial coordinate are considered, since they illustrate the nature of the phenomena and the required calculations without having to deal with unduly complicated vector systems.

Electrostatic Field. The basic equation describing the force acting on a charged particle in an electric field is the Lorentz equation

$$F = q\mathcal{E} \qquad \text{newtons} \qquad (4\text{-}1)$$

where q is the charge, in coulombs, and $\mathcal{E}$ is the electric-field intensity, in volts per meter. From mechanics it will be recalled that the time rate of change of momentum is also the force on a mass unit. This reduces to the familiar

$$F = ma \qquad (4\text{-}2)$$

as long as the particle velocity is less than one tenth that of the speed of light. Should the particle velocity approach the speed of light, then relativistic mass changes must be considered, and this obviously will complicate the problem. In conventional vacuum tubes, electron velocities of 10^6 to 10^7 m/sec (meters per second) are common, but, since the speed of light is 3×10^8 m/sec, no appreciable electron mass changes occur.

If the acceleration is a constant, the following formula gives the displacement of a particle as a function of time:

$$y = y_o + v_o t + \tfrac{1}{2}at^2 \qquad (4\text{-}3)$$

where y_o is the initial displacement, in meters; v_o is the initial velocity, in meters per second; and a is the acceleration in meters per second squared.

If the electron is considered to have only kinetic energy, and if the velocity is again considerably less than the speed of light, the energy is

$$W = Ve = \tfrac{1}{2}mv^2 \qquad (4\text{-}4)$$

where V is the voltage potential difference through which the electron travels; and e, m, and v are the electron charge, in coulombs; mass, in kilograms; and velocity magnitude, in meters per second, respectively.

From the preceding four relations it is possible to calculate the transit time of an electron from one electrode to another, and the velocity and energy of the electron at any point in a uniform electrostatic field.

EXAMPLE 4-1.   Two large, parallel conducting electrodes are 8 cm (centimeters) apart, with a potential difference of 32 volts between them (see Fig. 4-1). If

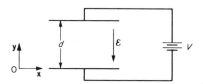

FIG. 4-1. Diagram for Example 4-1.

an electron is released from the negative electrode with an initial energy of 15 ev (electron volts)[1] normal to the electrode, find a) the time of travel to the positive electrode and b) the energy and final velocity of the electron upon impact.

Solution:

$$\text{Electron charge } e = -1.6 \times 10^{-19} \text{ coulomb}$$

$$\text{Electron mass } m = 9.1 \times 10^{-31} \text{ kg (kilogram)}$$

By Eq. 4-4,

$$v_o = \sqrt{\left|\frac{2Ve}{m}\right|}$$

$$= \sqrt{\frac{2 \times 15 \times 1.6 \times 10^{-19}}{9.1 \times 10^{-31}}}$$

$$= 2.29 \times 10^6 \text{m/sec in the } +y \text{ direction}$$

[1] An electron volt is a unit of energy and is equal to 1 (volt) $\times$ e (coulomb) = 1.6×10^{-19} joule.

From Eq. 4-3,

$$y = v_o t + \tfrac{1}{2} a t^2 \qquad \text{meters}$$

where

$$a = \frac{e}{m} \mathcal{E} \qquad \text{meters/sec/sec}$$

using Eqs. 4-1 and 4-2. From consideration of Fig. 4-1

$$\mathcal{E} = \frac{V}{d} (-\mathbf{y}_1) \qquad \text{volts/meter}$$

where V = 32 volts
$\qquad d$ = 0.08 m
$\qquad \mathbf{y}_1$ = a unit vector in $+y$ direction

Note that the electric field intensity $\mathcal{E}$ is constant between the two plates and directed from the positive plate to the negative plate.

$$0.08 = 2.29 \times 10^6 t + \frac{1}{2} \times 1.76 \times 10^{11} \times \frac{32}{0.08} t^2$$

or

$$3.52 \times 10^{13} t^2 + 2.29 \times 10^6 t - 0.08 = 0$$

Upon using the quadratic formula

$$t = -3.26 \times 10^{-8} \pm 5.72 \times 10^{-8} \text{ seconds}$$

and, since only positive times are of interest, a) t = 24.6 nanosec, where nano = 10^{-9}.

The total energy upon impact is simply the sum of the initial energy of the electron and the energy gained in passing through the potential difference between the electrodes or b) W_T = 32 + 15 = 47 ev or 47 $\times$ 1.6 $\times$ 10^{-19} = 75.2 $\times$ 10^{-19} joule.

The final velocity can be found by using the usual kinetic-energy relation

$$v_{\text{final}} = \sqrt{\left| 2 \frac{e}{m} V \right|}$$
$$= \sqrt{2 \times 1.76 \times 10^{11} \times 47}$$
$$= 4.02 \times 10^6 \text{m/sec in the } +y \text{ direction}$$

or one can use

$$v_{\text{final}} = v_o + at$$
$$= 2.29 \times 10^6 + 7.04 \times 10^{13} \times 24.6 \times 10^{-9}$$
$$= 4.04 \times 10^6 \text{ m/sec} \qquad (check)$$

Magnetostatic Fields. The behavior of an electron in a static magnetic field is a bit more difficult to visualize than in the electrostatic case. The basic observed law that describes this situation can be derived from the familiar Ampère's rule

$$\mathbf{F}_m = \mathbf{ll} \times \mathbf{B} \tag{4-5}$$

where $\mathbf{F}_m$ is the force exerted on a current-carrying conductor of length l situated in a magnetic field whose flux density is $\mathbf{B}$ webers/sq m. This law was determined by Ampère before it was known that electrons existed. To apply the law to an electron, define $i = dq/dt = e/\Delta t$ for a single electron, and velocity $v = \Delta l/\Delta t$. Upon making these substitutions in equation (4-5) one obtains

$$\mathbf{F}_m = \frac{e}{\Delta t}\mathbf{l} \times \mathbf{B} = e\mathbf{v} \times \mathbf{B} \tag{4-6}$$

As the vector cross product indicates, the force on the electron is always normal to the direction of motion, and thus no work is done by the magnetic field. When a moving charge enters a region in which there is a magnetic field, it is necessary that the charge be deflected from its original path, since the magnetic force is producing an acceleration normal to the direction of motion. Since the charge does not gain or lose energy, the only possible path that the particle can take is a circle.[2] This circular motion, in turn, causes an outward centrifugal force to be exerted on the

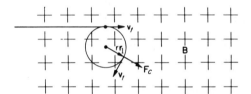

FIG. 4-2. Circular motion of a constant-velocity electron in a uniform magneto-static field.

particle (see Fig. 4-2). From dynamics this force is

$$\mathbf{F}_c = mr\dot{\theta}^2\mathbf{r}_1 = mr\left(\frac{v_t}{r}\right)^2\mathbf{r}_1 = m\frac{v_t^2}{r}\mathbf{r}_1 \tag{4-7}$$

where $\mathbf{r}_1$ is a unit vector directed radially outward, r is the radius of curvature, $\dot{\theta}$ is the angular velocity, and v_t is the tangential velocity. Since v_t is considered to be normal to the magnetic flux density, Eqs. 4-6 and 4-7 can be equated as follows:

$$\left|Bev_t\right| = \left|\frac{mv_t^2}{r}\right|$$

---

[2]Should the charged particle have a component of velocity parallel to the magnetic flux lines, the path will be a helix of constant radius.

since

$$\mathbf{F}_c + \mathbf{F}_m = |\mathbf{F}_c|\,\mathbf{r}_1 - |\mathbf{F}_m|\,\mathbf{r}_1 = 0$$

for equilibrium. Thus the radius is

$$r = \left|\frac{m\,\mathrm{v}_t}{e\,B}\right| \qquad \text{meters} \qquad (4\text{-}8)$$

The angular velocity in terms of the tangential velocity is

$$\omega = \frac{\mathrm{v}_t}{r}$$

$$= \left|\frac{Be}{m}\right| \qquad \text{radians/seconds} \qquad (4\text{-}9)$$

The time for one revolution is

$$T = \frac{1}{f} = \frac{2\pi}{\omega}$$

$$= \left|\frac{2\pi m}{Be}\right| \qquad \text{seconds} \qquad (4\text{-}10)$$

Equations 4-8 and 4-10 show that although a high-velocity electron travels a path of larger radius than does a low-velocity electron, both travel their respective circular paths in the same amount of time.

EXAMPLE 4-2.   An electron with an initial velocity of 4×10^6 m/sec enters a uniform magnetic field with a density of 0.006 weber/sq m at an angle of 60° to the field. Find the radius of its circular orbit and the time required for the electron to traverse one circular path.

Solution:

$$r = \left|\frac{m}{e}\,\frac{\mathrm{v}_t}{B}\right| = \frac{1}{1.76 \times 10^{11}} \times \frac{4 \times 10^6}{0.006} \times \sin 60°$$

$$= 3.27 \times 10^{-3}\,\text{m}$$

$$T = \left|\frac{2\pi m}{Be}\right| = \frac{2\pi \times 9.1 \times 10^{-31}}{0.006 \times 1.6 \times 10^{-19}}$$

$$= 5.95\ \text{nanosec}$$

The importance of understanding the behavior of electrons under the influence of electric and magnetic fields will be more apparent after one reads the two sections on the cathode ray oscilloscope and the cyclotron that appear at the end of this chapter.

4-2. Solid-State Electronics

The study of solid-state phenomena is inherently a more difficult task than the study of particles in a gas or vacuum simply because of the close proximity of the molecules in a solid. In a vacuum the behavior of a particle can be described by the well-known laws of particle dynamics, where each particle is treated as a separate entity and not as being influenced by other particles (or at least by not more than two other particles). In a solid this is not possible, as the individual particles are definitely influenced in their behavior by the other particles.

Crystal Structure. The structure of a solid can be either crystalline or amorphous in nature. Amorphous solids, such as glass, do not have an easily understood atomic or molecular array and will not be considered here. Suitable models which describe adequately the crystalline solids are presently available, but a detailed presentation of these models is beyond the scope of this text.[3] Briefly, the crystalline solids are composed of a latticelike regular array of the atoms which make up the solid. At the intersections of the lines which make up the lattice, that is, the lattice points, are situated the individual atoms. In a substance like common table salt (NaCl) the lattices have alternate sodium and chlorine ions arranged in a cubelike array. This particular array is called a face-centered cubic (FCC) crystal-lattice structure (see Fig. 4-3).

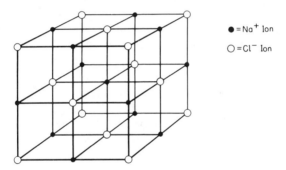

● $= Na^+$ Ion

○ $= Cl^-$ Ion

FIG. 4-3. The face-centered cubic lattice structure of NaCl.

An important fact to remember is that the ions remain essentially fixed at the lattice points, although there is a small thermal vibration about the equilibrium point.

Nonconductors. In a nonconductor almost all the electrons in the atoms remain bound to their atoms. Very few electrons are allowed to drift through the crystal; thus an applied electric field produces almost no current. The energy-level diagram of Fig. 4-4a illustrates the phenomenon more clearly. The filled energy band means that all of the allowed energy states in the crystalline structure are filled by electrons. In order for an

[3]See Kittel, Charles, *Introduction to Solid State Physics*, 3d ed., John Wiley & Sons, Inc., New York, 1966.

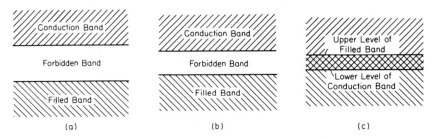

F‍IG. 4-4. Energy level diagram for a) a nonconductor, b) a semiconductor, and c) a good conductor.

electron to be available for conduction purposes, it must jump the energy gap represented by the forbidden band. Just how these bands come about is satisfactorily explained by quantum mechanical principles and the Pauli exclusion principle. For the nonconductor this forbidden energy band is several electron volts.

Conductors. Figure 4-4c shows that the filled and conduction bands in effect overlap for the good conductor. The conduction band is partially filled at all temperatures which means that an abundance of free electrons are available for conduction purposes. Normally, the free electrons drift about within the material in a random manner, but, if an external electric field is applied, the free electrons drift against the field since they are negatively charged.

Semiconductors. As the name implies, the semiconductor is neither a good nonconductor nor a good conductor. The variation between the conductivity of a good conductor and a good nonconductor is about 10^{22}, which is indeed an extremely large range. Figure 4-4b shows the energy diagram of a semiconductor material, and, again, the same three bands of Fig. 4-4a appear. However, in this case the forbidden band is much narrower, being only 1 ev or so. Thus, at low temperatures, the semiconductor actually becomes a good nonconductor, but at room temperatures the thermal vibrations of the lattice ions impart sufficient energy to some of the filled-band electrons to enable these electrons to jump the forbidden band and to enter the conduction band.

It is the semiconductor that is of interest in electronics, for this material has opened up a large, fast-growing industry in solid-state electronics. Germanium and silicon are particularly important in the practical application of semiconductors to the solid-state electronics field. Germanium has a valence of 4, and each of these valence electrons forms a covalent, or electron-pair, bond with an electron of a neighboring atom. Whenever an electron is removed from a covalent bond, a hole remains. This vacancy has a positive charge, since the germanium atom was previously neutral. The freed electron may drift through the lattices in a random manner unless an external electric field is applied. The hole also drifts in a random fashion, but the individual ions or atoms do not move from their fixed places in the lattice structure. The holes tend to drift because one ion takes away an electron from an adjacent atom, thus creating a new ion and an apparent drifting of the positive charge.

Pure germanium and silicon are not useful in making semiconductor devices as their conductivity is too low; to improve this, impurities can be added to the pure material. It has been found that certain impurities tend to enhance the number of free conduction electrons, whereas others tend to cause a predominance of holes. This is a happy consequence which makes possible semiconductor diodes and transistors.

If the impurity has a valence of 5, such as antimony, phosphorus, or arsenic, then 4 of the electrons of the impurity atom form a covalent bond with the germanium atom. This allows the fifth electron to drift about within the crystal and become a conduction electron. The entire crystal is electrically neutral, but, as with good conductors the current carriers are electrons. Semiconductor materials with electrons as current carriers are called *N-type semiconductors.* The impurities which create N-type semiconductors are called *donor impurities.*

In a similar fashion, impurities with a valence of 3, such as indium, boron, aluminum, or gallium, combine with the germanium atoms to form 3 covalent bonds. To complete the fourth covalent bond, the indium atom must borrow an electron from a neighboring germanium atom. This creates a deficiency of 1 electron, or a hole. As this process of borrowing an electron to complete a covalent bond continues, the hole appears to move about the crystal lattice. The dominant current carriers in the semiconductor are now holes, and the impurities which create this *P-type semiconductor* are called *acceptor impurities.*

As will be shown in later chapters, the P- and N-type semiconductors are used to fabricate diodes and transistors.

4-3. Electron Emission

Simple electrostatic coulombic forces seem to offer a plausible explanation as to how crystalline solids are held together. The free electrons in a crystal are prevented from leaving the surface of the crystal because of the electrostatic forces of attraction present at the surface. If an electron did escape, then the crystal body would no longer be electrically neutral, and the force of attraction between the positively charged crystal and the electron would normally cause the electron to return. However, the electron can be given sufficient energy to overcome the influence of the forces of the crystal. The energy required to remove completely an electron from the attractive influence of a material is called the *work function* of the material. Although the preceding statements greatly simplify the really complex field configuration at the boundary of a material, they do give a good approximation to the solution of the problem.

Usually, electrons receive energy for emission from either thermal or photon sources, but secondary and field emission are also important. Thermal sources cause the temperature of the atoms and electrons to increase uniformly, whereas photon sources can impart energy solely to the electrons. Photons are packets or bundles of radiant electromagnetic energy. The photon concept is normally applied to the study of visible light, X-rays, and other similar high frequencies. The phasor, sinusoidal concept of Chapter 12, which is used in circuit analysis, is useful only at the lower radio frequencies.

Thermal Emission. First, let us consider the thermal, or thermionic, emission of electrons. As a material is heated, the atoms and electrons begin to vibrate more and more furiously. If the electrons near the surface have enough energy, and if their motion is directed away from the surface, they may actually leave the surface and escape, since only the electrostatic surface forces tend to prevent their escape. The mean energy of all the electrons in a material, even at high temperatures, may be considerably below the work-function energy of the material, but, from a statistical standpoint, *some* electrons have a large energy whereas others have only a small amount; for instance, at 1000 K the mean energy of an electron is only 0.17 ev, whereas the work function of metals falls between 1 and 7 ev. Measurable emission is noted, however, even at this rather low emission temperature. Tungsten, thoriated tungsten, and oxide-coated metals are most frequently used for emitters in vacuum tubes. Tungsten is valuable because of its high melting point, and the oxide-coated metals have a low work function. Nickel coated with barium and strontium oxides is a commonly used emitter material having a work function of about 1 ev.

Photoelectric Emission. When electrons near the surface of a material receive energy from photons of radiant energy, photoelectric emission may occur. Einstein, in 1905, formulated this phenomenon by

$$\tfrac{1}{2}m\mathrm{v}^2 = hf - hf_w \qquad (4\text{-}11)$$

where $\tfrac{1}{2}m\mathrm{v}^2$ is the kinetic energy of the released electrons, hf is the energy of the incident photons, and hf_w is the work-function energy of the material under bombardment by the photons. For electrons to escape, the incident photon energy must be greater than the work-function energy. This equation describes the maximum kinetic energy that an electron may have, but most of the released electrons will have smaller kinetic energies owing to internal collisions before leaving the material.

The relationship of the velocity of light to the frequency and wavelength of a photon in free space is

$$c = \lambda f \qquad (4\text{-}12)$$

where c is the velocity of light (3×10^8 m/sec); λ is the wavelength, in meters; and f is the frequency, in hertz, of the photon.

EXAMPLE 4-3.   An X-band radar operates at a frequency of 10,000 MHz or 10 gHz. What is the wavelength of the radar signal, expressed in centimeters?

Solution: Using Eq. 4-12

$$\lambda = \frac{c}{f} = \frac{3 \times 10^{10}}{10^4 \times 10^6}$$

$$= 3 \text{ cm}$$

EXAMPLE 4-4.   A metal surface is exposed to radiant energy whose wavelength is 5100 A. If the work function of the metal is known to be 1.6 ev, what is the maximum velocity of a released electron?

Solution: First, two new constants need to be defined:

$$h \text{ (Planck's constant)} = 6.6 \times 10^{-34} \text{ joule-sec}$$

$$A \text{ (angstrom)} = 10^{-10} \text{m}$$

The energy of the incident photons is

$$hf = 6.6 \times 10^{-34} \times \frac{c}{\lambda} = 6.6 \times 10^{-34} \times \frac{3 \times 10^8}{5100 \times 10^{-10}}$$

$$= 3.87 \times 10^{-19} \text{ joule or } 2.43 \text{ ev}$$

Upon employing Einstein's equation,

$$\tfrac{1}{2}mv^2 = 2.43 - 1.6 = 0.83 \text{ ev}$$

$$v^2 = \frac{2 \times 0.83 \times 1.6 \times 10^{-19}}{9.1 \times 10^{-31}} = 29.2 \times 10^{10}$$

or

$$v = 5.41 \times 10^5 \text{ m/sec}$$

Field and Secondary Emission.   Field and secondary emission are also important when considering the ejection of electrons from a material. The removal of electrons from the surface of a material by the application or presence of an external electric field is called *field emission.* For emission to occur, the external electric-field intensity must overcome the local electric field tending to hold the electrons within the material. *Secondary emission* of electrons is caused when the surface of a material is bombarded with high-velocity or high-energy particles.   In vacuum-tube studies, the bombarding particles are usually electrons, and the target is usually the plate of the tube.   The high-energy impinging electrons transfer their energy to one or more electrons within the material being bombarded, with the result that perhaps 5 or 10 secondary electrons may be released for every primary or bombarding electron. The ratio of secondary to primary electrons depends on the type of material, the surface preparation, and the energy of the primary electrons.

4-4. The Frequency Spectrum

Broadly speaking, the study of electrical engineering consists of the study of the concepts, devices, and techniques of problem solution within a band of frequencies. The familiar power spectrum used in the study of motors, generators, transmission lines, and so forth, is generally limited to 0 to 1000 Hz with 60 Hz being the most common frequency. The radio AM (amplitude-modulated) broadcast band is limited to 500 to 1500 KHz. In this text, electronic devices operating above 1 MHz[4] or so will not be mathematically investigated. Figure 4-5 shows a broad breakdown of the

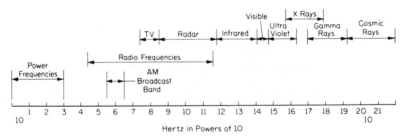

FIG. 4-5. The frequency spectrum.

entire frequency spectrum from 0 to $f \rightarrow \infty$ Hz. Prior to World War II, the electrical engineer confined most of his investigations to frequencies under 10^9 Hz, while the physicist confined his to above 10^{14} Hz. The infrared region was a sort of no-man's-land, since neither generators nor detectors existed for the investigation of this portion of the frequency spectrum. The advent of radar and associated equipment helped to change this picture and now frequencies of 3×10^{11} Hz, or 300 gigahertz (gHz), are being both generated and received. The recent development of maser-type devices promises to extend this frequency range even beyond the visible light frequencies (10^{14} to 10^{15} Hz). While the electrical engineer is finding it expedient to understand and use the higher frequencies such as X-ray frequencies, so the physicist finds microwaves of 10^{10} Hz or so very useful as a diagnostic tool in measuring microscopic particle behavior. Chapter 17, which deals with radio propagation and modulation, presents a rather interesting aspect of the frequency spectrum in relation to the earth's atmospheric envelope.

---

[4]The reason for this limitation is so that only resistive circuit models need be considered.

4-5. The Cathode-Ray Oscilloscope

The cathode-ray oscilloscope, also called CRO, oscilloscope, or scope, is one of the most valuable instruments now used for electronic measurements. It is a highly versatile and flexible instrument which enables one to make many types of electrical measurements. Its greatest appeal to the experimenter is because it provides a visual display of signal waveforms as they actually occur in time. The waveforms of signals in the frequency range from d-c to over 1000 megahertz can be observed (with sampling techniques). The basic oscilloscopes typically allow the observation of input waveforms with frequencies up to a few hundred kilohertz or a few megahertz without a distorted display of the input waveform.

The heart of the CRO is the cathode-ray tube which is very similar to the television picture tube. The visual display on the CRO and TV screen is made by bombarding a fluorescent coating on the inside face of the tube with a beam of electrons. The electrons typically have an energy of 5 to 20 kiloelectron-volts (kev) before striking the screen. Upon striking the phosphors the electrons give up their kinetic energy and in the process excite electrons in the atomic structure of the phosphor atoms into excited energy states. As these electrons drop from a higher excited energy state to another, photons (light waves or packets) are emitted. The emitted photons produce the bright visible spot or line of light at the point on the screen where the bombarding electron beam is impinging. The CRO tube is constructed so that the electron beam can be focused and directed onto the screen. By moving the beam along the screen the image of a line, curve, etc. can be formed. If this image is formed repetitively, then the persistence of photon emission from the phosphors and the persistence of vision in the eye will give the impression of a stationary image on the scope screen. The intensity and positioning of the electron beam is controlled by electronic circuits exterior to the CRO tube. Some of these circuits as well as the CRO tube will be discussed in subsequent sections.

The Cathode-Ray Tube. The vacuum cathode-ray tube (CRT) is the more proper name for all of the picture tubes now used in TV sets, radarscopes, and oscilloscopes. There are three basic parts to a CRT: 1) the electron gun, which furnishes a continuous stream of moderately high-energy electrons (a few thousand electron volts); 2) the deflection plates, which can move the electron stream by means of electrostatic forces; and 3) the evacuated glass envelope, which has a thin layer of phosphor material deposited on one end.

The electron gun is shown in Fig. 4-6. In order to explain the operation of the electron gun, it is necessary to assume that the entire assembly is situated in a vacuum so that the electrons will not collide with gas atoms, thereby impeding their progress.

The indirectly heated cathode emits electrons using thermal or

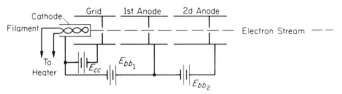

FIG. 4-6. Electron-gun assembly.

thermionic emission which was mentioned earlier. The grid, first anode, and second anode are usually constructed of a hollow, cylindrically shaped piece of metal with a metallic disk placed near the center of the cylinders. A small hole is drilled in the center of the disk in order to allow the electron stream to pass through, thus forming the "barrel" of the gun.

As shown in Fig. 4-6, the three cylinders are aligned axially, and different potentials are applied to the cylindrical electrodes. Using the cathode as the reference electrode, a negative voltage is applied to the grid. The purpose of the grid is to control the number of electrons that are allowed to flow through the grid orifice. The electrons that manage to pass through the grid orifice are rapidly accelerated by the rather large (about 500 volts) positive potential that is applied to the first accelerating anode. Any electrons that stray from the axial path strike the anode structure and are lost as far as the electron stream is concerned. The remaining axial electrons pass through the first anode orifice and are further accelerated by the potential of the second anode (1000 volts or higher). Those electrons that stay on the straight-line path pass through the orifice of the second accelerating anode as a rather fine, densely packed stream of fast-moving electrons. The diameter of the electron stream (or beam, as it is often termed) can be controlled, to some extent, by varying the potential that is applied to the first accelerating anode. This is known as *focusing* the electron beam. Since the electrons have a tendency to repel each other, there is a practical limit to the minimum beam diameter. Normally, an electron stream that is about the diameter of a pencil lead can be obtained.

To complete the CRT, it is necessary to add the deflection plates and a phosphorescent screen. The deflection plates move the electron stream from its axial path, and the phosphorescent screen enables one to see where the electron stream is being directed. Figure 4-7 shows the complete CRT assembly. Almost all small-diameter CRT's use electrostatic deflection, but magnetic deflection is usually necessary for large-diameter screens such as those used in TV sets. The electrostatic deflection of the electron beam is accomplished by passing the electron beam between two sets of charged parallel plates. One set of plates is used for vertical de-deflection and the other for horizontal deflection. The charge is placed on

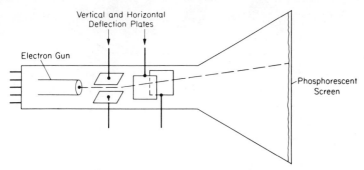

Fig. 4-7. A typical cathode-ray tube.

the deflection plates by means of an externally applied voltage, and, if the applied voltage is varied, the position of the electron beam is also varied. Since electrons have such a small mass, the useful frequency range of the CRT covers a bandwidth of many megahertz. No mechanical recording device can even remotely challenge this upper-frequency limitation (about 5000 Hz is the upper-frequency limit of mechanical recorders).

To deflect the electron beam magnetically, it is necessary to pass magnetic flux lines through and normal to the axial path of the electron beam. To accomplish this, current-carrying coils are situated on the outside neck of the CRT. The current comes from an externally applied signal source, and the current, in turn, produces the magnetic flux lines which deflect the electron beam.

The impact of the electron beam on the phosphors which cover the inner surface of the screen of the CRT excites the phosphors and causes visible light to be emitted. The color and persistence of the visible light depend on the type of phosphor material that is used to coat the screen. Phosphors that emit electromagnetic radiation corresponding to all colors from red to blue are now obtainable.

The Basic Cathode-Ray Oscilloscope. As discussed earlier the cathode-ray oscilloscope (CRO) uses a CRT to present a visual picture (the output) of any electrical signal that is applied to the input of the oscilloscope. The basic operation of the video portion of a TV set is really quite similar to the basic operation of the CRO. A very simple but basic CRO is shown in Fig. 4-8.

The CRT and a power supply for the electron gun are the same as shown in Figs. 4-6 and 4-7. The only two additions that have been made are the X-axis (horizontal) sawtooth generator and Y-axis (vertical) amplifier. The purpose of the vertical amplifier is simply to boost the voltage level of a small Y-input signal to a value which, when applied to the vertical deflection plate, will produce a sizable deflection of the electron beam. About 100 volts must be applied to the deflection plates of

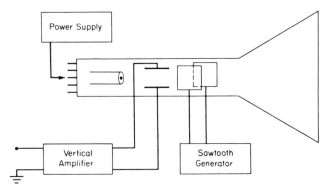

FIG. 4-8. A basic cathode-ray oscilloscope.

most CRT's in order to cause an appreciable deflection of the electron beam.

The oscilloscope is normally used to show a voltage as a function of time, that is, a Cartesian or rectangular plot with a magnitude of the desired voltage plotted as the ordinate and time plotted as the abscissa. The X-axis or abscissa can be calibrated in terms of time if a linear time-varying voltage is applied to the horizontal deflection plates; that is, $v_{x\,axis} = kt$, which is the equation of a straight line. Such a voltage is shown in Fig. 4-9a, except that the voltage is repeated every t_1 sec. It is

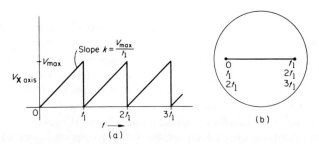

FIG. 4-9. a) A sawtooth voltage used for a linear sweep on an oscilloscope. b) Time base as shown on the face of an oscilloscope.

relatively easy to construct an electronic voltage generator that will generate this sawtooth wave form. If a sawtooth generator is connected to the horizontal deflection plates (see Fig. 4-7) and if the maximum voltage V_{max} is adjusted so that the electron beam is horizontally deflected a sizable distance across the face of the oscilloscope, then every t_1 sec the electron beam traces a line representing a linear time base across the face of the CRT. Figure 4-9b shows that at time = t_1, $2t_1$, $3t_1$, and so on, the horizontal sweep voltage suddenly jumps from V_{max} to zero volts, which causes the electron beam to start again at the origin and retrace the linear time base. Since the glow of the phosphors tends to persist, the trace appears to remain as a continuous line. The time interval t_1 is adjustable over a rather wide range in most oscilloscopes.

To illustrate the use of the oscilloscope, let us consider Fig. 4-10a, b,

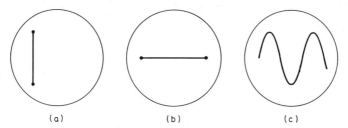

(a)　　　　　　(b)　　　　　　(c)

FIG. 4-10. a) A sine wave applied to the vertical axis. No x-axis voltage. b) A sawtooth voltage applied to the x axis. No y-axis voltage. c) Voltages of parts a and b applied simultaneously.

and c and assume that part a represents a sinusoidally varying voltage that is applied to the vertical deflection plates, but with no voltage applied to the horizontal deflection plates. The trace, although varying sinusoidally, merely appears as a straight line, and it would be difficult to tell just what sort of voltage was being applied to the vertical deflection plates. Figure 4-10b shows a sawtooth voltage applied to the horizontal deflection plates, with no voltage applied to the vertical deflection plates. Figure 4-10c shows the net result if the situations depicted in parts a and b occur simultaneously. It is easy to see that the applied Y-axis voltage is indeed sinusoidal in nature, and, if the time base is accurately calibrated, the frequency of oscillation of the sine wave can be determined. Also, if the vertical axis is calibrated for voltage measurements, the peak value of the sine wave can be determined.

There are many ways to utilize an oscilloscope. It is one of the most useful and versatile test instruments ever developed. If desired, a photograph can be taken so that the trace is permanently recorded.

4-6. The Cyclotron

The cyclotron, developed in 1931 by E. O. Lawrence and M. S. Livingston, is a positive-ion particle accelerator used to generate a stream of high-energy charged particles. The cyclotron principle can be explained on the basis of Eqs. 4-8 and 4-10. The basic construction of a cyclotron is shown in Fig. 4-11. A flat can is cut along a diameter, and the two sections are separated a short distance. These two sections are called *dees*.

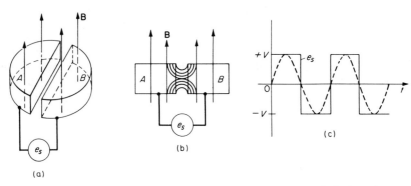

FIG. 4-11. a) Basic cyclotron structure. b) Electric and magnetic field configuration in a cyclotron. c) Voltage wave shape applied to the cyclotron dees.

A magnetic field of flux density B is directed axially through both dees, and a time-varying voltage e_s is applied to the metal dees. Because the dees are good conductors, almost all the electric flux lines appear at the gap separating the dees, and thus the interior of the dees is essentially free of any electric field (see Fig. 4-11b). Notice also that the electric field lines at the center of the gap are parallel to a circular cross section of the dees. This is an important factor because charged particles inserted anywhere along the diametric center of the gap are accelerated normal to, rather than parallel to, the axis of the dees. The dees are evacuated of all gases so that the accelerated charged particles will not collide with gas atoms.

If positive ions are released at the center of the dees and if e_s is a square wave, the cyclotron operation can be explained as follows (note Figs. 4-11c and 4-12):

As soon as a positive ion is released, let us assume that dee A becomes negative with respect to dee B. The positive ion is accelerated from

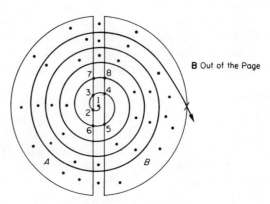

FIG. 4-12. Path of a positive ion in the cyclotron.

point *1* to point *2* and gains an energy of *Vq* ev. Upon reaching point *2*, the positive ion is thrust into the electric-field-free region of dee *A* and thus acquires no additional energy. The ion then travels a circular path of radius *r*, as determined by its velocity at point *2* and the magnetic flux density *B*. When the ion reaches point *3*, we let dee *B* become negative and dee *A* positive. The ion accelerates toward point *4* and gains an additional energy of *Vq* ev. At point *4* the ion enters the field-free region of dee *B* and again travels a circular path as defined by Eq. 4-8. This process is repeated perhaps thousands of times until the ion reaches the desired energy. At this point an auxiliary electric field forces the positive ion from its circular orbit and out of the evacuated chamber of the dees through a thin film of material, such as aluminum. The high-energy ions can then be used for any experimental purpose deemed necessary.

Because the travel time or period for each circular orbit is the same, the cyclotron can generate streams of high-energy particles rather than just one particle at a time. In a practical case, e_s is sinusoidal in nature rather than a square wave.

Since the positive ions approach the speed of light, relativistic mass change must be considered. This affects both the radius and the period of the ion path, and, to keep these factors constant, compensation is usually made by properly designing the magnetic-flux pattern that passes through the dees.

Problems

4-1. An electron having an initial velocity of 4×10^6 m/sec is ejected into the electric field between two parallel plates, as shown in the figure. How far does it travel horizontally before striking one of the plates? Assume no fringing of the electric field.

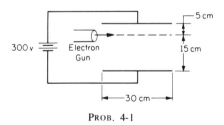

PROB. 4-1

4-2. Find the flux density *B* and the direction of *B* so as to prevent the electron from being deflected from its horizontal path in Prob. 4-1.

4-3. If an electron has an initial velocity of 6×10^7 m/sec, find its initial energy in a) electron volts, b) joules.

4-4. A positively charged particle consisting of 5 protons is ejected between and parallel to two plates. If a magnetic flux density of 0.4 weber/sq m is directed as shown in the figure, and the velocity $v_o = 5 \times 10^5$ m/sec. a) In which direction does the charged particle tend to move? b) What must be the voltage applied to terminals AB to keep the particle on a straight-line course? c) Sketch in a battery showing the correct polarity in part b.

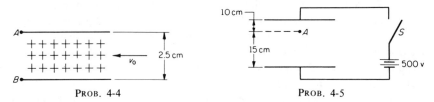

PROB. 4-4   PROB. 4-5

4-5. A charged particle with 3 net protons and a total mass of 6600 times that of an electron is at rest at point A in the figure. When the switch is thrown, how long will it take the particle to reach the correct plate?

4-6. An electron is moving in the direction shown in the figure. a) Find the vertical component of the force on the electron if $v = 8 \times 10^6$ m/sec and $B = 0.5$ weber/sq m. b) Is the force in part a up or down? c) Find the radius of the electron's orbit.

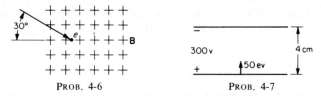

PROB. 4-6   PROB. 4-7

4-7. An electron with an initial energy of 50 ev is ejected from and normal to the positive plates, as shown in the figure. a) How much time will elapse before the electron velocity reaches zero, i.e., it stops? b) How far does the electron travel from the positive plate? c) How much energy, in joules, is delivered to the positive plate by the return of the electron?

4-8. An electron leaves point A with no initial energy. How long will it take the electron to reach point B? Assume a straight-line path and no fringing of the electric fields.

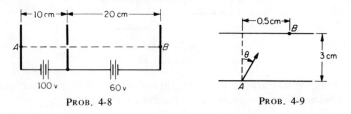

PROB. 4-8   PROB. 4-9

4-9. An electron leaves point A with an initial energy of 35 ev and at an angle θ with the vertical, as shown in the figure. If it strikes the upper plate at point B, find the angle θ. Also find the transit time.

4-10. An electron is released from point A with no initial energy, as shown in the figure. If a voltage $v = 50t$ is applied between the plates, how long will it take the electron to reach point B?

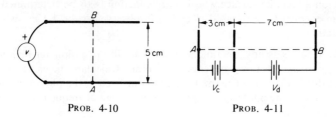

PROB. 4-10   PROB. 4-11

4-11. An electron is released from point A with an initial energy of 35 ev in the x direction, as shown in the figure. If V_d is 150 volts, what must V_c be in order to cause the electron to strike point B with a velocity of 2×10^6 m/sec? Neglect all fringing of the electric fields.

4-12. An electron enters a hole in plate A with an initial velocity of 8×10^6 m/sec and at an angle of 30°, as shown in the figure. What is the value of the battery voltage E if the electron reaches plate B with zero velocity in the y direction?

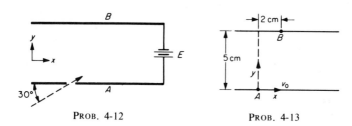

PROB. 4-12   PROB. 4-13

4-13. An electron is released from point A with an initial energy of 8 ev in the x direction, as shown in the figure. What voltage must be applied to the plates to cause the electron to strike the upper plate at point B, as indicated on the figure?

4-14. An electron leaves point A at an angle of 60° to the vertical with an initial velocity of 8×10^6 m/sec, as shown in the figure. Find the distance d that the electron travels along the x axis before striking the upper plate at point B.

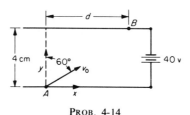

PROB. 4-14

4-15. What is the maximum wavelength of a photon that will release electrons from a metallic surface whose work function is 2.3 ev?

4-16. A monochromatic ultraviolet light source, whose wavelength is 3400 A, impinges on a metallic surface whose work function is to be determined. If the maximum energy of the released electrons is observed to be 1.1 ev, what is the work function?

4-17. Radar sets now operate at frequencies in the millimeter region. What frequency corresponds to a wavelength of 2 mm? Calculate the energy of a radar photon at this frequency. (The small energy contained in the photon wave packet is one reason that the photon concept is not employed at frequencies much below that of visible light.)

5 / *Diodes*

Historically, the diode, which was invented in 1897 by J. A. Fleming of England, was the forerunner of radio and electronics in general. Fleming developed his "valve" to detect radio signals and the diode has continued to serve this important purpose up to today. In addition to the detection of radio signals the rectifying and isolating properties of the vacuum-tube and solid-state diodes are extremely useful in all kinds of electronic devices and systems, such as radios, television sets, radar and microwave systems and the rapidly expanding computer field.

A diode is a unilateral electronic device because it is a good conductor in one direction but a poor conductor in the reverse direction. For most applications the solid state diode has replaced the vacuum-tube diode and so more attention will be focused on the solid state diode in this chapter and throughout the text. There are two applications, however, where the tube-type diode is still superior to the solid-state diode. The vacuum-tube diode is useful where high voltages are involved, such as in radar and microwave transmitters and television receivers. Gas-tube diodes, such as the mercury cathode rectifiers, are capable of handling thousands of amperes and are used in applications where large amounts of power must be transformed from alternating to direct current. Because progress is continually being made in solid-state technology the many medium voltage and current (a few thousand volts and hundreds of amperes) ratings that are now being successfully handled by solid state diodes will be extended even further in the future.

Only the vacuum and semiconductor diodes will be discussed in this chapter. Additional comments on the gas diode will be found in Chapter 18.

5-1. The Vacuum Diode

The vacuum diode consists of two electrodes which are enclosed in a glass or metal envelope in order to maintain the high vacuum required for proper operation. The two electrodes are called the anode, or plate, and the cathode, and, since the cathode emits the electrons, the study of it is of particular interest.

The cathode is electrically heated to a temperature that will produce the thermionic emission of electrons. The operating temperature of a cathode varies from 1000°K for the oxide-coated cathodes to 2400°K for tungsten cathodes. The equation that formulates the thermionic-emission law is attributed to either Richardson or Dushman and is as follows

$$J = AT^n e^{-b_o/T} \qquad \text{amps/square meter} \qquad (5\text{-}1)$$

where J = emitted current density in amperes/square meter

A = a universal constant of metals and equals 120.5×10^4

b_o = a constant dependent upon the work function of the material

T = temperature of the cathode material, in degrees Kelvin

n = a constant which equals $\frac{1}{2}$ for Richardson's equation and 2 for Dushman's equation

In practice, A is about one half of the theoretical value, but otherwise the law expresses quite well the experimental shape of the emission curve. Because the exponential term is dominant in Eq. 5-1 it makes little difference in the shape of the curve whether $n = \frac{1}{2}$ or 2. Figure 5-1 shows

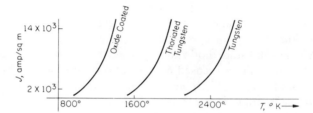

FIG. 5-1. Emission characteristics of the three most common emitting materials.

a comparison of the characteristics of three common thermionic-emission materials used in the construction of the cathode. These materials are tungsten, thoriated tungsten, and oxide-coated metals.

Cathode Construction. Cathode construction is determined, to a large extent, by whether the cathode is to be heated directly or indirectly. If tungsten is used, it is much more efficient to heat the cathode directly. This can best be accomplished by passing current through the filament wire making up the cathode. Figure 5-2 shows types of construction commonly used for directly heated cathodes.

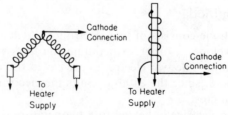

FIG. 5-2. Two types of construction for directly heated cathodes.

Tungsten cathodes are very rugged and are used whenever high voltages must be applied to the tube. In all vacuum tubes, some gas atoms are present, and if some of these atoms become ions (say by collision with one of the thermally emitted electrons), the cathode, which is usually at a negative potential, attracts these ions. In some tube applications the ions may fall through a potential difference of several thousand volts before striking the cathode. This high-energy ion bombardment is sufficient to destroy either the oxide-coated or the thoriated-tungsten cathode surface, but the tungsten itself can withstand the bombardment.

The indirectly heated cathode generally consists of a small-diameter nickel cylinder coated with the oxide material. Inside the cylinder is a tungsten heater which is electrically insulated from the cathode. Figure 5-3 shows a typical indirectly heated cathode construction.

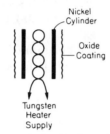

Nickel
Cylinder

Oxide
Coating

Tungsten
Heater
Supply

FIG. 5-3. Indirectly heated cathode construction.

Indirectly heated cathodes are useful and, indeed, necessary where a common filament or heater source is used but where the cathodes themselves are at different potentials. Also, it is economically more feasible to use a-c rather than d-c heater currents, and the indirectly heated cathode tends to mimimize objectionable 60 Hz hum or pickup. (Almost all a-c power is supplied at a frequency of 60 Hz in this country.)

Space-Charge Phenomena. Under normal operating conditions the cathode in a vacuum tube is surrounded by a virtual sea or cloud of electrons. This excess of electrons is necessary to provide smooth operation of the tube, and, at the same time, this negative space charge limits the current obtainable at the anode. To clarify this, Fig. 5-4 shows the effect of space charge on the potential distribution in a diode with parallel plate construction of the electrodes. Curve a, the straight-line potential distribution, shows that a constant electric-field intensity exists between the plates with the cathode too cold to emit electrons. Curve b shows the effect of a large number of cathode-emitter electrons. Note that from K to M the slope of the potential curve is negative, and from M to P it is positive. This means that low-energy emitted electrons are repelled back to the cathode, while those electrons with an energy (usually ex-

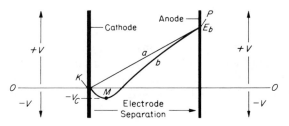

FIG. 5-4. Potential distribution in a diode. Curve *a*, without space-charge effects and curve *b*, with space-charge effects.

pressed in electron volts) greater than eV_c electron volts will overcome the repelling space-charge force and progress rapidly toward the plate. It is rather interesting to note that the electron space charge, or cloud, causes a point away from the cathode (point *M*) to be at a lower potential $(-V_c)$ than the cathode itself.

Figure 5-5 is a current-voltage diagram of a diode. For a given plate voltage e_b, the plate current i_b from 0 to *A* is seen to be the same,

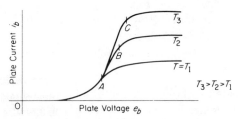

FIG. 5-5. Current-voltage diagram of a vacuum diode for various cathode temperatures.

regardless of the temperature of the cathode, but, for higher plate voltages, a considerable divergence is noted. The curve from 0 to *A* (for temperature T_1) represents the normal operating region of the diode where the space-charge effect is dominant; that is, a large cloud of electrons surrounds the cathode. The flatter portions of the curves represent the saturation region where the plate collects almost all the electrons emitted by the cathode. Since any vacuum tube is designated to operate at a given temperature, assume that a given diode has a normal operating temperature of T_2 as shown in Fig. 5-5. The tube operates under space-charge conditions up to point *B*. The equation for the curve for the region 0 to *B* for temperature T_2 has been theoretically formulated and is known as the Langmuir-Child's law

$$i_b = Ke_b^{3/2} \tag{5-2}$$

where K is a constant for any given geometrical electrode configuration. Here, note that the plate current varies with the three-halves power of the plate voltage.

5-2. The Solid-State Diode

The concept of the N- and P-type semiconductor materials was introduced in Chapter 4. As will be seen the combination of an N- and P-type semiconductor produces a diode, whereas the proper combination of two N's and a P or two P's and an N can produce a transistor. The rather involved explanation of the operation of the transistor is deferred to a later chapter, but the behavior of the semiconductor diode is relatively easy to explain from a qualitative standpoint. Assume that a P-type and a N-type semiconductor block are placed together so that atomic or microscopic bonding takes place. Actually the junction of a P and N region is usually accomplished by a carefully controlled doping and crystal growing process. At the instant of contact the P and N regions remain unchanged as shown in Fig. 5-6a. Because there is an abundance of holes in the P region and an abundance of electrons in the N region density gradients are now created at the junction between the P and N regions. Note that no density gradients exist in the P and N regions as long as the two regions are separate blocks of semiconductor material. The natural law of diffusion comes into play as soon as the two blocks form an intimate atomic bond. Since the holes and electrons are con-

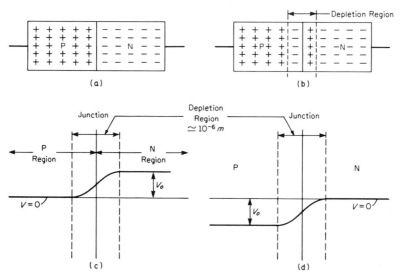

FIG. 5-6. a) PN junction at the instant of contact between the P and N blocks. b) PN junction in equilibrium. c) Potential energy barrier for holes. d) Potential energy barrier for electrons.

stantly in motion, it is only natural that some of the holes from the P region will drift into the N region and conversely electrons from the higher density N region will drift into the P region. The effect of this drifting is shown in Fig. 5-6b. Since the P and N regions were initially electrically neutral, the movement of charges across the boundary means that the two regions are electrically charged (the entire PN block is still electrically neutral). From Coulomb's law it is apparent that these displaced charges cause an electric field to exist across the junction or boundary of the P and N regions tending to counteract the diffusion motion. When the forces on the charges due to the electric field equal the diffusion forces, equilibrium is attained and the net diffusion of charges ceases.

The diffusion of charges across the boundary occurs over a rather small distance of about 10^{-6} meters. This region is variously called the space-charge region, the depletion region, or the transition region and the electric field across this region creates what is termed a potential-energy barrier, which is called either the contact or diffusion potential. The potential-energy barrier V_o for holes is shown in Fig. 5-6c and for electrons in Fig. 5-6d. For doped germanium the potential energy barrier is about 0.2 electron-volt and for doped silicon the barrier is about 0.7 electron-volt.

If a voltage is applied to the two terminals at the end of the diode in such a manner that this diffusion flow of electrons and holes is aided, the diode is said to be forward-biased (see Fig. 5-7a) and the result is an

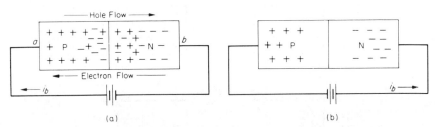

FIG. 5-7. a) A forward-biased PN junction. b) A reverse-biased PN junction.

appreciable current flow. In this situation the electrons enter the diode at point *b* and move to the left, whereupon, near the junction, they combine with the holes. At point *a* an electron is removed so that a new hole is created to replace the one that has just been filled. A similar argument can be made for the hole flow. It is well to remember that current can be considered a flow of positive and/or negative charges. Figure 5-7b shows the diode reverse-biased. In this configuration very little current is evidenced, as the holes and electrons are prevented from crossing the junction by the action of the externally applied electric field.

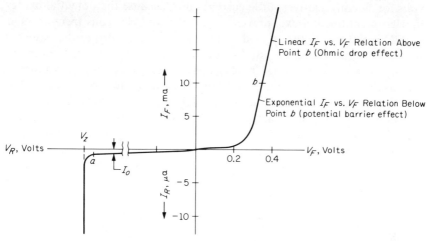

FIG. 5-8. Typical germanium PN-junction current-voltage characteristic. Note the change in scale for I_F and I_R.

The ratio of the forward current to the reverse current is commonly found to be a factor of several thousand for germanium diodes and a million or so for silicon diodes.

A typical germanium PN-junction current-voltage characteristic is shown in Fig. 5-8. The scale factors on this curve are typical and will be very useful in explaining the behavior of the transistor (see Chapter 6). From point a to point b of Fig. 5-8, the current-voltage characteristic is described by the equation

$$I = I_o(e^{V/nV_T} - 1) \qquad (5-3)$$

Although this equation can be derived from a consideration of solid state physical behavior, such a derivation is somewhat beyond the intent and scope of this text. Should the reader be interested there are several excellent texts[1] that elaborate on this equation. The constants in Eq. 5-3 are identified as follows:

I_o = reverse leakage or saturation current (see Fig. 5-8)
n = 1 for germanium
n = 2 for silicon
V_T = $T/11,600$

thus at

$$T = 300°K \qquad V_T \simeq 26 \text{ mv}$$

The diode is forward-biased for $V = V_F > 0$ (in Fig. 5-8). As the forward-bias voltage V_F is increased from 0 to 0.2 or so, volts, (0 to 0.7

[1]See for example, "Electronic Devices and Circuits," by J. Millman and C. C. Halkias, McGraw Hill Book Company, New York, 1967.

volts for silicon) relatively little current flows because the potential-energy barrier, mentioned above, prevents the holes and electrons from crossing the PN junction. As the applied voltage V_F increases further the potential barrier is largely overcome and the forward current I_F rises exponentially.

As the applied voltage is increased even more the current continues to increase, but not at an exponential rate. This is because of another effect that is not accounted for by Eq. 5-3; namely, the fact that at high currents, the ohmic, or $I_F R_D$, drop across the entire diode material is more dominant than the potential-barrier effect. In the low current region the ohmic drop is negligible compared to the potential-barrier effect.

The effect of negative values of V from the origin to point a in Fig. 5-8 is also described adequately by Eq. 5-3. As V decreases, $V = V_R < 0$, the exponential term quickly becomes negligibly small compared to unity, thus $I = I_R = -I_0$. The reverse-biased diode characteristic indicates that the reverse current is almost independent of the reverse voltage, up to a certain point. In this situation the back-biasing voltage is aiding rather than opposing the potential barrier, and thus it seems logical that only a few electrons and holes will be able to cross the potential barrier. However, as the reverse-bias is increased an abrupt increase in the current is noted. Although all semiconductor diodes (and transistors) exhibit this characteristic, the actual point of voltage breakdown can be controlled by proper doping of the diode during manufacture. The breakdown phenomenon is due to the Zener and the avalanche effects. Zener breakdown is associated with high internal electric field emission where electrons in the valence band of the P material are pulled across the junction into the conduction band of the N region by means of the reverse voltage V_R (which creates the electric field). Avalanche breakdown occurs when minority carriers (holes in N material and electrons in P material) acquire sufficient energy from the externally applied electric field to cause ionization when these carriers collide with neighboring atoms. Each ionization produces another electron-hole pair and if this ionizing continues, a manifold increase in carriers (and hence current) is evidenced. Both the Zener and the avalanche effects are present in breakdown diodes with the Zener effect dominant for lower voltage diodes and the avalanche effect dominant in the higher voltage diodes. As used in this text the terms breakdown diode, reference diode, regulator diode, avalanche diode and Zener diode have similar meanings; i.e., there is a reverse voltage phenomenon at which the diode voltage is relatively constant for a wide variation in diode current. The diode is usually not damaged by the breakdown phenomenon unless too much current is passed through the diode, thereby causing excessive heat dissipation in the breakdown region. Later it will be demonstrated how this effect can be put to practical use.

For germanium diodes the reverse current I_0 is generally a few microamperes but for silicon diodes the reverse current is much smaller, usually

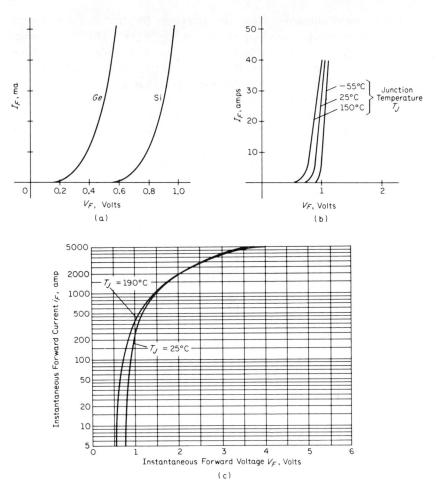

FIG. 5.9. Diode characteristics. a) Forward ampere-volt characteristic for a typical germanium or silicon diode. b) 1N3660 silicon-rectifier diode (peak reverse voltage − 100 volts). (Courtesy Motorola, Inc.) c) Forward instantaneous i_F versus v_F relation for a 1N4045 high power silicon rectifier diode. (Courtesy Westinghouse Electric Corp.)

a few nanoamperes. A comparison of a typical germanium diode and a silicon diode of comparable rating is given in Fig. 5-9a. Note that the shape of the two curves is similar but that they are displaced about 0.4 volt from each other. The fact that I_0 for silicon is 1,000 times smaller than I_0 for germanium is the major reason for the separation between these two curves. The forward voltage at which the diode current begins to increase visibly is called the threshold or cut-in voltage. This threshold voltage for germanium and silicon is about 0.2 and 0.6 volts, respectively.

The forward ampere-voltage characteristic for the medium current 1N3660 silicon rectifier diode is shown in Fig. 5-9b. The three curves indicate how the characteristic behavior changes with diode temperature. The forward characteristics for the 1N4045 high power silicon-rectifier diode is given in Fig. 5-9c. The 1N4045 diode has a rated average current of 275 amperes and can operate at a junction temperature of 190°C. Although the average current through the diode is limited to 275 amperes, an instantaneous forward current of up to 5,000 amperes is allowed. Two points need to be emphasized.

1. The principal reason for the current limitation of 275 amperes is because the heat generated by the diode power loss, $V_F I_F$, at the junction cannot be dissipated rapidly enough to prevent diode damage. This means that a proper heat sink must be provided so that the diode is cooled to limit the temperature rise. At an average current of 275 amperes $V_F = 0.95$ volts so that the power in the diode that must be dissipated is

$$V_{\text{DIODE}} = V_F I_F = 0.95 \times 275 = 261 \text{ watts}$$

2. Note how the current I_F begins to cease increasing at an exponential rate as V_F approaches and exceeds one volt. As explained earlier, Eq. 5-3 is valid only for values of V_F up to about one volt. As V_F increases further the ohmic drop in the semiconductor material dominates the I_F versus V_F characteristics. This phenomenon is evident by the bending of the curve as shown in Fig. 5-9c.

The maximum allowable reverse voltage V_R before the breakdown or avalanche (or Zener) effect is noted is 100 volts for the 1N4045. Another diode in this series, the 1N4056 has a breakdown V_R of 1,000 volts. Other diodes in this same series have breakdown voltages from between 50 to 1,000 volts. Diodes manufactured for utilization of their breakdown voltage value are generally termed Zener or regulator diodes and are available with breakdown values ranging from two volts to hundreds of volts within a specified degree of tolerance.

5-3. Diode Applications

Although diodes were and still are used for their ability to rectify alternating voltages to direct voltages in order to detect radio waves or to make ac to dc power supplies, diodes are being used in a variety of applications in the electronic circuit designs of today. Vast quantities of diodes are used in digital computer circuits, logic circuits, in radio, television, radar and microwave systems, and in a host of home appliance, industry and military electronic and electrical applications.

Rectifiers. First consider the use of diodes as a device to rectify alternating currents or voltages into unidirectional currents or voltages.

A simple series circuit consisting of a sinusoidal voltage source, a diode, and a resistance, is shown in Fig. 5-10a. The conventional circuit symbols for a vacuum and a semiconductor diode are displayed in Fig. 5-10b. The diode is considered to be ideal, that is, a short circuit when the anode voltage is positive with respect to the cathode and an open circuit when the anode voltage is negative as shown in Fig. 5-10c. In Fig. 5-10d, the relationship of the three voltage waveforms: $v_s = v_{ad}$, the applied voltage; $v_D = v_{ab}$, the voltage drop across the diode; and $v_R = v_{bd}$, the voltage drop across the resistance, and the current waveform i are displayed.

When v_s causes the anode to be positive, the diode is a short circuit and v_s is applied across the resistance giving rise to the current flow i. When v_s causes the anode to go negative, the diode is an open circuit and v_s is dropped across the diode, since there is no current flow and hence no iR drop across the resistance.

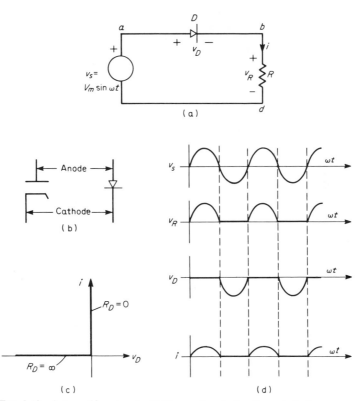

FIG. 5-10. a) A rectifier circuit. b) Conventional circuit symbols for vacuum and semiconductor diodes. Arrows point in the direction of conventional current flow; that is, the arrows indicate the anode, and the straight lines the cathode side of the diode. c) Ideal diode characteristic. d) Instantaneous voltage waveforms for $v_s = v_{ad}$, $v_R = v_{bd}$, $v_D = v_{ab}$ and current waveform i.

Although an actual diode behaves like an ideal diode in series with a small resistance R_p, for many practical circuits it is possible to neglect this series resistance. The diode circuit of Fig. 5-10 is known as a half-wave rectifier and in essence this circuit "wastes" half of the applied voltage as only the upper portion of the applied sine wave reaches the output. Since the primary goal of the usual power rectifier is to convert the alternating voltage or current into a unidirectional voltage or current, it is much more efficient if the entire alternating voltage can be converted into a unidirectional voltage at the output. This can be rather easily accomplished by means of a bridge rectifier as shown in Fig. 5-11. Note that

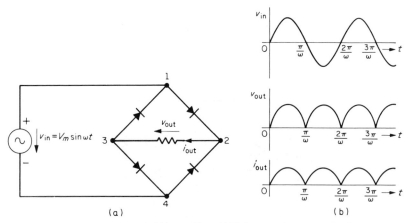

FIG. 5-11. a) Full wave bridge rectifier. b) Voltage and current waveforms.

when node 1 is positive with respect to node 4 the conduction path through the bridge is 1234. When node 4 is positive with respect to node 1 the conduction path for current is 4231; thus it is evident that node 2 is always positive with respect to node 3. If the 4 diodes are connected opposite to the way shown in Fig. 5-11a, then node 3 is positive with respect to node 2. The full wave rectifier utilizes the entire input waveshape as is shown by the output voltage and current waveforms in Fig. 5-11b.

The most important reason for rectifying an alternating voltage is to obtain a voltage whose average value is not zero. By inspection, the average value of a sine wave over a full cycle is zero since the areas above and below the time axis are identical. The average value of the full wave rectified sine wave is $(2/\pi) V_m$ (proved in Chapter 15) and the average value of a half wave rectified sine wave is $\frac{1}{2}$ of that value. The conversion of the rectified sine wave from a pulsating unidirectional voltage to a steady direct voltage is discussed in the next section.

Regulator Diodes. For purposes of rectification the forward characteristic of the diode is utilized, while for regulation the reverse characteristics of a diode are important. Two types of regulator diode char-

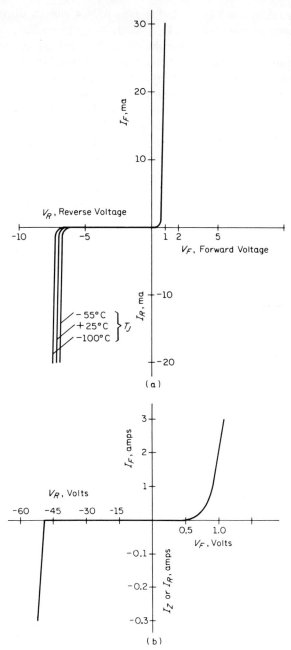

FIG. 5-12. a) Type 653C4 alloyed silicon-reference diode; breakdown voltage is within ±5 percent of indicated value. b) Type 1N2997 diffused silicon-power regulator, 10 watts maximum dissipation. (Courtesy Texas Instruments.)

acteristics are shown in Fig. 5-12. The type 653C4 alloyed silicon-reference diode is shown in Fig. 5-12a. This diode is more useful as a reference source than as a regulator since the allowable current capacity is only 20 or so milliamperes. Such reference diodes are used in electronic power supplies, which can maintain a very stable output voltage over a range of currents. The breakdown voltage is a function of temperature although the variation of voltage with junction temperature T_j is remarkably small.

A true regulator diode characteristic is shown in Fig. 5-12b. The type 1N2997, 10-watt diffused silicon power regulator has a reverse breakdown voltage of 51 volts. This diode belongs to a class of power regulators having nominal breakdown voltages, V_z, from 6.8 to 200 volts. It is noted that the reverse characteristic is not exactly a vertical line. The nominal values of the breakdown voltage and rated current are taken at a point such that their product is equal to the continuous power dissipation capability. Of course, as with any manufactured component there is a percent tolerance attached to these voltage and current values. Table 5-1 lists five of the many diodes belonging to this class of 10-watt regulators.

TABLE 5-1
DIFFUSED SILICON POWER REGULATORS
10 WATTS MAXIMUM DEVICE DISSIPATION

Type	V_z(Volts) Nominal Breakdown Voltage	$I_{z\,max}$ Nominal	$I_{z\,max}$ Maximum d-c Current ma
1N2970	6.8	370	1500
1N2984	20	125	480
1N2997	51	50	185
1N3005	100	25	91
1N3015	200	13	46

An effective and typical application for this regulator diode is its use in a shunt regulator as shown in Fig. 5-13a. Note the symbol $(\rightarrow\!\!\!\!\!\!\top)$ used to indicate the diode is employed as a Zener or regulator diode, i.e., its reverse breakdown characteristic is to be utilized. It is also common to consider the Zener current, I_z, and the Zener voltage, V_z, as the diode current and the diode voltage drop in the reverse direction so that the Zener region of the diode can be labeled with positive values. Admittedly, these arbitrary conventions instituted for ease of use by the experienced designer are often confusing to the beginning student but must be employed if the literature on the subject is to be intelligible to the reader. It is desired to convert a sinusoidal voltage of $v_{in} = 120\sqrt{2}\sin(2\pi \times 60)t$ volts into a d-c output voltage of approximately 20 volts with a current range of 0 to 480 milliamperes. Assuming ideal rectifier diodes in the bridge network, v_1 varies from 0 to $120\sqrt{2}$ volts in a pulsating unidirectional manner as shown in Fig. 5-11c. Thus it is not possible to use the

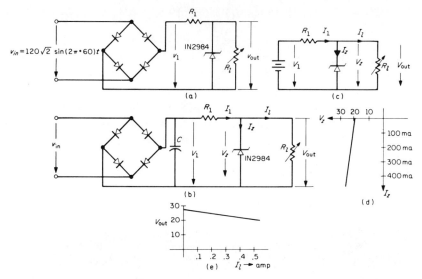

FIG. 5-13. Shunt voltage regulator.

circuit of Fig. 5-13a to achieve the desired goal because when v_1 drops below 20 volts it is impossible for the Zener diode to maintain 20 volts across the output.

Instead let us consider the circuit shown in Fig. 5-13b. Although at this point it is beyond the scope of this chapter to explain how the capacitor C affects the circuit, suffice it to say that the capacitor maintains the voltage V_1 in the neighborhood of 150 volts d-c. The circuit to be analyzed thus reduces to that shown in Fig. 5-13c which is a circuit driven by a d-c source V_1. The analysis now proceeds as follows.

The purpose of R_1 is to limit the Zener current I_z to a safe value. It is assumed that the load current I_l can vary from zero to some maximum value as yet to be determined. Since the maximum allowable Zener current I_z is 480 milliamperes assume that $I_z = 480$ ma and $I_l = 0$ to determine R_1.

$$V_1 = I_1 R_1 + V_z \qquad (5\text{-}4)$$

From Fig. 5-13d, which shows a typical current-voltage characteristic of a 1N2984 diode, $V_z = 27$ volts at $I_z = 480$ milliamperes. Using Eq. 5-4

$$150 = 480 \times 10^{-3} \times R_1 + 27$$

$$R_1 = \frac{123}{0.480} = 256 \text{ ohms}$$

Now let I_z drop to 50 milliamperes and see how the circuit behaves. For $I_z = 50$ ma, $V_z = 19.5$ volts according to Fig. 5-13d. Using Eq. 5-4

$$150 = I_1 \cdot 256 + 19.5$$

$$I_1 = \frac{130.5}{256} = 0.510 \text{ amp}$$

This means that I_l is now

$$I_l = I_1 - I_z$$

$$= 510 - 50 = 460 \text{ ma}$$

The shunt regulator circuit shown in Fig. 5-13b is often called a d-c power supply in that it converts an alternating voltage into a direct voltage. Because the output voltage of the power supply V_{out} is relatively constant (it varies from 27 to 19.3 volts) over an output current variation of 0 to 510 milliamperes as shown in Fig. 5-13e, it is possible to use such a power supply to replace a battery. The regulator circuit shown in this example would make a battery charger for batteries with a nominal voltage

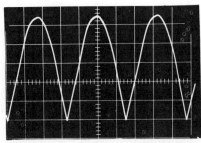

The input voltage $V_{in} = 120\ 2 \sin 377t$ as a full wave rectified voltage V_1
Scale: 30 volts per major division

(a)

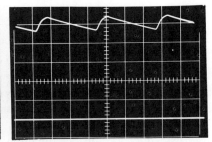

The voltage V_1 vs. t with $C = 100\ \mu f$, $I_z = 50$ ma, $I_l = 360$ ma (refer to Fig. 5-13b)
Scale: 30 volts per major division

(b)

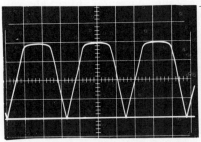

The output voltage V_{out} vs. t with $C = 0$, $I_z = 50$ ma and $I_l = 360$ ma
Scale: 5 volts per major division

(c)

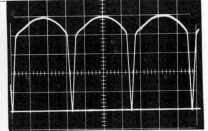

The output voltage V_{out} vs. t with $C = 0$, $I_z = 300$ ma and $I_l = 50$ ma
Scale: 5 volts per major division

(d)

FIG. 5-14. Photographs of v_1 and v_{out} for various Zener currents I_z, load currents I_l and capacitor values C for the circuit shown in Fig. 5-13.

of 22 to 25 volts. If one wished to design a 12-volt charger for a car battery, simply choose a regulator diode with a 9 to 11 volt breakdown voltage and reevaluate R_1 to limit the maximum allowable value for I_Z to a safe value. The battery would replace R_l in Fig. 5-13b. Since the charging current for a battery should initially be high (the terminal voltage of the battery is initially low), and since the terminal voltage of a battery increases as the battery accepts a charge (it is good practice to reduce the charging current as the battery is being charged), the curves shown in Figs. 5-13d and e indicate that this is exactly how the shunt regulator operates. To reduce the power loss in the current-limiting resistor R_1 due to the large difference between V_i and V_{out} it would be wise to reduce v_{in}. This can be done by using a transformer.

Actual laboratory tests on the shunt regulator of Fig. 5-13 are shown in Fig. 5-14. Figure 5-14a shows the full wave rectified input voltage,

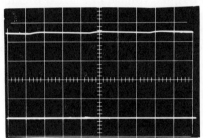

The output voltage V_{out} vs. t with
$C = 100\,\mu f$, $I_z = 300$ ma, $I_l = 50$ ma
Scale: 5 volts per major division
(e)

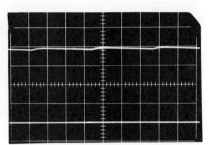

The output voltage V_{out} vs. t with
$C = 100\,\mu f$, $I_z = 50$ ma, $I_l = 360$ ma
Scale: 5 volts per major division
(f)

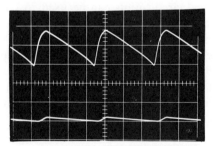

The ripple voltage portion of V_{out} with
$I_z = 50$ ma, $I_l = 360$ ma; the upper
curve occurs when $C = 100\,\mu f$ and the
lower curve is for $C = 750\,\mu f$
Scale: 0.2 volt per major division
(g)

FIGURE. 15-14 (*continued*)

which is the v_1 of Fig. 5-13a. Figure 5-14b shows the effect on v_1 of a shunt capacitor of 100 μf. Note how the capacitor smooths out the pulsating a-c voltage so that a fairly constant voltage $V_1 \simeq 150$ volts is obtained. A peak-to-peak ripple voltage of 24 volts is rather evident.

The next two photographs show the pulsating output voltage v_{out} for two extremes of average load current I_l. Note that the Zener diode clamps the output voltage between 20 to 25 volts but because there is no input capacitor the voltage varies from 0 up to the Zener voltage. This pulsating d-c output would be totally unacceptable for many power supply applications.

The last three photographs show the effects of adding a capacitor across the output terminals of the bridge rectifier. Note the marked improvement in the constancy of the output voltage v_{out}. Although a ripple in the output voltage is evident, it is rather small and for many power supply applications is negligible. If the ripple voltage in Fig. 5-14f is too large, it can be reduced much more by the addition of a larger shunt capacitor. As shown in Fig. 5-14g, the output ripple voltage is reduced from a peak-to-peak value of 0.36 volts to 0.05 volts by increasing the capacitor from 100 μf to 750 μf.

A more detailed discussion of the shunt capacitor as a filter or smoothing device is given in Chapter 15 on Power Supplies and Filters.

5-4. Logic Circuits

The two state nature (conducting and nonconducting) of diodes makes them ideal elements for logic circuits, switching matrices and digital computer applications. This section will show how diodes can be used to implement electronically certain logical functions and Chapter 11 will show how these logical circuit units can be used to construct the fundamental elements of a digital computer. The two logical functions considered in this section are the AND and the OR functions.

The AND logical function simply stated is "if proposition A is true, AND proposition B is True, AND proposition C is true, ..., then some specific result T is true." Each proposition can be in one of two possible states, namely, it is either true or false with no ambiguity. The symbols A, B, C, . . ., T are termed logical variables. An example illustrates the definition.

EXAMPLE 5-1.
Proposition A: temperature is above 70°F
Proposition B: humidity is below 85 percent
Result T: test is to proceed if and only if proposition A AND proposition B are *both* satisfied
Statement: If A is true AND B is true, then T is true
Mathematical Formulation in terms of Boolean algebra: $T = A \cdot B$; ($\cdot$) signifies logical AND connective

Conditions	State of *A*	State of *B*	(*A* AND *B*) Implication on *T*
Temperature = 65°F Humidity = .86	false	false	false
Temperature = 70°F Humidity = .80	false	true	false
Temperature = 81°F Humidity = .92	true	false	false
Temperature = 85°F Humidity = .73	true	true	true

Each proposition can be represented by any device that can assume only two possible states. Devices exhibiting such a characteristic might be identified by contrasting adjectives used in their description, e.g., on-off, high-low, positive-negative, positive-zero, negative-zero, etc. Consider the ideal diode circuit shown in Fig. 5-15a. If a proposition is satisfied (true), indicate this state electronically by applying a positive 5 volts to the appropriate terminal and if a proposition is not satisfied (false), indicate this state electronically by applying 0 volts (ground) to the appropriate terminal. The dotted lines show how such a circuit could be implemented. In practice this circuit symbolism is not used. If a point is labeled with a literal (logical variable), it is assumed the point may take on either of two different voltage potentials (in this case 0 or 5 volts) depending on the state of the logical variable represented by the literal. Thus the inputs to the circuit in Fig. 5-15a are merely labeled *A* and *B*. If a point is held at a fixed potential, it is simply labeled with the numerical value (including polarity). It is understood that there must be a power supply furnishing the potential. The 5-volt potential applied to the resistor in the circuit of Fig. 5-15a is such a point. Restating the AND

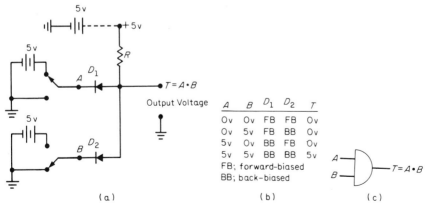

Fig. 5-15. a) Diode implementation of the positive-logic AND gate. b) Biasing and output conditions for all possible input conditions. c) Standard logical AND gate symbol.

logical function in terms of its electronic implementation is that the output of the AND gate is to be +5 volts *if and only if* both *A* AND *B* are +5 volts. The term "gate" refers to the fact that only under certain conditions will the logic circuit allow a signal to pass to the output.

Even though the circuit shown in Fig. 5-15a has two inputs only, its electrical behavior is independent of the number of inputs and therefore this basic circuit configuration can be used to implement an AND gate with an arbitrary number of logical inputs. Note that as long as any *one* logical variable is false, the corresponding circuit input is held electrically at 0 volts, the associated diode is forward-biased, and neglecting the diode's small forward voltage drop, the output of the gate is held at 0 volts. The resistor *R* limits the source current drain when one or more of the diodes is forward-biased. If *all* logical variables are true, all inputs are electrically held at 5 volts, each associated diode is back-biased, hence no current flows from the source, therefore, the resistor voltage drop is 0 and the output is at 5 volts potential.

A table is constructed in Fig. 5-15b listing all possible voltage combinations of the input logical variables, the resulting states of the diodes and the resulting voltage level of the output logical variable. Such tables, often called truth or logic tables, are useful in establishing the logical operation implemented by a given logic circuit. As usual in the treatment of electronic systems, if the circuit details are unimportant or of no concern to the designer of a larger system, a symbol is adopted to represent the circuit. The standard symbol for an AND logic gate is given in Fig. 5-15c. The dot (·) represents the logical AND connective. It should be pointed out the same symbol represents the AND logic gate independent of how it is implemented; electrically, mechanically, etc.

If a high voltage (+5 v) represents the true state of a logical variable and a low voltage (0 v) represents the false state of a logical variable (high and low are used relatively), the implementation is termed *positive-logic*. In Fig. 5-16, an implementation is shown using a negative voltage supply. It is left to the reader to show by way of a table similar to Fig. 5-15b, that this circuit also electronically implements the logical AND function if *negative-logic* is used. Negative-logic represents the true state

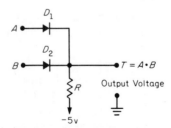

FIG. 5-16. Diode implementation
of the negative-logic AND gate.

of a logical variable by a low voltage and the false state of a logical variable by a high voltage (relatively speaking).

The use of positive- or negative-logic is based on many factors not the least of which is economic. Availability of power supply type, price discounts for quantity purchases, etc., all have a role to play in the choice. A mixture may also be used to realize a given switching system more efficiently.

The second logical function is the OR function. The OR logical function simply stated is "if proposition A is true, OR proposition B is true, OR proposition C is true, ..., then some specific result T is true." It directly follows that for an electronic OR gate using positive-logic, the output is to be high if any one or more of the inputs is high. The electronic implementation of the OR gate is shown in Fig. 5-17a where the

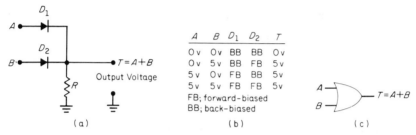

A	B	D_1	D_2	T
0 v	0 v	BB	BB	0 v
0 v	5 v	BB	FB	5 v
5 v	0 v	FB	BB	5 v
5 v	5 v	FB	FB	5 v

FB; forward-biased
BB; back-biased

(a)   (b)   (c)

FIG. 5-17. a) Diode implementation of the positive-logic OR gate. b) Biasing and output conditions for all possible input conditions. c) Standard logic OR gate symbol.

true state of a logical variable is represented by $+5$ volts and the false state represented by 0 volts. The truth table corresponding to the OR gate is given in Fig. 5-17b while the symbol used to represent the OR logic gate is given in Fig. 5-17c. The symbol $(+)$ is used to indicate the logical OR connective in the mathematical formulation of Boolean algebra. The negative-logic OR gate is shown in Fig. 5-18 for -5 and 0 volts.

If in the four circuits investigated (Figs. 5-15, 5-16, 5-17, and 5-18) opposite type of logic is assumed, i.e., positive instead of negative or

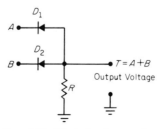

FIG. 5-18. Diode implementation of the negative-logic OR gate.

negative instead of positive, then the former AND gate implementation becomes an OR gate implementation and the former OR gate implementation becomes an AND gate implementation. The reader is asked to verify this statement in Prob. 5-7. This demonstrates that considerable flexibility is available to the logic circuit designer.

The diode gates can be used in either timed-pulse logic circuits or in level logic circuits. Pulse logic circuits utilize pulse representations of the states of the logical variables. These pulses arrive at the gate inputs during specified time intervals and the gate output has meaning only during the same time interval. Consider the pulse trains given in Fig. 5-19

FIG. 5-19. Input and output pulse trains for an OR gate.

applied to the OR gate of Fig. 5-17 which would result in the output pulse train as shown. The true state is represented by a positive pulse and the false state is represented by a zero volt pulse (absence of a pulse). Of course, timing becomes extremely important in pulse logic circuits. No inference is made to the state of a logical variable between pulse intervals. A timing or synchronizing pulse is used frequently in these systems but will not be considered here.

In level logic circuits, constant voltage levels represent the states of the logical variables and the only circuit ambiguity that can exist is when a variable is changing states (transient mode). The output is monitored during the so-called steady state-mode, i.e., all logical variables at rest in some state.

Logical functions can also be implemented by relays, switches and other mechanical devices. Diode circuit implementation offers advantages in speed, simplicity and economy while the main advantage of a mechanical device would be its nonambiguity in state recognition, e.g., a contact is either open or closed. In contrast the electrical voltage drops due to the loading effects on a gate output (current might have to be furnished to or absorbed from the inputs of succeeding gates), or due to the non-ideal nature of the diodes, etc. For the examples considered, this could cause the output voltage to drop from the ideal 5 volts. As an extreme, if the output voltage of a gate was 2.5 volts, it would be difficult to state with certainty whether it was meant to be 0 or 5 volts. In Chapter 11, it will be shown that a diode-transistor gate can be used to solve this problem since the transistor will be able to furnish the energy necessary to compensate for the diode drops, loading effects, etc. A general rule is that only two levels of diode logic can be used before an active (energy supplying) logic element must be used to remove the degradation from the signals and return the signal amplitudes to their proper levels.

Diodes are also used extensively in switching matrices. Encoding and decoding networks are good examples of such use. Consider the network given in Fig. 5-20. The four input logical variables consist of a pair of

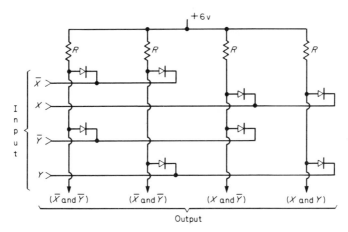

FIG. 5-20. Many to one diode decoder matrix.

variables and their complements. If X is high, then $\overline{X}$ is low since $\overline{X}$ is defined to be in the true state if and only if the variable X is in the false state, and $\overline{X}$ is termed the complement of the logical variable X. Note the logical variable $\overline{X}$ is always in the opposite state of the logical variable X and $(\overline{X}) = X$. It follows then for the system under consideration, two of the input variables must be high, say 5 volts as in previous examples, and two of them must be low, say 0 volts. The four output lines give all possible AND combinations of the input variables. Any input variable that is in the low state, 0 volts, will clamp the two associated output lines at zero volts independent of the state of the other variable which is diode-connected to the same output line. Only the output line diode-connected to the two input lines whose logical variables are in the high state, 5 volts, will be high, thus identifying the current state of the system. This system is termed a "many to one" decoder since one and only one output line out of many is energized at any given time. Further applications of diodes to logic and digital circuitry are given in Chapter 11.

5-5. Diode D-C Circuit Analysis

The circuits in the previous sections contained diodes in connections such that one could determine by inspection whether a given diode was in the conducting or nonconducting state. For the logic circuits, the circuit solution could be rather easily determined for each of the configurations since the source voltages could take on only one of two distinct values. Consider the circuit in Fig. 5-21 containing an ideal diode. The

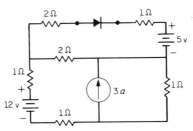

FIG. 5-21. Circuit containing ideal diode.

state of the diode is not apparent from visual observation of the circuit. Must one force a nonlinear analysis of this circuit as outlined in Chapter 3 to effect a solution? For a single diode in a circuit, the answer is no. One can assume a state of the diode, i.e., conducting or nonconducting, and then replace it by a short or open circuit respectively. This then gives a linear circuit with no diode which can be solved by any technique for linear circuit analysis. Upon obtaining the solution, one must then check the original assumption. If the diode had been assumed conducting and replaced with a short circuit, the solution current in the short must be *positive* in the forward direction of the replaced diode. If the diode had been assumed nonconducting and replaced with an open circuit, the solution voltage across the open must have a polarity that would *back-bias* the replaced diode. If the original assumption as to the state of the diode proves to be incorrect, then the opposite state is the correct state. Of course, another solution will have to be obtained for the correct state since all voltages and currents will change when the short circuit representation of the diode is changed to the open circuit representation or vice versa. These latter two statements are correct as long as the circuit does not contain dependent or controlled energy sources. Circuits containing such sources and nonlinear elements (such as diodes) may be astable, i.e., oscillate from one state to another state continuously. Such circuits will not be considered in the remarks to follow.

Assume the diode in the circuit of Fig. 5-21 is conducting and therefore replaceable by a short circuit as shown in Fig. 5-22a. Using the loop currents I_1, I_2, and I obtain

$$5I_1 - 2I_2 = -5$$
$$-2I_1 + 5I_2 = 12 - 3 \cdot 1$$

from which

$$I_1 = \frac{\begin{vmatrix} -5 & -2 \\ 9 & 5 \end{vmatrix}}{\begin{vmatrix} 5 & -2 \\ -2 & 5 \end{vmatrix}} = \frac{-7}{21} = -\frac{1}{3} \text{ amp}$$

Since the current I_1 is the current through the short circuit in the forward direction of the replaced diode and is negative, the original assumption of

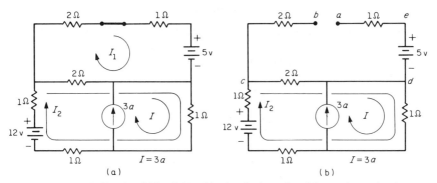

FIG. 5-22. a) Circuit of Fig. 5-21 with the diode replaced by a short circuit. b) Circuit of Fig. 5-21 with the diode replaced by an open circuit.

the diode being in the conducting state is incorrect. It must be noncon-ducting and is replaceable by an open circuit as shown in Fig. 5-22b. As a check, the describing equation is

$$5I_2 = 12 - 3 \cdot 1$$

from which

$$I_2 = \frac{9}{5} \text{ amp}$$

Now if the diode is back-biased as assumed, V_{ab} must be positive. Con-sider

$$V_{ab} = V_{ae} + V_{ed} + V_{dc} + V_{cb}$$

or

$$V_{ab} = 0 + 5 + 2(-I_2) + 0 = 5 - \frac{18}{5} = +\frac{7}{5} \text{ volts}$$

which verifies the assumed nonconducting state of the diode.

If a circuit contains more than one diode, this solution technique be-comes more difficult since there are 2^N different state combinations for N diodes. Even if all states but one are seemingly chosen correctly on the basis of the circuit solution there is no guarantee that reversing the incor-rect state of the one diode will then lead to a correct solution. All voltages and currents will change in the new solution and other assumed diode states may now be incorrect. One notes that the process is an iterative one which can be long and time consuming. Digital programs can be used to advantage in solving such circuits containing many diodes. Such circuits will not be pursued in this text. It is also obvious that a careful visual analysis of the circuit leading to an intelligent estimate of the states of the various diodes reduces the amount of computational effort. This completes the presentation on the diode and the next chapter turns attention to the transistor.

PROBLEMS

5-1. In the accompanying two circuits the diodes are ideal, $R = 1000$ ohms and $v_i = 100 \sin \omega t$. Sketch $v_o(t)$. Give a name that best approximates the behavior of the output voltage of these two circuits.

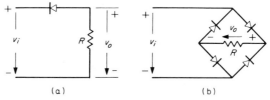

(a)   (b)

PROB. 5-1

5-2. In the accompanying two circuits $R = 1000$ ohms, the diodes are ideal, $V_1 = 10$ volts and v_{in} is as shown. Sketch $v_o(t)$.

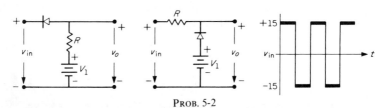

PROB. 5-2

5-3. In the accompanying circuit consider the diodes to be ideal with $R = 1000$ ohms, $V_1 = 10$ volts, $V_2 = 20$ volts. Determine v_o for $v_i = 30 \sin \omega t$ and sketch v_o versus v_i over this range. Such a circuit is termed a limiter or clipper.

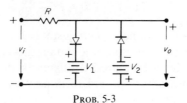

PROB. 5-3

5-4. In the accompanying circuit consider the diodes to be ideal with R = 1000 ohms, V_1 = 15 volts, V_2 = 10 volts. Determine v_o for v_i = 30 sin ωt and sketch v_o versus v_i over this range. This circuit is a dead band simulator or threshold circuit.

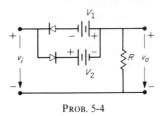

PROB. 5-4

5-5. An avalanche or Zener diode can be used to prevent damaging a sensitive meter movement due to overloading. The accompanying figure shows a d'Arsonval meter movement which reads 50 volts full scale. The internal meter resistance is 520 ohms and the meter current for full scale deflection is 400 microamperes. If the diode is a 39-volt Zener find R_1 and R_2 such that when V_{in} > 50 volts the Zener diode will shunt the overload current away from the meter movement.

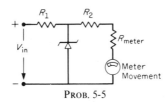

PROB. 5-5

5-6. A certain avalanche or Zener diode rated at 30 volts will maintain this constant voltage over a range of diode current from 5 to 150 ma. If V_1 = 180 volts calculate R_1 necessary to allow a voltage regulation for all load currents from I_L = 0 to $I_L = I_{L\,max}$. What is $I_{L\,max}$ and what is the wattage rating necessary for R_1? (*Hint:* Keep in mind that V_o remains fixed at 30 volts.)

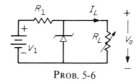

PROB. 5-6

5-7. Assuming negative logic, show that the circuit in Fig. 5-15a implements the OR logical function and the circuit in Fig. 5-17a implements the AND logical function. Assuming positive logic, show that the circuit in Fig. 5-16 implements the OR logical function and the circuit in Fig. 5-18 implements the AND logical function.

5-8. a) Using positive logic, diode implement the logical function given by $T = A \cdot (B + C)$. b) Can you deduce the logical function that results if negative logic is assumed for the implementation found in (a)?

5-9. a) Using negative logic, diode implement the logical function given by $T = A + (B \cdot C)$. b) Can you deduce the logical function that results if positive logic is assumed for the implementation found in (a)?

5-10. In the accompanying circuit diodes D_1 and D_2 are to be considered ideal. Determine the correct state of both diodes and verify your answer in terms of the bias conditions.

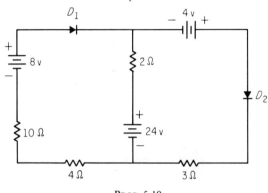

PROB. 5-10

5-11. Using the figure accompanying Prob. 2-6, insert a diode in series with the 4Ω resistor between nodes c and d. a) If the anode of the diode is facing node c determine V_{cd}. b) If the anode of the diode is facing node d determine V_{cd}.

6 / *The Transistor and Its Use in Practical Design*

The transistor is a three-terminal (or more) solid-state device. (The two-terminal solid-state device is called a diode.) Transistors are classified as either point-contact or junction types. Since the junction transistor is easier to manufacture and is more dependable in operation, practically all of the transistors now in use are the junction type.

As compared to the vacuum tube, the transistor 1) requires no filament power and therefore no warm-up time before it begins to operate, 2) is usually much smaller in size, 3) is more rugged mechanically, 4) usually has a much longer operational lifetime, and 5) usually requires lower operating potentials. Since miniaturization and compactness of equipment are necessities in the space age, the small size and low weight, ruggedness, and low heat dissipation of the transistor are important qualities.

Some of the disadvantages of transistors as compared to vacuum tubes, are that 1) they are more easily damaged when electrically overloaded, 2) their operating characteristics vary a great deal with temperature, 3) their operational frequency range is more limited, and 4) they are more susceptible to damage by X-ray and other high-energy radiation effects. Although special-purpose vacuum tubes are designed to operate in the microwave region (microwaves are roughly defined as frequencies above 1000 MHz), at the present time no transistor operates satisfactorily at frequencies above a few gigahertz.

The transistor discussed in this chapter is often called the bipolar junction transistor, or BJT. Another transistor, called the field effect transistor (FET) is discussed in Chapter 8.

6-1. The Bipolar Junction Transistor

The bipolar junction transistor consists of a thin [0.001-in. (inch) or less] layer of N or P semiconductor material sandwiched between two P or two N semiconductor layers, respectively. This forms a PNP or a NPN junction transistor. Leads are then soldered or welded to each separate region (three in all), and the transistor is ready to be hermetically encapsulated.

The manufacture of the PNP or NPN junction can be accomplished by placing a seed crystal of germanium in a molten bath of pure germanium (the purity may be as high as 10^9 germanium atoms for each impurity atom) and slowly withdrawing the crystal. By carefully adjusting the temperature of the molten bath and the rate of withdrawal, a solid crystal of the pure germanium is formed. If the molten bath is alternately doped with P- and then N-type materials, it is possible to produce a crystal with alternating P and N regions. This large crystal can then be sliced and diced to form either PNP or NPN transistor crystals. Other transistor fabrication methods will be discussed in Sec. 6-3.

Junction-Transistor Behavior. Figure 6-1a shows a NPN junction transistor. The three semiconductor regions are labeled emitter, base, and collector. As mentioned above, the base region is very thin, for it is this

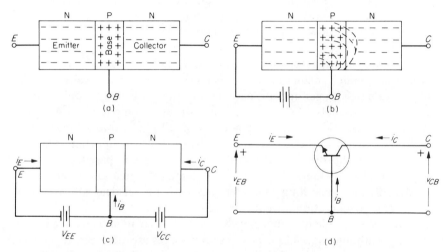

FIG. 6-1. The junction-transistor. a) Pictorial diagram. b) With the base-emitter diode section forward biased. c) With both the base-emitter and collector-emitter diodes properly biased. d) Schematic diagram.

thinness that makes the transistor behave properly. The upper limit of the transistor frequency response or usefulness is directly proportional to the thinness of the base region.

All transistors (and grid-type vacuum tubes) used in circuit applications must be properly biased. Biasing means the application of voltages across and/or currents through a transistor or tube so that the transistor or tube is ready to operate properly. Transistors and tubes tend to be unilateral, nonlinear devices (current flows easily in one direction, but a high resistance to current flow is offered in the opposite direction), but biasing allows them to operate as bilateral, linear elements as far as the desired incremental input and output signals are concerned. Biasing can be thought of as "setting up" the circuit for operation so that the desired electrical signals are able to pass through the electronic system from input to output with the output signal waveform being an amplified, faithful reproduction (linear amplification) of the input signal waveform. The necessity for linear operation can be fully justified in many applications — such as in high-fidelity monaural or stereophonic sound systems. Figure 6-1b shows the transistor emitter biased in the forward direction in that the base-emitter diode section is biased so that there is a sizable base-emitter current. Some of the electrons that flow from the emitter into the base region pass into the collector region. This diffusion is caused by the saturation of the thin base region, even with only a relatively small number of electrons, which creates a large electron density, thereby causing many of the emitter electrons to diffuse or be repelled into the collector region.

If the collector-base NP junction is reverse-biased, so that the collector terminal is positive, as shown in Fig. 6-1c, many of the emitter electrons that diffuse through the base region are attracted to the collector terminal. In practice, it is found that a large percentage of the emitter electrons pass through the base region and continue to the collector terminal rather than returning to the base terminal. The ratio of the output current to the input current, which in this case is the ratio of the collector current to the emitter current, is one of a useful number of parameters in transistor theory. This ratio is defined as

$$\alpha_{fb} \triangleq - \left. \frac{\text{rate of change of output or collector current}}{\text{rate of change of input or emitter current}} \right|_{\substack{\text{collector-to-base voltage held} \\ \text{constant}}}$$

or mathematically

$$\alpha_{fb} \triangleq - \frac{\partial i_C}{\partial i_E} = - \left. \frac{\Delta i_C}{\Delta i_E} \right|_{v_{CB} = \text{ a constant}} \tag{6-1}$$

where α_{fb} is called the forward short-circuit current transfer ratio or gain. The reason for the term "short circuit" will be seen later. The subscript f stands for forward, and b signifies a common-base configuration. The partial derivatives indicate that all other circuit variables except i_C and i_E are held constant. This is more evident in the corresponding incremental expression (the rightmost term in Eq. 6-1) for α_{fb}, where it is noted that v_{CB}, the collector to base voltage, is a constant. The minus sign is introduced so that α_{fb} will be a positive quantity. As will be seen later, the incremental expression will be useful when working with the transistor graphical characteristics in that by inspection a visual interpretation of α_{fb} will be available. As is indicated by Fig. 6-1 the collector current i_C is the emitter current i_E minus a small amount of the emitter current that flows out the base lead. From this argument it should be evident that the collector current is always somewhat smaller than the emitter current, and thus the magnitude of α_{fb}, from Eq. 6-1, is always less than unity. Other α parameters for other transistor-circuit configurations will be defined later in this chapter.

Another important relation can be established by inspection of Fig. 6-1c. By Kirchhoff's second law the summation of currents entering the transistor (which is the same as the currents leaving node B) must equal zero or

$$i_E + i_C + i_B = 0$$

and, solving for i_E,

$$i_E = -(i_C + i_B) \tag{6-2}$$

Figure 6-1d is the conventional schematic representation for a transistor, with the various voltages and currents noted. The direction of the arrows marked on the emitter terminals denotes the type of transistor being used. The arrow points in the direction of conventional current or hole flow and thus is directed toward the base for a PNP transistor and away from the base for an NPN transistor.

Terminology. As with any electrical or electronic device, many symbols are necessary to distinguish the various voltages and currents present in a transistor circuit. Some of these symbols are defined as follows:

i_B = total instantaneous base current

i_b = instantaneous value of the a-c component of the base current

I_B = average or d-c value of the quiescent base current

I_b = effective or rms (root-mean-square) value of the a-c component of the base current

$\left.\begin{array}{l} i_C \\ i_c \\ I_C \\ I_c \end{array}\right\}$ = same as above, except that the word *collector* is substituted for base

$\left.\begin{array}{l} i_E \\ i_e \\ I_E \\ I_e \end{array}\right\}$ = same as above, except that the word *emitter* is substituted for base

v_{EB} = total instantaneous emitter-base voltage drop

v_{eb} = instantaneous value of the a-c component of emitter to base voltage drop

V_{EB} = average or d-c value of emitter to base voltage drop

V_{eb} = effective or rms value of the a-c component of emitter to base voltage drop

Definitions exist for the base-emitter, collector-emitter, and collector-base, voltages and so on, that are similar to those defined by the preceding four emitter-base voltage relations. In addition

V_{CC} = collector-circuit d-c supply voltage
V_{BB} = base-circuit d-c supply voltage
V_{EE} = emitter-circuit d-c supply voltage
e_s = instantaneous a-c component of the signal or input voltage
v_{out} = the instantaneous a-c component of the output voltage

As is usual in circuit theory, the capital I's and V's represent either rms or d-c values, while lowercase i's and v's represent instantaneous values. Uppercase subscripts represent total or average values, and lowercase subscripts indicate the a-c component only. It is extremely important that the reader familiarize himself with these terms, as they will be used frequently in this and the next chapter, not to mention other articles concerning transistors.

6-2. Common-Base Input and Output Characteristics

A family of common-base input (or emitter) characteristics and output (or collector) characteristics for a PNP or NPN transistor can be obtained by using the circuits given in Figs. 6-2a or 6-2b respectively. The

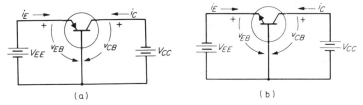

(a)   (b)

Fig. 6-2. Circuit that can be used to obtain the common-base characteristics of a) the type 2N404 PNP alloy junction germanium transistor, and b) the type 2N3704 NPN epitaxial planar silicon transistor.

transistor is said to be connected in the common-base configuration because the base terminal is common or is connected to both an input and an output terminal. Characteristic curves such as displayed in Fig. 6-3 are obtained by varying the input and output voltages and currents over a range of values. A comparison between the forward-biased emitter-base P-N junction of a 2N404 transistor and the N-P junction of a 2N3704 transistor is shown in Fig. 6-3a. Note the characteristic 0.4 volt separation that occurs between germanium and silicon P-N junctions. Note also that the curves have almost identical shapes. If the collector-base voltage, V_{CB}, is greater than zero, the curves shift to the left about 0.05 volt. This effect will be shown on a later figure concerning the common emitter con-

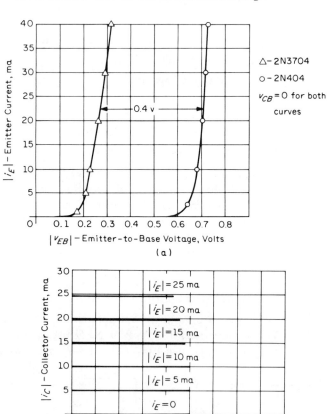

Fig. 6-3. a) Input or emitter-base characteristics for the 2N404 and 2N3704 transistors in common base mode. b) Output or collector-base characteristics for the 2N404 and 2N3704 transistors in common base mode.

figuration. It should be noted that i_E and v_{EB} have negative values for PNP transistors (like the 2N404) and positive values for NPN transistors (like the 2N3704). The input characteristic is important because it gives the dynamic input resistance R_{ib} which is defined as the reciprocal of the slope of the input current $|i_E|$ versus the input voltage $|v_{EB}|$ curve at any point of interest for the transistor in the common base mode. Since the 2N404 or 2N3704 transistors are usually operated beyond the knee of the curve ($|i_E| > 5$ ma), the dynamic input resistance $R_{ib} \triangleq \Delta v_{EB} / \Delta i_E \simeq 2$ ohms. Because this is such a small resistance it can very often be neglected in the other circuit analysis calculations. As contrasted to the dynamic input resistance, which is defined as the slope of a curve, the d-c or static input resistance, R_{in} (static), is simply the ratio of the input voltage v_{EB} to the input current i_E, as measured from the origin, at any particular point on the $|i_E|$ versus $|v_{EB}|$ curve. The value of the static input resistance for the common base connection varies from about 10 to 100 ohms, which is considerably larger than the dynamic input resistance.

The curves shown in Fig. 6-3b are the average output characteristics for a 2N404 or 2N3704 transistor in a common-base mode. These curves are somewhat idealized since there is a slight difference in the behavior of a PNP and NPN transistor in the vicinity of $v_{CB} = 0$. The primary purpose in displaying these curves is to show that the emitter current, i_E, is slightly greater than the collector current, i_C. The difference between i_E and i_C is the base current, i.e., $i_E - i_C = i_B$. Also note that a family of curves is obtained for i_C versus v_{CB} with i_E considered the variable that controls what the transistor output collector current will be.

6-3. Transistor Fabrication Techniques

The two transistors emphasized in this text are the 2N404 PNP alloy junction germanium and the 2N3704 NPN epitaxial planar silicon types. Although little emphasis is placed on the physics of semiconductors in this text, some words of explanation concerning the descriptive meaning of "alloy junction" and "epitaxial planar" are in order. In addition to the previously mentioned grown junction process two other popular methods of producing junctions are by alloying or by diffusion processes.

Alloy Process. As the name implies the 2N404 is fabricated by the alloy process, which is illustrated in a general way in Fig. 6-4a. The alloy junction transistor is fabricated by placing a small pellet of dopant, such as indium, boron, arsenic, antimony, etc, on each side of a thin wafer of germanium or silicon of opposite type conductivity. The unit is then heated until the dopant melts (some 500°C or so). At the interface between the liquified dopant and the semiconductor an alloying between the dopant and semiconductor takes place thus forming a semiconductor

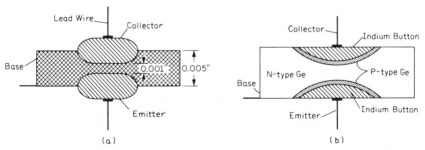

FIG. 6-4.   a) Pictorial diagram of a typical alloy junction transistor. b) Detailed
diagram of a PNP alloy junction transistor.

region whose conductivity is opposite to that of the original wafer of semiconductors. A more detailed diagram illustrating this process is shown in Fig. 6-4b, where a wafer of N-type germanium is alloyed with two pellets of indium. During the heating process a portion of the liquified indium combines with the N-type germanium and creates a region of P-type germanium between the pure indium and the N-type germanium wafer. Of course the temperature must be controlled closely so that the P-type impurity of the indium atoms can overcome the N-type impurity in the original germanium wafer and at the same time allow the new P-type germanium to recrystallize with the same crystal orientation as the parent germanium wafer, the remaining pure indium merely serves as an ohmic contact between the newly formed P-region and the lead wires that are soldered to the indium buttons.

Diffusion Process. The terms planar and diffusion are commonly employed when one discusses the BJT or the MOS (Metal-Oxide-Semiconductor) transistor (which is discussed in Chapter 8). Planar means that the transistor is built-up layer by layer and diffusion means that the impurity atoms are placed inside of a silicon or germanium wafer or layer by the laws controlling the diffusion process. The planar process is particularly adaptable to mechanized or automated fabrication techniques.

A typical NPN epitaxial planar silicon BJT is shown in Fig. 6-5. The fabrication begins with a P-type silicon wafer that has been cut, polished

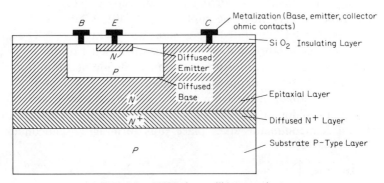

FIG. 6-5. NPN planar silicon transistor.

and cleaned to about 0.006 in. thick. This wafer (called the substrate) is then placed in a high temperature oven (800°C to 1200°C) and exposed to an N-type dopant such as arsenic. At these high temperatures the dopant is in a gaseous state and the lattice atoms of the substrate are highly agitated. The arsenic impurity atoms tend to diffuse into the substrate because of the existing concentration gradient of arsenic atoms at the surface of the substrate. The arsenic atoms thus tend to occupy the vacancies in the substrate lattice structure due to the wanderings of the agitated silicon atoms. The result is a heavily doped N^+ region.

Next, a gas containing an N-type impurity deposits an epitaxial layer on top of the N^+ region (note that this is not a diffusion process). Another diffusion process with a P-type impurity creates the base region, then a third diffusion process creates the N-type emitter region within the P-type base region. A layer of silicon oxide for insulation purposes and three appropriately placed metal contacts complete the fabrication of the basic transistor. Note that the base is tied to a P region, the emitter to a N region and the collector to a N region thus forming a NPN transistor.

6-4. Common-Emitter Characteristics

During the early period (1948–1952) of transistor development, the common-base characteristics were used extensively, because it was thought that the common-base connection was the logical way to use the transistor. Since these characteristics are still referred to in the literature, it is perhaps necessary to consider the common-base characteristics, even though today the common-base configuration is rarely used in circuits. One major reason that the common-base configuration is not desirable is because the emitter and collector currents are almost equal. This means that since the current gain α_{fb} is always less than unity, a voltage or power gain is attained only if the output or load resistance is greater than the transistor input resistance. However, such a difference in input and output resistances may be undesirable as far as other circuit characteristics or behaviors are concerned. If the base is used as the input node and the emitter as the common node, a considerable collector-to-base current gain can be realized; i.e., the output current is greater than the input current.

Since the common-emitter circuit configuration is presently so widely used, it is important to consider the common-emitter transistor characteristics. The common-emitter circuit, using a 2N3704 NPN epitaxial planar silicon transistor, as shown in Fig. 6-6a, can be used to obtain the input or base and the output or collector characteristics for the common emitter mode, as shown in Figs. 6-6b and 6-6c, respectively.

Since Fig. 6-6b is a plot of base current i_E versus base voltage v_{BE}, it is evident that the slope of the curves can be used to define the *dynamic* base input resistance of the transistor. This input resistance varies from about 50 to 1000 ohms for normal operation of the various transistors, and is larger than the emitter input resistance for the common-base connection (which was about 10 ohms). As an example, above the knee of the curve (see Fig. 6-4b for $i_B > 50$ μa) the slope of the i_B versus v_{BE} curve is

$$\frac{\Delta i_B}{\Delta v_{BE}} = \frac{200 \times 10^{-6}}{0.695 - 0.63} = 3.08 \times 10^{-3} \text{ mhos}$$

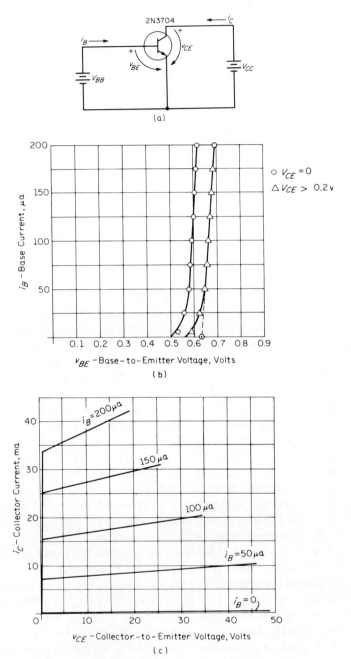

FIG. 6-6. Type 2N3704 NPN epitaxial planar silicon transistor. a) Circuit that can be used to obtain the static common-emitter characteristics for an NPN transistor. b) Common-emitter input characteristics. c) Common-emitter output characteristics. (Experimental curves by R. Lueg.)

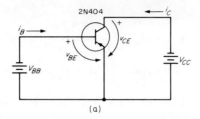

(a)

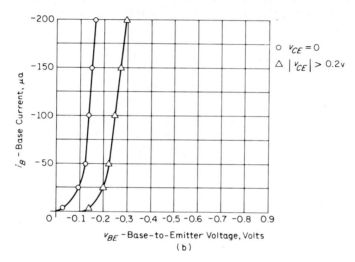

v_{BE} – Base-to-Emitter Voltage, Volts

(b)

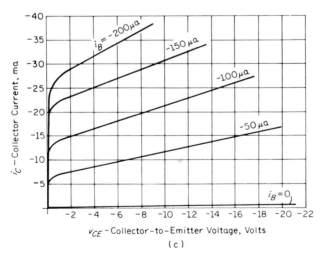

v_{CE} – Collector-to-Emitter Voltage, Volts

(c)

FIG. 6-7. Type 2N404 PNP alloy junction germanium transistor. a) Circuit that can be used to obtain the static common-emitter characteristics for a PNP transistor.   b) Common-emitter input characteristics. (Experimental curves by R. Lueg.) c) Common-emitter output characteristics.   (Courtesy Texas Instruments.)

or

$$R_{in} = R_{ie} = \frac{1}{3.08 \times 10^{-3}} = 325 \text{ ohms}$$

where R_{ie} is the transistor input resistance for the common-emitter configuration.

Recall that the d-c or *static* resistance, is defined simply as the ratio of the input voltage to input current. For instance for a base current of $i_B = 50$ μa and $V_{CE} > 0.2$ volts, $v_{BE} = 0.65$ volt according to Fig. 6-4b, and the static input resistance, R_{in} (static), is

$$R_{in}(\text{static}) = \frac{0.65}{50 \times 10^{-6}} = 13,000 \text{ ohms}$$

Figure 6-7a, b, and c show the circuit, the base, and collector characteristics, respectively, for a type 2N404 PNP alloy junction germanium transistor in the common emitter mode. The explanation for the shape of these common-emitter characteristics is so similar to that used for the common-base characteristics that no further discussion will be given here.

Figures 6-6a and 6-7a indicate that a new output-to-input current ratio or α factor can be defined. For the common-emitter circuit

$$\alpha_{fe} \triangleq \frac{\text{rate of change of output or collector current}}{\text{rate of change of input or base current}} \bigg|_{\substack{\text{collector-to-emitter voltage held} \\ \text{constant}}}$$

or mathematically

$$\alpha_{fe} = \frac{\partial i_C}{\partial i_B} = \frac{\Delta i_C}{\Delta i_B} \bigg|_{v_{CE} = \text{a constant}} \tag{6-3}$$

where α_{fe} is the common-emitter forward short-circuit current transfer ratio or gain. α_{fe} can easily be much greater than unity since i_B is normally only a fraction of i_C. [i_B is usually measured in microamperes (μa) and i_C in milliamperes (ma).] Quite often the symbol β (beta) is used to indicate α_{fe}. A sample calculation for α_{fe} or β will be given later. A β of 50 to 300 is typical of many transistors. The values of R_{ie} and R_{in} (static) given here for the common-emitter mode should be compared by the reader with the corresponding quantities for the 2N3704 transistor in the common-base mode.

6-5. Common-Emitter Amplifier

Figure 6-8a shows the circuit diagram for the basic common-emitter amplifier using an NPN transistor. Figure 6-8a is similar to Fig. 6-6a, which was used to obtain the common-emitter characteristics, but there are two important differences. First, an a-c voltage generator, signified by e_s, and its equivalent internal resistance, signified by R_g, are added in series with the base-emitter input circuit. The reader should not be con-

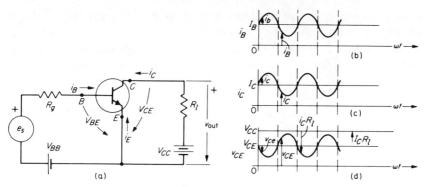

FIG. 6-8. a) Basic common-emitter amplifier configuration. b), c), and d) Some important phase relations for the circuit of part a.

cerned about the appearance of this "mysterious" a-c signal source and resistance. It must be realized that this basic amplifier is but one of many circuits that might be contained in an overall electronic device or system. As will be more apparent after one reads the material in Chapter 7, this is no more than the Thévenin equivalent circuit of all of the preceding transistor amplifiers (or stages, as they are also commonly called), but, since it is desirable to start with the simplest circuit configuration possible, the reduction of the preceding stages to the Thévenin equivalent will be left for the next chapter. The second major difference between the two circuits is the addition of the load resistance R_l. (R_l is purely resistive in this circuit, but, beginning with Chapter 12, Z_l, a complex impedance, will replace R_l.) This load resistance includes the input or driving-point resistance of the stage following the one shown in Fig. 6-8a. Again, for purposes of simplification, it is desirable to consider R_l simply as some fixed resistance that is placed in the circuit.

In Fig. 6-8b, c, and d are shown some of the important phase or time relations for the circuit of Fig. 6-8a. Before discussing these, several preparatory comments are in order. The d-c voltage V_{CC} supplies most of the energy to this circuit, that is, to the load resistance R_l and the collector-emitter circuit of the transistor. The d-c voltage V_{BB} is considered the base-emitter biasing voltage. Both V_{CC} and V_{BB} set up the transistor so that it can behave like a bilateral, linear device. When the signal voltage $e_s = 0$, the transistor amplifier settles down to its equilibrium or quiescent state. This quiescent state is achieved when $i_B = I_B$, $i_C = I_C$, and $v_{CE} = V_{CE}$; that is, i_B, i_C, and v_{CE} have only a d-c value. This condition is represented by the three straight lines displaced from the abscissa or ωt axis on Figs. 6-8b, c, and d. If the base-to-emitter transistor input resistance is small compared to R_g, then the transistor input resistance can be neglected. In many transistor-circuit designs the biasing resistance—which, in this particular circuit, is R_g—is made large so that the inherent variations of the transistor input resistance with temperature, operating point, frequency, and so on, will not be noticeable. Under the assumption of a large base-biasing resistor, the base-biasing current is obtained by writing a KVL (Kirchhoff's voltage law) equation around the base-emitter mesh or

$$I_B R_g + V_{BE} - V_{BB} = 0$$

Thus

$$I_B = \frac{V_{BB} - V_{BE}}{R_g} \tag{6-4}$$

If $R_g \gg R_{in}$ (static) then $V_{BB} \gg V_{BE}$ ($V_{BE} \simeq 0.6$ volt for a silicon transistor and 0.2 volt for a germanium transistor), and Eq. 6-4 reduces to

$$I_B = V_{BB}/R_g$$

The relationship between the collector-to-emitter voltage V_{CE} and the collector current I_C is obtained by writing a KVL equation around the collector-emitter mesh

$$V_{CE} = V_{CC} - I_C R_l \tag{6-5}$$

where I_C is obtained from a graphical analysis. In practice, the values of both I_C and V_{CE} are normally obtained by a graphical analysis, by setting up the circuit and measuring the quiescent values, or from knowing β and using equivalent circuits.

EXAMPLE 6-1. Using the circuit of Fig. 6-8a and the 2N3704 input and output characteristics of Figs. 6-6b and c, respectively, calculate I_B, I_C, and V_{CE} if $V_{BB} = 5$ volts, $V_{CC} = 30$ volts, $R_l = 1000$ ohms, and $R_g = 50,000$ ohms.

Solution: Assuming $R_g \gg R_{in}$ (static) for the transistor, Eq. 6-4 can be used to calculate the quiescent base current

$$I_B = \frac{5}{50,000} = 1 \times 10^{-4} \text{ amp} = 100 \text{ } \mu a$$

To find the value of the quiescent collector current, we note that the collector characteristics of Fig. 6-6c are straight, reasonably parallel, sloping lines, so that a good estimation of I_C for a base current of 100 μa is between 15 and 20 ma. Taking the average

$$I_C = 17.5 \text{ ma}$$

V_{CE} can be calculated by using Eq. 6-5:

$$V_{CE} = 30 - 17.5 \times 10^{-3} \times 1000 = 30 - 17.5 = 12.5 \text{ volts}$$

The reader should confirm that R_g is indeed much greater than R_{in} (static).

Another method of solution, using a d-c load line, will be presented in a later section of this chapter.

The next step is to consider what happens when the signal voltage e_s takes on some value other than zero. Since a vast majority of the signals sent through electronic circuits are sinusoidal or at least repetitive in nature (such as in audio and video signal amplification), we can assume that $e_s = E_m \sin \omega t$. Since only linear circuits are being considered, the other a-c voltages and currents are all sinusoidal in nature and have the same frequency as the input voltage e_s. As the sinusoidal voltage e_s, which causes both v_{be} and i_b, becomes more positive, i_B tends to increase because e_s and V_{BB} are additive in the positive sense. The net result is shown in Fig. 6-8b. From previous discussions it is known that when i_B changes, i_C changes in the same manner. Thus i_c increases and varies sinusoidally as does i_b, as is shown in Fig. 6-8c. Finally, since $v_{CE} = V_{CC} - i_C R_l$, it is evident that when i_C increases, V_{CE} decreases. This effect is shown in Fig. 6-8d.

Before stating the final conclusions, the definition of the output voltage v_{out} must be restated. By definition, the output voltage v_{out} is the a-c component of the collector-to-emitter voltage, or

$$v_{out} \triangleq v_{ce} \tag{6-6}$$

for this particular circuit. The output voltage v_{out} is always defined as the instantaneous or a-c component of the output voltage, but $v_{out} \neq v_{ce}$ in every common-emitter circuit. (An example of this will be given in the next chapter.)

The conclusions that should be remembered from this analysis are: 1) i_b and i_c are in phase or in exact time relation to each other and 2) e_s and v_{out} are 180° out of phase or in opposite time relation to each other for the common-emitter amplifier with a purely resistive load. Although a NPN-type transistor was used in the analysis, a PNP-type transistor produces identical phase relations. Proof of this is left as an exercise for the reader (see Prob. 6-1).

6-6. Graphical Circuit Analysis

In the preceding section it was stressed that the transistor is usually operated within the linear portion of its characteristic curves. If the transistor is operated in its nonlinear region a graphical approach is frequently desirable, but it is a bit premature to discuss nonlinear graphical-analysis methods in detail. (This subject will be taken up in Chapter 14). Instead, a linear problem will be solved by graphical means as an introduction to graphical analysis. The technique of graphical analysis, which was introduced in Chapter 3, is straightforward and easy to use.

D-C Load Lines. Let us consider the following example.

EXAMPLE 6-2. A 2N3704 NPN transistor is used in the circuit of Fig. 6-9. a) Find the Q point and the values of I_B, I_C, and V_{CE}. b) If $e_s = 2.5 \sin \omega t$ volts, find v_{out}, V_{out}, I_c/I_b, V_{out}/E_s, and $V_{out}I_c/E_sI_b$.

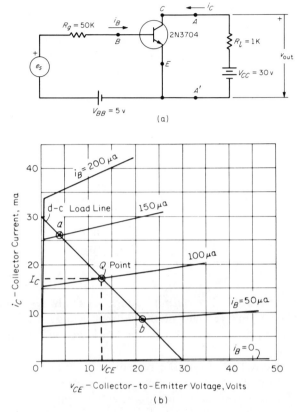

FIG. 6-9. a) Circuit to be used for Example 6-2.
b) Graphical analysis using the common-emitter
output characteristics for the 2N3704 transistor.

Solution: The collector characteristics for the 2N3704 transistor are repeated in Fig. 6-9b. These characteristics define the behavior of the voltage across
and the current entering terminals AA' on Fig. 6-9a. The circuit to the right of
terminals AA' is a linear circuit consisting of a resistance R_l and a voltage source
V_{CC}. This is exactly the same situation as was outlined in the graphical-analysis
section of Chapter 3. The first step in the graphical analysis is to draw the voltage-
ampere characteristic of the equation $V_{AA'} = V_{CC} - I_C R_l$ on the transistor collector characteristics. The standard way of doing this is first to let $V_{AA'}$, which is
the same as V_{CE}, equal zero. The current through R_l is then $V_{CC}/R_l = 30/1000 =$
30 ma. This is plotted as the point $v_{CE} = 0$, $i_C = 30$ ma, on the ordinate of the
collector characteristics. (The reason that the total instantaneous values equal
the d-c quiescent values is because $e_s = 0$ for these calculations.) For the second
calculation it is assumed that the current through R_l is equal to zero. Since there
is no IR drop across R_l, $V_{AA'} = V_{CE} = 30$ volts. This second point is plotted as
$v_{CE} = 30$, $i_C = 0$ on the abscissa of the collector characteristics. A straight line is

then drawn through these two points, and it can be seen that the slope of the line is $(-30 \times 10^{-3})/30 = -1/1000 = -1/R_l$. This straight line is usually referred to as the *d-c load line*.

The next step in the analysis is to determine the Q or quiescent point. Since there is a family of curves drawn of i_C versus v_{CE} for various values of i_B, it is necessary to find the correct value of i_B. This is easily done if it can be assumed that R_g, which happens to be the base-biasing resistor in this particular circuit, is much greater than the transistor input resistance. The d-c or static (as contrasted with the a-c or dynamic) input resistance of the 2N3704 transistor is in the neighborhood of 5000 ohms. For example, referring to the common-emitter input characteristics of Fig. 6-6b, we note that, for an estimated collector voltage v_{CE} of 20 volts and an assumed base current of 100 μa, a base voltage v_{BE} of about 0.66 volt is obtained. Thus $R_{in} = v_{BE}/i_B = 0.66/(100 \times 10^{-6}) = 6600$ ohms. If R_g is selected to be 50,000 ohms or greater, then the assumption $R_g \gg R_{in}$ is valid. Under this assumption, $I_B = V_{BB}/R_g = 5/50,000 = 100 \mu a$.

a) The Q point is then determined by the intersection of the d-c load line and the $i_B = 100 \mu a$ curve on the collector characteristics. The values of I_B, I_C, and V_{CE} are $I_B = 100 \mu a$, $I_C = 17$ ma, and $V_{CE} = 13$ volts. These are the quiescent (no a-c signal present) voltages and currents that simply set up the transistor so that, if desired, it can operate like a bilateral, linear element. The transistor is now ready to receive the a-c signal e_s and perform the process called amplification.

b) If $e_s = 2.5 \sin \omega t$, then $i_b = e_s/R_g = 2.5 \sin \omega t/50,000$ amp $= 50 \sin \omega t$ μa. The assumption $R_g \gg R_{in}$ is even more valid in the case of the a-c signal because the dynamic input resistance R_{ie} is only 1000 ohms or so, as has already been established in an earlier section.

Since $i_B = I_B + i_b = 100 + 50 \sin \omega t$ μa, the base current varies sinusoidally between $i_B = 150$ and $i_B = 50$ μa. As the loci of the collector voltages and currents are forced simultaneously to satisfy the voltage and current requirements of the load resistance R_l as well as the transistor characteristics, we proceed up the d-c load line to $i_B = 150$ μa and then down the load line to $i_B = 50$ μa. These two points are marked a and b on Fig. 6-9b. The quantities v_{out}, V_{out}, I_c/I_b, and V_{out}/E_s can now be calculated.

The quantity v_{out} is the instantaneous variation of the collector voltage v_{CE}. The peak-to-peak variation of v_{CE} is

$$v_{CE \text{ peak-to-peak}} = v_{CE_b} - v_{CE_a} = 22 - 4 = 18 \text{ volts}$$

Since linear operation is assumed (by observation the transistor appears to be operating in a linear portion of the characteristic curves), the peak-to-peak variation in v_{CE} causes an output voltage whose peak-to-peak value is also 18 volts, or

$$v_{out} = -\frac{18}{2} \sin \omega t = -9 \sin \omega t \tag{6-7}$$

where the division by 2 is necessary to obtain only the maximum value of v_{out}, and the minus sign is attached because an analysis in a previous section showed that v_{out} and e_s are 180° out of phase.

The quantity V_{out} is the rms value of v_{out}, which for a sine wave is

$$V_{out} = \frac{v_{out_{max}}}{\sqrt{2}} = \frac{9}{\sqrt{2}} \text{ volts} \qquad (6\text{-}8)$$

The rms value of the collector current I_c can be found by dividing the rms value of the output voltage by the load resistance R_l, or

$$I_c = \frac{V_{out}}{R_l} = \frac{9}{\sqrt{2} \times 1000} = \frac{9}{\sqrt{2}} \text{ ma} \qquad (6\text{-}9a)$$

The same value can be obtained graphically from Fig. 6-9b:

$$I_c = \frac{i_{c_a} - i_{c_b}}{2\sqrt{2}} = \frac{26.5 - 8.5}{2\sqrt{2}} = \frac{9}{\sqrt{2}} \text{ ma} \qquad (check) \qquad (6\text{-}9b)$$

The ratio I_c/I_b is called the current gain A_i of the transistor amplifier and for this example is

$$A_i = \frac{I_c}{I_b} = \frac{\dfrac{9}{\sqrt{2}} \times 10^{-3}}{2.5/(\sqrt{2} \times 50{,}000)} = 180 \qquad (6\text{-}10)$$

The ratio V_{out}/E_s is called the voltage gain A_v of the transistor amplifier and for this example is

$$A_v = \frac{V_{out}}{E_s} = -\frac{9/\sqrt{2}}{2.5/\sqrt{2}} = -3.6 \qquad (6\text{-}11)$$

where the minus $(-)$ sign is added to indicate that V_{out} and E_s are $180°$ out of phase.

The ratio $|V_{out} I_c/E_s I_b|$ is called the power gain A_p of the transistor amplifier and is equal to the product of the voltage and current gains

$$A_p = \left| \frac{V_{out} I_c}{E_s I_b} \right| = 180 \times 3.6 = 648 \qquad (6\text{-}12)$$

6-7. A General Graphical Analysis Procedure

The purpose of Example 6-2 was to give the reader a rapid overview of how a common emitter transistor amplifier behaves. This example used a circuit in which the d-c and a-c signals had the same current paths. In general the a-c and d-c signals "see" different circuits. One of the most confusing concepts in electronic circuit behavior lies in understanding how the a-c and d-c currents combine and separate within the circuit. It cannot be emphasized too strongly that the purpose of the d-c voltages and currents is to bias or "set up" the electronic devices so that the circuit is ready to receive the a-c or time varying signals (voltages and/or currents). The reader should also keep in mind the fact that the d-c sources provide all of the energy for converting low-energy-level a-c signals into high-energy-level a-c signals.

In an effort to clarify the behavior of a typical, basic electronic amplifier an exhaustive, step-by-step analysis of a common-emitter transistor amplifier will next be presented. Although the amplifier presented here is somewhat simplified as compared to amplifiers that will be studied in later sections, the actual analysis procedure varies little from one amplifier type to another. Fundamental ideas concerning the sinusoidal voltages and currents will be introduced as needed. Of necessity some coverage of material presented earlier will be included in the following development.

Consider the amplifier given in Fig. 6-10 and let it be desired to find the voltage, current and power gain associated with the a-c signal, e_s, being applied to the input.

As in Example 6-2 the first task is to find the quiescent or Q point, which is also known as the equilibrium point, operating point or zero a-c point. This is accomplished by considering the d-c solution only. The objective is to insure that the transistor has been biased properly with V_{BB} and V_{CC}. For accurate amplification; i.e., the waveshape of the output voltage and current being faithful reproductions of the sinusoidal input voltages and currents, the transistor should have a quiescent point that is in the "linear region" of the output characteristics. As long as the a-c operation about this quiescent point is maintained within this linear region where the transistor output characteristics are equally spaced with respect to the base current, this linearity between the output and input is maintained.

The role of the capacitor shown in the output circuit is to distinguish between d-c and a-c. The impedance[1] (analogous to resistance) that the capacitor C offers to current flow is $1/\omega C$ where ω is the radian frequency of the signal under consideration. Note that as $\omega \rightarrow 0$ (d-c), the impedance of the capacitor $1/\omega C \rightarrow \infty$ (open circuit). Conversely, as $\omega \rightarrow \infty$ (high frequency), the impedance $1/\omega C \rightarrow 0$ (short circuit). Of course this latter condition can never be realized since the frequency must be finite. However, if the capacitor is in *series* with a resistor, e.g., such as R_i in Fig. 6-10, then as the frequency is increased, a point is reached where $1/\omega C \ll R_i$ and the current in the branch is essentially controlled by the resistor alone. For frequencies greater than this frequency, the capacitor can be considered a short circuit and plays no role in the a-c analysis. The mathematically precise role of the capacitor is treated in Chapter 12 for circuits where the above frequency separation assumption is no longer valid. Naturally any a-c source makes no contribution to the d-c solution and therefore can be replaced by its internal resistance. For the purposes

---

[1] Impedance is a generic parameter that describes the voltage to current ratio in a circuit. Strictly speaking the impedance quality of a capacitor C is termed a capacitive reactance $1/\omega C$, the impedance quality of an inductor L is termed an inductive reactance ωL, and a resistor R offers an impedance called resistance R. The units of impedance, reactance, and resistance are ohms.

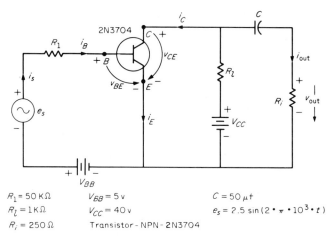

$R_1 = 50 \text{ K}\Omega$ $V_{BB} = 5 \text{ v}$ $C = 50 \, \mu f$
$R_l = 1 \text{ K}\Omega$ $V_{CC} = 40 \text{ v}$ $e_s = 2.5 \sin(2 \cdot \pi \cdot 10^3 \cdot t)$
$R_i = 250 \, \Omega$ Transistor - NPN - 2N3704

FIG. 6-10. Common-emitter amplifier.

of simplifying the amplifier circuit in order to make a d-c analysis, the following rules may be employed:

1. Replace a-c voltage sources by their internal resistance. If none is indicated, replacement is by a short circuit.
2. Replace a-c current sources by their internal conductance. If none is indicated, replacement is by an open circuit.
3. Remove any branch containing a capacitor since the capacitor blocks d-c current flow and such a branch will not enter into the d-c analysis.

For the circuit in Fig. 6-10, no internal resistance is indicated for the a-c voltage source. One could argue that it has been lumped together with R_1 or that R_1 itself is the internal resistance of the a-c signal source so that no generality has been lost. The a-c voltage source is therefore replaced with a short circuit. The d-c circuit to be analyzed is given in Fig. 6-11 with suitable annotation in terms of the d-c components of the various voltages and currents. Note the branch containing the capacitor has been deleted as per rule 3. Consider the equation describing Kirchhoff's voltage law around the input or base-emitter loop:

$$I_B R_1 + V_{BE} - V_{BB} = 0 \qquad (6\text{-}13)$$

This equation contains two unknown variables, I_B and V_{BE}, and cannot be solved directly. Consider the equation describing Kirchhoff's voltage law around the output or collector-emitter loop:

$$I_C R_l + V_{CE} - V_{CC} = 0 \qquad (6\text{-}14)$$

This equation contains two unknowns, I_C and V_{CE}, and cannot be solved directly. Quite obviously, no further independent loop equations can be

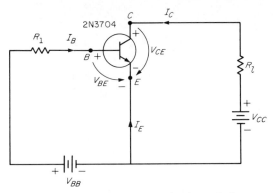

FIG. 6-11. Amplifier circuit for d-c analysis.

written. One must therefore rely on the characteristics of the transistor itself to be able to solve Eqs. 6-13 and 6-14.

Attention is directed toward Fig. 6-12a. Note that the base voltage is in the neighborhood of 0.6–0.7 volts for the entire range of base current values given on the graph. One of two approximations could be made and they are as follows:

1. Assume that V_{BE} = 0.65 volt for the d-c analysis since it varies very little from this value for a wide range of base current and collector-to-emitter voltage values.

2. Assume that V_{BE} = 0 volt since $V_{BE} \ll V_{BB}$ and little error will be introduced into the solution of the input Eq. 6-13 for the base current, I_B.

Of course assumption 1 leads to a slightly more accurate solution for I_B than assumption 2. However, one must remember that many component values may have tolerances of ± 20 percent and the transistor characteristic is for a class of transistors any one of which will have a characteristic which differs some from that of the published characteristic. In this respect, the use of assumption 2 is justified in most cases and leads to the simple expression

$$I_B = \frac{V_{BB}}{R_1}$$

and in this particular example

$$I_B = \frac{5 \text{ volts}}{50 \text{ K}\Omega} = 100 \,\mu\text{a} \tag{6-15}$$

Now consider the output Eq. 6-14 written in the form

$$I_C = -\frac{1}{R_l} V_{CE} + \frac{1}{R_l} V_{CC} \tag{6-16}$$

or more generally

$$I_C = -\frac{1}{R_{DC}} V_{CE} + \frac{1}{R_{DC}} V_{CC} \qquad (6\text{-}17)$$

It is recognized that Eq. 6-17 is linear in terms of the unknown variables I_C and V_{CE} and hence would be graphically represented by a line. It should be noted that for a more complicated bias circuit on the output, one could argue that a Thévenin equivalent circuit could be determined to the right of the collector-emitter terminals. The resulting Thévenin equivalent resistance would be R_{DC} and the Thévenin equivalent voltage would be the "effective" V_{CC} used in Eq. 6-17. In the example under consideration $R_l = R_{DC}$ and so substitution into Eqs. 6-16 or 6-17 obtains

$$I_C = -\frac{1}{1000} V_{CE} + \frac{1}{1000} \cdot 40 \qquad (6\text{-}18)$$

Now this equation must be satisfied simultaneously with the transistor characteristics. The straight line representing Eq. 6-18 must be plotted on Fig. 6-12b. This is easily accomplished by determining two points on the line from Eq. 6-18 as was done in Chapter 3:

Point 1: $I_C = 0 \rightarrow V_{CE} = 40$ volts
Point 2: $V_{CE} = 0 \rightarrow I_C = 40$ ma

The first point is on the abscissa of the output characteristic while the second point is on the ordinate. These two points determine the so-called d-c load line as shown in Fig. 6-12b. To satisfy Kirchhoff's law, the d-c solution must be on this line. Therefore, the intersection of this line with the characteristic given by the previously determined value of the base current, $I_B = 100$ μamps, gives the quiescent ("Q") point. If the base current value is between two of the given transistor characteristics, then linear interpolation would be used to determine the position of the Q point on the d-c load line. With the Q point determined, it is a simple matter to read the d-c components of the collector-to-emitter voltage, $V_{CE} = 22$ volts, (point 3) and of the collector current, $I_C = 18$ ma (point 4). It is noted that the Q point is located in the linear region of the characteristics thereby assuring faithful reproduction of the input signal. With the value of V_{CE} ascertained, the Q point can also be determined on the input characteristic as shown in Fig. 6-12a. This completes the d-c analysis for locating the quiescent point. It should be pointed out that the procedure would be reversed in a synthesis or design situation. One would establish the Q point in an appropriate part of the linear region and then back-calculate values of R_1, R_l, V_{BB} and V_{CC} to give the desired Q point. The same objective could also be met by an iterative procedure of modifying the values of R_1, R_l, V_{BB} and V_{CC}, until the approximate location of the Q point was attained.

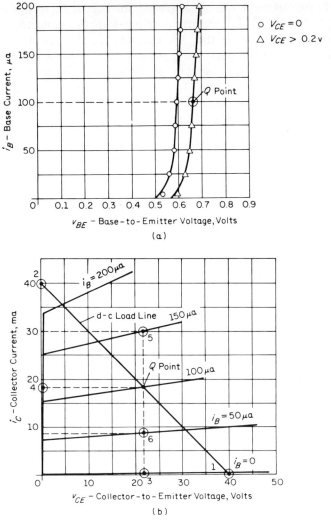

FIG. 6-12. Characteristics of the 2N3704 transistor in a common-emitter mode. a) Input characteristics. b) Output characteristics.

At this point the forward current transfer ratio can be determined. Recall

$$\beta = \alpha_{fe} \triangleq \left. \frac{\partial i_C}{\partial i_B} \right|_{v_{CE} = \text{constant}} = \left. \frac{\partial i_C}{\partial i_B} \right|_{v_{CE} = V_{CE}} \sim \left. \frac{\Delta i_C}{\Delta i_B} \right|_{v_{CE} = V_{CE}} \qquad (6\text{-}19)$$

Keeping v_{CE} constant is equivalent to having the a-c component zero so that v_{CE} is made up of a d-c component only. Since only the graphical characteristics of the device are available, the partial derivative is approximated by the ratio of incremental changes in the collector current i_C to incremental changes in the base current i_B. The variation is sought along a vertical line through the Q point since $v_{CE} = V_{CE}$ on this vertical line. The variation should also be confined to the linear region to give valid results for the transistor. It is common practice to take the points at base current characteristics adjacent to the Q point. Such points are labeled 5 and 6 in Fig. 6-12b. The ratio is given by

$$\beta = \alpha_{fe} = \frac{i_{C_5} - i_{C_6}}{i_{B_5} - i_{B_6}} = \frac{30\,\text{ma} - 8\,\text{ma}}{150\,\mu\text{a} - 50\,\mu\text{a}} = 220$$

This represents the current gain of the transistor itself and is the upper limit of the current gain that can be achieved in an amplifier using this transistor.

At this stage, analysis of operation away from the Q point due to the a-c signal source, e_s, will be undertaken. A brief overview of the steps follow.

1. Replace each d-c voltage source by its internal resistance. If none is indicated, replace d-c source by a short circuit.
2. Check impedances of all capacitors. If the impedance of any capacitor is much smaller (factor of 10) than a series resistor, neglect impedance of capacitor, i.e., replace capacitor by a short circuit.
3. Determine a-c load line, i.e., the Thévenin equivalent resistance associated with the circuitry to the right of the collector-to-emitter terminals.
4. Analytically determine the a-c base current swing, i.e., the limits of base current variation from the Q point value.
5. Graphically determine the a-c collector current swing and the collector-to-emitter voltage swing corresponding to the a-c base current swing.
6. Analytically determine the a-c voltage and current swings in the load resistor. This load resistor may or may not be a resistor per se. It simply represents the device in which the energy from the amplifier is utilized. In many cases it may represent the input resistance of the succeeding device.
7. Analytically determine the voltage, current and power gain of the amplifier.

The a-c circuit that must be analyzed is given in Fig. 6-13 with the annotation in terms of the a-c components of the various voltages and currents. The capacitor impedance is determined by

$$\frac{1}{\omega C} = \frac{1}{2\pi f C} = \frac{1}{2 \cdot \pi \cdot 10^3 \cdot 50 \cdot 10^{-6}} = 3.2\ \text{ohms}$$

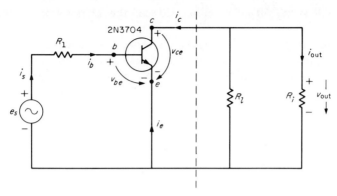

FIG. 6-13. Amplifier circuit for a-c analysis.

which is negligible in comparison to the series resistor, $R_i = 250$ ohms and therefore the capacitor has been replaced by a short circuit in the a-c circuit.

Applying Kirchhoff's voltage law around the input or base-emitter loop gives

$$R_1 i_b + v_{be} - e_s = 0 \qquad (6\text{-}20)$$

Again, two unknown variables appear, namely, the a-c components of the base current, i_b, and base to emitter voltage, v_{be}. However, observing the transistor input characteristic in Fig. 6-12a shows that the base to emitter voltage varies little for any reasonable base current swing away from the Q point since the characteristics are almost vertical lines. Thus negligible error is introduced by assuming the a-c component of the base-to-emitter voltage, v_{be}, to be zero. Solving for i_b in Eq. 6-20 with $v_{be} = 0$ obtains

$$i_b = \frac{e_s}{R_1} = \frac{2.5 \sin (2 \cdot \pi \cdot 10^3 \cdot t)}{50 \times 10^3} = 50 \times 10^{-6} \sin (\omega t) \, \text{amp} \qquad (6\text{-}21)$$

Note that the frequency value itself is not needed since with proper biasing linearity between transistor output and input is assumed, i.e., in this a-c analysis, all voltages and currents are assumed sinusoidal with a frequency of 10^3 Hz. A plot of i_b is given in Fig. 6-14 where it is noted that the base

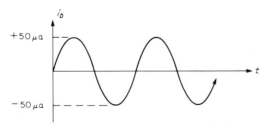

FIG. 6-14. Sinusoidal component of the base current i_b.

current will swing 50 μa on either side of the Q point in a sinusoidal manner.

The next step is to determine the a-c collector-to-emitter a-c resistance. In other words, the Thévenin equivalent resistance of the circuitry to the right of the dashed line in Fig. 6-13. In this example, it is made up of the resistor R_l in parallel with the load resistor R_i. The a-c resistance is

$$R_{ac} = \frac{R_l \cdot R_i}{R_l + R_i} = 200 \text{ ohms} \qquad (6\text{-}22)$$

and the output circuit could be redrawn as shown in Fig. 6-15. The describing equation is given by

$$R_{ac}i_c + v_{ce} = 0 \qquad (6\text{-}23)$$

This equation contains two unknown variables, i_c and v_{ce} and can be rearranged to give

$$i_c = -\frac{1}{R_{ac}} v_{ce} \qquad (6\text{-}24)$$

which is a linear equation and is graphically represented by a line having a slope of $-1/R_{ac}$. As Eq. 6-24 stands, it suggests a line through the origin

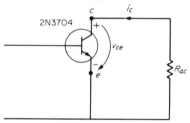

FIG. 16-5. Output circuit for a-c
analysis.

of the output characteristics in Fig. 6-16 with a slope of $-1/R_{ac}$. However, recall that with the a-c signal set to zero, the d-c components of the voltage and current will force the amplifier to be at the Q point. In other words, the a-c solution must be superimposed on the previously determined d-c solution. This is easily accomplished by insisting that the line described by Eq. 6-24 called the a-c load line, *must* pass through the Q point. To get a line with the proper slope, consider

$$\frac{\Delta i_C}{\Delta v_{CE}} = -\frac{1}{R_{ac}} \rightarrow |\Delta v_{CE}| = R_{ac}|\Delta i_C| \qquad (6\text{-}25)$$

One can pick a convenient value of Δi_C, say 25 ma, and locate this point 1 on the ordinate of the output characteristic as in Fig. 6-16. From Eq. 6-25, the corresponding value of Δv_{CE} is 5 volts and this point 2 is located on the abscissa of the output characteristic. A line through these two points has a slope of $-1/R_{ac}$ as desired. However, it was demon-

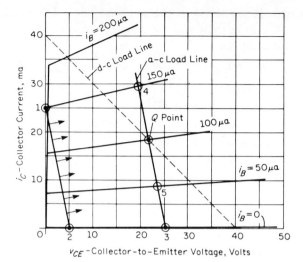

FIG. 6-16. Locating a-c load line on output characteristics.

strated that this a-c load line must pass through the Q point so it is graphically moved parallel until it goes through the Q point as shown in Fig. 6-16.

Another technique of directly locating the a-c load line is to determine the intercept on the abscissa, point 3. Recall that from the equation of a line $y = mx + b$, that the x-axis intercept in terms of the slope of the line and a point (x_1, y_1) through which the line must pass is

$$x \text{ (intercept)} = x_1 - \frac{y_1}{m}$$

so analogously,

$$v_{CE_3} = V_{CE} - \frac{I_C}{\left(-\dfrac{1}{R_{ac}}\right)} = V_{CE} + I_C R_{ac} \qquad (6\text{-}26)$$

Using Eq. 6-26, point 3 is computed as

$$v_{CE_3} = 22 + (18 \text{ ma})(0.2 \text{ K}\Omega) = 25.6 \text{ volts}$$

and the a-c load line can be drawn directly through point 3 and the Q point. Either technique can be used or one can be used as a check on the other. Now for an a-c swing of the base current, the transistor is constrained to move along the a-c load line. Whenever the a-c signal is zero the transistor is at its equilibrium or Q point.

Consider the a-c component of the base current to be at its maximum value of 50 μa. The total base current is

$$i_B = I_B + i_b \rightarrow i_{B_{max}} = I_B + i_{b_{max}} = 100 \ \mu a + 50 \ \mu a = 150 \ \mu a$$

and the transistor is at point 4 on the a-c load line. The corresponding values of the collector current and the collector-to-emitter voltage can be determined graphically, i.e.,

$$i_{B_{max}} = 150 \, \mu a \rightarrow i_{C_{max}} = 29 \, ma \qquad \text{and} \qquad v_{CE_{min}} = 19 \text{ volts}$$

Now consider the a-c component of the base current to be at its minimum value of $-50 \, \mu a$. The total base current is

$$i_B = I_B + i_b \rightarrow i_{B_{min}} = I_B + i_{b_{min}} = 100 \, \mu a - 50 \, \mu a = 50 \, \mu a$$

and the transistor is at point 5 on the a-c load line. The corresponding values of the collector current and the collector-to-emitter voltage are

$$i_{B_{min}} = 50 \, \mu a \rightarrow i_{C_{min}} = 9 \, ma \qquad \text{and} \qquad v_{CE_{max}} = 23 \text{ volts}$$

Since operation is in the linear region of the characteristics and the base current variation is sinusoidal (Eq. 6-21), the collector current and collector-to-emitter voltage variations are also sinusoidal. Therefore, the peak-to-peak variation of these two variables must be the difference between their maximum and minimum values. One can thus describe them mathematically as

$$i_c = \frac{i_{C_{max}} - i_{C_{min}}}{2} \sin (2 \cdot \pi \cdot 10^3 \cdot t) \tag{6-27}$$

and

$$v_{ce} = - \frac{v_{CE_{max}} - v_{CE_{min}}}{2} \sin (2 \cdot \pi \cdot 10^3 \cdot t) \tag{6-28}$$

Note that the expression for v_{ce} is negative. This results from the fact that v_{CE} is increasing when i_B is decreasing and vice-versa. These phase or relative time relationships are shown in Fig. 6-17. Substituting in the appropriating values Eqs. 6-27 and 6-28 become

$$i_c = 10 \times 10^{-3} \sin \omega t \text{ amp} \tag{6-29}$$

and

$$v_{ce} = -2 \sin \omega t \text{ volts} \tag{6-30}$$

As a check on the graphical accuracy the ratio of v_{ce}/i_c should give the negative of R_{ac}.

In this case,

$$\frac{v_{ce}}{i_c} = \frac{-2 \sin \omega t}{10 \times 10^{-3} \sin \omega t} = -200 \text{ ohms}$$

which checks exactly. That this is so should be evident from the fact that the maximum and minimum values of the a-c components are taken off of the a-c load line whose slope is $-1/R_{ac}$. One could have first determined the collector current swing and then used this ratio to determine the corresponding collector-to-emitter voltage swing. This is useful for near-vertical a-c load lines since the voltage swing will be small and difficult to evaluate graphically.

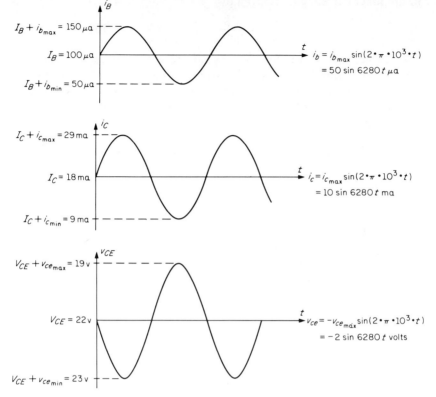

$I_B + i_{b_{max}} = 150\,\mu a$

$I_B = 100\,\mu a$

$I_B + i_{b_{min}} = 50\,\mu a$

$i_b = i_{b_{max}} \sin(2\cdot\pi\cdot10^3\cdot t)$
$= 50 \sin 6280\,t\,\mu a$

$I_C + i_{c_{max}} = 29\,ma$

$I_C = 18\,ma$

$I_C + i_{c_{min}} = 9\,ma$

$i_c = i_{c_{max}} \sin(2\cdot\pi\cdot10^3\cdot t)$
$= 10 \sin 6280\,t\,ma$

$V_{CE} + v_{ce_{max}} = 19\,v$

$V_{CE} = 22\,v$

$V_{CE} + v_{ce_{min}} = 23\,v$

$v_{ce} = -v_{ce_{max}} \sin(2\cdot\pi\cdot10^3\cdot t)$
$= -2 \sin 6280\,t\,volts$

FIG. 6-17.   Relative time relationships.

Now returning to the amplifier circuit for a-c analysis given in Fig. 6-13, it is evident that the amplifier output voltage v_{out} is identical to the collector-to-emitter voltage v_{ce} given by Eq. 6-30.   The amplifier voltage gain then is given by

$$A_v = \frac{v_{out}}{e_s} = \frac{v_{ce}}{e_s} = \frac{-2 \sin(\omega t)}{2.5 \sin(\omega t)} = -0.8$$

where the negative sign indicates the phase reversal between the amplifier output voltage and the input signal voltage.

It is evident from Fig. 6-13 that the amplifier output current through R_i is *not* the collector current.   However, with the two resistors in parallel, the current splitting formula may be used which gives

$$i_{out} = -i_c \times \frac{R_l}{R_l + R_i} = -10 \times 10^{-3} \sin(\omega t) \frac{1000}{1000 + 250}$$

$$= -8 \times 10^{-3} \sin(\omega t)\,amp$$

$$= -8 \sin \omega t\,ma$$

The amplifier current gain is given by

$$A_i = \frac{i_{\text{out}}}{i_s} = \frac{i_{\text{out}}}{i_b} = \frac{-8 \times 10^{-3} \sin \omega t}{50 \times 10^{-6} \sin \omega t} = -160$$

where the negative sign indicates the phase reversal between the amplifier output current and the input signal current. Some observations follow:

1. The amplifier output current usually will not be equal to the collector current but is always expressible in terms of the collector current.
2. The amplifier output voltage may not be equal to the collector-to-emitter voltage but is always expressible in terms of the collector-to-emitter voltage.
3. The base current may not be equal to the signal current but is always expressible in terms of the signal current.
4. The amplifier voltage gain may be less than one, i.e., the amplifier output voltage is less than the signal voltage.
5. The amplifier current gain is less than the transistor forward current transfer ratio, β. This ratio is the limiting value of the amplifier current gain.

The power gain incorporates both the voltage and current gains to show if more a-c energy is being provided to the load than contained in or extracted from the a-c signal source. The power gain is given by

$$A_p = \left| \frac{p_{\text{out}}}{p_s} \right| = \left| \frac{v_{\text{out}} i_{\text{out}}}{e_s i_s} \right| = \left| \frac{v_{\text{out}}}{e_s} \frac{i_{\text{out}}}{i_s} \right| = |(-.8)(-160)| = 128$$

This calculation shows that the load is provided with a-c energy 128 times that extracted from the signal source. This completes the analysis for determining the gains of an amplifier through use of the transistor characteristics.

The following rules are a résumé of the graphical-analysis approach:

1. Identify the d-c path in the collector circuit and draw the d-c load line on the collector characteristics.
2. Knowing V_{BB}, the biasing voltage, and the biasing resistance, which is R_1 in the circuit of Fig. 6-10, calculate the biasing current I_B and locate the Q point.
3. Calculate the a-c collector load resistance R_{ac}.
4. Realizing that the a-c and d-c load lines intersect at a common point for $e_s = 0$, draw the a-c load line through the Q point.
5. Knowing e_s, calculate i_B and, after denoting the maximum and minimum variations of i_B on the collector characteristics, find whatever values are required, that is, $V_{\text{out}}, I_c, I_b, A_i, A_v,$ and/or A_p.

6-8. Transistor Power Relationships

In the preceding example, the amplifier power gain was determined by comparing the a-c power developed in the load resistor to that furnished by the a-c signal source. Since the load resistor is being furnished much more a-c power than provided by the a-c signal source, the question arises as to the source of this additional a-c power. Two approaches to answering that question can be made. One is to examine the a-c and d-c power conditions in the transistor separately in much the same way as the a-c and d-c analyses were conducted in the previous example and then insist on an energy balance to relate the a-c and d-c powers in answering the question. Another more elegant approach is to apply a mathematical analysis to the combined a-c and d-c solutions and let the mathematical expressions provide the answer. This section will take the former approach while material in Appendix E takes the latter approach. The preceding example will serve as the model for obtaining the power relationships. Recall that the power lost between two points 1 and 2 in a circuit is given by

$$p_{12}(t) = v_{12}(t) \cdot i_{12}(t) \tag{6-31}$$

In the case of a d-c voltage across or current through a resistor, located between points 1 and 2, the power lost is non-time-varying and is given by

$$P_{12} = V_{12}I_{12} = (I_{12})^2 R = (V_{12})^2 / R \tag{6-32}$$

where Ohm's law ($V_{12} = I_{12} \cdot R$) has been used to get alternative forms for the d-c power lost in a resistor. Consider a resistor through which an a-c current

$$i_{12}(t) = I_m \sin \omega t \tag{6-33}$$

is flowing so that the voltage is

$$v_{12}(t) = i_{12}(t) \cdot R = I_m \cdot R \sin \omega t = V_m \sin \omega t \tag{6-34}$$

where

$$V_m = I_m \cdot R \tag{6-35}$$

The power lost in the resistor is given by Eq. 6-31, i.e.,

$$p_{12}(t) = v_{12}(t)i_{12}(t) = V_m I_m \sin^2 \omega t = I_m^2 \cdot R \sin^2 \omega t = \left(\frac{V_m}{R}\right)^2 \sin^2 \omega t \tag{6-36}$$

Examination of Eq. 6-36 shows that unlike the d-c case, the power dissipated in the resistor continually varies in time when generated by an a-c voltage or current. Plots of these quantities appear in Fig. 6-18. Some observations can be made from these plots. The voltage across and the current through the resistor are in phase as they should be according to Ohm's law. Although the power varies continually in time, it is always equal to or greater than zero which agrees with the fact that a resistor is a dissipative element, i.e., it can neither store nor furnish energy to the circuit. The obvious answer to the question "What is the a-c power dissipated in the resistor?" must be "At what instant of time is it desired?" How can time be removed from the analysis without leading to significant

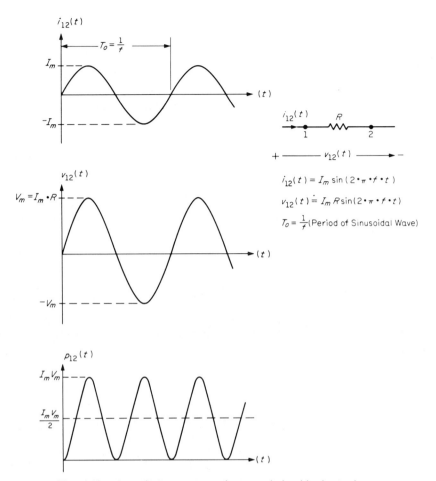

FIG. 6-18.   A-c voltage, current, and power relationships in a resistor.

error or a loss of information about the a-c behavior? One notes that the power behavior is periodic, repeating itself regularly at a frequency double that of the voltage or current waveforms. An *average power per cycle*, P_{avg}, can therefore be calculated which is time independent. If the total energy dissipated in the resistor over a time interval ΔT is desired, it can be approximated by

$$\text{Energy dissipated} \simeq P_{avg} \cdot \Delta T \qquad (6\text{-}37)$$

with a maximum error of approximately

$$\text{Error} = \frac{1}{\Delta T \cdot f \cdot 2} \qquad (6\text{-}38)$$

where f is the cyclic frequency of the voltage and current waves. Note that for 60 Hz and one second operation, the error is less than 1 percent. In essence, Eqs. 6-37 and 6-38 indicate that if the circuit has been operating for an interval of time much greater than the period $(1/f)$ of the sinusoidal voltage and current, then neglecting a partial cycle and making the energy calculation from the average power over the full time interval will introduce negligible error. Therefore, the average power per cycle, P_{avg}, is a good index of the a-c power being lost and is independent of time.

The next step is to find an expression for P_{avg}, in terms of the a-c voltages and currents. Eq. 6-36 can be written as

$$p_{12}(t) = V_m I_m \left(\frac{1 - \cos 2\omega t}{2} \right) = \frac{V_m I_m}{2} - \frac{V_m I_m}{2} \cos 2\omega t \quad (6\text{-}39)$$

In this form, it is composed of two components, the first being nontime-varying and the second varying as a cosine function in time. Now the average value of a cosine or sine wave over an *integral* number of cycles is zero since there is always as much negative area as positive area. The average value of a number (independent of time) over any interval of time is simply the number so that from Eq. 6-39, it can be concluded that

$$P_{12_{avg}} = \frac{V_m I_m}{2} = \frac{I_m^2 \cdot R}{2} = \frac{V_m^2}{R \cdot 2} \quad (6\text{-}40)$$

One will get the same result (see Chapter 12) by obtaining the mathematical average from the equation

$$P_{12_{avg}} = \frac{1}{T_o} \int_0^{T_o} p_{12}(t)\, dt$$

The only unsatisfying note about Eq. 6-40 is that the expressions are not quite the same as for the d-c case given by Eq. 6-32. A new index or measure of an a-c voltage and current will now be defined. Let the *effective* value of an a-c voltage be defined so that

$$\frac{V_{eff}^2}{R} \triangleq P_{12_{avg}}$$

and, therefore, be of the same form as the d-c power expression. But from Eq. 6-40

$$P_{12_{avg}} = \frac{V_m^2}{R \cdot 2} = \frac{(V_{eff})^2}{R}$$

so that

$$V_{eff} = \frac{V_m}{\sqrt{2}} \quad (6\text{-}41)$$

In a similar manner, define the *effective* value of an a-c current by

$$(I_{eff})^2 \cdot R \triangleq P_{12_{avg}} = \frac{I_m^2 \cdot R}{2}$$

so that

$$I_{\text{eff}} = \frac{I_m}{\sqrt{2}} \tag{6-42}$$

with the further result that

$$P_{12_{\text{avg}}} = \frac{V_m I_m}{2} = \frac{V_m}{\sqrt{2}} \frac{I_m}{\sqrt{2}} = V_{\text{eff}} I_{\text{eff}} \tag{6-43}$$

The expressions for the average a-c power in terms of the effective values of the voltages and currents are

$$P_{\text{avg}} = V_{\text{eff}} I_{\text{eff}} = I_{\text{eff}}^2 \cdot R = \frac{(V_{\text{eff}})^2}{R}$$

which are identical in form to those for the d-c case given by Eq. 6-32. *Furthermore, Kirchhoff's voltage and current laws can be applied to the effective values of sinusoidal quantities rather than using the instantaneous expressions when dealing with resistive circuits.* (This is demonstrated analytically in Chapter 12.) This makes a-c analysis of resistive circuits identical to d-c analysis.

A power analysis of the common-emitter amplifier given in Fig. 6-10 will make use of the expressions just developed.

The d-c analysis will utilize Fig. 6-11. The d-c power dissipated in R_1 is given by

$$P_{R_1} = (I_B)^2 \cdot R_1 = (100\,\mu a)^2 \cdot (50\,\text{K}) = 500\,\mu\text{watts}$$

The d-c power entering the transistor between the base and emitter terminals is given by

$$P_{BE_{\text{in}}} = V_{BE} I_B$$

which depends on how V_{BE} is handled. If it has been assumed to be zero, then of course the d-c power is zero. If it has been assumed to be 0.6–0.7 volts then $P_{BE_{\text{in}}}$ is 60 to 70 μ watts which is around 10 percent of the power lost in R_1 and is of small consequence. This is especially true in view of the much higher power levels existing on the output or collector-emitter side of the transistor as will be determined later. The power lost in the base bias supply V_{BB} is given by

$$P_{BB} = V_{12} I_{12} = (V_{BB})(-I_B) = (5)(-100\,\mu a) = -500\,\mu\text{watts}$$

This negative power loss or circuit gain exactly balances the loss in R_1 and if V_{BE} had not been set to zero, then V_{BB} would also account for $P_{BE_{\text{in}}}$.

On the output side, the d-c power loss in R_l is given by

$$P_{R_l} = (I_C)^2 \cdot R_l = (18\,\text{ma})^2 \cdot (1\,\text{K}) = 324\,\text{mwatts}$$

Neglecting the base current the d-c power entering the transistor between the collector and emitter terminals is given by

$$P_{CE_{\text{in}}} = V_{CB} I_C + V_{BE} I_C \tag{6-44}$$

But

$$V_{CE} = V_{CB} + V_{BE}$$

so that

$$P_{CE_{in}} = V_{CE}I_C = (22\text{ v})(18\text{ ma}) = 396\text{ mwatts}$$

The power lost in the collector supply is given by

$$P_{CC} = V_{32}I_{32} = V_{CC}(-I_C) = (40\text{ v})(-18\text{ ma}) = -720\text{ mwatts}$$

This negative power loss (or circuit power gain) is identically equal to $P_{R_1} + P_{CE_{in}}$. This exact correspondence was somewhat fortuitous since both V_{CE} and I_C came from a graph but it emphasizes that the graphical analysis does indeed give quite accurate and satisfactory results. At this point, it should be pointed out that in the absence of an a-c signal, the transistor must dissipate $P_{CE_{in}}$ as heat and for large power transistors this can be a challenging thermal problem. This is due to the fact that transistors are inherently small with the major volume being the packaging rather than the active electrical material. In essence, something approaching a point heat source can be envisioned, so that heat conduction away from the source is most important if the source temperature is not to rise to a destructive value.

Consider now the a-c circuit given in Fig. 6-13. The effective values are to be used in the power analysis. The correspondence becomes

$$e_s = 2.5 \sin \omega t \Rightarrow E_{s_{eff}} = \frac{2.5}{\sqrt{2}}\text{volts} = E_s$$

$$i_s = i_b = 50 \times 10^{-6} \sin \omega t \Rightarrow I_{b_{eff}} = \frac{50 \times 10^{-6}}{\sqrt{2}}\text{amps} = I_b = I_s$$

$$i_c = 10 \times 10^{-3} \sin \omega t \Rightarrow I_{c_{eff}} = \frac{10 \times 10^{-3}}{\sqrt{2}}\text{amps} = I_c$$

$$v_{ce} = -2 \sin \omega t \Rightarrow V_{ce_{eff}} = \frac{-2}{\sqrt{2}}\text{volts} = V_{ce}$$

where it is now understood that an a-c quantity is the effective value unless otherwise indicated. The a-c power lost in R_1 is given by

$$P_{R_{1_{a-c}}} = (I_b)^2 \cdot R_1 = \left(\frac{50 \times 10^{-6}}{\sqrt{2}}\right)^2 \cdot (50\text{ K}) = 62.5\text{ }\mu\text{watts}$$

The a-c power entering the transistor between the base and emitter terminals is given by

$$P_{be_{a-c_{in}}} = V_{be}I_b = 0$$

which is zero since v_{BE} is essentially constant at 0.6–0.7 volts, and therefore, has no a-c component. The power lost in the a-c signal source is given by

$$P_s = V_{12_{a-c}} I_{12_{a-c}} = E_s(-I_s) = \left(\frac{2.5}{\sqrt{2}}\right)\left(\frac{-50 \times 10^{-6}}{\sqrt{2}}\right) = -62.5\,\mu\text{watts}$$

This negative power loss or circuit power gain is totally dissipated in R_1 so that the a-c signal source only provides an a-c current into the base of the transistor but virtually no a-c power. The a-c power lost in R_l is given by

$$P_{R_{l_{a-c}}} = (V_{ce})^2/R_l = (-2/\sqrt{2})^2/1\,\text{K} = 2\,\text{mwatts}$$

The a-c power used in the output or load resistor R_i is given by

$$P_{R_{i_{a-c}}} = (V_{ce})^2/R_i = (-2/\sqrt{2})^2/(0.25\,\text{K}) = 8\,\text{mwatts}$$

As a check on the calculations, note that the power gain

$$\frac{P_{R_{i_{a-c}}}}{|P_s|} = \frac{8\,\text{mwatts}}{62.5\,\mu\text{watts}} = 128$$

is the same as before as it should be. Again assuming $I_b \ll I_c$ the a-c power lost in the collector to emitter terminals is given by

$$P_{ce_{a-c_{in}}} = V_{ce}I_c + V_{be}I_c = V_{ce}I_c + 0$$

Substituting in the effective values gives

$$P_{ce_{a-c_{in}}} = V_{ce}I_c = \left(\frac{-2}{\sqrt{2}}\right)\left(\frac{10 \times 10^{-3}}{\sqrt{2}}\right) = -10\,\text{mwatts}$$

This negative power loss or circuit power gain exactly balances the a-c power lost in R_l and R_i as it should, since the output of the transistor must be the source of this a-c power on the output side. The key question, "Where does this a-c power come from since relatively zero a-c power entered the transistor on the input side?" There is only one source; namely, of the 396 mwatts of d-c power going into the collector to emitter terminals, 10 milliwatts is transformed within the transistor and returned as a-c power. Another observation would be that under a-c operation, the transistor actually dissipates *less* heat than it does when the a-c signal source is removed or is zero. Another way of stating this phenomenon is that the transistor itself operates cooler with an a-c signal applied. Furthermore, the d-c source V_{CC} is the ultimate source of a-c power, although the transistor provides the conversion function. This analysis demonstrates how power amplification truly comes about through the transistor action. This completes the power analysis of a typical common emitter amplifier. All amplifier power analyses can be conducted in a similar manner.

6-9. A-C, D-C, and RMS Concepts

The preceding section discussed many of the power relations for the basic common-emitter amplifier circuit of Fig. 6-10. It is appropriate to extend this discussion to a somewhat more complicated and more typical circuit. Not only can there sometimes be confusion over what is alternating and what is direct current in an electronic circuit, but it is frequently necessary to combine the two terms. In addition to the a-c and d-c terms, we can refer to the total rms value of the voltages and currents. The relationship among these three is

$$I_{\text{eff}}^2 = I_{\text{rms}}^2 = I_{\text{a-c}}^2 + I_{\text{d-c}}^2 \tag{6-45}$$

where the a-c and d-c terms are dimensionally rms² quantities. Equation E-24 illustrates the derivation of Eq. 6-45 (refer to Appendix E).

EXAMPLE 6-3. As an example of the application of these concepts, consider the circuit shown in Fig. 6-19. This circuit is similar to the one of Fig.

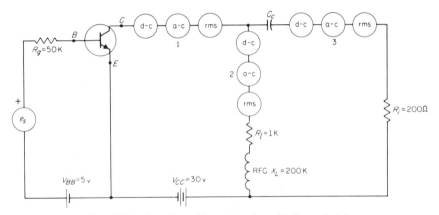

FIG. 6-19. Circuit used in conjunction with Example 6-3.

6-10 with the addition of the sets of milliammeters at points *1*, *2*, and *3*, and the RFC (radio-frequency choke) in series with the d-c load resistance. The choke offers no impedance to the flow of d-c current but offers an inductive reactance of 200,000 ohms to the a-c current. In this circuit the a-c and d-c load lines have the same slopes as the respective load lines shown in Figs. 6-12b and 6-16. This is because the d-c and a-c load resistances are again 1000 and 200 ohms; respectively. (The reader should verify this approximation.)

The problem in Example 6-3 is to find the reading on the three sets of milliammeters. Each set contains a meter that reads direct current only, a meter that

---

[2]The abbreviation rms stands for *root-mean-square* and is named for the mathematical formulation used to obtain said values. The terms *effective* (eff) and root-mean-square are synonomous.

reads the rms value of alternating current only, and a meter that reads total rms current. The argument as to what meters will read is as follows:

1. The d-c path is through the 1000-ohm resistor, the RFC, and the collector terminal, since the capacitor blocks the d-c current.
2. The resistive-inductive impedance is so much greater than the resistive-capacitive impedance that almost all of the a-c current flows through the 200-ohm resistance and the capacitor.

Table 6-1 tabulates the readings of the various milliammeters in each set.

TABLE 6-1
TABULATION OF AMMETER READINGS FOR EXAMPLE 6-3

Meter Type	Meter Set Number		
	1	2	3
I_{d-c}	18 ma	18 ma	0
I_{a-c}	$10/\sqrt{2}$ ma	0	$10/\sqrt{2}$ ma
I_{rms}	$\sqrt{(18)^2 + (10/\sqrt{2})^2}$ ma	18 ma	$10/\sqrt{2}$ ma

There is a good reason for being able to determine the various currents passing through elements, for, in order to complete properly the design of an electronic circuit, it is necessary not only to stipulate the ohmic value of the circuit resistances but also to state the energy-dissipation requirement. Resistors of the same ohmic value are made in many different physical sizes, and usually the physical size varies in direct proportion to the wattage rating.

EXAMPLE 6-4. If 1000- and 200-ohm resistors are available in $\frac{1}{100}$, $\frac{1}{50}$, $\frac{1}{10}$, $\frac{1}{4}$, $\frac{1}{2}$, and 1-watt sizes, find the minimum-wattage resistors that can be safely used in the circuit of Fig. 6-19 of Example 6-3.

Solution: The load resistance R_l has a total current of 18 ma flowing through it.

$$P_{R_l} = I^2 R_l = (18 \times 10^{-3})^2 \times 1000 = 0.324 \approx \frac{1}{3} \text{ watt}$$

Thus R_l should be rated at least $\frac{1}{2}$ watt. A similar calculation for R_i yields

$$P_{R_i} = I^2 R_i = \left(\frac{10}{\sqrt{2}} \times 10^{-3}\right)^2 \times 200 = 0.01 = \frac{1}{100} \text{ watt}$$

Thus R_i could safely be rated at $\frac{1}{50}$ watt or larger. Of course the resistors with the smaller wattage ratings might cost more than those with the larger ratings, so it might be economically feasible to choose a $\frac{1}{4}$-watt instead of a $\frac{1}{50}$-watt resistor for R_i.

6-10. Common-Emitter Biasing Techniques

The common-emitter amplifier shown in Fig. 6-20 is the same as the amplifier shown in Fig. 6-9a except for the biasing arrangement. In prac-

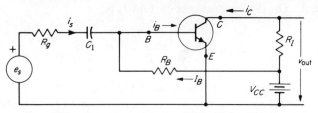

FIG. 6-20. Common-emitter amplifier with fixed biasing.

tice very few transistor circuits are designed with a separate d-c-bias supply source V_{BB}. Many times the collector supply source V_{CC} can be used to provide the necessary biasing current I_B, which eliminates the need for an expensive, separate power supply.

Fixed Bias. The method of biasing shown in Fig. 6-20 is called *fixed base-current bias*. Fortunately, in the common-emitter configuration the collector supply voltage V_{CC} and the base-biasing voltage V_{BB} are of the same polarity, so it is possible for V_{CC} to supply the base-biasing current I_B. Using Kirchhoff's first law, the base-biasing current I_B is found to be

$$I_B = \frac{V_{CC} - V_{BE}}{R_B} \qquad (6\text{-}46)$$

but, since V_{CC} is usually 20 or 30 volts while the average base-emitter voltage V_{BE} is only 0.5 or so volt (see Fig. 6-6b), it is convenient to neglect V_{BE} in making the calculation for the base-biasing current, or

$$I_B \simeq \frac{V_{CC}}{R_B} \qquad (6\text{-}47)$$

The capacitor C_1 is a very necessary part of this biasing scheme because it blocks the d-c current from the branch containing the source and thus forces I_B to flow through the transistor base-emitter junction but at the same time allows the a-c signal input current $i_b = e_s/R_g$ to pass through to the base of the transistor.

Self-Bias. If R_B is connected between the collector and base terminals, instead of the collector supply V_{CC} and base terminal, the transistor is said to be self-biased. Figure 6-21 illustrates this type of self-biasing

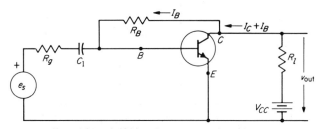

FIG. 6-21. Self-biased common-emitter circuit.

arrangement. The d-c or quiescent current I_B is

$$I_B = \frac{V_{CB}}{R_B} = \frac{V_{CE} + V_{EB}}{R_B} \simeq \frac{V_{CE}}{R_B} \qquad (6\text{-}48)$$

where V_{BE} is again assumed to be so small that $V_{CE} \gg V_{BE}$.

The reason for choosing self-biasing over fixed biasing is apparent if we realize that the collector current I_C tends to increase as the temperature of the transistor increases. This effect is much more pronounced in germanium transistors as compared with silicon transistors. If fixed biasing is used, the base current I_B remains constant even if the collector current I_C increases. This may produce two undesirable effects: 1) the transistor may become biased either in or close to a nonlinear region of operation, and 2) the increased collector current may produce a further, and perhaps damaging, temperature rise in the transistor. If self-biasing is used, and if I_C increases, V_{CE} decreases (owing to the relation $V_{CE} = V_{CC} - I_C R_l$) with the result that I_B also decreases (see Eq. 6-48). This decrease in I_B, in turn, decreases I_C (note the common-emitter collector characteristics in Fig. 6-6c). As a consequence, I_C does not increase as much with a self-biasing arrangement as with fixed biasing. The overall result is a greater stability in the operation of the transistor.

Unfortunately, the self-biased circuit of Fig. 6-21 allows a portion of the a-c collector-emitter voltage v_{ce} to be fed back to the base-emitter voltage v_{be}. Earlier in this chapter it was established that the a-c input and output voltages were 180° out of phase for the common-emitter circuit. To avoid this undesirable degeneration (called *negative feedback* since the output a-c signal is fed back out of phase and tends to decrease the input a-c signal) the biasing resistance is divided into two resistances R_{B_1} and R_{B_2} and then a capacitor C_2 is tied to the node between these resistances and ground, as shown in Fig. 6-22. If $1/\omega C_2 \ll R_B$, the capacitor effectively bypasses the a-c signals to ground and prevents the a-c feedback signal from reaching the base terminal. The subject of feedback is discussed in some detail in Chapter 21.

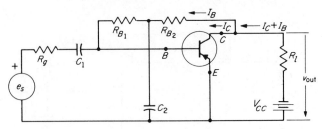

Fig. 6-22. Self-biasing scheme for the common-emitter amplifier to avoid undesirable degeneration.

EXAMPLE 6-5.   In Fig. 6-20, $V_{CC} = 30$ volts, $R_l = 5000$ ohms, and I_B is chosen to be 100 μa. What is the proper value of R_B?

Solution: $R_B = V_{CC}/I_B = 30/(100 \times 10^{-6}) = 300{,}000$ ohms.

EXAMPLE 6-6.   a) If the circuit shown in Fig. 6-22 uses a 2N404 transistor where $V_{CC} = 15$ volts, $R_l = 500$ ohms, and I_B is chosen as -100 μa, what is the proper value for R_B where $R_{B_1} + R_{B_2} = R_B$?

Solution: Since $R_B = V_{CE}/I_B$ in this self biased circuit, it is first necessary to find V_{CE}. A good way to find V_{CE} is to use the accompanying 2N404 common-emitter collector characteristics (Fig. 6-23). Using the given data, the Q point a is

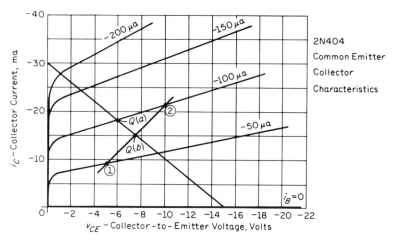

FIG. 6-23.   2N404 common-emitter collector characteristics.

seen to be located at $V_{CE} = -6$ volts, $I_C = -18$ ma. Thus

$$R_B = \frac{-6}{-100 \times 10^{-6}} = 60{,}000 \text{ ohms}$$

Any reasonable combination of resistors can be used to divide R_B into R_{B_1} and R_{B_2}. If possible it is desirable to make $R_{B_1} > 10R_{ie}$ and $R_{B_2} > 10R_l$ to avoid loading either the input or output circuits. The latter constraint assures that the d-c and a-c load lines are essentially the same. Quite often dividing R_B into two equal valued resistors works nicely. For this example

$$R_{B_1} = R_{B_2} = \frac{R_B}{2} = 30{,}000 \text{ ohms}$$

b) Suppose R_B is changed from 60,000 to 100,000 ohms. Now where is the Q point?

Solution: This problem is a little more difficult to solve than the problem posed in part a. The best way to solve this new problem is to construct what is known as a base bias line. Again recognizing that

$$V_{CE} = I_B R_B$$

if V_{BE} is assumed negligibly small, choose two or three convenient values for I_B and calculate the resulting values for V_{CE}. For $I_B = -50\,\mu a$

$$V_{CE} = -50 \times 10^{-6} \times 100{,}000 = -5\,\text{volts}$$

For $I_B = -100\,\mu a$

$$V_{CE} = -100 \times 10^{-6} \times 100{,}000 = -10\,\text{volts}$$

Now plot $I_B = -50\,\mu a$, $V_{CE} = -5$ volts as point 1 on the 2N404 characteristics. Plot $I_B = -100\,\mu a$, $V_{CE} = -10$ volts as point 2 and draw a straight line between these points. The intersection of the d-c load line and the base bias line gives the solution for the Q point b as indicated in Fig. 6-23.

A Superior Self-Bias Circuit. As mentioned in the foregoing sections the problem of the shifting of the Q point due to temperature variations in the transistor can be quite detrimental to the satisfactory operation of a transistor circuit. Although self biasing is far superior to fixed biasing the emitter self-biasing circuit (herein called the hybrid-biasing circuit) shown in Fig. 6-24a is much superior to the two previously mentioned biasing circuits.[3] To avoid cluttering the diagram only the essential features of the biasing circuit are emphasized in this figure. Note that because the top of R_1 is attached to V_{CC} there is no need for a bypass capacitor (point a is at a-c ground or zero potential).

The problem is to determine R_E, R_1, and R_2, if V_{CC}, R_l, V_{CE}, V_{BE}, I_C, and I_B are given. In other words given the desired Q point determine the biasing resistors. To proceed with the analysis it is desirable to replace the circuit of Fig. 6-24a with the rearranged circuit shown in Fig. 6-24b. There is absolutely no difference in the operation of the circuits shown in Figs. 6-24a or 6-24b. The voltage value of both V_{CC} power supplies shown in Fig. 6-24b is the same since physically there is only one power source. To simplify the analysis next replace the bias circuit (to the left of terminals AA') with its Thévenin equivalent as shown in Fig. 6-24c. By inspection it should be evident that

$$V_o = \frac{R_2}{R_1 + R_2}\,V_{CC} \qquad (6\text{-}49)$$

$$R_o = \frac{R_1 R_2}{R_1 + R_2} \qquad (6\text{-}50)$$

Summing voltage drops around the collector circuit obtains

$$I_C R_l + V_{CE} + (I_C + I_B) R_E - V_{CC} = 0 \qquad (6\text{-}51)$$

Note that all of the quantities in Eq. 6-51 are known except R_E for the problem under consideration; thus, one can solve for R_E.

---

[3]Proof of this statement can be found in many more advanced texts on electronics such as *Electronic Devices and Circuits*, by Millman and Halkias, McGraw-Hill, 1967.

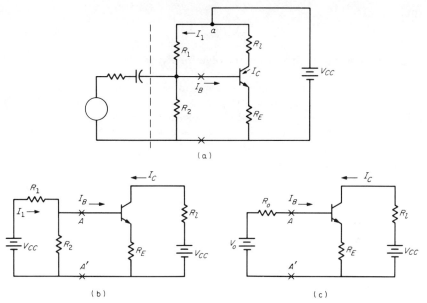

FIG. 6-24. a) Highly temperature-stabilized hybrid self-bias circuit. b) Re-arranged bias circuit. c) Thévenin's equivalent of bias circuit.

It should be apparent that any number of values for V_o and R_o will produce the desired I_B; hence, many values exist for R_1 and R_2 subject to two practical constraints: 1) V_o cannot be greater than V_{CC} if only one battery is to supply both the base and collector circuits; 2) R_1 and R_2 should not be so small as to produce an undue current drain, I_1, on V_{CC}. (Or as will be seen later, to attenuate unduly the applied a-c input signal.) On the other hand R_1 and R_2 should be as small as possible to maximize the temperature stability of the circuit. A compromise between a small I_1 and a small R_o must be made. Generally acceptable[4] values for R_o are

$$0.5R_E \leq R_o \leq 3R_E \tag{6-52}$$

Summing voltage drops around the base circuit of Fig. 6-24c obtains

$$I_B R_o + V_{BE} + (I_B + I_C) R_E - V_o = 0 \tag{6-53}$$

By choosing R_o subject to the constraint of Eq. 6-52, V_o can be solved for. Recall the constraint that V_o must be less than V_{CC}; thus if $V_o \geq V_{CC}$, choose a smaller value for R_o.

From Eqs. 6-49 and 6-50

$$R_1 V_o = \frac{R_1 R_2}{R_1 + R_2} V_{CC} = R_o V_{CC} \tag{6-54}$$

---

[4]A thorough discussion of thermal stability is beyond the scope of this text. See op. cit.

thus

$$R_1 = \frac{V_{CC}}{V_o} R_o \qquad (6\text{-}55)$$

Equation 6-50 can then be used to solve for R_2.

EXAMPLE 6-7. Assume that it is desired to establish the Q point of a 2N3704 NPN silicon transistor at $I_B = 100$ μa, $V_{CE} = 20$ volts. Using the 2N3704 collector characteristics in Fig. 6-12b it is seen that $I_C = 18$ ma at this Q point. Solve for the necessary values of R_1, R_2, and R_E to establish the Q point. Let $V_{CC} = 35$ volts and $R_l = 400$ ohms.

Solution: Assuming the circuit configuration shown in Fig. 6-24 first find R_E by using Eq. 6-51

$$\begin{aligned} R_E &= \frac{V_{CC} - V_{CE} - I_C R_l}{(I_C + I_B)} \\ &= \frac{35 - 20 - 18 \times 10^{-3} \times 400}{(18 \times 10^{-3} + 10^{-4})} \\ &= 430 \text{ ohms} \end{aligned}$$

To minimize the current drain on V_{CC} use the largest permissible value for R_o as given in Eq. 6-52

$$R_o = 3R_E = 1290 \text{ ohms}$$

Next determine V_o from Eq. 6-53

$$\begin{aligned} V_o &= I_B R_o + V_{BE} + (I_B + I_C) R_E \\ &= 10^{-4} \times 1290 + 0.6 + (10^{-4} + 18 \times 10^{-3}) 430 \\ &= 8.53 \text{ volts} \end{aligned}$$

From Eq. 6-55

$$R_1 = \frac{V_{CC}}{V_o} R_o = \frac{35}{8.53} \times 1290$$
$$= 5290 \text{ ohms}$$

From Eq. 6-50

$$1290 = \frac{5290 R_2}{5290 + R_2}$$

thus

$$\begin{aligned} R_2 &= \frac{1290 \times 5290}{5290 - 1290} \\ &= 1700 \text{ ohms} \end{aligned}$$

6-11. The D-C Equivalent Circuit for the Transistor

Although graphical circuit analysis methods are useful, instructive and important, the designer would be quite limited if he had to rely on graphical methods alone. The equivalent, linearized circuit approach, which complements the graphical method, will be presented in this section.

The most important step in analyzing an electronic circuit by means of an equivalent circuit is modeling the physical device (such as a diode or transistor) so that conventional, linear circuit analysis methods can be used. This is not always an easy task and if one chooses an improper or inadequate circuit model then the accuracy of the analysis is subject to question. The reader might ask if circuits with transistors can be analyzed using the linear analysis methods of Chapters 1, 2, and 3. The answer is yes, if one can model or describe the transistor in terms of resistors, voltage sources and current sources. In this section we shall develop the obvious dc circuit model of the transistor and show how this model is useful in analyzing circuits using transistors.

In the previous sections the base input resistance of the transistor was discussed in terms of a static resistance, R_{in} (static), and a dynamic resistance, R_{ie}. The static resistance is the ratio of V_{BE}/I_B for a transistor in the common-emitter mode and as noted earlier it varies greatly as the Q point is moved about. To illustrate this fact, a plot of R_{in} (static) versus I_B for $V_{CE} > 0.2$ volt is shown in Fig. 6-25. Note that

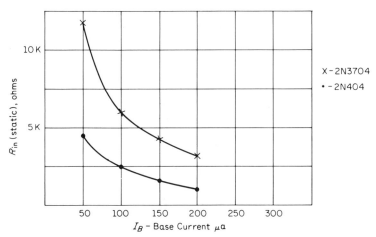

Fig. 6-25.   Static input resistance, R_{in} (static), versus base current for the 2N3704 and 2N404 transistors.

the static input resistance varies rather markedly as a function of base current and that the static input resistance of a germanium transistor is about 1/3 that of a silicon transistor.

Because of the wide variation of R_{in} (static) versus base current a better model of the base-emitter input characteristics is desired. First consider the two straight lines $0A$ and AB shown in Fig. 6-26a which approximate the actual input characteristics of the transistor. The voltage at point A, V_A, is either about 0.2 or 0.7 volt depending on whether the transistor is made of germanium or silicon and the voltage polarity depends on whether the transistor is NPN or PNP. From analytic geometry the equation of the line AB is

$$v_{BE} = V_A + R_{ie}i_B \qquad \text{for} \qquad v_{BE} \geq V_A \qquad (6\text{-}56a)$$

or if only d-c voltages and currents are considered

$$V_{BE} = V_A + R_{ie}I_B \qquad (6\text{-}56b)$$

The d-c equivalent circuit for the input of a transistor is shown in Fig. 6-26b and follows directly from inspection of Eq. 6-56a or b. R_{ie} is the

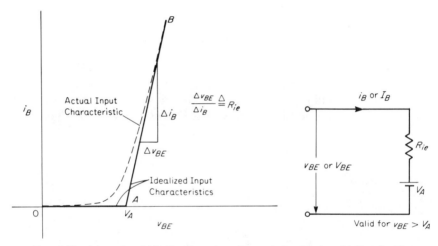

FIG. 6-26. a) Actual and idealized transistor input characteristics. b) D-c electric circuit model of the transistor input.

slope of line AB expressed as

$$R_{ie} \triangleq \frac{\Delta v_{BE}}{\Delta i_B}$$

Next let us turn our attention to the output portion of the transistor to see what model might be useful in describing the output behavior. Keep in mind that we are considering only the linear or idealized behavior of the transistor; thus only the straight line portion of the output characteristics are of interest to us in obtaining the circuit model for the transistor. Using a bit of judicious foresight let's obtain a relationship between i_B and i_C from Fig. 6-27a. Any vertical line such as the one shown in Fig. 6-27a will produce the desired results both because the collector characteristics are almost horizontal to the v_{CE} axis and are almost parallel to

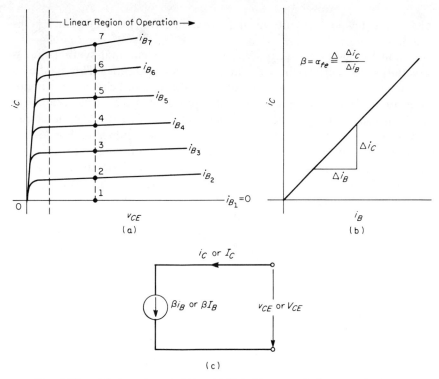

$$\alpha_{fe} = \beta \triangleq \frac{\Delta i_c}{\Delta i_B}\bigg|_{V_{CE} = \text{a constant}}$$

FIG. 6-27. a) Output characteristics. b) Plot of i_C versus i_B for a constant v_{CE}. c) D-c circuit model of the transistor output.

each other. The resulting plot of i_C versus i_B (for a particular v_{CE}) is shown in Fig. 6-27b. The slope of this line is the short circuit current gain α_{fe} or β

$$\alpha_{fe} = \beta \triangleq \frac{\Delta i_c}{\Delta i_B}\bigg|_{V_{CE} = \text{a constant}}$$

The corresponding electric circuit model for the output side of the transistor is simply a current generator[5] as shown in Fig. 6-27c. The com-

[5]This observation is entirely correct only if the i_C versus v_{CE} curves are horizontal to the v_{CE} axis. Although this is a good assumption for many transistors the exact d-c equivalent circuit should be a current generator in parallel with a resistor where R_{oe} is the slope of the i_C versus v_{CE} curve.

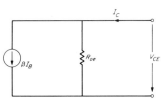

plete d-c equivalent circuit which replaces the transistor symbol is shown in Fig. 6-28b. Fig. 6-28a shows the conventional symbol that is generally used in electronic schematic or circuit diagrams. This symbol is adequate for a graphical analysis of transistor circuit behavior but is not very useful if one wishes to use the powerful linear circuit analysis methods such as loop currents or node voltages. The d-c electric circuit shown in Fig. 6-28b, which replaces the symbol shown in Fig. 6-28a, contains the types of electric elements or models that can be used with loop currents, node voltages or any other linear electric circuit analysis technique. Note that the circuit variables V_{BE}, I_B, V_{CE}, and I_C are exactly the same for both Fig. 6-28a and b. An example will help to clarify the usefulness of the transistor model of Fig. 6-28b.

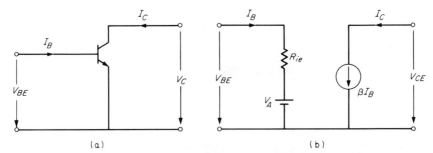

FIG. 6-28. a) Conventional circuit symbol for a NPN transistor. b) D-c electric circuit or model for the transistor.

EXAMPLE 6-8. Determine the Q point of the circuit of Example 6-6 using the d-c equivalent circuit.

Solution: Redrawing Fig. 6-22 so that the transistor symbol is replaced by the d-c equivalent circuit for the transistor results in Fig. 6-29a. Both R_{ie} and V_A can be obtained by showing the straight line approximation to the i_B versus v_{BE} input characteristic as shown in Fig. 6-29b. By inspection

$$V_A = -0.2 \text{ volts}$$

and

$$R_{ie} = \frac{\Delta V_{BE}}{\Delta i_B} = \frac{-0.3 - (-0.2)}{(-200 - 0) 10^{-6}}$$

$$= 500 \text{ ohms}$$

The short circuit current gain β can be obtained from Fig. 6-29c where the vertical line is chosen between points 1 and 2 simply for convenience in making the calculation. The value for β will be about the same (a 30 to 40 percent variation is to be expected) regardless of where one chooses to place the vertical line

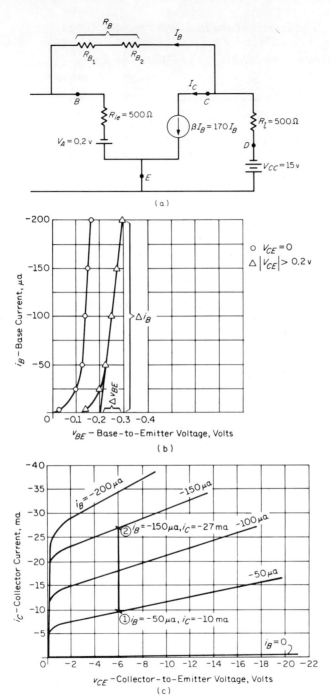

(a)

(b)

(c)

FIG. 6-29. a) D-c equivalent circuit for Fig. 6-22. b) 2N404 common-emitter input characteristics from which R_{ie} and V_A can be obtained. c) 2N404 common-emitter output or collector characteristics from which β can be obtained.

as long as one chooses $|v_{ce}| > 1$ volt. The value for β is

$$\beta = \frac{\Delta i_C}{\Delta i_B}\bigg|_{v_{CE} \,=\, \text{a constant}} = \frac{[-27 - (-10)]\,10^{-3}}{[-150 - (-50)]\,10^{-6}}$$

$$= 170$$

a) The problem posed in Example 6-6a stipulated that one should find the necessary value for R_B if $I_B = -100\ \mu a$. Writing a KVL equation around loop $CBEDC$ obtains

$$R_B I_B + 500\,I_B - 0.2 + 15 + (170 + 1)\,I_B \times 500 = 0$$

and since $I_B = -100\ \mu a$

$$-R_B \times 10^{-4} - 0.05 - 0.2 + 15 - 171 \times 10^{-4} \times 500 = 0$$

$$R_B = \frac{15 - 0.25 - 8.55}{10^{-4}}$$

$$= 62,000\ \text{ohms}$$

which compares favorably with the 60,000 ohms found by graphical means.

b) The second part of Example 6-6 required that the Q point be determined if $R_B = 100,000$ ohms. Using the same KVL equation around loop $CBEDC$ as before results in

$$10^5 I_B + 500\,I_B - 0.2 + 15 + 171 \cdot 500 \cdot I_B = 0$$

where

$$186,000\,I_B = -14.8$$

$$I_B = -79\ \mu a$$

Since the Q point is usually expressed in terms of I_C and V_{CE} note that $I_C = 170 I_B$ and

$$V_{CE} = -(I_C + I_B)\,R_l - V_{CC}$$

Therefore the Q point is located at

$$I_C = 170(-79) = -13400\ \mu a$$

$$= -13.4\ \text{ma}$$

and

$$V_{CE} = -(-13,400 - 79)\,10^{-6} \cdot 500 - 15$$

$$= 6.74 - 15$$

$$= -8.26\ \text{volts}$$

From the graphical analysis of Example 6-6 the Q point was found to be located at

$$I_C = -15\ \text{ma} \qquad V_{CE} = -7.5\ \text{volts}$$

which is reasonably close to the equivalent circuit approach given in this example.

The previous example indicates both how to obtain the d-c equivalent circuit for the transistor and its usefulness in analyzing or designing circuits using transistors. The somewhat more sophisticated a-c equivalent circuit for the transistor is presented in Chapter 7. Although the input resistance R_{ie}, the short circuit current gain α_{fe} or β, and the output resistance R_{oe}, are the same for both the d-c and low frequency a-c equivalent circuits, one other rather important parameter is necessary to describe adequately the overall a-c or small-signal equivalent circuit.

Minimizing Q Point Variation. In Sec. 6-9 it was pointed out that the collector current tends to increase with increasing temperature. To compensate for this effect the self-biased and hybrid-biased temperature-stabilized biasing networks were introduced. Since the physics of semiconductors is only very briefly treated in this text, the reader is asked to accept the reasoning presented here or refer to the many texts that treat this subject in much more detail in regard to the temperature effects in transistors.

The thermal effect is much more pronounced with germanium than with silicon transistors. In fact the thermal effect can almost be considered negligible if one uses only silicon transistors. Usually of much greater concern in transistor circuit design is the fact that there is a large variation in β for transistors of the same type. Although the use of the characteristic curves given in this text might indicate that all 2N404 or 2N3704 transistors have similar collector characteristics, such is far from being true. Both industry and the professions have accepted the fact that present day manufacturing techniques cannot produce large quantitites of the same type of transistor with identical (or even nearly identical) collector characteristics. This lack of uniformity in the characteristics has been accepted for two reasons:

1. Transistors can be produced for an extremely cheap price if one is not too fussy about uniform characteristics.
2. The electronic circuit designer has learned to design circuits that behave very well in spite of this nonuniformity in characteristics.

The same hybrid biasing circuit that produces a temperature stabilizing effect also makes the circuit behavior relatively insensitive to changes in β. There is a price to pay for this good effect; namely, the overall circuit voltage current and power gains are reduced—sometimes drastically. Because transistors have rather large betas and because they are relatively inexpensive, the temperature and beta desensitizing circuits are considered a must in almost all of the mass produced electronic devices, circuits and systems of today.

To illustrate the stabilizing effect that the hybrid biasing circuit has on the Q point let us consider Fig. 6-24c with the transistor replaced by its d-c equivalent circuit. The new circuit is shown in Fig. 6-30 where only symbols are shown for the circuit parameters and the transistor is assumed to be NPN.

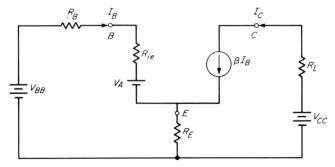

Fɪɢ. 6-30. D-c equivalent circuit for a hybrid-biased transistor circuit.

Writing a KVL equation around the base emitter loop obtains

$$I_B(R_B + R_{ie}) + V_A + (\beta + 1) I_B R_E - V_{BB} = 0$$

and

$$I_B = \frac{V_{BB} - V_A}{R_B + R_{ie} + (\beta + 1)R_E} \qquad (6\text{-}57)$$

Recognizing that $I_C = \beta I_B$ multiply Eq. 6-57 by β and divide numerator and denominator by $\beta + 1$, such that

$$\beta I_B = I_C = \frac{\dfrac{\beta}{\beta + 1}(V_{BB} - V_A)}{\dfrac{R_B + R_{ie}}{\beta + 1} + R_E}$$

For $\beta > 10$

$$\frac{\beta}{\beta + 1} \simeq 1$$

thus I_C can usually be expressed as

$$I_C = \frac{V_{BB} - V_A}{\dfrac{R_B + R_{ie}}{\beta + 1} + R_E} \qquad (6\text{-}58)$$

If

$$\frac{R_B + R_{ie}}{\beta + 1} \ll R_E$$

then Eq. 6-58 reduces to the simple expression

$$I_C = \frac{V_{BB} - V_A}{R_E} \qquad (6\text{-}59)$$

which shows that the collector current is indeed independent of β if the two previously stated conditions

$$\frac{\beta}{\beta + 1} \simeq 1 \tag{6-60}$$

and

$$\frac{R_B + R_{ie}}{\beta + 1} \ll R_E \tag{6-61}$$

are satisfied. Let us return to Example 6-7 and see if the circuit parameter values as chosen for temperature stabilizing purposes also meet the requirements for stabilizing the Q point for variations in β. In Example 6-7 R_E = 430 Ω, R_B = 1290 Ω (R_B is the same as R_o), and assuming β = 100 and R_{ie} = 500 ohms for the transistor results in the conditions of Eq. 6-60

$$\frac{100}{100 + 1} \simeq 1$$

and Eq. 6-61

$$\frac{1290 + 500}{100 + 1} = 17.7 \ll 317$$

being met very adequately. Thus even if β varies from 50 to 250 the Q point value of I_C (and therefore V_{CE}) will not vary significantly. The reader is encouraged to check this statement (see Prob. 6-22). Design Example 6-5 in the next section illustrates that amplifier gains are also held almost constant by the aforementioned circuit; i.e., the voltage, current, and power gains of an amplifier are not dependent on the beta of the transistors.

6-12. Design Problems

The purpose of this section is to introduce the concept of design in electronic circuits and to show that many simple, yet practical, applications can be appreciated by the beginning student in electronics. Hopefully, the examples developed in this section will encourage the interested reader to pursue the design of other electronic circuits.

The first four design problems will emphasize the use of graphical analysis techniques even though today almost all electronic circuits are designed by using equivalent circuit methods. The reason that graphical design is not widely used is because of the large variation in transistor characteristics even with transistors of the same type. In the opinion of the authors a beginning student should master the graphical approach before proceeding to the use of equivalent circuits. The reason for this is simple—the graphical method allows one to have a visual display of the circuit behavior. This provides a valuable reinforcement to the analytical procedures that are used with the graphical method. In the equivalent circuit approach one must rely almost solely on analytical methods to obtain a solution.

The design problems presented here are practical and have been proven to be workable by many students. In practice the student is asked to measure the characteristics of all of the devices used in the design. This means that in part he must obtain the characteristics of the particular transistor he is going to use in the design of the electronic amplifier. A Tektronix 575 or 576 (or equivalent) transistor curve tracer allows one to obtain the collector characteristics of a specific transistor easily and quickly.

Design Examples 6-1 and 6-2 emphasize direct coupled (d-c) amplifier design. As the name implies direct coupled amplifiers are used to amplify d-c or slowly varying (with respect to time) voltages or currents.

Design Examples 6-3 and 6-4 emphasize the design of a-c (alternating current) amplifiers. If the signal to be amplified varies relatively rapidly with respect to time (say above 10 to 20 Hz) then an a-c amplifier is generally used in preference to the d-c amplifier. The main reason for a-c amplifiers being preferred over d-c amplifiers is because the a-c amplifiers isolate the d-c biasing voltages and currents from the a-c signals. Even with the stabilizing effects of feedback (recall the hybrid biasing circuit), there is still a tendency for the d-c quiescent voltages to vary or "drift." If several direct-coupled stages are cascaded or connected in series, these drift voltages can cause the amplifiers to produce an unacceptable error in the output or even cause the amplifiers to saturate[6] and effectively become inoperative as far as the incoming signal is concerned.

As mentioned above Design Examples 6-1 through 6-4 emphasize the graphical approach to design. Design Example 6-5 shows how to design an amplifier using only equivalent circuit methods. In addition this latter example shows how to design a circuit that is practically insensitive to changes in transistor beta. In designing circuits that are to be mass-produced it is mandatory that the design take into account the fact that transistors of the same family or batch have betas that vary by as much as a factor of 3.

DESIGN EXAMPLE 6-1. The control of a large amount of power by a small sensing element or transducer is often encountered in the control of various processes. One of the simplest types of controllers is the bang-bang or on-off circuit where an electromechanical relay is often used as the primary "power

---

[6]Saturation occurs when there is no longer a reasonably linear relationship between i_B and i_C. Recall that I_C stabilizes at a constant value for large values of I_B; e.g., refer to Fig. 6-9b and note that I_C remains fixed at 29 ma for $I_B > 175$ μa.

amplifier." In this design one is asked to develop a circuit that will produce a satisfactory regulation of the temperature in a chemical bath. To be specific it is necessary to hold the temperature T in a chemical bath that would normally be below $0°C$ to a range $0°C < T < 20°C$. Although it is desirable to narrow the temperature variation as much as possible, it is mandatory that the temperature neither drop below $0°C$ nor go above $20°C$.

Solution: In almost any design situation there are usually many ways to solve a given problem. Since the above problem is stated rather loosely, several quite different solutions could produce satisfactory results. Since this text emphasizes electronic circuitry, a solution to the problem involving electronic or electrical devices will be presented. Let us consider the diagram shown in Fig. 6-31. The

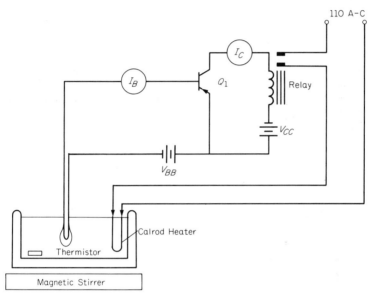

FIG. 6-31. Automatically temperature regulated chemical bath. The relay is shown with the relay contacts open.

general operation of the circuit is rather simple. A relay is used to open and close an electric circuit supplying power to a heater element. Since the temperature of the chemical bath tends to be less than $0°C$, it is necessary only that heat be added to the bath and so no provision to cool the bath is necessary. From a conceptual standpoint the addition of a refrigeration cycle would present no particular problem if this were necessary. The relay is driven by a transistor, which in turn is controlled by the thermistor temperature-sensing transducer. Even though the reader is presented with a partially completed design, several questions must be asked. What components are needed? What voltage, current, and power ratings must the components have? What supply voltages are necessary? The two most important points to consider before choosing specific components are: 1) component specifications; and 2) component compatibility. These two points are interrelated. For instance, if it is assumed that a 200 watt heater is sufficient to keep the chemical bath at the correct temperature, then the relay contacts should be capable of handling about 2 amperes (the line voltage is 120 volts so the actual current through the relay contact is $I = P/V = 200/120 = 1.67$ amps). A glance at any electronics parts catalog (such as, Allied Radio or Newark Electronics) shows that there are literally hundreds of relays that have electrical contacts that will handle 2 amperes or more. There are also hundreds or even thousands of transistors from which to choose. The choice of thermistors is a bit more limited.

At this point the careful design engineer usually has to make several trial-and-error choices before arriving at a selection of components that are compatible and that will produce a satisfactory operation of the circuit. Naturally, cost of the components is an important factor in making the final choice especially if mass production is required.

To continue with the design, let us choose a single-pole double-throw relay that has a coil resistance of 600 ohms, a pickup current of 22 ma and a drop-out current of 15 ma. For those readers who may be unfamiliar with relays the term pick-up means that the relay is energized or activated when 22 ma passes through the relay coil. In other words, the relay contacts move and either open or close with $I_{coil} = 22$ ma. As the coil current is reduced from 22 ma to 15 ma the relay contacts remain fixed in position, but when the coil current drops below 15 ma the relay is deactivated and the relay contacts move to their former position. This is why the 15 ma is called the drop-out current. Another word of explanation is in order concerning the relay contacts. Relays generally have a set or sets of normally open and normally closed electrical contacts. The word *normally* as used here means that when the relay coil current is zero the relay is operating in its *normal* mode.

Since the 2N404 transistor has been discussed at some length in this text, it will be chosen to be used in the amplifier of Fig. 6-31. The reader has a right to ask if the 2N404 is the best transistor to use in this application. The answer is a probable no, but until the circuit specifications are more clearly stated the 2N404 should be a reasonably satisfactory choice.

The type of thermistor to use is the next choice to be made. A glance at the collector characteristics for the 2N404 transistor shows that when $I_c = 22$ ma (the relay pick up current), I_B is between 120 and 140 μa. This means that a thermistor should be chosen that will have a safe current-carrying capacity of over 150 μa. To achieve a maximum sensitivity it is desirable to choose a thermistor that has a large ohmic resistance variation with temperature. The JP45J2 disc thermistor, which is manufactured by Fenwal Electronics, has all of the desired characteristics. The resistance versus temperature characteristic of this thermistor is shown in Fig. 6-32.

Now that all of the components that have been chosen are known to be reasonably compatible (this is sometimes referred to as an in-the-ballpark-design) it is time to begin the detailed design to see if the temperature regulator will really meet or exceed the desired specifications. Let us start the analysis by choosing a reasonable value for V_{CC}. Since the transistor collector current must be able to pass at least 22 ma to pick up the relay the very minimum value for

V_{CC} is

$$V_{CC}(\text{minimum}) = 22 \times 10^{-3} \times R_{\text{coil}}$$
$$= 22 \times 10^{-3} \times 600$$
$$= 13.2 \text{ volts}$$

In order to allow for change of the transistor characteristics due to aging, damage, or replacement it would be well to choose a V_{CC} of 15 volts or more.

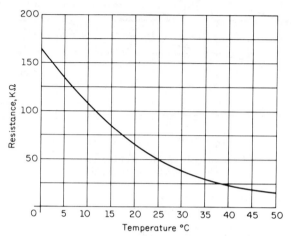

FIG. 6-32. Resistance-temperature characteristics of type JP45J2 disc themistor.

To be on the safe side let us choose a value of 18 volts for V_{CC}. The d-c load line for this choice of V_{CC} is shown on the 2N404 collector characteristics of Fig. 6-33.

Next note the pick-up and drop-out relay currents of 22 ma and 15 ma, respectively, on the d-c load line. The only task remaining is to choose V_{BB}, such that (hopefully) the relay will pick-up and drop-out within the 0°C to 20°C temperature range. It is evident from Fig. 6-32 that because the thermistor has a negative resistance versus temperature characteristic that the thermistor current (which is the same as the transistor base current) is a minimum at low temperatures and increases with increasing temperature. This means that the relay will drop-out at low temperatures and pick-up at the high temperatures. The relay contacts therefore must be normally closed to apply heat when the relay drops out at low temperatures and to remove the heat source when the relay picks up at high temperatures.

From Fig. 6-32 the resistance of the thermistor R_T is about 165 KΩ at 0°C and 70 KΩ at 20°C. The base current is equal to

$$I_B = \frac{-V_{BB} + 0.2}{R_T}$$

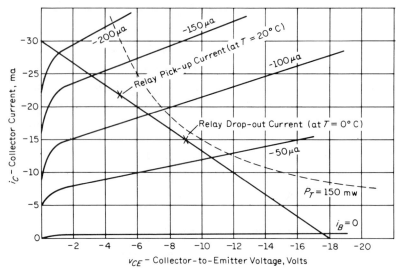

FIG. 6-33. Graphical analysis of the temperature regulator using the 2N404A collector characteristics.

where the 0.2 volt is the typical forward-biased voltage drop for a germanium PN junction. The two sets of conditions that must be satisfied are

$$\text{at } 0°C \quad R_T \simeq 165 \text{ K}\Omega \quad \text{from Fig. 6-32}$$
$$I_B \simeq -70 \ \mu a \quad \text{from Fig. 6-33}$$
$$\text{at } 20°C \quad R_T \simeq 70 \text{ K}\Omega \quad \text{from Fig. 6-32}$$
$$I_B \simeq -130 \ \mu a \quad \text{from Fig. 6-33}$$

Using the 0°C data first

$$-70 \times 10^{-6} = \frac{-V_{BB} + 0.2}{135,000}$$

or

$$V_{BB} = 9.65 \text{ volts}$$

Now see if I_B is at least $-130 \ \mu a$ using the 20°C data and $V_{BB} = 9.65$ volts

$$I_B = \frac{-9.65 + 0.2}{70,000}$$
$$= -135 \ \mu a$$

which is larger in magnitude than the required $-130 \ \mu a$. The circuit should work satisfactorily.

Some Observations.   Any relay that has a pick-up current from 5 ma to 30 ma and a drop out current of $15/22 = 0.7$ times the pickup current should work satisfactorily in this circuit (this is simply an observation based on assumed linearity of the transistor characteristics).   If another relay is chosen it is only necessary to change V_{BB} to the new required base current I_B. An adjustment of V_{CC} might also be desirable. Note that the d-c load line is inside of the allowable transistor dissipation curve marked $P_T \simeq V_{CE}I_C = 150$ mw on Fig. 6-33. This is a very important point to emphasize as the transistor can be permanently damaged or destroyed if it is overheated.[7]

One might question the feasibility of allowing the rather large temperature variation of 20°C in the chemical bath. Suppose only a 5°C variation were acceptable—what then? If the temperature regulator is to operate satisfactorily within a narrower allowable temperature range one or more changes must be made in the circuit. One must choose a relay with a much narrower range between pick-up and drop-out currents or a thermistor with a larger slope of resistance versus temperature or a transistor with a higher current gain. The first two suggestions allow only a limited promise of improvement in regulator performance, and even the third suggestion by itself does not offer a significant improvement in performance. But if more than one transistor is used to amplify the current changes due to thermistor resistance variation then a vast improvement in temperature regulation can be effected. The promise as well as the problems of adding a second transistor are illustrated in the next example problem.

DESIGN EXAMPLE 6-2. Repeat Design Example 6-1 except narrow the temperature variation from 0°C–20°C to 0°C–5°C.

Solution: As mentioned above it would be possible to choose more sensitive relays or thermistors to help achieve the desired goal but the most direct way is to put another stage of amplification in the circuit. In fact it is possible to achieve almost any desired temperature control characteristic by the judicious use of electronics.

Even though the same relay and thermistor will be used in the new circuit, there are several ways to go about designing the new electronic amplifier. It should be emphasized that a designer is limited only by his own ingenuity and innovativeness. In the authors' opinion the design that is to be presented is a good one but there are other good and perhaps even better designs that could be brought forth. It is suggested that the interested reader consider other ways to solve the problem.

Because the thermistor is excited by a d-c (direct current) source and be-

---

[7]The reader should note that fixed biasing is being used in this example problem. In an earlier section it was pointed out that fixed biasing is the worst form of biasing because of the thermal instability in transistors. The tendency of the collector current to increase with temperature is much more noticeable with germanium than with silicon transistors. Because of this it would be better to use a silicon PNP transistor in place of the 2N404 in this example problem, or one could use the 2N3704 NPN silicon transistor and obtain satisfactory operation of the temperature regulator. The 2N404 is used in this example primarily because both a PNP and a NPN transistor are needed in Design Example 2 and since the 2N404 is used extensively in this text it is felt that it is better to use this transistor rather than a silicon PNP with which the reader may be unfamiliar.

cause the temperature varies slowly with respect to time a d-c (direct coupled) amplifier will be utilized. One of many possible designs is shown in Fig. 6-34. There are several interesting features about this design: 1) Q_1 is a PNP transistor and Q_2 is an NPN transistor; 2) V_1 is the collector-emitter supply voltage for Q_1 and the emitter-collector supply voltage for Q_2; 3) V_{BB} could easily be replaced by a voltage divider network across V_1. This last point means that only one supply source V_1 is necessary to operate the entire circuit. This can be an attractive economic feature in any design. V_{BB} will be retained because it makes the operation of the circuit easier to explain.

The purpose of the circuit shown in Fig. 6-34 is to cause a large variation

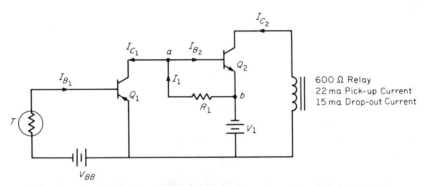

FIG. 6-34. Temperature regulator using a two-stage direct-coupled (d-c) amplifier.

in I_{C_2} for small variations in I_{B_1}. The reason that this circuit is better than the one shown in Fig. 6-31 is because I_{B_1} directly controls I_{C_1} which through I_{B_2} controls I_{C_2}. The result is a multiplying effect of the current gains of each individual stage; thus, although each stage might have a current gain of only 100, the overall current gain could be as high as 10,000. This means that the original temperature variation of 20°C using a one-stage amplifier might be reduced to as little as 0.2°C using a two-stage amplifier. Let us continue with the design and see what actually happens.

The 2N404 and the 2N3704 transistors will be chosen for Q_1 and Q_2, respectively, primarily because they have already been studied earlier in this chapter. In other words, other transistor selections could be as good or even better but these two choices should be satisfactory. The 2N3704 silicon transistor is chosen as the output transistor because it may have to dissipate more energy than the 2N404 germanium transistor is capable of handling. Silicon transistors have a higher power rating and less variation of characteristics with temperature than do their germanium counterparts.

The design of amplifier stages cascaded is really not too difficult if the individual stages are coupled capacitively (as will be evident in the next design problem) but direct coupled stages present a biasing problem. Remember that it is extremely importan. that a transistor be biased properly if it is to operate satisfactorily. This means that each transistor must have a quiescent collector current and voltage (and of course base current) such that the transistor is ready to amplify time varying voltages or currents. Stated in other terms, the Q point must be located somewhere within the linear region of operation of the transistor and such that the heat dissipation capability of the transistor is not exceeded. These are quite general stipulations and give the circuit designer a great deal of latitude in choosing values for the circuit components as well as the circuit current and voltage levels.

Perhaps the best way to continue with the design is to redraw the circuit given in Fig. 6-34 with its d-c equivalent circuit as shown in Fig. 6-35. Note

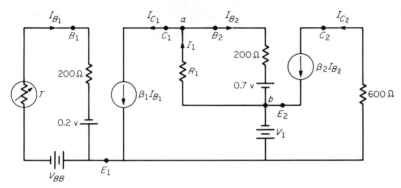

FIG. 6-35.   D-c equivalent circuit of Fig. 6-34.

that R_{ie} is chosen as 200 ohms for both transistors. This value is lower than the 400 to 500 ohms obtained earlier but is still an acceptable value and was chosen so as to emphasize that the choice of this parameter value is usually not critical. In addition to Fig. 6-35 the common-emitter collector characteristics of the two transistors as given in Fig. 6-36a and 6-36b will be used in the design-analysis procedure.

First let us pick a reasonable value for V_1. Since V_1 supplies the collector-emitter circuits of both transistors and provides the current through the relay some obvious constraints restrict our range of values for V_1. The first constraint lies in recognizing that if a relay pick-up current of 22 ma is to be realized then the absolute minimum value for V_1 is

$$V_1(\text{min}) = I_{\text{relay}}(\text{max}) R_{\text{relay}} + V_{C_2 E_2} = I_{C_2}(\text{max}) R_{\text{relay}} + V_{C_2 E_2}$$
$$= 22 \times 10^{-3} \times 600 + V_{C_2 E_2}$$
$$= 13.2 + V_{C_2 E_2} \text{ volts}$$

and since the minimum value for $V_{C_2 E_2}$ is approximately 0 volts (if I_{B_2} is large enough)

$$V_1(\text{min}) \simeq 13.2 \text{ volts}$$

The maximum value for V_1 is limited by the smaller maximum allowable value of collector-emitter voltage for the two transistors. The 2N404A transistor

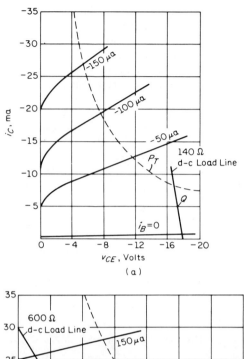

FIG. 6-36. Common-emitter collector characteristics for a) 2N404, and b) 2N3704 transistors.

has the smaller value of 24 volts (the 2N3704 allows a V_{CE} = 50 volts) so

$$V_1(\text{max}) \simeq 24 \text{ volts}$$

Armed with this information let us try an average value of V_1 = 18 volts and see if a satisfactory design results. The 600-ohm d-c load line with V_{CC_2} = V_1 is shown drawn on Fig. 6-36b. Note that the 2N3704 has a base current variation of about 180 μa between a fully cut-off ($V_{C_2 E_2}$ = 18 volts) and a fully cut-on ($V_{C_2 E_2} \simeq 0$ volts) condition.

Now check the effect of the choice for V_1 on the operation of the 2N404 stage. As mentioned earlier it is necessary to bias both transistors properly for satisfactory circuit operation. Recall that at 0°C the critical value of current to cause the relay to drop out is 15 ma. According to Fig. 6-36b a base current I_{B_2} of 90 μa is necessary to cause I_{C_2} to be 15 ma. From Fig. 6-35 it is evident that I_{B_2} is equal to

$$I_{B_2} = \frac{V_{ab} - 0.7}{200}$$

and that V_{ab} can be written in terms of $V_{C_1 E_1}$ and V_1 or

$$V_{ab} = V_{C_1 E_1} + V_1$$
$$= V_{C_1 E_1} + 18$$

Substituting this into the expression for I_{B_2}

$$I_{B_2} = \frac{V_{C_1 E_1} + 18 - 0.7}{200}$$

Now substituting I_{B_2} = 90 μa (which causes the relay to drop out) and solving for $V_{C_1 E_1}$ obtain

$$V_{C_1 E_1} = 200 \times 90 \times 10^{-6} - 18 + 0.7$$
$$= 0.018 - 18 + 0.7$$
$$= -17.282$$
$$\simeq -17.3 \text{ volts}$$

The 2N404A characteristics of Fig. 6-36a indicate that the transistor can operate with a $V_{C_1 E_1}$ of −17.3 volts up to a maximum value of −8 ma for I_{C_1} without exceeding the power dissipation rating of the 2N404A. Let's choose I_{C_1} = −5 ma for two reasons: 1) the transistor Q point is placed reasonably well into the region of linear operation without exceeding the allowable power dissipation of the transistor; and 2) the magnitude of I_{C_1} is so much larger than I_{B_2} (5000 μa compared to less than 200 μa) that the current through R_1 is essentially equal to I_{C_1}. This second reason allows us to calculate R_1 from the relation

$$V_{ab} = -I_{C_1} R_1 \simeq 0.7 \text{ volts}$$

and for I_{C_1} = −5 ma

$$R_1 = \frac{0.7}{5 \times 10^{-3}} = 140 \text{ ohms}$$

Next, a proper value for V_{BB} must be chosen. This is a straightforward calculation since from Fig. 6-36a I_{B_1} is seen to be approximately −17 μa for the Q point of stage 1 to be at $V_{C_1 E_1}$ = −17.3 volts and I_{C_1} = −5 ma. Using the previously noted fact that $R_{thermistor}$ = 165 K ohms at T = 0°C and Fig. 6-35

$$V_{BB} = -I_{B_1}(165,000) + 0.2$$
$$= +17 \times 10^{-6}(165,000) + 0.2$$
$$= 3 \text{ volts}$$

Finally, it is necessary to see what happens when the temperature of the bath (and hence the thermistor) rises above 0°C. I_{C_2} = 22 ma at relay pick-up. Referring to Fig. 6-36b it is seen that an I_{B_2} of about 130 μa is required to cause I_{C_2} to be 22 ma. This in turn means that

$$
\begin{aligned}
V_{ab} &= 200 I_{B_2} + 0.7 \\
&= 200 \times 130 \times 10^{-6} + 0.7 \\
&= 0.726 \text{ volts}
\end{aligned}
$$

and so

$$
\begin{aligned}
V_{C_1 E_1} &= V_{ab} - V_1 \\
&= 0.726 - 18 \\
&= -17.274 \text{ volts}
\end{aligned}
$$

Graphically it is practically impossible to distinguish -17.274 volts from the -17.282 volts which established the original Q point of stage 1 (see Fig. 6-36a). A crude estimate might be that I_{B_1} changes from -17 μa to -18 μa as $V_{C_1 E_1}$ changes from -17.282 to -17.274 volts, respectively, along the 140 ohm d-c load line (this ignores the shunting effect of the base-emitter diode of the 2N3704 transistor). The new value required for $R_{\text{thermistor}}$ would be

$$
\begin{aligned}
R_{\text{thermistor}} &= \frac{V_{BB} - 0.2}{-I_{B_1}} \\
&= \frac{3 - 0.2}{18 \times 10^{-6}} \\
&= 156{,}000 \text{ ohms}
\end{aligned}
$$

From Fig. 6-33 it is seen that when $R_{\text{thermistor}}$ = 156,000 ohms $T \simeq 2°C$. Thus, the new two stage amplifier narrows the temperature variation in the chemical bath from 20°C to less than 2°C. A third stage of amplification could narrow this temperature variation even further.

Since it is so difficult to read accurately small fluctuations on the graphical characteristics, it is desirable to use only the complete d-c equivalent circuit of Fig. 6-35 to compute the actual variation in I_{B_1} for a change in $V_{C_1 E_1}$ from -17.282 to -17.274 volts. To do this one must determine β_1 at the Q point in Fig. 6-36a.

$$
\begin{aligned}
\beta_1 &= \frac{\Delta I_C}{\Delta I_\beta}\bigg|_{V_{C_1 E_1} = -17.3 \text{ volts}} \\
&\simeq \frac{15 \times 10^{-3}}{50 \times 10^{-6}} = 300
\end{aligned}
$$

Since

$$
\Delta V_{C_1 E_1} = 17.282 - 17.274 = 0.008 \text{ volts}
$$

$$
\simeq \Delta I_{C_1} \times \frac{200 \times 140}{200 + 140} = 82 \Delta I_{C_1}
$$

where both the 140Ω and 200Ω resistors share the ΔI_{C_1} current change. The d-c batteries can be neglected when working with incremental variations. To illustrate this fact note that if two arbitrary values for $V_{C_1 E_1}$ are selected; i.e., $V_{C_{11} E_{11}} = -I_{11} R_1 - V_1$ and $V_{C_{12} E_{12}} = -I_{12} R_1 - V_1$ then by definition

$$\Delta V_{C_1 E_1} \overset{\Delta}{=} V_{C_{12} E_{12}} - V_{C_{11} E_{11}}$$
$$= - (I_{12} - I_{11}) R_1$$

and the constant V_1 term is seen to be eliminated from the incremental variation $\Delta V_{C_1 E_1}$. Thus returning to the expression for ΔI_{C_1} in terms of $\Delta V_{C_1 E_1}$.

$$\Delta I_{C_1} = \frac{0.008}{82} \simeq 100 \times 10^{-6} \, \text{amp}$$
$$= 100 \, \mu a$$

Next note that

$$\beta_1 \Delta I_{B_1} = \Delta I_{C_1}$$

or

$$\Delta I_{B_1} = \frac{100}{300} = 0.33 \, \mu a$$

This shows that the one microamp estimate for ΔI_{B_1} from the graphical characteristics was off by a factor of 3. In other words, the circuit should reduce the temperature variation of the chemical bath to well under 1°C.

Several factors can contribute to errors in the above analysis: 1) error in reading the graphical characteristics; 2) differences between the actual and assumed circuit parameters; and 3) the equivalent circuit or model (as shown in Fig. 6-35) is not an entirely accurate one. In other words, being off by a factor of 3 or even more is not really unusual in a graphical design such as this. Adverse effects of transistor parameter variation can be largely eliminated by proper design as will be evident when Design Example 6-5 is studied. The next step is to build the circuit and verify the theoretical design by experiment. The authors have done this and have indeed found the circuit to control the temperature to within 1°C.

A Practical Modification. Although the two-stage direct coupled amplifier discussed in Design Example 6-2 works quite nicely, the problem of thermal runaway in the 2N404 transistor is still a potential threat to satisfactory operation of the circuit because fixed biasing is used. Reference to Fig. 6-36a shows that it is possible for the Q point to shift up the 140 Ω d-c load line so that the Q point is above the maximum allowable power dissipation. If this happens thermal runaway occurs, which in this case means that the Q point continues up the d-c load line until the transistor is overheated and destroyed. One way to avoid this situation is to introduce another biasing voltage V_2 into the emitter circuit of Q_1. This is shown in Fig. 6-37a. If V_2 equals 10 volts then the effective collector

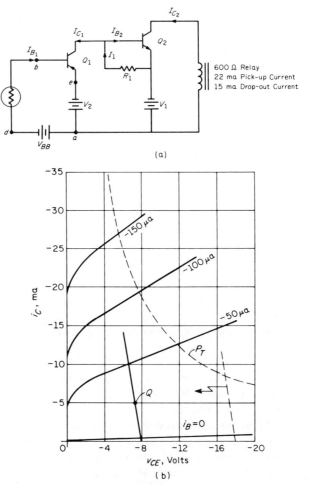

FIG. 6-37. Modified temperature controller. a) Modified circuit. b) Modified d-c load line.

supply voltage becomes

$$V_1 - V_2 = 18 - 10 = 8 \text{ volts}$$

Since R_1 still equals 140 ohms, the d-c load line is now shifted from an intercept of $V_{CE} = -18$ volts to $V_{CE} = -8$ volts as shown in Fig. 6-37b.

To maintain the Q point at $I_C = -5$ ma note that $V_{C_1 E_1}$ drops from -17.3 to -7.3 volts and I_{B_1} is increased from -17 to -25 μa. Thus the new V_{BB} for the 2N404 is obtained by writing a KVL equation around loop *aebda* or

$$
\begin{aligned}
V_{BB} &= V_2 + 0.2 - I_{B_1}(165{,}000) \\
&= 10 + 0.2 + 25 \times 10^{-6}(165{,}000) \\
&= 14.33 \text{ volts}
\end{aligned}
$$

The modified circuit assures satisfactory operation of both the 2N404 and 2N3704 transistor but at the expense of adding another power supply. Other methods can eliminate the need for the additional power supplies (such as the use of a Zener diode) but it is felt that further discussion is beyond the scope of this text.

Some introductory material will be presented before proceeding with Design Example 6-3. The purpose of this next design is to build a strain indicator by interfacing a strain gauge with a d-c milliammeter. The strain gauge is one of the most widely used transducers in engineering experimental work. Although simple in concept it took almost 10 years of intensive effort before an acceptable strain gauge was developed. The strain gauge operates on the principle that if a piece of wire is stretched or compressed the resistance varies (recall Eq. 1-12 or $R = \rho(l/A)$, which indicates that if the wire is stretched l increases and A decreases so that the resistance is increased while the reverse happens if the wire is compressed). The usual strain gauge (such as the SR-4) is about $\frac{1}{4}$ in. long and $\frac{3}{32}$ in. wide and consists of a length of resistance wire folded back and forth in a sawtooth fashion as shown in Fig. 6-38. The wire is cemented to

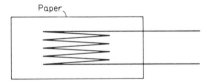

FIG. 6-38. Pictorial diagram of a typical strain gauge.

a piece of tough, parchment type paper so that when the gauge is cemented to the structural member under test, the stretching or compressing of the outer fibres of the structural member is transmitted directly to the strain gauge resistance wires.

Although the unstressed resistance of a typical strain gauge is 120 ohms, the actual variation of resistance under moderate stress is less than 1 ohm. The usual expression for strain ϵ of the gauge is

$$\epsilon = \frac{\Delta R/R}{\text{G.F.}} \text{ meters/meter} \qquad (6\text{-}62)$$

where $R = 120$ ohms (the nominal resistance of the gauge), G.F. is the gauge factor which is about 2 for many strain gauges, and ΔR is the variation of the gauge resistance from its nominal value of 120 ohms due to the stretching or compressing of the wires in the gauge.

Although the SR-4 strain gauge is still used, photo etching techniques now produce a foil gauge of very high quality and uniformity. In fact similar photo-reduction and etching processes are employed in the manufacture of integrated electronic circuits.

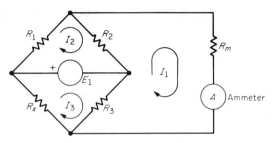

FIG. 6-39. Wheatstone bridge.

The best way to detect small changes in resistance is to use a Wheatstone bridge circuit as shown in Fig. 6-39. Using the loop current analysis techniques of Chapter 2 it can be shown that

$$I_1 =$$

$$\frac{(R_2 R_x - R_1 R_3)E_1}{R_3[R_1 R_2 + R_M(R_1 + R_2)] + R_x[R_1 R_2 + R_1 R_3 + R_2 R_3 + R_M(R_1 + R_2)]}$$

$$\text{amp} \qquad (6\text{-}63)$$

where R_1, R_2, R_3, R_M and E_1 are fixed values whereas R_x can vary. If R_x is the resistance of the strain gauge, the value for R_x varies between 119 and 121 ohms for a gauge with a nominal resistance of 120 ohms. Although many combinations of values for R_1, R_2, R_3, R_M and E_1 could be made, the resistance values will probably not exceed 10,000 ohms and the excitation source E_1 will be alternating rather than direct current. The reasons for choosing an a-c over a d-c excitation will be given later.

It can also be shown that with $R_M = 0$

$$I_3 = \frac{-R_1(R_2 + R_3)E_1}{R_1 R_2 R_3 + R_x(R_1 R_2 + R_1 R_3 + R_2 R_3)} \qquad (6\text{-}64)$$

which indicates that the current through the strain gauge does not change significantly as R_x varies from 119 to 121 ohms. On the other hand Eq. 6-63 indicates that where $R_2 R_x = R_1 R_3$, $I_1 = 0$. The bridge is nulled when this condition is reached. To illustrate the sensitivity of the bridge output (I_1) to changes in gauge resistance (R_x) let us choose some values for the bridge parameters and calculate some actual values for I_1. When the strain gauge is unstressed the nominal value of the gauge resistance is 120 ohms. Although it will not be proven here, good bridge sensitivity (in other words for reasonable variation of I_1 for variations in R_x) can be obtained without exceeding current ratings in the bridge resistors if $R_1 = R_2 = R_3 = 120$ ohms. To protect the strain gauge the current through the gauge should be limited to less than 30 ma. This means that E_1 must be limited to a value as defined by Eq. 6-64 or

$$E_1 = \frac{4 \times (120)^3 I_3}{2 \times (120)^2} = 2 \times 120 \times 30 \times 10^{-3} = 7.2 \text{ volts}$$

Instead of 7.2 volts use 6.3 volts, which is the rms value of a standard 60Hz vacuum tube filament voltage that is readily available in most electronic laboratories. With E_1 = 6.3 volts assume that the strain gauge is stretched so that the new value for R_x is 120.2 ohms. Substitution into Eq. 6-63 obtains

$$I_1 = \frac{(120 \times 120.2 - 120 \times 120)6.3}{(120)^3 + 120.5[3 \times (120)^2] + 240(240.5)R_M}$$

$$= \frac{24 \times 6.3}{481.5 \times (120)^2 + 5.76 \times 10^4 R_M}$$

$$= \frac{151}{6.93 \times 10^6 + 5.76 \times 10^4 R_M} \text{ amp}$$

$$= \frac{151}{6.93 + 0.0576R_M} \mu\text{a} \tag{6-65}$$

A 0.2 ohm variation in R_x represents a reasonable deformation or strain in the gauge. Assuming a gauge factor of 2.05

$$\epsilon = \frac{0.2/120}{2.05} = 8.13 \times 10^{-4} \text{ meters/meter}$$

$$= 813 \,\mu\text{meters/meter}$$

which is a typical strain encountered in a structural member undergoing tests or in a practical application.

DESIGN EXAMPLE 6-3. Build a strain indicator using the aforementioned strain gauge and a D'Arsonval-type milliammeter movement.[8] Suppose that both a 0.1 ma and a 5 ma full scale meter are available to us. It is required that a full scale deflection of the meter movement correspond to a strain of 1000 μ meters/meter.

Solution: The circuit shown in Fig. 6-40 is a suggested first attempt at solving this problem. Because compression as well as tension in the strain gauge is expected, it would be difficult (but not impossible) to use a d-c amplifier in this particular problem. Biasing as well as other circuit changes can cause annoying drift in the Q point of d-c amplifiers. Thus if a d-c amplifier was used in this particular application, it would be difficult to maintain a meter indication of zero for zero strain. Bias drift in capacitively coupled amplifiers (also known as a-c amplifiers) presents no particular problem since the d-c bias voltages cannot reach either the output or input stages of the circuit. Note how C_1 and C_3 perform this d-c isolation function in Fig. 6-40.

The design of the transistor amplifier is straightforward and is very similar to the self-biased amplifier presented earlier in this chapter. First let us estimate what a single stage amplifier using a 2N404 transistor can do. (Again the choice of the 2N404 is rather arbitrary. The main reason for choosing the 2N404 over the 2N3704 is because β is higher for the 2N404.) The maximum value for β for

---

[8]The D'Arsonval meter movement converts a direct electrical current into a mechanical rotation. A brief discussion of this meter is found in Appendix B.

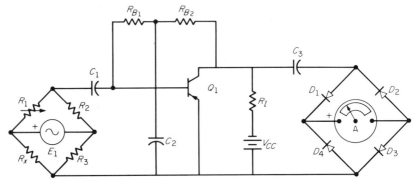

FIG. 6-40. Strain gauge indicator.

the 2N404 is about 300 so let's be conservative and choose an anticipated current gain on the conservative side of 100 for the amplifier. Since a previous calculation showed that the maximum bridge output would be about 20 μa (refer to Eq. 6-65 with R_M = 0) the choice between using a 100 μa or 5 ma full scale meter movement is obvious. In other words a maximum of 2 ma can be expected at the amplifier output so the 100 μa movement is the only one that can be driven to full scale.

The necessity for choosing the 100 microammeter movement is even more apparent when the base input resistance is considered in solving for I_1 from Eq. 6-63. For base currents between -20 and -100 μa the dynamic base input resistance varies between about 1500 and 500 ohms. This can be verified by referring to Fig. 6-7a. Let's choose an average value of 1000 ohms and since R_M of Eq. 6-65 is the same as the base input resistance, Eq. 6-65 becomes

$$I_1 = \frac{151}{6.93 + 57.6} = 2.34 \, \mu a$$

This calculation shows that the actual base current is reduced by almost a factor of 10 from the expected value for R_M = 0. Thus the maximum output current of the amplifier is in the vicinity of 200 μa instead of 2 ma which makes it even more mandatory to choose the 100 microammeter movement as the strain indicator.

Looking ahead further it should be noted that there is a diode bridge circuit connected to the output of the amplifier. As discussed in Chapter 5, the purpose of the diode bridge is to rectify or change the output voltage from a pulsating bidirectional to a pulsating unidirectional signal. This is necessary if a d-c instead of a-c milliammeter is used as the strain indicator. Each diode has a forward resistance of from 100 to 1000 ohms depending on the current passing through the diode and the milliammeter or microammeter has an internal resistance of another 500 to 1000 or more ohms depending on which manufacturer's type is used. In other words a resistance of from 700 to 3000 ohms can be expected to be in parallel with R_l as far as the a-c signal is concerned. This means that if R_l is set equal to this 700 to 3,000 ohms then $\frac{1}{2}$ of the 200 μa of a-c current mentioned above flows through R_l and the other half through the microammeter movement. Now only about 100 μa are available to drive the meter movement instead of the 2,000 μa (2 ma) that seemed at first to be available.

At this point the reader may be wondering if all the estimating done earlier isn't bordering on wild guesswork or black magic. The answer is an emphatic no! It must be emphasized that designing *any* circuit (or almost anything else for that matter) is not done in the straightforward fashion of analysis where all parameters are known and one simply solves for the prescribed variable. In a design problem there are usually any number of parameter combinations that can produce the desired variables. Where parameters (such as base input resistance, diode resistance, or meter resistance) are not known precisely it is quite reasonable to estimate the anticipated range of parameter values and proceed with the design accordingly.

Let us now continue with the design of the amplifier. Knowing that at least $\frac{1}{2}$ of the a-c collector current must pass through the meter movement choose

$$R_l \geq R_{meter} + 2 R_{diodes}$$
$$\geq 3,000 \text{ ohms}$$

A value of $R_l = 3300$ ohms will be selected as this is one of the standard resistor ohmic values that are in common manufacture.

Next select a suitable value for V_{CC}. In order to obtain a Q point I_C of about -5 ma so that the 2N404 is biased into its linear region of operation, let's try

$$V_{CC} = 21 \text{ volts}$$

Using these choices for V_{CC} and R_l results in the d-c load line shown on Fig. 6-41.

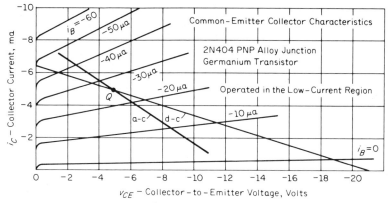

Fig. 6-41. Using the 2N404 low-current collector characteristics to establish the Q point for the strain gauge amplifier.

If the Q point is placed at $I_C = -5$ ma then $V_{CE} = -5$ volts and $I_B = -25$ μamps means that the base-biasing resistor must be

$$R_B = R_{B_1} + R_{B_2} = \frac{V_{CE}}{I_B} = \frac{-5}{-25 \times 10^{-6}} = 200,000 \text{ ohms}$$

where the 0.2 volt base-to-emitter voltage is assumed negligibly small compared to the 5 volts collector-to-emitter voltage. The usual assumption of letting $R_{B_1} = R_{B_2} = \frac{1}{2} R_B$ will be used here. The Q point could vary along the d-c load line between a V_{CE} of -3 and -8 volts or so and one would still expect satisfactory operation of the circuit. Assuming that the meter and diodes have a resistance of 3000 ohms draw an a-c load line with a slope of $-1/1570$ through the Q point, since this line represents the a-c resistance of 1570 Ω calculated from 3300 Ω in parallel with 3000 Ω.

In Chapter 15 it is shown that the relationship between the average and maximum values of a full wave rectified sine wave is

$$I_{avg} = \frac{2}{\pi} I_{max}$$

Thus if the meter is to read an average or d-c current of 100 μa full scale the maximum value of the full wave rectified sine wave must be

$$I_{max} = \frac{\pi}{2} \times 100 = 157 \,\mu a$$

Since the a-c collector current of the 2N404 is about twice that going into the diode-meter circuit the peak variation of I_c is

$$
\begin{aligned}
I_c(\text{peak}) &= 2 \times I_{max} \\
&= 2 \times 157 \\
&= 314 \,\mu a
\end{aligned}
$$

Although it is difficult to read accurately the variation of I_c of 314 μa on either side of the Q point it is easy to see that the expected values of base current lie between -20 and -30 μa and that the incremental variation of i_B required to produce a 100 μa deflection of the meter is less than 5 μa.

The complete circuit used in building the strain gauge indicator is shown in Fig. 6-42. There are two additions to the original circuit suggested in Fig. 6-40. One is a "gain" control in the base circuit of Q_1. The purpose of this potentiometer (often abbreviated pot) is to enable the user to adjust the gain of the amplifier so that a strain ϵ of 1000 μmeters/meter is equal to the full scale deflection of the 100 μa meter movement. The second addition is a 0.5 microfarad

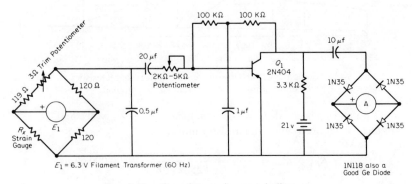

$E_1 = 6.3$ V Filament Transformer (60 Hz)

1N118 also a Good Ge Diode

FIG. 6-42. Complete strain gauge indicator.

capacitor connected across the bridge output. The purpose of this capacitor, which may or may not be needed, is to filter or short out any high-frequency components that may be picked up by the power supply or any of the other bridge components. Fluorescent light fixtures are a good source of high frequency electrical noise that could disturb the balancing of the bridge. The "equivalent" resistance of a 0.5 μfd capacitor at 60 Hz is

$$\frac{1}{\omega C} = \frac{1}{377 \times 0.5 \times 10^{-6}} = 5300 \text{ ohms}$$

so that little of the 60 Hz bridge output is shunted away from the input to Q_1. Note also that the d-c blocking-a-c passing capacitors C_1, C_2, and C_3 are 20, 1 and 10 μf respectively. At 60 Hz a 10 μf capacitor has an "equivalent" resistance of 260 ohms which is an order of magnitude smaller than the 3000 ohms of the diode-meter circuit and the 2600 ohms of C_2 is less than one tenth of the 100,000 ohms in the base circuit. Similarly the 130 ohms "equivalent" resistance of C_1 is about a factor of 10 less than the expected base input resistance of 1000 ohms or so (note that the addition of the 2000–5000 ohm "gain" control potentiometer will increase the effective base input resistance and preserve the order of magnitude criterion for choosing C_1 at 20 μf rather than a larger value).

It is almost mandatory to choose germanium rather than silicon diodes for D_1, D_2, D_3, and D_4 in the bridge rectifier because the germanium diodes have a forward voltage drop of less than $\frac{1}{3}$ of that of silicon diodes (about 0.2 volt compared to 0.7 volt). In other words the collector voltage V_{ce} must be at least 1 volt or so before any current will pass through the meter if silicon diodes are used (recall that 2 diodes are in series with the meter) whereas, a V_{ce} of only 0.3 volt or less is necessary to produce a meter deflection if germanium diodes are used. It should be emphasized that the circuit will work with silicon or germanium diodes but a more linear relationship between strain and meter deflection will be obtained with germanium diodes (such as the 1N35 or 1N118).

General Comments. The authors have found the circuit shown in Fig. 6-42 to be practical and to give good results in the laboratory. Since a sharp null in the bridge circuit is necessary to achieve satisfactory operation of the overall circuit, some practical pointers are in order. The strain gauge used is a C-12, 120 ohm, metallic foil gauge manufactured by the Budd Company. This gauge is temperature compensated for aluminum and it is recommended that a nominal gauge current of 25 ma be used. Higher currents can be used (30–45 ma) only if the gauge is affixed to a good heat sink.

The resistor used for balancing purposes (in this case R_1) should be a fixed resistor in combination with a trim pot of 2 to 5 ohms. The bridge null is quite sensitive and a fraction of an ohm variation in R_1 is detectable. Thus it is necessary to have a pot that produces only a small variation in R_1 for a given rotation of the wiper arm of the pot. The actual values of R_1, R_2, and R_3 can be anywhere between 100 and 150 ohms but they must be chosen such that the trim pot can produce a null in the circuit. In other words 1 percent, 5 percent or 10 percent resistors can be used so long as the bridge can be nulled. To minimize bridge capacitance (which can prevent a sharp null from being obtained when R_1 is varied) keep the interconnecting leads in the bridge as short as possible and use only fixed resistors to form R_1, R_2, and R_3. Decade resistance units add capacitance and tend to be noisy and so should not be used in the bridge.

The bridge excitation is perhaps best provided by any standard filament transformer (the authors used the filament supply on the Hewlett-Packard 711a Power Supply) as this provides a convenient low resistance source that is isolated from ground. To calibrate the strain gauge indicator the authors cemented the Budd C12 strain gauge to an aluminum bar as shown in Fig. 6-43. From mechanics the strain ϵ in the outer fibers

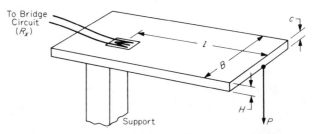

FIG. 6-43. Strain gauge cemented to an aluminum bar.

of an aluminum bar are

$$\epsilon = \frac{Mc}{EI} \text{ inches/inch}$$

where M is the bending moment $l \times P$; c is the distance from the neutral axis to the outer fibers in inches; E is the modules of elasticity (for aluminum $E = 10^7$ lbs/sq in.); I is the moment of inertia of the cross section of the beam or

$$I = \frac{BH^3}{12} \text{ inches}^4$$

Knowing the weight P, the length l, and all of the dimensions of the aluminum bar ϵ can be calculated. An appropriate M can be obtained and the 2K–5K pot in the base circuit of Q_1 adjusted so that the 100-microammeter movement reads the required 1,000 μmeters/meter (or μinch/inch).

DESIGN EXAMPLE 6-4. Suppose a 100-microammeter movement is not available but a 5 milliammeter is. Design a strain indicator using the 5 milliammeter movement.

Solution: Since the previous design works quite nicely for the 100-micro-ammeter movement, let's try another stage of electronic amplification and see what happens. Because a-c instead of d-c signals are being used it is possible to use capacitor coupling for the next stage of amplification. As will soon be evident this makes the design of the second stage much easier than in the case of Design Example 6-2 where a two-stage d-c amplifier had to be used.

The circuit shown in Fig. 6-44 emphasizes the second stage of amplification.

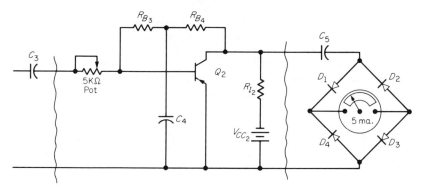

Fig. 6-44. Strain gauge indicator using a 5 milliammeter movement and a second stage of amplification.

Note that the "gain" control has been transferred to the input of the second stage rather than the first stage. The reason for this is because the first stage was never driven out of the linear region of operation by the output of the strain gauge bridge and so there is no need to limit the gain of the first stage. Thus it is logical to shift the gain control to the second stage.

Two questions need to be asked and answered in the preliminary stages of the design of this second stage. 1) Since V_{CC_1} was chosen as 21 volts can V_{CC_2} also be chosen as 21 volts? 2) What type of transistor should be used for Q_2? It should be obvious that for reasons of economy V_{CC_1} and V_{CC_2} should be the same supply voltage. This means that Q_2 must be a PNP transistor and Q_2 must be capable of handling at least 10 ma or more if it takes 5 ma to drive the strain indicator to full scale (remember that not all of the collector current is used to drive the meter-bridge circuit). Since the 2N404 transistor meets all of these qualifications let's choose Q_2 to be a 2N404.

To force as much of the collector current as possible into the meter circuit choose R_{l_2} to be as large as possible and still place the Q point of Q_2 into a favorable region of operation. Unfortunately these two criteria are conflicting because as R_{l_2} is increased the Q point is forced nearer the X axis which reduces the possible variation of I_{C_2}. A compromise must be made. The best way to proceed with the design is to determine just how much of a variation in i_C is required to produce a full scale output on the 5-milliammeter movement. Again recall that the average value of a full wave rectified sine wave is

$$I_{\text{avg}} = \frac{2}{\pi} I_{\text{max}}$$

If I_{avg} in the meter is to be 5 ma, then I_{C_2} must be at least 10 ma if it is assumed $I_{R_{l_2}}$ and I_{meter} are equal. Thus

$$I_{max} = \frac{\pi}{2} \times 10$$

$$= 16 \, ma$$

which means that the a-c load line must at least allow a 32 ma peak-to-peak swing on i_C.

Let's try an R_{l_2} of 500 ohms. This is not good because as shown in Fig. 6-45 it causes the d-c load line to pass through the region of maximum allowable power dissipation. Next, try $R_{l_2} = 1000$ ohms. This is okay as far as the transistor power dissipation is concerned but it may not allow a peak-to-peak swing of 32 ma on i_C. A value of 820 ohms (this is another resistance value of resistors in common manufacture) seems to be a satisfactory choice between 500 and 1000 ohms.

Now to place the Q point and then determine R_{B_3} and R_{B_4}. The Q point must be placed near $i_{C_2} = 16$ ma so for convenience try the Q point given by the intercept of the d-c load line with the $I_B = -100 \, \mu a$ curve. At this Q point $V_{CE} = -6$ volts, so that

$$R_{B_3} + R_{B_4} = \frac{-6}{-100 \times 10^{-6}} = 60,000 \text{ ohms}$$

The closest combination of two standard resistance values to 60,000 ohms is to let $R_{B_3} = 30,000$ ohms and $R_{B_4} = 30,000$ ohms. Using resistors with ± 5 percent tolerance a value of $R_{B_3} + R_{B_4}$ of between 57,000 ohms and 63,000 ohms can be expected. None of these resistance values will shift the Q point to an undesirable position.

The a-c load line must next be drawn through the chosen Q point. The resistance of the 5-milliampere movement should be about 10 to 20 ohms which is much less than that of the 100-microampere movement (which was about 1,500–2,000 ohms). This means that the diode-meter resistance is reduced to about 1,000 ohms, which is the forward resistance of two PN junction diodes. The slope of the a-c load line is approximately

$$\text{Slope of a-c load line} = -\frac{1}{\underset{1820}{1000 \times 820}} = -\frac{1}{450}$$

which is shown on Fig. 6-45. As can be seen the a-c load line intercepts the i_{C_2} axis at 32^+ ma. This means that the circuit should work but with very little margin of safety. In fact the Q point should be lowered so that $I_B = -80 \, \mu a$ or so in order that a uniform swing of 16 ma of collector current can be obtained. The complete circuit is shown in Fig. 6-46. The reason for choosing the capacitor values for C_4 and C_5 should be obvious.

Some redesign of stages 1 and 2 might be necessary to achieve a wider variation of i_{C_2} so as to drive the 5-milliampere movement to full scale without saturating the second stage of amplification. The most obvious thing to do is reduce V_{CC_2} and R_{l_2} a bit to increase the i_C intercept of the a-c load line. It might even be necessary to choose a PNP transistor with a larger allowable collec-

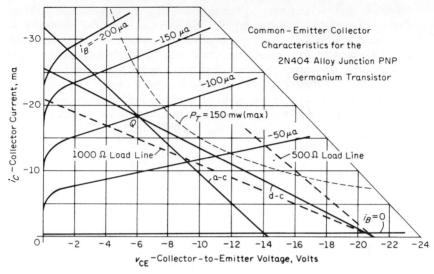

FIG. 6-45.   Using the 2N404 collector characteristics to design the second stage of amplification of the strain indicator.

tor dissipation as this is the limiting or troublesome factor in this particular design using the 2N404. Such a redesign is left as an exercise for the reader.

The authors have built and tested the circuit shown in Fig. 6-46 and found it to be satisfactory.

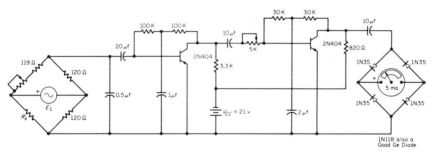

FIG. 6-46.   Complete strain gauge indicator using a two-stage amplifier and 5 milliammeter movement as the strain indicator.

DESIGN EXAMPLE 6-5.   To make the previous four designs work it is necessary to obtain the specific characteristics for each transistor used in the circuit. If these circuits were to be mass-produced, then the designs would not in general work satisfactorily since the scale on the ordinate axis (i_C) would vary by a factor of ± 3 or so for transistors of the same type.   In other words a beta variation from 40 to 360 might be expected (and considered acceptable) for a given transistor from a batch of mass-produced transistors.

Solution: Let us refer back to the hybrid-biased transistor circuit and see how this circuit could make the amplifier in Design Example 6-3 insensitive to changes in β. Although it was mentioned earlier that the overall gain of the amplifier was reduced if the hybrid-bias circuit was used, only a calculation on Q point stability was presented. Now it is desirable to see just how much v_{CE} actually varies for a given change in v_{in}. Referring back to the d-c load line on Fig. 6-41 the voltage gain can be calculated by arbitrarily choosing a Q point where $I_B = -20\ \mu a$ intersects the d-c load line. Next, letting I_B vary 10 μa on either side of this Q point and noting that $v_{in} = i_b R_{in}$ ($R_{in} \simeq 500\ \Omega$) obtains

$$A_V = \frac{\Delta v_{CE}}{\Delta v_{in}} = \frac{-11.5 - (-3.5)}{(-30 - (-10))\,10^{-6} \times 500} = -800$$

without the diode-bridge output circuit connected and with the internal resistance of the strain gauge bridge neglected. The new hybrid-biased circuit, which is shown in Fig. 6-47, needs to have about this same voltage gain if it is to operate

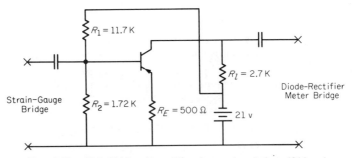

FIG. 6-47. Hybrid-biased amplifier that replaced the self-biased amplifier of Fig. 6-42.

satisfactorily. If we use the same Q point shown in Fig. 6-41, i.e., $V_{CE} = -5$ volts, $I_C = -5$ ma, and if we let $V_{CC} = 21$ volts as in Fig. 6-42, writing a KVL equation around the collector-emitter loop obtains

$$R_E = \frac{-V_{CC} - V_{CE} - I_C R_l}{(I_C + I_B)} = \frac{-21 + 5 + 5\cdot 10^{-3}(2.7 \times 10^3)}{(-5 \times 10^{-3} + 25 \times 10^{-6})}$$
$$= 500 \text{ ohms}$$

where it is seen that R_l had to be reduced from the original 3.3K to 2.7K if R_E is to have a reasonable value of several hundred ohms. If we now arbitrarily let $R_o = 3R_E$ (which is within the allowed range given by Eq. 6-52),

$$V_o \simeq V_{BE} + I_C R_E = -0.2 - 5 \times 10^{-3} \times 500 = -2.7 \text{ volts}$$

by neglecting the base current I_B in Eq. 6-53.

From Eq. 6-55 (note that V_{CC} reverses polarity between Figs. 6-24 and 6-47)

$$R_1 = \frac{-V_{CC}}{V_o} R_o = \frac{-21}{-2.7}\ 1500 = 11,700\ \Omega$$

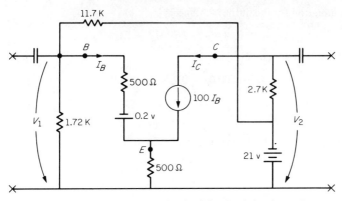

FIG. 6-48.   D-c equivalent circuit of the circuit in Fig. 6-47.

and from Eq. 6-50

$$R_o = \frac{R_1 R_2}{R_1 + R_2} \quad \text{or} \quad 1500 = \frac{11,700 R_2}{11,700 + R_2}$$

Solving for R_2 obtains

$$R_2 = 1,720 \ \Omega$$

To calculate the voltage gain of this new hybrid-biased amplifier let us consider the d-c equivalent circuit of the circuit shown in Fig. 6-47.[9] The new circuit
is shown in Fig. 6-48 and although it appears to be a bit complicated the analysis
for the voltage gain

$$A_V \triangleq \frac{\Delta V_2}{\Delta V_1}$$

is quite simple as long as β is greater than about 20.

[9]The reader should note that a graphical analysis was used to obtain the voltage gain
of the self-biased amplifier of Fig. 6-40 (and associated graphical characteristics of Fig.
6-41). On the other hand an equivalent circuit analysis is suggested for Fig. 6-47 because
it is not easy to explain how to use a graphical analysis to obtain the voltage gain for the
hybrid circuit. A moments reflection should indicate why a graphical analysis is troublesome. The accompanying figure shows that the a-c voltages V_{in} and V_{out} are referenced
to ground and not to the emitter, whereas the common-emitter graphical characteristics are
plotted in terms of the emitter being the reference node. If a capacitor is connected (as
shown by the dashed lines) across R_E then the hybrid-biased circuit appears to be a common-emitter amplifier as far as the a-c signals are concerned (as long as $1/\omega C \ll R_E$)
and the resulting circuit voltage, current and power gains can be easily determined
graphically.

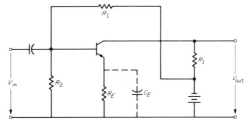

The equation for the output voltage V_2 is

$$V_2 = -2,700I_C - 21 \qquad (6\text{-}66)$$

and for the input voltage V_1 is

$$V_1 = 500I_B - 0.2 + (I_B + I_C)\,500 \qquad (6\text{-}67)$$

Using the inequality

$$I_C \gg I_B$$

which is valid for $\beta > 20$, allows us to write Eq. 6-67 as

$$V_1 = -0.2 + 500I_C \qquad (6\text{-}68)$$

Now, if we are interested only in the change or variation of V_2 as V_1 is varied Eqs. 6-66 and 6-68 become

$$\Delta V_2 = -2700\,\Delta I_C \qquad (6\text{-}69)$$

and

$$\Delta V_1 = 500\,\Delta I_C \qquad (6\text{-}70)$$

where the constants of 21 volts and 0.2 volts do not vary and hence are of no interest in these equations that show a variation of V_1, V_2, and I_C.

The symbol Δ simply means $(V_{2a} - V_{2b})$, $(I_{ca} - I_{cb})$, or $(V_{1a} - V_{1b})$ and implies a variation around some given operating or Q point.

Dividing Eq. 6-69 by Eq. 6-70 results in

$$A_v = \frac{\Delta V_2}{\Delta V_1} = \frac{-2,700\,\Delta I_C}{500\,\Delta I_C} = -5.4$$

The voltage gain A_v of the self-biased amplifier is -800 but drops to -5.4 for the hybrid-biased amplifier. This would not be particularly desirable except for the fact that the voltage gain of the hybrid biased amplifier is almost totally independent of the transistor beta. As long as $\beta > 20$ any 2N404 transistor (or its equivalent) should give satisfactory results.

The problem is how to increase the voltage gain of the hybrid-biased amplifier from 5.4 to 800. The answer is to cascade, or connect in series, enough hybrid-biased amplifiers to achieve a voltage gain of 800. Four cascaded stages results in a voltage gain of

$$A_v = (5.4)^4 = 860$$

which is more than sufficient to meet the need. The resulting multiple stage hybrid-biased amplifier circuit that replaces the single stage self-biased circuit is shown in Fig. 6-49.

Even though the circuit complexity is increased and it might appear that the 4-stage amplifier is costlier than its single-stage counterpart, the advantage of making the circuit behavior almost totally independent of β is highly desirable where mass-produced circuits are concerned. Almost all micro or integrated circuits (see Chapter 8) are designed to operate satisfactorily as long as the transistor beta is greater than 20 or so.

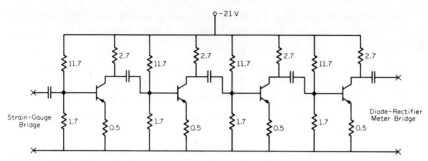

Fɪɢ. 6-49. Complete four-stage hybrid-biased amplifier circuit that replaces the single stage self-biased amplifier for Fig. 6-26. All resistor values are in K ohms.

In all fairness to the reader it should be pointed out that a more exhaustive analysis should be undertaken if the circuit in this design problem were to be mass-produced. The authors have chosen to concentrate on the voltage gain of the hybrid-biased circuit and ignore the current or power gains. Also the loading effect of the biasing resistors was ignored. These resistors, particularly the 1.7 K resistors, will reduce the overall gain of the circuit. The authors feel that a complete analysis of this circuit is beyond the scope of this text and would probably be rather tedious to read if presented. Further analysis might reveal that 3 or 5 stages would be sufficient or necessary to give satisfactory circuit operation. If at all possible the designer should build and test his design in the laboratory to see how it really behaves.

It is hoped that these five design examples will give the reader greater insight as to how electronic devices can be designed and will encourage him to develop other designs of his own choosing.

PROBLEMS

6-1. Using the accompanying figure, sketch curves similar to those of Figs. 6-8b, c, and d, showing the phase relations among e_s, i_B, i_C, and v_{CE}, if $e_s = E_m \sin \omega t$. What is the relative phase shift between a) i_B and i_C, and b) e_s and v_{out}?

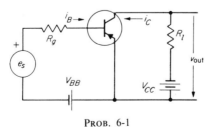

Pʀᴏʙ. 6-1

6-2. If the transistor shown in Prob. 6-1 is a type 2N404, and if V_{BB} = 5 volts, R_g = 5 × 10^4 ohms, V_{CC} = 16 volts, and R_l = 400 ohms, find V_{CE}, I_C, and I_B at the Q point.

6-3. A common-emitter amplifier using a 2N3704 transistor has a V_{CC} of 30 volts and a d-c load resistance of 1000 ohms. Plot the collector current i_C versus the base current i_B. Over what range can the curve be considered fairly linear?

6-4. Repeat Prob. 6-3 using a 2N404 transistor, a V_{CC} of 20 volts, and a d-c load resistance of 2000 ohms.

6-5. Using the circuit of Prob. 6-1 and the data of Prob. 6-2 with e_s = 5 sin ωt volts find a) v_{out}, b) V_{out}, c) the current gain I_c/I_b, d) the voltage gain V_{out}/E_s, and e) the power gain.

6-6. Use graphical techniques to analyze the circuit of the accompanying figure to find a) v_{out}, b) the current gain I_{out}/I_b, c) the voltage gain V_{out}/E_s, and d) the power gain.

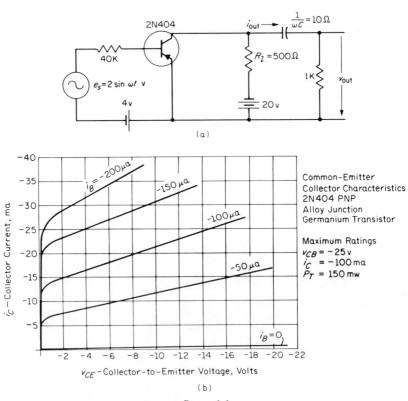

PROB. 6-6

6-7. Using the accompanying figure and graphical characteristics and if the Q point is known to be at V_{CE} = 10 volts, I_B = 20 μa, determine a) A_v = V_{out}/V_{in}, b) A_i = I_c/I_{in}, and c) the power gain A_p.

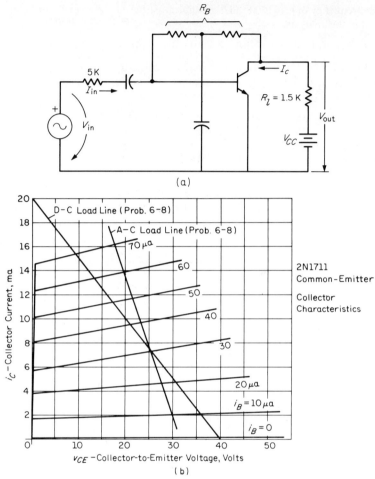

PROB. 6-7. 2N1711 graphical characteristics to be used in conjunction with Probs. 6-7 and 6-8.

6-8. Using the accompanying circuit diagram, and graphical characteristics and load lines shown in Prob. 6-7 determine a) V_{CC}, b) R_B, c) R_L, and d) R_i.

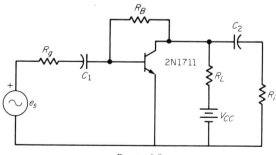

PROB. 6-8

e) If the frequency of e_s is 1000 rad/sec what must be the value of C_2 to be able to ignore it in the analysis?

6-9. Repeat Prob. 6-6 with V_{CC} = 10 volts. Comment on the results.

6-10. A 2N3704 transistor is used in the circuit of Fig. 6-20. If V_{CC} = 9 volts, find the value of R_B and R_l to place the Q point at I_B = 20 μa, V_{CE} = 5 volts. Do not neglect V_{BE} in making the calculations.

6-11. Using the self-biasing circuit shown in Fig. 6-22, find R_{B_1} and R_{B_2} if V_{CC} = 20 volts, R_l = 1000 ohms, and I_B is to be a) 100 μa, b) 75 μa, c) 50 μa, and d) 25 μa. Use a 2N3704 transistor and comment on which value of bias current would be most desirable if i_b = 15 sin ωt μa.

6-12. Using the accompanying circuit determine a) R_B if the Q point is located at I_B = 100 μa, b) the minimum value for C_2 if C_2 is to be neglected in the analysis, c) the minimum value for C_1 if C_1 is to be neglected in the analysis, d) V_{out}, e) v_{out}, f) wattage rating of the 1K resistor using standard values, g) wattage rating of the 3K resistor using standard values, and h) the Q point if R_B = 250 K.

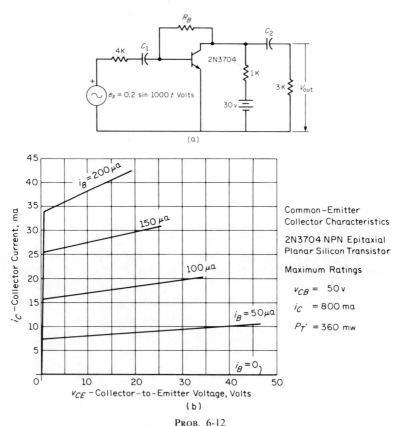

(a)

(b)

PROB. 6-12

6-13. In the circuit of Prob. 6-6, find the power loss in each resistor and in the transistor, and the power supplied by V_{BB}, e_s, and V_{CC}.

6-14. Using the circuit of Prob. 6-2 with e_s = 0, find the power loss in each resistor and in the transistor, and the power supplied by V_{BB} and V_{CC}.

6-15. Find the direct currents and the rms values of the alternating and total currents in each branch of the circuit in Prob. 6-6.

6-16. Using a 2N404 transistor determine the minimum value of the d-c collector load resistance that can be used in a fixed- or self-biased common-emitter amplifier if a) V_{CC} = 8 volts; b) V_{CC} = 10 volts, c) V_{CC} = 14 volts, and d) V_{CC} = 20 volts. (Hint: Recall that the maximum power dissipation of the transistor is 150 mw. Also assume that the Q point may be situated anywhere on the d-c load line.)

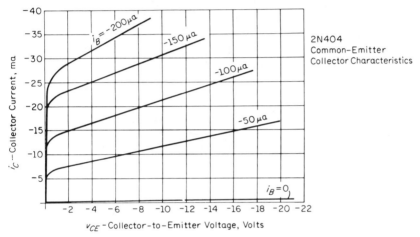

PROB. 6-16

6-17. It is desired to fix the Q point of a 2N404 transistor used in an amplifier at V_{CE} = −8 volts, I_B = −50 μa, and V_{CC} = 12 volts. Determine the values of a) the d-c load resistor R_1, b) R_B for fixed biasing, c) R_B for self-biasing, and d) R_1, R_2, and R_E for emitter self-biasing. Sketch the d-c biasing circuit for each part of this problem. Assume the maximum allowable values for R_1 and R_2 for acceptable temperature stability.

6-18. If V_{CC} = 10 volts and R_L = 250 Ω in a common-emitter amplifier using a 2N404 transistor determine the Q point for a) R_B = 80 KΩ in a fixed-bias circuit, b) R_B = 60 KΩ in a self-bias circuit, and c) R_E = 150 Ω, R_1 = 200 KΩ, R_2 = 50 KΩ in an emitter self-bias circuit. d) Repeat part (c) if R_1 = 20 KΩ and R_2 = 5 KΩ.

6-19. Using the accompanying figures and the 2N404 graphical characteristics. a) Determine (using figure a) the Q point values for I_C and V_{CE}. b) (Using figure b) the Q point is located at the intersection of I_C = −15 ma and the d-c load line, determine R_B. c) Determine (using figure c) the Q point values for I_C and V_{CE}.

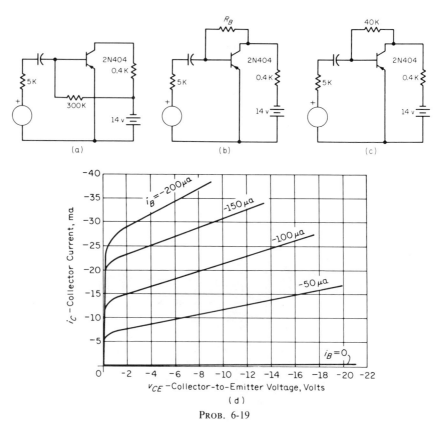

PROB. 6-19

6-20. Use the d-c equivalent circuit for the 2N404 transistor with $\beta = 70$, $R_{ie} = 500\ \Omega$, $V_A = -0.2$ v and repeat Prob. 6-17.

6-21. Use the d-c equivalent circuit for the 2N404 transistor with $R_{ie} = 500\ \Omega$, $V_A = -0.2$ v and repeat Prob. 6-17 with a) $\beta = 250$, b) $\beta = 100$, and c) $\beta = 50$. Comment on the Q point stability for each biasing circuit.

6-22. Calculate the variation in Q point (I_C and V_{CE}) using the circuit values given in Example 6-7 and the d-c equivalent circuit for the transistor. Assume a 2N3704 transistor is used and that the following betas can be expected a) $\beta = 250$, b) $\beta = 150$, and c) $\beta = 50$. [*Hint:* For the hybrid biasing circuit note that $V_{CE} = V_{CC} - I_C(R_l + R_E)$.]

6-23. Using the accompanying circuit and collector characteristics of the 2N3704 transistor determine a) R_1 and R_2 to place the Q point at $I_B = 30\ \mu a$, b) the voltage, current, and power gains, and c) state appropriate values for C_1, C_2, and C_3 if the lowest frequency for e_s is 100 Hz.

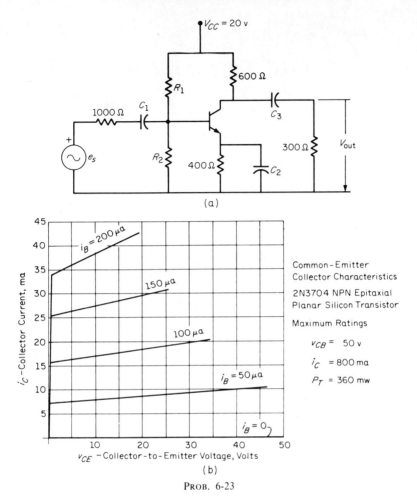

(a)

Common-Emitter
Collector Characteristics

2N3704 NPN Epitaxial
Planar Silicon Transistor

Maximum Ratings

$v_{CB} = 50$ v

$i_C = 800$ ma

$P_T = 360$ mw

(b)

PROB. 6-23

6-24. Repeat Design Example 6-1 using a 2N3704 transistor in place of the 2N404.

6-25. Repeat Design Example 6-2 using a relay with a pick-up current of 10 ma and a drop-out current of 5 ma. What is the temperature variation allowed by this new circuit?

6-26. Repeat Design Example 6-3 using a 2N3704 transistor in place of the 2N404.

6-27. Repeat Design Example 6-4 using 2N3704 transistors in place of the 2N404s.

7 / The Transistor Used as a Linear Device

In Chapter 6 the graphical bipolar junction transistor (BJT) characteristics were studied. Using the graphical characteristics, the transistor circuit was analyzed using graphical methods. In this chapter the h parameters, as obtained either from the graphical characteristics or from experimental measurements, will be used in the equivalent-circuit approach to transistor-circuit analysis. When the transistor is considered to be a linear element, conventional two-port network-analysis methods, as discussed in Chapter 3, can be used in the overall analysis. The two-generator equivalent circuit that was developed in Chapter 3 will be used in the analysis throughout this and the remaining chapters. In addition, only the common-emitter transistor amplifier and the emitter-follower circuit (using the h parameter equivalent circuit) will be discussed. Admittedly, these restrictions narrow the scope of the overall field of transistor analysis, but the material in this and the two preceding chapters should lay a good foundation for further investigation by the reader.

7-1. The A-C Equivalent Circuit of the Common-Emitter Amplifier

If the d-c currents and voltages bias the transistor in the linear portion of its characteristic curves and if the a-c signal current i_b is relatively small (so that the maximum and minimum variations of i_C and v_{CE} remain within the linear region of the collector characteristics), the transistor can be replaced by an a-c equivalent circuit. When the physical transistor circuit is replaced by the a-c equivalent circuit, all information about the internal d-c voltages and currents is lost. However, since the circuit designer is usually interested only in the a-c signals passing into and out of an electronic system, the replacement of the physical transistor by its a-c equivalent circuit allows the designer to utilize the familiar and well-developed tools of conventional a-c circuit analysis. If the total power in the electronic circuit is to be calculated, then the original circuit, which includes the d-c and a-c currents and voltages, must be analyzed.

The replacement of the physical transistor circuit by its a-c equivalent circuit closely follows the development of the equivalent circuit for the two-port linear network which was presented in Chapter 3. Rewriting Eq. set 3-23

$$V_1 = h_{11}I_1 + h_{12}V_2 \\ I_2 = h_{21}I_1 + h_{22}V_2 \Bigg\} \tag{7-1}$$

and redefining the symbols so that

$$\left. \begin{array}{ll} V_1 = V_{be} & h_{11} = R_{ie} \\ I_1 = I_b & h_{12} = \mu_{re} \\ V_2 = V_{ce} & h_{21} = \alpha_{fe} \\ I_2 = I_c & h_{22} = 1/R_{oe} \end{array} \right\} \tag{7-2}$$

Eq. set 7-1 becomes

$$V_{be} = R_{ie}I_b + \mu_{re}V_{ce} \tag{7-3a}$$

$$I_c = \alpha_{fe}I_b + \frac{1}{R_{oe}}V_{ce} \tag{7-3b}$$

Although the hybrid parameters are, in general, complex quantities (for frequencies approximately 50,000 Hz or greater), in this chapter only low-frequency application will be assumed, and thus the hybrid parameters can be considered purely real.

The physical circuit of a common-emitter amplifier circuit is shown in Fig. 7-1a and its a-c equivalent circuit is shown in Fig. 7-1b. Figure 7-1b is similar to the two-generator equivalent circuit presented in Chapter 3, but the symbols have been redesignated in accordance with Eq. set 7-2 and

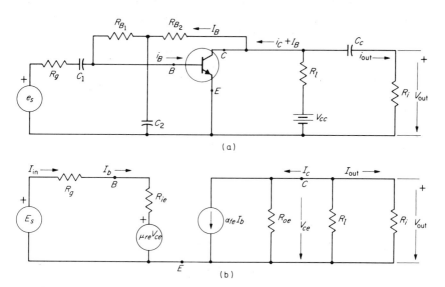

(a)

(b)

FIG. 7-1.   a) *R-C*-coupled common-emitter amplifier with self-biasing.   b) A-c equivalent circuit for part a.

7-3. The reader should note that no hint of the d-c voltages and currents appears in the circuit of Fig. 7-1b. The bias resistances R_{B_1} and R_{B_2} are not included in the equivalent circuit because it is assumed that both of these resistances are much greater in value than any of the other resistances in the circuit, and the capacitances C_1, C_2, and C_c are not included because it is assumed that the frequency ω of the signal voltage e_s is high enough so that the capacitive reactances $1/\omega C$ are negligibly small.

One of the big questions to be answered is how does one obtain the a-c equivalent circuit of the transistor as shown in Fig. 7-1b? The answer to this very important and pertinent question will be given in the next section.

7-2. Transistor Parameters

The input and output characteristics for the common-emitter transistor circuits bear further observation. First of all, except for the smaller values of voltage and current, the sets of curves are very regular and uniform. This uniformity[1] of the characteristics means that the transistor can be used as a linear element in electronic circuits.

In Chapter 3 the subject of linear two-port network analysis was introduced. Figure 7-2 shows a common-emitter transistor in a standard two-port configuration. Since two of the terminals are common or shorted together (the emitter terminal), the circuit might be considered a special or simplified two-port network. To employ two-port network theory, it is assumed that the transistor is a linear device. Just how the transistor is properly biased to operate as a linear device was explained in Chapter 6. As pointed out in Chapter 3, any two-port linear network can be completely described in terms of the two external currents and two external voltages. Of the six sets of parameters that can be used to describe the two-port network, only the hybrid or h parameters will be used here. Repeating Eq. set 3-22 in terms of instantaneous quantities, and substituting $v_{BE} = v_1, i_B = i_1, v_{CE} = v_2, i_C = i_2$,

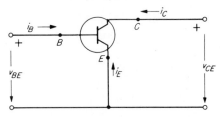

FIG. 7-2. Transistor in a two-port circuit configuration.

[1]Uniformity herein means that the family of curves of i_C versus v_{CE} as a function of i_B are approximately straight, evenly spaced, and parallel lines.

$$\left.\begin{array}{l} v_{BE} = h_{11} i_B + h_{12} v_{CE} \\ i_C = h_{21} i_B + h_{22} v_{CE} \end{array}\right\} \qquad (7\text{-}4)$$

but, since the transistor voltages and currents are a combination of d-c and a-c values, we write the following relations for v_{BE}, v_{CE}, i_B, and i_C (refer to Figs. 6-6a, b, c, and d):

$$\begin{array}{l} v_{BE} = V_{BE} + v_{be} \\ v_{CE} = V_{CE} + v_{ce} \\ i_B = I_B + i_b \\ i_C = I_C + i_c \end{array}$$

and substitute these relations into Eq. set 7-4

$$\left.\begin{array}{l} V_{BE} + v_{be} = h_{11}(I_B + i_b) + h_{12}(V_{CE} + v_{ce}) \\ I_C + i_c = h_{21}(I_B + i_b) + h_{22}(V_{CE} + v_{ce}) \end{array}\right\} \qquad (7\text{-}5)$$

Definition. An arbitrary but important definition of the two-port network parameters (such as the h, z, or y parameters) must now be stated. The two-port network parameters are defined only for linear networks and must be independent of the voltages and currents in the network; however, the parameters can be dependent on other variables such as frequency, temperature, and so forth. In Eq. set 7-5, the d-c voltages and currents are not linearly related (refer to the graphical input or output characteristics). If the transistor is properly biased in its linear region of operation, the a-c voltages and currents in Eq. set 7-5 are linearly related. Thus the only way that the h parameters can be defined properly is to consider only the a-c voltage and currents in Eq. set 7-5 and to eliminate the d-c terms, or

$$\left.\begin{array}{l} v_{be} = h_{11} i_b + h_{12} v_{ce} \\ i_c = h_{21} i_b + h_{22} v_{ce} \end{array}\right\} \qquad (7\text{-}6)$$

Although the definition of the h parameters could be obtained from Eq. set 7-6, it is perhaps better to consider another method of solution. Equation set 7-5 can be rewritten in a functional form as

$$\begin{array}{l} v_{BE} = v_{BE}(i_B, v_{CE}) \\ i_C = i_C(i_B, v_{CE}) \end{array}$$

If the total derivative[2] of v_{BE} and i_C is now taken,

$$\left.\begin{array}{l} dv_{BE} = \dfrac{\partial v_{BE}}{\partial i_B} di_B + \dfrac{\partial v_{BE}}{\partial v_{CE}} dv_{CE} \\[3mm] di_C = \dfrac{\partial i_C}{\partial i_B} di_B + \dfrac{\partial i_C}{\partial v_{CE}} dv_{CE} \end{array}\right\} \qquad (7\text{-}7)$$

---

[2]The authors feel that the use of derivatives at this point is desirable in order to explain the development of suitable expressions for the transistor hybrid parameters. If unfamiliar with derivative functions, the reader can skip to Eq. set 7-10 without any real handicap in understanding subsequent material.

Equation 7-7 can be used to provide a better understanding of the h parameters. The definition of the h parameters is

$$
\left.
\begin{aligned}
h_{11} &= \frac{\partial v_{BE}}{\partial i_B} = \frac{\Delta v_{BE}}{\Delta i_B}\bigg|_{v_{CE} \,=\, \text{a constant}} \\[2mm]
h_{12} &= \frac{\partial v_{BE}}{\partial v_{CE}} = \frac{\Delta v_{BE}}{\Delta v_{CE}}\bigg|_{i_B \,=\, \text{a constant}} \\[2mm]
h_{21} &= \frac{\partial i_C}{\partial i_B} = \frac{\Delta i_C}{\Delta i_B}\bigg|_{v_{CE} \,=\, \text{a constant}} \\[2mm]
h_{22} &= \frac{\partial i_C}{\partial v_{CE}} = \frac{\Delta i_C}{\Delta v_{CE}}\bigg|_{i_B \,=\, \text{a constant}}
\end{aligned}
\right\} \tag{7-8}
$$

where the partial derivative ∂ and the total derivative d can be replaced by the incremental Δ variations, since linear operation is assumed. The Δ variations imply that a finite amplitude variation of the transistor voltages and currents occurs about some operating point (the Q point). For these variations to have any meaning, the Q point must be situated in the linear region of operation of the transistor. Furthermore, these Δ variations of the total voltage and currents (v_{BE}, v_{CE}, i_B, and i_C) can be considered as identically equal to the a-c components only (v_{be}, v_{ce}, i_b, and i_c) or

$$
\left.
\begin{aligned}
\Delta v_{BE} &\equiv v_{be} \\
\Delta v_{CE} &\equiv v_{ce} \\
\Delta i_B &\equiv i_b \\
\Delta i_C &\equiv i_c
\end{aligned}
\right\} \tag{7-9}
$$

Equation set 7-7 is seen to be identical to Eq. set 7-6 if the total derivative symbol d is also replaced by the incremental Δ and Eq. sets 7-8 and 7-9 are substituted into Eq. set 7-7. Furthermore, Eq. sets 7-6 and 7-7 can be written in a more useful form if *rms* instead of instantaneous quantities are employed:

$$
\left.
\begin{aligned}
V_{be} &= h_{11} I_b + h_{12} V_{ce} \\
I_c &= h_{21} I_b + h_{22} V_{ce}
\end{aligned}
\right\} \tag{7-10}
$$

Using Eq. set 7-10, the h parameters can also be defined as

$$
\left.
\begin{aligned}
h_{11} &= \frac{V_{be}}{I_b}\bigg|_{V_{ce} = 0} & \qquad h_{21} &= \frac{I_c}{I_b}\bigg|_{V_{ce} = 0} \\[2mm]
h_{12} &= \frac{V_{be}}{V_{ce}}\bigg|_{I_b = 0} & \qquad h_{22} &= \frac{I_c}{V_{ce}}\bigg|_{I_b = 0}
\end{aligned}
\right\} \tag{7-11}
$$

Although the h parameters as defined by Eq. sets 7-8 and 7-11 are the same, Eq. set 7-11 shows more clearly that the h parameters can be obtained from appropriately measured voltages and currents. The h parameters defined in Eq. set 7-11 are identical to those defined in Eq. set 3-26.

The constraint of v_{CE} or i_B being a constant in Eq. set 7-8 is easily interpreted if it is recalled that $v_{CE} = V_{CE} + v_{ce}$ and $i_B = I_B + i_b$. Thus, when v_{CE} or i_B = a constant, v_{ce} and i_b (the a-c components of v_{CE} and i_B, respectively) are equal to zero.

The reader should carefully study the preceding material (Eq. sets 7-4 through 7-11) because one of the most difficult concepts in electronics for the new student lies in being able to see how the a-c and d-c components of the voltages and currents can be separated and treated independently of each other. *We are not using the principle of superposition in separating the a-c and d-c components, since the d-c voltages and currents are not linearly related. We are assuming that the incremental variations of the transistor voltages and currents occur only over a linear portion of the characteristic curves.* The a-c input signals must be small enough so that the incremental voltage and current variations are always linearly related, otherwise the definition of the h parameters indicated in Eq. sets 7-8 or 7-11 is meaningless. If the incremental variations are not linearly related—that is, the transistor is not operated in the linear region of its characteristics—then graphical circuit-analysis techniques must be employed in the problem solution.

Graphical Determination. The question as to how these parameters are determined will now be considered. Although experimental measurements offer a quick and simple solution, the graphical input and output characteristics will be employed to obtain these h parameters. Theoretically, the h parameters can be defined at any desired d-c operating point, but, practically speaking, the d-c operating point (or Q point) must be situated in the linear region of the characteristic curves (this is because the Δ variations are usually taken as relatively large quantities). As an example, we use the NPN characteristics for the 2N3704 transistor of Figs. 6-4b and c repeated here in Fig. 7-3. We choose an operating point such that $V_{CE} = 20$, $I_B = 100$ μa, $V_{BE} = 0.66$ volt, and $I_C = 18$ ma. The operating point (called Q for quiescent) is shown on each part of the figure. To calculate h_{11}, use Fig. 7-3a, and hold v_{CE} constant and vary v_{BE} and i_B around this point. A moment's reflection should indicate that this is the same as finding the slope of a line drawn tangent to the curve $v_{CE} = 20$ at the Q point. Thus

$$h_{11} = \left. \frac{\Delta v_{BE}}{\Delta i_B} \right|_{v_{CE}=20} = \frac{0.695 - 0.63}{(200 - 0)10^{-6}} = 325 \text{ ohms}$$

The next hybrid parameter h_{12} can be obtained from the same input characteristics. This time, i_B must be held constant, as indicated by the horizontal line drawn through the Q point. Unfortunately, it is difficult to obtain an accurate value of h_{12} from these curves (notice the very unequal spacing between the curves as there is a noticeable separation as v_{CE} is varied from 0 to 0.2 volt but from 0.2 to 20 volts there is no discernible

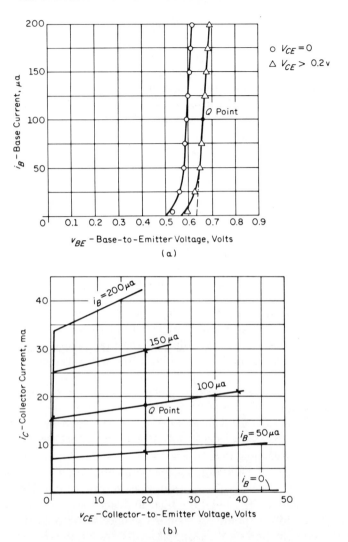

FIG. 7-3. Common-emitter input and output characteristics for the 2N3704 transistors.

separation between curves), but a rough approximation is

$$h_{12} = \frac{\Delta v_{BE}}{\Delta v_{CE}}\bigg|_{i_B = 100} = \frac{0.66 - 0.59}{20 - 0} = 0.0035 = 3.5 \times 10^{-3}$$

However, a value of an order of magnitude less than this is more typical; that is, $h_{12} \cong 3 \times 10^{-4}$.

The two input parameters h_{11} and h_{12} are usually measured by placing the transistor in a circuit designed for this purpose, and no attempt is made to obtain them by graphical means as was done above. However, it is at least instructive to see how they might be obtained graphically if an accurate (and enlarged) set of input characteristics was available.

A graphical determination of the output parameters h_{21} and h_{22} can be made quite accurately by using the common-emitter output or collector characteristics of Fig. 7-3b. The Q point again governs the point on the curves at which the h parameters are to be determined. To find h_{21}, we hold v_{CE} constant at 20 volts and vary i_B and i_C around this point. It is usually convenient to extend the Δ variation to the adjacent i_B curves on either side of the Q point. In this case the vertical line corresponding to $v_{CE} = 20$ is extended until it intersects $i_B = 150$ and 50 μa. Reading the corresponding values of i_C on the vertical axis, we can calculate

$$h_{21} = \left. \frac{\Delta i_C}{\Delta i_B} \right|_{v_{CE}=20} = \frac{(29 - 8)10^{-3}}{(150 - 50)10^{-6}} = 210$$

The last hybrid parameter h_{22} is found by keeping i_B constant. The slope of a line drawn tangent to the $i_B = 100$ μa curve and passing through the Q point yields h_{22}, or

$$h_{22} = \left. \frac{\Delta i_C}{\Delta v_{CE}} \right|_{i_B=100} = \frac{(21.5 - 15.5)10^{-3}}{40 - 0} = 150 \ \mu mhos$$

For purposes of greater accuracy in making the calculation, the tangent line is shown extended from the y-axis to $v_{CE} = 40$ volts.

It would be convenient if it could be assumed that the h parameters were constant for a given transistor or a given set of transistors of the same type. Unfortunately, this is not true; not only do the h parameters for a given transistor vary with a choice in the Q point, but quite often there is a wide variation among transistors of the same type. In other words, the characteristic curves shown in Figs. 6-5 and 7-3 are merely typical and do not necessarily reflect the actual characteristics of a given 2N404 or 2N3704 transistor. This means that, when developing a circuit, the electronic-circuit designer must, if at all possible, allow for these parameter variations.

Much confusion and annoyance may result if the reader now refers to other literature on transistors, because a great variety of symbolism and of definitions of terms will be encountered. Even though there is an IEEE standard on semiconductor symbols, many early and even some later articles and texts employ different notations. The authors have attempted to use only those symbols accepted by the IEEE Standards.

For the common-emitter transistor circuit the hybrid parameters h_{11}, h_{12}, h_{21}, and h_{22} become

$h_{11} = h_{11e} = R_{ie}$ the small-signal (or a-c) common-emitter short-circuit input impedance or resistance

$h_{12} = h_{12e} = \mu_{re}$ the small-signal (or a-c) common-emitter open-circuit reverse-voltage transfer ratio

$h_{21} = h_{21e} = \alpha_{fe}$ the small-signal (or a-c) common-emitter forward short-circuit transfer ratio or gain

$h_{22} = h_{22e} = \dfrac{1}{R_{oe}}$ the small-signal (or a-c) value of the open-circuit output admittance or conductance

$$\left. \right\} \quad (7\text{-}12)$$

Since the e terms in the subscripts in Eq. set 7-12 mean common emitter, many authors drop the e when it is understood that only the common-emitter circuit is under consideration. (The same could be said of the common-base or common-collector circuits.) This is sanctioned by the IEEE Standards, so one must be very careful, when reading articles or texts on transistors, that no confusion arises over terminology.

The reader should note that Eq. sets 7-10 and 7-12 combine to form Eq. 7-3.

7-3. A-C or Small Signal Analysis of Transistor Circuits

Current, Voltage, and Power Gain. In Chapter 6 the current gain A_i, the voltage gain A_v, and power gain A_p were introduced but were not specifically defined. Any circuit gain in a two-port network, whether it is a current, voltage, or power gain, is usually defined as the ratio of an output to an input quantity. Unfortunately, there does not seem to be any definite standard as to how the inputs and outputs are to be defined. The reader is thus again cautioned to be careful when studying not only this text but other material on transistor circuits. This text defines the circuit gains (for the specific circuit of Fig. 7-1b) as follows:

The current gain A_i is

$$A_i \triangleq \frac{I_{out}}{I_{in}} = \frac{I_{out}}{I_b} \qquad (7\text{-}13)$$

but some authors define A_i as $-I_c/I_b$, which is not always the same as Eq. 7-13.

The voltage gain A_v is defined as

$$A_v \triangleq \frac{V_{out}}{E_s} \qquad (7\text{-}14)$$

but some authors define A_v as V_{out}/V_{be} or V_{ce}/V_{be} or V_{ce}/E_s, which are not necessarily the same as Eq. 7-14.

The power gain A_p is defined as

$$A_p \triangleq \frac{V_{out}I_{out}}{E_s I_b} \qquad (7\text{-}15)$$

which is the product of the current and voltage gains as defined in Eqs. 7-13 and 7-14.

Now that the various gains have been defined, it is desired to derive the analytical expression for these gains by using the circuit of Fig. 7-4,

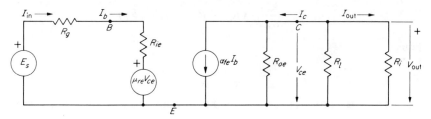

FIG. 7-4. Figure 7-1b repeated.

which is the same as Fig. 7-1b. The derivation of the gain expressions is straightforward and using network reduction methods proceed as follows:

The output voltage V_{out} is

$$V_{out} = V_{ce} = -(\alpha_{fe} I_b) R_T \qquad (7\text{-}16)$$

where R_T is the parallel equivalent of R_{oe}, R_l, and R_i, or

$$R_T = \frac{1}{\dfrac{1}{R_{oe}} + \dfrac{1}{R_l} + \dfrac{1}{R_i}} \qquad (7\text{-}17)$$

The output current I_{out} is

$$I_{out} = -(\alpha_{fe} I_b) \frac{R_b}{R_b + R_i} \qquad (7\text{-}18)$$

where R_b is the parallel equivalent of R_{oe} and R_l, or

$$R_b = \frac{R_{oe} R_l}{R_{oe} + R_l} \qquad (7\text{-}19)$$

Equation 7-18 makes use of the well known current splitting formula, which was developed in Chapter 2.

To solve for the current amplification A_i, we solve for the ratio I_{out}/I_b in Eq. 7-18, or

$$A_i = \frac{I_{out}}{I_b} = -\alpha_{fe} \frac{R_b}{R_b + R_i} \qquad (7\text{-}20)$$

To solve for the voltage amplification A_v, we first substitute V_{ce} from Eq. 7-16 into Eq. 7-3a

$$V_{be} = I_b(R_{ie} - \mu_{re} \alpha_{fe} R_T) \qquad (7\text{-}21)$$

Next, we write an expression for E_s in terms of I_b and V_{be}, or

$$E_s = I_b R_g + V_{be} \qquad (7\text{-}22)$$

Now, substituting V_{be} from Eq. 7-21 into Eq. 7-22,

$$E_s = I_b(R_g + R_{ie} - \mu_{re}\alpha_{fe}R_T) \tag{7-23}$$

As a final step, we substitute I_b from Eq. 7-16 into Eq. 7-23 and solve for V_{out}/E_s

$$E_s = -\frac{V_{out}}{\alpha_{fe}R_T}(R_g + R_{ie} - \mu_{re}\alpha_{fe}R_T)$$

Thus

$$A_v = \frac{V_{out}}{E_s} = \frac{-\alpha_{fe}R_T}{R_g + R_{ie} - \mu_{re}\alpha_{fe}R_T} \tag{7-24}$$

The power gain A_p is defined as the product of the voltage and current gains, or

$$A_p = \frac{V_{out}I_{out}}{E_s I_b} = \frac{(\alpha_{fe})^2 R_b R_T}{(R_g + R_{ie} - \mu_{re}\alpha_{fe}R_T)(R_b + R_i)} \tag{7-25}$$

Input and Output Resistance. Two other circuit parameters are of the utmost importance—the input resistance R_{in} and the output resistance R_{out} of the amplifier shown in Fig. 7-4. By definition, the input resistance R_{in} will be defined at this time as the driving-point resistance looking into the base-emitter terminals, or

$$R_{in} \triangleq \frac{V_{be}}{I_b} \tag{7-26}$$

where R_g is considered a part of the previous stage or amplifier and hence is not a part of the circuit or stage under consideration. Before calculating this resistance, we redraw Fig. 7-4 as Fig. 7-5 but substitute R_T for the

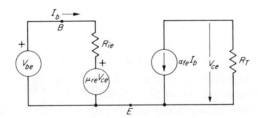

FIG. 7-5. Circuit used for the calculation of the transistor-circuit input impedance R_{in}.

parallel combination of R_{oe}, R_l, and R_i (see Eq. 7-17) and replace E_s and R_g with a new voltage source V_{be}. Because the voltage source $\mu_{re}V_{ce}$ and the current source $\alpha_{fe}I_b$ are either directly or indirectly dependent upon the input current I_b, these two sources must remain in the circuit for the calculation of R_{in}. Equation 7-21 gives the relationship between V_{be} and

I_b so that

$$R_{\text{in}} = \frac{V_{be}}{I_b} = R_{ie} - \mu_{re}\alpha_{fe}R_T \qquad (7\text{-}27)$$

To calculate the output resistance R_{out}, Fig. 7-4 must be redrawn. Figure 7-6a is the same as Fig. 7-4, except that an independent source

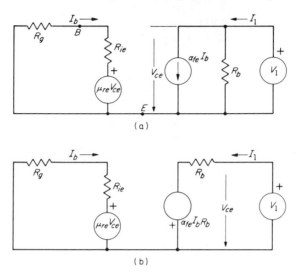

(a)

(b)

FIG. 7-6.   a) Circuit used for the calculation of the transistor-circuit output resistance R_{out}. b) Same as part a, except that a Thévenin generator replaces the Norton generator.

voltage V_1 is assumed to be applied to the output terminals, and the resistance R_i is taken out of the circuit (this is done because R_i is usually considered to be the input resistance of the next stage and is therefore not a part of the stage under consideration) just as R_g was not included in the calculation for the input resistance. The voltage source E_s is independent of V_1 and must be replaced by a short circuit, but the voltage source $\mu_{re}V_{ce}$ and the current source $\alpha_{fe}I_b$ are still either directly or indirectly dependent upon V_1 (note that $V_1 = V_{ce}$ in this circuit) and must remain in the circuit. The output resistance R_{out} is defined as

$$R_{\text{out}} \triangleq \frac{V_1}{I_1} \qquad (7\text{-}28)$$

One way to approach the solution for R_{out} is to replace the Norton generator (consisting of the current source $\alpha_{fe}I_b$ and the shunt resistance R_b) with a Thévenin generator, as in Fig. 7-6b. Next, we use I_b and I_1 as loop currents and write the following two loop equations:

$$(R_g + R_{ie})I_b + \mu_{re}V_{ce} = 0 \qquad (7\text{-}29a)$$

$$R_b I_1 - \alpha_{fe} R_b I_b = V_1 \qquad (7\text{-}29b)$$

Now we substitute I_b from Eq. 7-29a into Eq. 7-29b and also note that $V_1 = V_{ce}$

$$R_b I_1 - \alpha_{fe} R_b \left(\frac{-V_1 \mu_{re}}{R_g + R_{ie}} \right) - V_1 = 0$$

or

$$R_b I_1 = -\left(\frac{\alpha_{fe} \mu_{re} R_b}{R_g + R_{ie}} - 1 \right) V_1$$

Finally, we solve for V_1 / I_1

$$R_{out} = \frac{V_1}{I_1} = \frac{R_b (R_g + R_{ie})}{R_g + R_{ie} - \alpha_{fe} \mu_{re} R_b} \qquad (7\text{-}30)$$

Fortunately, there are two simplifying assumptions that can often (but not always) be made in regard to the circuit parameters. Usually, μ_{re} is quite small ($\mu_{re} \simeq 10^{-4}$ for many transistors) and R_{oe} is usually quite large ($R_{oe} > 5000$ ohms for many transistors). If the assumptions that $\mu_{re} = 0$ and $R_{oe} = \infty$ are made, the gain and resistance equations become

$$A_i = \frac{I_{out}}{I_b} = -\alpha_{fe} \frac{R_l}{R_l + R_i} \qquad (7\text{-}31)$$

$$A_v = \frac{V_{out}}{E_s} = \frac{-\alpha_{fe} R_l R_i}{(R_g + R_{ie})(R_l + R_i)} \qquad (7\text{-}32)$$

$$A_p = A_i A_v = \frac{(\alpha_{fe})^2 R_l^2 R_i}{(R_g + R_{ie})(R_l + R_i)^2} \qquad (7\text{-}33)$$

$$R_{in} = \frac{V_{be}}{I_b} = R_{ie} \qquad (7\text{-}34)$$

$$R_{out} = \frac{V_1}{I_1} = R_l \qquad (7\text{-}35)$$

EXAMPLE 7-1. A 2N3704 NPN transistor is used in the circuit of Fig. 7-1a. A previous graphical solution for the 2N3704 hybrid parameters yielded the following results:

$$h_{11e} = R_{ie} = 325 \text{ ohms} \simeq 300 \text{ ohms}$$
$$h_{12e} = \mu_{re} = 3 \times 10^{-4}$$
$$h_{21e} = \alpha_{fe} = \beta = 210$$
$$h_{22e} = \frac{1}{R_{oe}} = 150 \times 10^{-6} \text{ mho} \quad \text{or} \quad R_{oe} \simeq 7{,}000 \text{ ohms}$$

If $R_g = 2000$ ohms, $R_l = 500$ ohms, and $R_i = 1{,}000$ ohms, find the current, voltage, and power gains and the input and output resistances of the transistor circuit of Fig. 7-1a.

Solution: Since μ_{re} is small and R_{oe} is large (relative to the other resistors on the collector side of the circuit), we use the simplified Eqs. 7-31 through 7-35 for the desired calculations:

$$A_i = -210\frac{500}{500 + 1000} = -70$$

$$A_v = -210\frac{500 \times 1000}{2300 \times 1500} = -30.4$$

$$A_p = A_i A_v = -70(-30.4) = 2130$$

$$R_{in} = 300 \text{ ohms}$$

$$R_{out} = 500 \text{ ohms}$$

7-4. Cascading of Amplifiers

A complete electronic circuit usually consists of many elementary units or stages tied together. The overall circuit behavior can be very complex, and the solution for the various gains can require advanced techniques in analysis. Although these advanced techniques are firmly based on Kirchhoff's two laws and on Ohm's law, a great reduction in time and confusion can be obtained by employing these advanced analytic methods. The purpose of this section is to give the student some "feel" for analyzing a simple cascading of amplifiers using only the more basic analytic methods studied in Chapter 3.

The specific problem to be solved is the finding of the overall circuit gains. Any number of network-analysis methods could be used in obtaining the solution for the circuit gains, but Thévenin's theorem will be chosen as the primary method of analysis. In developing the analysis, we always start at the circuit input and proceed in an orderly manner to the circuit output. The circuit of Fig. 7-7a is a two-stage *R-C* (resistance-capacitance)-coupled common-emitter amplifier with fixed bias. The coupling resistor is R_{l_1}, and the coupling capacitor is C_c. The frequency ω of the sinusoidal input voltage e_s is high enough so that the series capacitive reactances $1/\omega C$ of C_1, C_2, and C_c are negligibly small compared to their associated resistors (for audio amplifiers this usually means frequencies above 100 Hz). Also, the biasing resistances R_{B_1} and R_{B_2} are large enough to be neglected in the equivalent circuit of Fig. 7-7b.

A further and quite helpful simplifying assumption can be made if the voltage generator $\mu_{re} V_{ce}$ can be neglected. Since μ_{re} is usually quite small this assumption is valid as long as $I_b R_{ie} \gg \mu_{re} V_{ce}$ (recall that $V_{be} = I_b R_{ie} + \mu_{re} V_{ce}$). The resistance R_{oe} could also be neglected in many cases because usually $R_{oe} \gg R_l$, but since the elimination of R_{oe} does not materially simplify the circuit, it will not be removed.

After the equivalent circuit is drawn, we must decide where to begin the analysis. Usually, the two nodes separating the first stage from the remaining stages is the obvious choice for "breaking" the circuit. These two

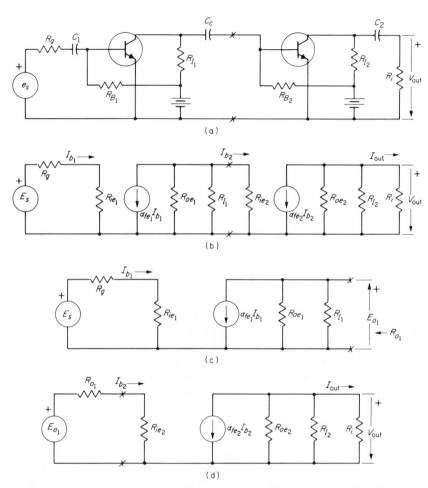

FIG. 7-7. a) Two-stage R-C-coupled common-emitter amplifier. b) Somewhat simplified ($\mu_{re} = 0$, R_{B_1}, and $R_{B_2} = \infty$, $1/\omega C = 0$ for C_1, C_2, and C_c) a-c equivalent circuit for part a. c) First stage of part b isolated in order to calculate the Thévenin equivalent circuit at terminals XX. d) Part b repeated, except that the first stage is replaced by an equivalent Thévenin generator.

nodes are marked XX in Figs. 7-7a and b. Thévenin's theorem will be employed as the primary analytic tool. To use Thévenin's theorem, we remove the circuit to the right of terminals XX and leave the circuit to the left, as shown in Fig. 7-7c. Next, we calculate the open-circuit voltage E_{o_1} and the output resistance R_{o_1}.

Since $\mu_{re} V_{ce}$ has been neglected, the output resistance R_{o_1} is obtained by removing the current source $\alpha_{fe} I_b$, which is now independent of V_{ce} since I_b is independent of V_{ce}, and calculating the resistance seen "looking into" the open-circuited terminals XX, or R_{o_1} is seen to be R_{oe_1} and R_{l_1} in parallel

$$R_{o_1} = \frac{R_{oe_1}R_{l_1}}{R_{oe_1} + R_{l_1}} \qquad (7\text{-}36)$$

where $R_{o_1} = R_b$ of stage 1 as defined by Eq. 7-19. The term R_{o_1} is used simply to emphasize that Thévenin's equivalent circuit is being used.

The open-circuit terminal voltage E_{o_1} is the product of the current source $\alpha_{fe_1}I_{b_1}$ and the parallel combination of R_{oe_1} and R_{l_1}, or

$$E_{o_1} = -(\alpha_{fe_1}I_{b_1})\frac{R_{oe_1}R_{l_1}}{R_{oe_1} + R_{l_1}}$$
$$= -(\alpha_{fe_1}I_{b_1})R_{o_1} \qquad (7\text{-}37)$$

where the negative sign is attached simply because of the way that E_{o_1} is defined. The reader should remember that the choice of current directions and many of the voltage polarities is completely arbitrary when beginning a problem, but, after the choices have been made, strict adherence to the voltage-polarity signs and current directions must be maintained throughout the analysis.

The current I_{b_1} in Eq. 7-37 can be replaced by

$$I_{b_1} = \frac{E_s}{R_g + R_{ie_1}} \qquad (7\text{-}38)$$

and

$$E_{o_1} = \frac{-\alpha_{fe_1}R_{o_1}}{R_g + R_{ie_1}}E_s \qquad (7\text{-}39)$$

Thus E_{o_1} can be expressed in terms of either the input current I_{b_1} (Eq. 7-37) or the input voltage E_s (Eq. 7-39).

The Thévenin equivalent circuit of the network to the left of terminals XX is now placed across these terminals, as shown in Fig. 7-7d. The circuit of Fig. 7-7d is similar in form to the circuit of Fig. 7-7b, except that a Thévenin generator has replaced the first stage. The output voltage V_{out} is

$$V_{\text{out}} = -\alpha_{fe}I_{b_2}R_T \qquad (7\text{-}40)$$

where R_T is the parallel equivalent of R_{oe_2}, R_{l_2}, and R_i as defined in Eq. 7-17.

The input current I_{b_2} is

$$I_{b_2} = \frac{E_{o_1}}{R_{o_1} + R_{ie_2}} \qquad (7\text{-}41)$$

The output current I_{out} is

$$I_{\text{out}} = -\alpha_{fe_2}I_{b_2}\frac{R_{b_2}}{R_{b_2} + R_i} \qquad (7\text{-}42)$$

where

$$R_{b_2} = \frac{R_{oe_2} R_{l_2}}{R_{oe_2} + R_{l_2}}$$

Equation 7-42 is identical in form to Eq. 7-18.

The current, voltage, and power gains of the two-stage R-C-coupled circuit can now be calculated. The current gain A_i is defined as

$$A_i = \frac{I_{\text{out}}}{I_{b_1}} \tag{7-43}$$

To evaluate A_i, we substitute E_{o_1} from Eq. 7-37 into Eq. 7-41, or

$$I_{b_2} = \frac{-\alpha_{fe_1} R_{o_1} I_{b_1}}{R_{o_1} + R_{ie_2}} \tag{7-44}$$

Then we substitute I_{b_2} from Eq. 7-44 into Eq. 7-42

$$I_{\text{out}} = \frac{\alpha_{fe_2} \alpha_{fe_1} R_{o_1} R_{b_2}}{(R_{o_1} + R_{ie_2})(R_{b_2} + R_i)} I_{b_1} \tag{7-45}$$

and thus the current gain A_i is

$$A_i = \frac{I_{\text{out}}}{I_{b_1}} = \frac{\alpha_{fe_1} \alpha_{fe_2} R_{o_1} R_{b_2}}{(R_{o_1} + R_{ie_2})(R_{b_2} + R_i)}$$

$$= \frac{-\alpha_{fe_1} R_{o_1}}{R_{o_1} + R_{ie_2}} \times \frac{-\alpha_{fe_2} R_{b_2}}{R_{b_2} + R_i} \tag{7-46a}$$

$$A_i = A_{i_1} A_{i_2} \tag{7-46b}$$

where Eq. 7-46 shows that the overall current gain of the two stages is the product of the individual current gains of each stage.

The voltage gain A_v is defined as

$$A_v = \frac{V_{\text{out}}}{E_s} \tag{7-47}$$

To derive the expression for A_v, we substitute I_{b_2} from Eq. 7-41 into Eq. 7-40

$$V_{\text{out}} = \frac{-\alpha_{fe_2} R_T}{R_{o_1} + R_{ie_2}} E_{o_1} \tag{7-48}$$

Next, we substitute E_{o_1} from Eq. 7-39 into Eq. 7-48

$$V_{\text{out}} = \frac{\alpha_{fe_2} \alpha_{fe_1} R_T R_{o_1}}{(R_{o_1} + R_{ie_2})(R_g + R_{ie_1})} E_s \tag{7-49}$$

and solving for V_{out}/E_s,

$$A_v = \frac{V_{out}}{E_s} = \frac{\alpha_{fe_1}\alpha_{fe_2}R_{o_1}R_T}{(R_{o_1} + R_{ie_2})(R_g + R_{ie_1})} = \frac{-\alpha_{fe_1}R_{o_1}}{R_g + R_{ie_1}} \times \frac{-\alpha_{fe_2}R_T}{R_{o_1} + R_{ie_2}} \quad (7\text{-}50a)$$

$$A_v = A_{v_1}A_{v_2} \quad (7\text{-}50b)$$

where Eq. 7-50b shows that the overall voltage gain of the two stages is the product of the individual voltage gains of each stage.

The power gain A_p is the product of A_i and A_v, or

$$A_p = A_iA_v = \frac{(\alpha_{fe_1}\alpha_{fe_2})^2 R_{o_1}^2 R_{b_2} R_T}{(R_{o_1} + R_{ie_2})^2(R_{b_2} + R_i)(R_g + R_{ie_1})} \quad (7\text{-}51)$$

Note that, because of the two stages of amplification, both the voltage and the current gains have a zero-degree phase shift between the input and the output.

EXAMPLE 7-2. Two 2N3704 transistors are used in the circuit of Fig. 7-7a. If $R_{l_1} = R_{l_2} = R_i = 1000$ ohms, $R_g = 800$ ohms, $R_{ie} = 300$ ohms, $\alpha_{fe} = 200$, and $R_{oe} = 7000$ ohms, find the current, voltage, and power gains of the two-stage amplifier.

Solution: The current gain A_i can be obtained from Eq. 7-46a

$$A_i = \frac{(\alpha_{fe})^2 R_{o_1} R_{b_2}}{(R_{o_1} + R_{ie_2})(R_{b_2} + R_i)}$$

where

$$R_{o_1} = R_{b_2} = \frac{7000 \times 1000}{7000 + 1000} = 875 \text{ ohms}$$

for this particular circuit. Thus

$$A_i = \frac{(200)^2(875)^2}{(875 + 300)(875 + 1000)}$$

$$= 13,900$$

The voltage gain can be obtained from Eq. 7-50a and from the relation

$$R_T = \frac{1}{\dfrac{1}{R_{oe_2}} + \dfrac{1}{R_{l_2}} + \dfrac{1}{R_i}} = \frac{1}{\dfrac{1}{7 \times 10^3} + \dfrac{1}{1 \times 10^3} + \dfrac{1}{1 \times 10^3}} = 470 \text{ ohms}$$

$$A_v = \frac{(200)^2 \times 875 \times 470}{(800 + 300)(875 + 300)}$$

$$= 12,700$$

The power gain is the product of A_i and A_v, or

$$A_p = A_iA_v$$

$$= 177 \times 10^6$$

This illustrates that a very small signal can be amplified to a quite large and useful signal with only a few stages of amplification. For instance, if the input signal level is measured in a few microwatts (as is usually found to be true for the output of a phonograph pickup cartridge), the output is a sizable signal level measured in watts, which, when fed to a speaker, would produce a loud audible signal.

7-5. The Emitter-Follower or Common-Collector Circuit

Although the BJT is most commonly used in the common-emitter configuration, the transistor can be connected in the common-base or common-collector mode also. As mentioned in Chapter 6, the transistor is seldom used in its common-base configuration simply because there are more advantages in using the BJT in its common-emitter or common-collector configuration. The common-emitter circuit takes advantage of the inherent current amplification property of the transistor when used in this mode, but the common emitter amplifier has the somewhat undesirable characteristics of a low input resistance coupled with a rather high output resistance.

The common-collector circuit as shown in Fig. 7-8a (often termed the emitter follower) greatly increases the input resistance V_i/I_b of the transistor circuit and greatly reduces the output resistance; however, the voltage gain of V_{out}/V_{in} of the emitter follower is less than unity, which

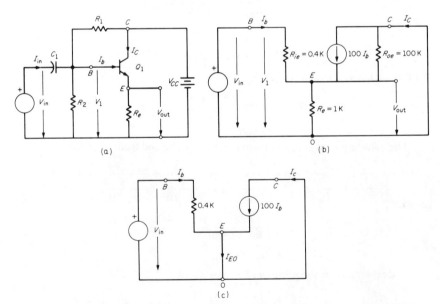

Fig. 7-8.   a) Emitter-follower circuit to accompany Example 7-3.   b) Simplified a-c equivalent of part a with values given to the transistor parameters and R_e. c) Part b modified for the calculation of the short-circuit current.

has to be considered an undesirable quality. Thus, the emitter follower complements the common-emitter amplifier very nicely. The circuit of Fig. 7-8a is considered to be in the common-collector mode because the collector is at the ground or zero (or reference) potential as far as the a-c signals are concerned. An example problem should illustrate the salient properties of the emitter follower.

EXAMPLE 7-3. Using the emitter-follower circuit shown in Fig. 7-8a determine a) the voltage gain $A_v = V_{out}/V_{in}$, b) the input resistance defined as V_1/I_b, c) the current gain $A_i = I_c/I_b$, and d) the output resistance.

Solution: First substitute the a-c equivalent circuit for Q_1, where $1/\omega C_1$ is considered small and R_1 and R_2 are considered large. The simplified equivalent circuit with typical values given to the various parameters is given in Fig. 7-8b.
Using the ground node 0 as a reference and noting that node 0 is the same as node C and $V_{EO} = V_{out}$ write a node voltage equation at node E

$$\frac{V_{out} - V_{in}}{0.4} + \frac{V_{out}}{1} - 100 I_b + \frac{V_{out}}{100} = 0 \qquad (7\text{-}52)$$

where it is recognized that all currents are in milliamperes since the resistors are expressed in Kohms. Also,

$$V_1 = V_{in}$$

and

$$I_b = \frac{V_{in} - V_{out}}{0.4} \text{ ma} \qquad (7\text{-}53)$$

Substituting in the value for I_b from Eq. 7-53 into Eq. 7-52 and solving for V_{out} in terms of V_{in} obtains

$$\left(\frac{1}{0.4} + \frac{1}{1} + \frac{1}{100}\right) V_{out} - \frac{V_{in}}{0.4} - \frac{100(V_{in} - V_{out})}{0.4} = 0$$

$$\left(\frac{100}{0.4} + \frac{1}{0.4} + \frac{1}{1} + \frac{1}{100}\right) V_{out} = \left(\frac{100}{0.4} + \frac{1}{0.4}\right) V_{in}$$

$$253.5\, V_{out} = 252.5\, V_{in} \qquad (7\text{-}54)$$

a) The voltage gain is

$$A_v = \frac{V_{out}}{V_{in}} = \frac{252.5}{253.5} = 0.996$$

b) To solve for the input resistance substitute V_{out} from Eq. 7-54 into Eq. 7-53 and solve for

$$\frac{V_{in}}{I_b} = \frac{V_1}{I_b}$$

$$I_b = \frac{V_{in} - \frac{252.5}{253.5} V_{in}}{0.4} = \frac{1.0}{253.5 \times 0.4} V_{in}$$

thus

$$R_{in} \triangleq \frac{V_{in}}{I_b} = \frac{101.4}{1.0} = 101.4 \text{K}\Omega$$

At first glance the input resistance of the circuit shown in Fig. 7-8b might appear to be $0.4 + 1 = 1.4\text{K}\Omega$; hence the value of $101.4\text{K}\Omega$ might be quite a surprise. This calculation emphasizes that many interesting and useful results can be achieved with the use of the active circuits of electronics. The emitter follower is a good example of feedback, which is a subject unto itself and is discussed in more detail in Chapter 21.

c) The current gain

$$A_i = \frac{I_c}{I_b} \simeq 100$$

if the current through R_{oe} is assumed to be negligibly small. The reader should prove that this assumption is warranted. [*Hint:* $I_c = 100I_b - (V_{out}/100,000)$ amp.]

d) The output resistance R_o can be determined in two ways. One method is by placing a voltage excitation across the V_{out} terminals $E\text{-}O$, replacing the excitation V_{in} with a short circuit, and then solving for the driving point resistance "looking into" the output terminals $E\text{-}O$. Instead, since the open-circuit voltage V_{out} has already been solved for, the second method is quicker. To use the second method the short-circuit collector current must be obtained since it will be recalled from the material in Chapter 3 that the output or Thévenin equivalent resistance R_o is

$$R_o = \frac{V_{E\text{-}O} \text{ (open circuit)}}{I_{E\text{-}O} \text{ (short circuit)}} \tag{7-55}$$

To find the short-circuit output current place a short across terminals $E\text{-}O$ and determine the resulting current through the short. The new circuit for this calculation is shown in Fig. 7-8c (R_{oe} is neglected because the short circuit placed across terminals $E\text{-}O$ also puts a short circuit across terminals CE). Using K.C.L. at node E

$$I_{EO} \text{(short circuit)} = I_b + 100I_b = 101I_b$$

where

$$I_b = \frac{V_{in}}{0.4} \text{ ma}$$

Thus

$$I_{EO} \text{(short circuit)} = \frac{101}{0.4} V_{in} \text{ ma} \tag{7-56}$$

Combining Eqs. 7-54 and 7-56 with Eq. 7-55 obtains

$$R_o = \frac{\dfrac{252.5}{253.5} V_{in}}{\dfrac{101}{0.4} V_{in}} = \frac{101}{253.5 \times 101} K\Omega$$

$$\simeq 4 \text{ ohms}$$

Although it may not be apparent, the ability of the emitter follower to match resistances between two electronic circuits (such as 101.4 KΩ to 4 Ω) has some very valuable applications, such as, matching the output resistance of a power amplifier to a 4 Ω speaker in order to transfer maximum power from the amplifier into the speaker. (Recall the maximum power transfer theorem presented in Chapter 3.)

Problems

7-1. Determine the common-emitter h parameters for a 2N404 transistor with $V_{CE} = 10$ volts, $I_B = -50 \, \mu a$. What are the values for I_C and V_{BE}?

7-2. Determine the common-emitter h parameters for a 2N3704 transistor with $V_{CE} = 15$ volts and $I_B = 100 \, \mu a$.

7-3. Plot $\alpha_{fe} = \beta$ versus V_{CE} for $I_B = 50 \, \mu a$ for the 2N3704 transistor. Comment on the results.

7-4. Plot R_{ie} versus I_B for $V_{CE} > 2$ volts for the 2N3704 transistor. Comment on the results.

7-5. Repeat Probs. 7-3 and 7-4 for the 2N404 transistor. Make appropriate sign changes for the currents and voltages.

7-6. In the circuit of Fig. 7-1a, the circuit and NPN transistor parameters are as follows:

$R_{ie} = 400$ ohms	$R_g = 1000$ ohms
$\mu_{re} = 10^{-3}$	$R_l = 2500$ ohms
$\alpha_{fe} = 55$	$R_i = 4000$ ohms
$R_{oe} = 50,000$ ohms	$V_{CC} = 28$ volts

If $I_C = 4$ ma, $I_B = 100 \, \mu a$, and the capacitive reactances are negligible, find a) the sum of the two resistors $R_{B_1} + R_{B_2}$, b) the current gain A_i, c) the voltage gain A_v, d) the power gain A_p, e) the output voltage V_{out} if $e_s = 0.2 \sin \omega t$, f) the input impedance Z_{in}, and g) the output impedance Z_{out}. (Recall $Z = R$ in resistive circuits.)

7-7. In the common-emitter amplifier circuit of Fig. 6-21, the circuit and NPN transistor parameters are as follows:

$R_{ie} = 800$ ohms	$R_g = 1500$ ohms
$\mu_{re} = 10^{-4}$	$R_l = 3000$ ohms
$\alpha_{fe} = 60$	$C_1 = 10 \, \mu f$
$R_{oe} = 100,000$ ohms	$\omega = 10^5$ radians/sec

If $R_B = 300,000$ ohms and $I_B = 80 \, \mu a$, find a) V_{CC}, b) the current gain A_i, c) the voltage gain A_v, d) the power gain A_p, e) the input impedance Z_{in}, and f) the output impedance Z_{out}. (Recall $Z = R$ in resistive circuits.) Cite all assumptions made in the analysis.

7-8. If $e_s = 0.3 \sin \omega t$ in Prob. 7-7, find a) the output voltage v_{out}, b) the a-c power in R_l, and c) the power in R_g.

7-9. Find the current, voltage, and power gains of the two-stage R-C-coupled common-emitter amplifier shown in Fig. 7-7a if $R_{B_1} = R_{B_2} = 200$ K, $R_g = R_{l_1} = R_{l_2} = R_i = 1800$ ohms, $X_{C_c} = X_{C_1} = X_{C_2} = 5$ ohms, and the transistors are identical with $\mu_{re} = 2 \times 10^{-4}$, $R_{ie} = 300$ ohms, $\alpha_{fe} = 40$, and $R_{oe} = 50$ K. Cite all assumptions made in the analysis.

7-10. Find the current, voltage, and power gains of the two-stage R-C-coupled common-emitter amplifier shown in Fig. 7-7a if $R_{B_1} = R_{B_2} = 100$ K, $R_g = R_{l_1} = R_{l_2} = R_i = 1500$ ohms, X_{C_1}, X_{C_c} and $X_{C_2} < 10$ ohms. The first transistor has a $\mu_{re} = 2 \times 10^{-4}$, $R_{ie} = 900$ ohms, $\alpha_{fe} = 35$, and $R_{oe} = 120$ K. The second transistor has a $\mu_{re} = 10^{-4}$, $R_{ie} = 1200$ ohms, $\alpha_{fe} = 115$, and $R_{oe} = 60$ K. Cite all assumptions made in the analysis.

7-11. A 2N334 NPN transistor is used in the circuit of the accompanying figure. If $R_{ie} = 400$ ohms, $\alpha_{fe} = 46$, $R_{oe} = 115$ K, and $e_s = 0.2 \sin 10^5 t$, find a) the current gain, b) the voltage gain, and c) the output voltage V_{out}. (The "equivalent resistance" for an inductor is ωL ohms and for a capacitor is $1/\omega C$ ohms.) Cite all assumptions made in the analysis.

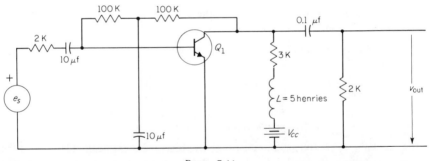

PROB. 7-11

7-12. Find the input and output resistance of the circuit used in Prob. 7-11. (*Hint:* The "equivalent resistance" of an inductor is ωL ohms and for a capacitor is $1/\omega C$ ohms. Note how the capacitors can be replaced by a short circuit and the inductors can be removed or replaced by an open circuit.)

7-13. Neglecting the effects of R_1, R_2 and C in the circuit shown in Fig. 7-8a, determine the symbolic expression for A_v, R_{in}, A_i and R_{out} for parts a, b, c, and d of Example 7-3. Use the simple model for the transistor by neglecting both μ_{re} and R_{oe}. Use R_{ie}, β, and R_e in the derivation of the expression or formula.

7-14. Determine the driving point resistance as seen by V_{in}, i.e., $R_{in} =$

V_{in}/I_{in}, in Example 7-3 if $R_1 = 60$ KΩ, $R_2 = 30$ KΩ and the remainder of the circuit parameters are as given in Example 7-3.

7-15. Calculate the current gain $A_i = I_{meter}/I_1$ for the single-stage thermistor circuit given in Design Example 6-1 of Chapter 6. Use a-c equivalent circuit-analysis methods and compare with the results obtained by graphical means.

7-16. Repeat Prob. 7-15 for the two-stage amplifier given in Design Example 6-2 of Chapter 6.

7-17. The accompanying emitter follower (or a common-collector amplifier because the collector is at ground or zero potential as far as the a-c signals are concerned) has transistor h parameters for Q_1 as follows:

$$R_{ie} = 1500\Omega \qquad \beta = 50$$

$$\mu_{re} = 2 \times 10^{-4} \qquad \frac{1}{R_{oe}} = 8\,\mu\text{mhos}$$

Determine a) the current gain A_i, b) the voltage gain A_v, c) the input resistance $R_{in} = V_{in}/I_{in}$, and d) the output resistance.

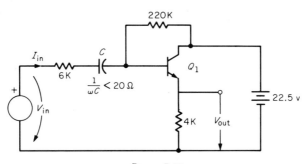

PROB. 7-17

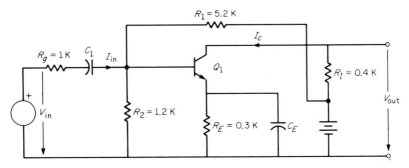

PROB. 7-18

7-18. A 2N3704 transistor is used for Q_1 in the circuit of the accompanying figure. If R_{ie} = 325 ohms, μ_{re} = 2 × 10^{-4}, β = 200 and R_{oe} = 60 K determine: a) $A_v = V_{out}/V_{in}$; b) $A_i = I_c/I_{in}$; c) A_p; d) $R_{in} = V_{in}/I_{in}$; e) R_{out}. f) If the minimum expected frequency for V_{in} is 100 radians/sec, what are the minimum capacitive values for C_1 and C_E if their respective reactances are to be neglected in the analysis and the amplifier is to be operated in the common-emitter mode? Cite all assumptions made in the analysis.

7-19. Repeat Prob. 7-18 if the capacitor C_E is removed from the circuit. Although the reader will no doubt observe the differences in the gains and resistances between Probs. 7-18 and 7-19, it should be pointed out that the temperature compensating effects of the hybrid-biasing circuit remain essentially the same for the two circuits.

8 / Field-Effect Transistors, MOSFET's, and Integrated Circuits

One of the major objections to the characteristics of the bipolar junction transistor studied in Chapters 6 and 7 is the low base-to-emitter input resistance. This objection was much more strongly voiced in the early period of transistor application than it is now because then all electronic circuit designers were accustomed to designing circuits with vacuum tubes, which have a very high input resistance of many megohms. The vacuum tube will be discussed in Chapters 9 and 10.

The field-effect transistor (FET) also has a very high input resistance of 10^{12} to 10^{14} ohms which is the same order of magnitude as the input resistance to the vacuum tube. The FET and the BJT thus complement each other very nicely. It must be emphasized that both the BJT and FET have areas of application in which one device is clearly superior over the other but there are other applications where the decision to use one or the other is not so easy to make.

There are two general categories of FET's; namely, the junction and the metal-oxide-semiconductor types. The conventional junction field-effect transistor is known as the JFET, or more simply the FET. The metal-oxide semiconductor type is called the MOSFET and is the most widely used field-effect transistor. The advent of large-scale integrated (LSI) circuits is the primary reason for all of the interest in MOSFET devices. It is now possible to achieve a denser arrangement of MOSFET devices over BJT devices in integrated circuits by almost an order of magnitude. The higher placement density coupled with lower manufacturing costs make the MOSFET a very attractive transistor, particularly where LSI is concerned because LSI circuits are used in increasing quantities in the very compact (and hopefully popular) digital computers.

8-1. FET Characteristics

Basically there is very little difference in the behavior of the JFET and the MOSFET. Both transistors operate on the principle of using electric fields to control the flow of current as opposed to using carrier (electrons or holes) injection as in the case of the BJT. In other words, there is no input current to a field-effect transistor which means that the FET is a voltage-controlled device as compared to the BJT, which is a current-controlled device.

The pictorial diagram of the JFET is shown in Fig. 8-1. The main difference between the JFET and the MOSFET is that the latter device

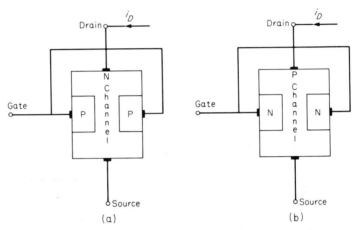

FIG. 8-1. a) N channel and b) P channel junction field-effect transistor.

has a very thin deposit of insulating material of some form of silicon oxide between the N or P gate and the P or N channel regions. (The channel is not formed in the MOSFET until the proper biasing voltages are applied.) This insulation assures that no current can flow between the gate and channel regardless of the voltage polarity on the gate. With the JFET one must be careful to apply the proper voltage polarity to the gate to avoid overheating and destroying the transistor. For the N or P channel JFET the forward bias should never exceed 0.5 or −0.5 volts, respectively (the reader should recall that the forward bias or point at which conduction occurs in the PN silicon junction is a little above 0.5 volts).

The schematic circuit symbols for the JFET and MOSFET are shown in Fig. 8-2. The insulating of the gate from the channel is emphasized by the separation of the gate node and channel on the symbol for the MOSFET.

The MOSFET is manufactured in general for either depletion-mode or enhancement-mode operation. The mode simply refers to whether the applied gate voltage is used to deplete or enhance the concentration of carriers in the channel. Because it now appears that usage of the enhance-

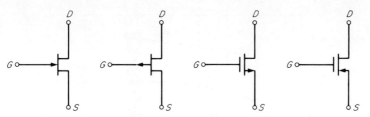

N–Channel JFET   P–Channel JFET   N–Channel MOSFET   P–Channel MOSFET

FIG. 8-2. Schematic diagram for the JFET and MOSFET.

ment-mode MOSFET is going to predominate over the JFET or the depletion-mode MOSFET, the enhancement-mode MOSFET will be emphasized in this chapter. The interested reader is referred to several of the excellent texts on FET's for a detailed discussion of the internal physics of the device.

Terminology. There is little real difference between the BJT and the FET concerning the terminology for the instantaneous and average values for the various circuit variables of current and voltage. A respective substitution of drain, gate, and source for collector, base and emitter in the list of terminology for the BJT given in Chapter 6 gives the appropriate voltage-current terminology for the FET.

JFET. The circuit shown in Fig. 8-3 can be used to obtain the JFET common-source drain characteristics. The drain characteristics for

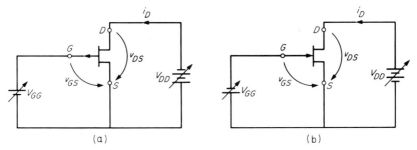

(a)   (b)

FIG. 8-3. a) Proper biasing for the P-Channel JFET. b) Proper biasing for the N-Channel JFET.

the 2N2498 P-Channel diffused planar silicon FET and for the 2N3819 N-Channel planar silicon FET are shown in Fig. 8-4a and 8-4b, respectively. Several comments are in order regarding these characteristic curves. The 2N2498 characteristics are a bit more complete than the 2N3819 characteristics and so will be used to illustrate the important features of the JFET. 1) Note that the gate to source voltage V_{GS} must be a few volts (about 2.5 volts in magnitude for both the 2N2498 and 2N3819) before the drain current, i_D is cut off. 2) Note that the gate-to-

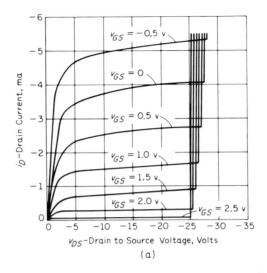

(a)

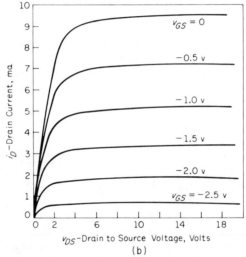

(b)

FIG. 8-4. Common source drain characteristics for the a) P-Channel 2N2498
JFET, b) N-Channel 2N3819 JFET. (Courtesy of Texas Instruments, Inc.)

source voltage can vary on either side of zero, but must not go below
-0.5 volts or else damage to the transistor can result. 3) Note the
rapid increase in drain current when V_{DS} reaches -25 volts. One should
avoid letting V_{DS} exceed 20 volts in magnitude to assure that no damage is
done to the JFET by the rapid rise in current. 4) The incremental
variation of drain current i_D for an incremental variation in gate voltage

V_{GS} is not uniform. Because of these nonuniform variations of i_D versus V_{GS}, the output of the JFET tends to have a high harmonic content as compared to the BJT. In other words, if V_{GS} varies sinusoidally about a given Q point, V_{DS} will contain harmonics (integer multiples of the fundamental) as well as the fundamental frequency of the input sinusoid. This phenomenon is discussed in greater detail in Chapter 13.

8-2. Graphical Analysis of a JFET Amplifier

Because the circuits (see Figs. 8-5a and b) and the graphical analysis of the common-source JFET amplifier are so similar to the common-emitter BJT amplifier, an example problem will be used to explain the analysis procedure.

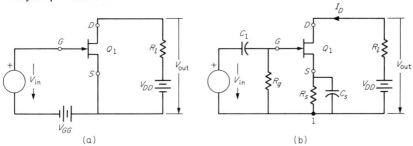

(a)   (b)

FIG. 8-5. Basic common source N-Channel JFET amplifier using a) external fixed-biasing, b) using source self-biasing.

EXAMPLE 8-1. Using the circuit shown in Fig. 8-5b and with Q_1 a 2N3819 JFET determine a) the biasing resistor R_s if $V_{DD} = 16$ volts, $R_l = 2$ KΩ and the Q point is to be located at $V_{GS} = -1.5$ volts; b) suitable values for R_g, C_1, and C_2; c) the voltage gain $A_v = V_{out}/V_{in}$.

Solution: a) The self-biasing scheme employed in Fig. 8-5b is similar to, but not the same as, the temperature compensating emitter-bias circuit presented in Chapter 6. Note that the voltage supply (called a bias voltage) in the gate circuit V_{GG} is opposite in sign to the drain-supply voltage V_{DD}. This means that V_{DD} cannot be used directly to furnish a bias voltage to the JFET as V_{CC} was used to furnish a bias current to the BJT. To see how the resistor R_S in Fig. 8-5b replaces the bias battery V_{GG} in Fig. 8-5a write a voltage equation around mesh $GS1G$ or

$$V_{GS} + V_{S1} + V_{1G} = 0$$

Because the gate current I_G is zero there is no d-c current flowing in R_g, therefore

$$V_{GS} = -V_{S1} = -I_D R_S \tag{8-1}$$

which shows that a negative gate-to-source voltage can indeed be obtained by placing a resistor in the source circuit.

The reader may question the necessity of using C_1, R_g, and C_s in the circuit. From a practical standpoint it is desirable to use capacitive coupling between one stage to the next if at all possible so to eliminate any stray d-c voltages that can disturb the biasing of the transistor, hence the need for C_1 to isolate the source V_{in} from the gate of the JFET. As soon as C_1 is inserted in the circuit it is necessary to include R_g so as to provide the gate with a d-c path to the source. The reason for this is not obvious but stems from the fact that there is a very small reverse-bias leakage current for PN junctions. Thus, if R_g is re-moved ($R_g = \infty$) or is too large the small d-c leakage current can cause a signifi-cant build-up of voltage (by charging capacitor C_1) across the gate-to-source. This can change the bias voltage V_{GS} and cause the Q point to move to an un-desirable location. Acceptable values for R_g vary from 1 to 10 megohms. The capacitor C_s is used to bypass the a-c signal voltages around R_s. This improves the voltage gain A_v markedly as will be seen in the later sections on a-c equivalent circuit analysis.

To continue with the analysis draw a d-c load line on the drain characteristics. The total resistance in the path of the quiescent drain current I_D is $R_l + R_s$, so that the y-axis intercept should be

$$I_D \mid_{V_{DS} = 0} = \frac{V_{DD}}{R_l + R_s} \tag{8-2}$$

but since R_s is not yet known initially assume it to be zero, so that

$$I_D \mid_{V_{DS} = 0} = \frac{16 \text{ v}}{2 \text{ K}\Omega} = 8 \text{ ma}$$

The resulting d-c load line is shown sketched on the 2N3819 drain character-istics of Fig. 8-6.

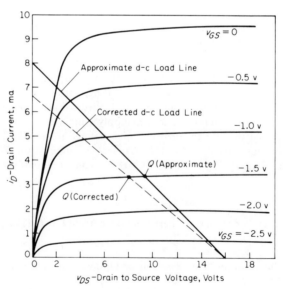

FIG. 8-6. The 2N3819 drain characteristics used to analyze Example 8-1.

The intercept of the d-c load line with a $V_{GS} = -1.5$ volts corresponds to a drain current $I_D = 3.3$ ma. Employing Eq. 8-1

$$R_s = \frac{-(-1.5\ \text{v})}{3.3 \times 10^{-3}\ \text{a}} = 455\ \Omega$$

Since R_s is not insignificant compared to R_l let us substitute this value of R_s into Eq. 8-2

$$I_D\big|_{V_{DS} = 0} = \frac{16\ \text{v}}{2.455\ \text{K}\Omega} = 6.53\ \text{ma}$$

The corrected d-c load line is shown as a dashed curve on Fig. 8-6. As can be seen, neither the Q point nor the quiescent drain current change to any significant degree, but to be more precise the dashed d-c load line will be used to complete the analysis.

Although strictly speaking the asked for solution to part a is complete a word of caution is in order at this point. The drain characteristics for the 2N3819 of Fig. 8-6 are only the *average* characteristics of many such transistors. The actual characteristics of any given transistor may vary by a factor of two, three, or even more and the transistor can still be a "good" operating transistor. (This same precautionary statement was made in reference to the BJT.) The effect of such a variation of the drain characteristics on the location of the Q point is depicted in Fig. 8-7. Using the value of $V_{GS} = -1.5$ volts as the desired Q point results in

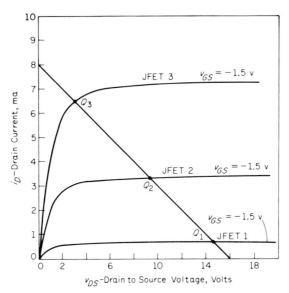

FIG. 8-7. Possible variations of drain characteristics (for one value of V_{GS}) for three different 2N3819 transistors.

widely varying values for R_s. For

JFET 1 $\qquad R_s = \dfrac{-V_{GS}}{I_D} = \dfrac{1.5 \text{ v}}{0.75 \text{ ma}} = 2 \text{ K}\Omega$

JFET 2 $\qquad R_s = \dfrac{1.5 \text{ v}}{3.3 \text{ ma}} = 0.455 \text{ K}\Omega$

JFET 3 $\qquad R_s = \dfrac{1.5}{6.3 \text{ ma}} = 0.238 \text{ K}\Omega$

Although Q_1 and Q_2 are acceptable quiescent points, Q_3 results in a very non-linear or distorted output for a linear variation of V_{GS}. In general Q_3 would represent an unacceptable bias point. The reader should affirm that the same conclusions will be reached if R_s is considered in drawing the d-c load line (Eq. 8-2).

Now suppose one had designed the amplifier under discussion using the characteristics of JFET 1 but JFET 2 was substituted into the circuit. The resulting analysis is shown in Fig. 8-8. The new Q point is obtained by first

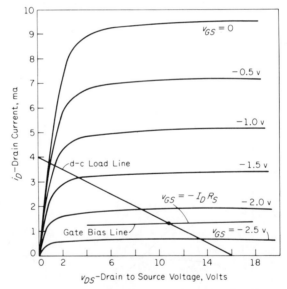

FIG. 8-8. Circuit of Fig. 8-5 analyzed with $R_s = 2$ K.

drawing a d-c load line with a slope of $-1/(R_s + R_l) = -1/(2 \text{ K} + 2 \text{ K}) = -1/4$ K and then drawing a gate bias line. The gate bias line, which is similar to the grid bias curve of the triode (see Chapter 9) or the base bias curve of the BJT, is a bit difficult to find due to the extreme flatness of the drain characteristics but after a few trials it should be evident that the line shown represents the gate bias line defined by $V_{GS} = -I_D R_s$. The Q point is located at $I_D = 1.2$ ma which means that

$$V_{DS} = V_{DD} - (R_s + R_l) I_D$$
$$= 16 - (4 \text{ K}) 1.2 \text{ ma}$$
$$= 11.2 \text{ volts}$$

which is a good check since $V_{DS} \simeq 11$ volts at the Q point according to Fig. 8-8.

The new value for V_{GS} is approximately -2.25 volts which is different from the -1.5 volts when $R_s = 455$ ohms. Although this seems to be a reasonably satisfactory Q point as long as v_{gs} is kept under 0.3-volt peak, it may not be satisfactory if the original circuit allowed v_{gs} to go to 0.7-volt peak or so. The reason for this assessment is that a 0.7-volt peak signal on the gate produced only a slightly nonlinear output voltage v_{ds} for $R_s = 455$ ohms, but for $R_s = 2000$ ohms a considerable amount of nonlinearity in v_{ds} is observed for $v_{gs} = 0.7 \sin \omega t$ volts.

After a rather lengthy aside (or word of caution) let us now consider the solution to the remainder of the problem.

b) A suitable value for R_g is 5 megohms and for input frequencies above 10 Hz C_1 should be such that $1/\omega C_1 \leq 0.1 R_g$ which results in the relationship

$$\omega R_g C_1 \geq 10$$

or

$$C_1 \geq \frac{10}{2\pi \times 10 \times 5 \times 10^6}$$
$$\geq 31.8 \text{ } \eta\text{f (nanofarad)}$$

for satisfactory operation. By a similar calculation let

$$\omega R_s C_s \geq 10$$

$$C_s \geq \frac{10}{2\pi \times 10 \times 455}$$
$$\geq 350 \times 10^{-6} \text{ f}$$
$$\geq 350 \text{ } \mu\text{f}$$

in order to make the self-biasing circuit a satisfactory replacement for the bias voltage source V_{GG}.

c) By letting the gate voltage vary an arbitrary but convenient ± 0.5 volts about the approximate Q point on Fig. 8-6 one obtains

$$A_v = \frac{V_{out}}{V_{in}} = \frac{\Delta v_{DS}}{\Delta v_{GS}} = \frac{11.1 - 4.1}{-2 - (-1)} = \frac{7}{-1} = -7$$

8-3. Enhancement-Mode MOSFET Characteristics

The circuit shown in Fig. 8-9 can be used to obtain the MOSFET drain characteristics of i_D versus v_{DS} with v_{GS} as a parameter. Because of the high gate resistance there is no need to plot the input characteristics of i_G versus v_{GS} for the MOSFET (i_G is considered to be zero in all of the discussion that follows). The same was true of the JFET.

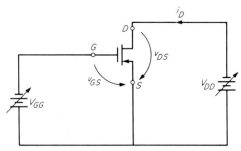

FIG. 8-9. Circuit used to obtain MOSFET
drain characteristics.

The common-source drain characteristics for the 3N160 P-channel insulated gate planar silicon FET, otherwise termed an enhancement-type
MOSFET, are given in Fig. 8-10. Several comments about the curves are
in order. 1) Note that the gate-to-source voltage v_{GS} must be about -4
volts before the transistor begins to show any drain current. This is characteristic of all enhancement-type MOSFET's since the gate-to-source
voltage is used to form the interconnecting P-channel between the P
regions of the drain and source and about -4 volts is needed to form the
channel. In the case of the JFET the channel is formed during the fabrication of the transistor. Once the channel is formed the MOSFET begins
to conduct. As the gate-source voltage v_{GS} increases in a negative sense

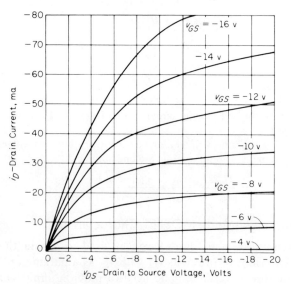

FIG. 8-10. Common-source drain characteristics for
the 3N160 MOSFET. (Courtesy of Texas Instruments, Inc.)

the *P*-interconnecting-channel widens and the transistor drain current increases as expected. 2) Note the extended knee of the curves before drain-current saturation is reached. With the BJT the collector current quickly reached saturation even for quite small collector-emitter voltages. The extended knee of the MOSFET drain characteristics is not desirable as it leads to a greater tendency toward nonlinear behavior. 3) The incremental variation of i_D for an incremental variation of v_{GS} is not as uniform for the MOSFET as it is for the incremental variation of i_C for an incremental variation of i_B for the BJT. This is not a desirable characteristic of the MOSFET as it tends to contribute even further to the nonlinear behavior of the transistor. In other words, the conditions mentioned by points 2 and 3 cause the output voltage v_{DS} not to follow the input voltage v_{GS} in a linear fashion. Thus, the MOSFET output voltage often contains distortion components. This is developed further in Chapter 13. It should be recalled that the JFET has limitations similar to the MOSFET, but because the MOSFET has an insulated gate it can withstand gate voltages that would permanently damage the JFET. For instance, an N-channel JFET can be damaged if $v_{GS} > 0.5$ volts but the N-channel MOSFET is not affected since the silicon oxide insulation prevents any appreciable current. However, the MOSFET is much more susceptible to permanent damage from momentary application of excessive gate-to-source voltages than the JFET, because if the silicon-oxide-insulating layer (also called a dielectric layer) is broken down (by excessive voltages), the MOSFET is permanently damaged. Even static electricity can cause the MOSFET to be so damaged. A loop or band of metal is usually placed around the gate-source-drain leads of the MOSFET to prevent damage by static electricity during handling.

Because the graphical analysis of the common-source MOSFET amplifier is so similar to the common-drain JFET amplifier, an example problem will be used here to explain the analysis.

EXAMPLE 8-2.   Using the circuit shown in Fig. 8-11 determine a) R_1 and R_2 if Q point is to be placed at $V_{GS} = -8$ volts, b) V_{out} if $v_{in} = 2 \sin \omega t$ volts, and c) A_v.

Solution: The d-c load line and the corresponding Q point are shown on the 3N160 MOSFET characteristics of Fig. 8-12. Since $V_{GS} = -8$ volts and the gate current can be assumed equal to zero, V_{GS} is equal to

$$V_{GS} = \frac{-18}{R_1 + R_2} R_2 = -8$$

or

$$\frac{R_2}{R_1 + R_2} = \frac{8}{18}$$

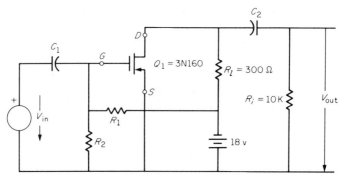

FIG. 8-11. Figure to accompany Example 8-2.

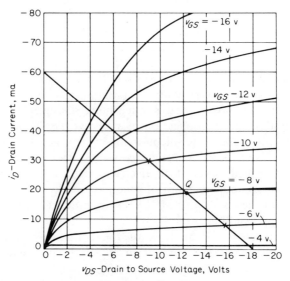

FIG. 8-12. Graphical analysis of Example 8-2.

a) Assuming that it is desired to keep the input resistance of the amplifier circuit high choose R_2 = 5 megohms and solve for R_1

$$5 \times 18 = 8(R_1 + 5)$$

$$R_1 = \frac{50}{8} = 6.25 \text{ megohms}$$

Of course many other combinations of values for R_1 and R_2 would be quite satisfactory for many circuit applications.

b) Since R_i = 10,000 ohms is so much larger than R_l = 300 ohms, let us assume that the a-c and d-c load lines are the same since the parallel equivalent would be essentially 300 ohms. The peak-to-peak value of v_{GS} varies from -10 to -6 volts so that the peak-to-peak variation of v_{DS} is read as approximately -16

to −9 volts, so that

$$v_{out} = \frac{-16 - (-9)}{2} \sin \omega t = -3.5 \sin \omega t \text{ volts}$$

or

$$V_{out} = \frac{3.5}{\sqrt{2}} \text{ volts}$$

c) $A_V = \dfrac{V_{out}}{V_{in}} = \dfrac{3.5/\sqrt{2}}{2/\sqrt{2}} = 1.75$

The voltage gain is rather small because R_l is small. If R_l was selected as large as 3000 ohms the voltage gain would be larger by almost a factor of 10. (The Q point would have to be changed from $V_{GS} = -8$ volts to about -6 or so volts to move the Q point from the saturation region.

The MOSFET as a Load Resistor. As mentioned earlier the MOSFET is an ideal device to use in conjunction with present integrated circuit technology for a variety of reasons (refer to Chapter 11 on Digital Computers and the end of this chapter for a discussion of these reasons). So far all of the emphasis in using BJT's, FET's, or MOSFET's, has been placed on using these devices as amplifiers. Another quite interesting and useful application of the MOSFET is its use as a variable load resistor. One configuration for obtaining the resistance of a MOSFET, such as the 3N160, is to use the circuit of Fig. 8-13a and obtain the static resistance between the drain and source terminals as

$$R_{DS} \triangleq \frac{V_{DS}}{I_D} \text{ ohms}$$

By referring to Fig. 8-13b, which shows the drain characteristics of the 3N160, note that the value of drain current is affected by the gate-source voltage V_{GS}. Let us arbitrarily choose $V_{DD} = 12$ volts in Fig. 8-13a (this means that $V_{DS} = -12$ volts), and draw a vertical line at $V_{DS} = -12$ volts on Fig. 8-13b. Now choose various values of V_{GS} and plot R_{DS} versus V_{GS} as shown on Fig. 8-13c. Between a V_{GS} of −4 volts to −16 volts R_{DS} varies from about 12,000 to 150 ohms. For a magnitude of V_{GS} less than 4 volts the value of R_{DS} increases tremendously (up to several megohms).

Although any MOSFET can be used as a driver (another name for an amplifier) device as well as a load (resistor) device, it is usually good practice to fabricate the two devices somewhat differently when using integrated circuit technology. For instance, the separation between the source and drain for a driver device is about 0.2 mil and is about 10 times this for a load device.

EXAMPLE 8-3. Consider the circuit shown in Fig. 8-14 where the gate of the load MOSFET is connected to its drain. It is desired to find V_{out} if V_{in} is a 2-volt peak-to-peak sine wave and the Q point of the driver MOSFET is to be located at $V_{D_2 S_2} = -8$ volts.

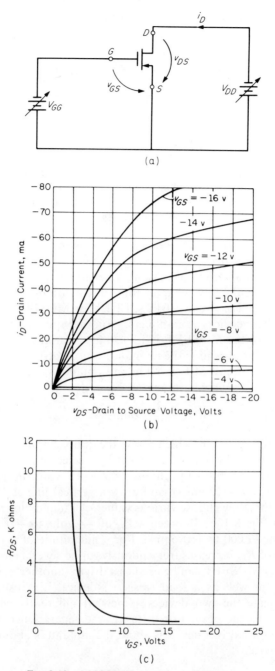

FIG. 8-13. a) MOSFET circuit to determine the static R_{DS}. b) Drain characteristics for a 3N160 P-Channel MOSFET with locus of $V_{DS} = -12$ volts plotted on characteristics. c) R_{DS} versus V_{GS} for $V_{DS} = -12$ volts.

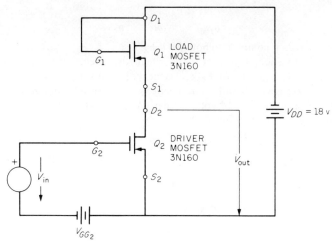

FIG. 8-14. An amplifier using a MOSFET driver and a MOSFET load device.

Solution: First sketch the locus of $V_{D_1 S_1} = V_{G_1 S_1}$ on a set of 3N160 characteristics as shown in Fig. 8-15. This establishes the resistance of the 3N160 used

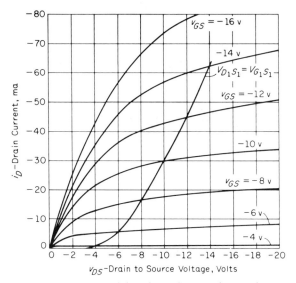

FIG. 8-15. Determining the resistance characteristics of the 3N160 with $V_{DS} = V_{GS}$.

as a load device. Next transfer this load characteristic to the drain characteristics of the 3N160 used as the driver device or amplifier. This is accomplished by recognizing that

$$V_{D_1 S_1} + V_{D_2 S_2} + V_{DD} = 0$$

or

$$V_{D_1 S_1} = -V_{DD} - V_{D_2 S_2}$$

and noting that I_D is common to both MOSFET's.

Now select various values of $V_{D_2 S_2}$ which then determines $V_{D_1 S_1}$ and $V_{G_1 S_1}$ (recall $V_{D_1 S_1} = V_{G_1 S_1}$). Until $V_{D_2 S_2}$ reaches -14 volts $V_{D_1 S_1} > -4$ volts which means $I_{D_1} = I_{D_2}$ is very small. The result is an essentially horizontal load line between $-18 < V_{D_2 S_2} < -14$ volts. When $V_{D_2 S_2} = -12$ volts, $V_{D_1 S_1} = -18 + 12 = -6$ volts and so from Fig. 8-15

$$I_{D_1} = -6 \text{ ma} = I_{D_2}$$

For $V_{D_2 S_2} = -10$ volts, $V_{D_1 S_1} = V_{G_1 S_1} = -8$ volts, and from Fig. 8-15

$$I_{D_1} = -17 \text{ ma} = I_{D_2}$$

For $V_{D_2 S_2} = -6$ volts, $V_{D_1 S_1} = V_{G_1 S_1} = -12$ volts, and from Fig. 8-15

$$I_{D_1} = -45 \text{ ma} = I_{D_2}$$

Now plot these points $(V_{D_2 S_2}, I_{D_2})$ on Fig. 8-16 to obtain the equivalent of a d-c load line for Q_1. Even though the resulting load line is somewhat nonlinear

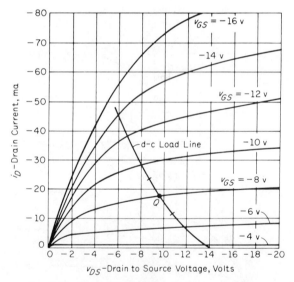

FIG. 8-16. Load characteristics of Q_1 transferred to the driver characteristics of Q_2.

the amplifier output is really quite linear for small variations about a given Q point.

$V_{G_2 G_2}$ was chosen as -8 volts so that the Q point of the amplifier is as indicated on Fig. 8-16. The remainder of the analysis proceeds in a straightforward manner. By varying $V_{in} = \pm 1$ volt about the Q point the peak-to-peak variation of V_{out} (which is the same as $V_{s_1 d_1}$ or $V_{s_1 s_2}$) is seen as about 10.8–8.6 or 2.2 volts.

Thus

$$A_V = \frac{2.2}{2} = 1.1$$

This isn't much of a voltage gain but a considerable power gain is realized from this circuit.

There are many other interesting and useful applications of using MOSFETs as driver and load devices but it is hoped that this one example will be sufficient to encourage the interested reader to pursue the matter further on his own.

8-4. Small-Signal Analysis

In Chapters 6 and 7 it was seen how electronic circuits employing the BJT could be analyzed or designed using both graphical and equivalent circuit methods. By now the similarities of the graphical analysis method between circuits employing the BJT or the FET should be apparent. As might be expected a small-signal a-c equivalent circuit can be found for the FET which allows us to replace the FET symbol with appropriate electronic circuit symbols so that one can use conventional linear circuit analysis methods to effect a solution of the circuit.

Recall that the concept of the equivalent circuit was discussed in Chapter 3 and that in Chapter 7 the h-parameter equivalent circuit for the BJT was utilized. Although it might not be obvious, it is not feasible to try to use an h-parameter equivalent circuit for the FET. This is because the gate current of the FET is small (and in fact is normally assumed to be zero) which means that both h_{11} and h_{21} would not be easy to determine. Instead, the y-parameter equivalent circuit is the natural two-port equivalent circuit to use to describe the FET. The symbol for a JFET is shown in Fig. 8-17 where it is assumed that the proper d-c or quiescent biasing voltages and currents are already applied to the JFET. The biasing point is assumed to be situated within the region of linear operation of the transistor. Although proper biasing is very im-

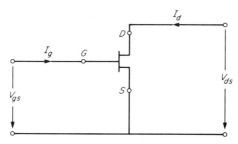

FIG. 8-17. Basic two-port configuration for a JFET.

portant, the a-c voltages and currents that fluctuate about the Q point are of primary interest in an a-c or small-signal analysis of a transistor circuit.

Repeating Eq. set 3-22 in terms of the a-c voltages and currents of the JFET shown in Fig. 8-17 obtains

$$\left.\begin{array}{l} I_g = y_{11} V_{gs} + y_{12} V_{ds} \\ I_d = y_{21} V_{gs} + y_{22} V_{ds} \end{array}\right\} \tag{8-3}$$

where the y-parameters are most easily obtained by making appropriate short circuit tests or

$$\left.\begin{array}{l} y_{11} = \dfrac{I_g}{V_{gs}}\bigg|_{V_{ds}\,=\,0} \\[2em] y_{12} = \dfrac{I_g}{V_{ds}}\bigg|_{V_{gs}\,=\,0} \\[2em] y_{21} = \dfrac{I_d}{V_{gs}}\bigg|_{V_{ds}\,=\,0} \\[2em] y_{22} = \dfrac{I_d}{V_{ds}}\bigg|_{V_{gs}\,=\,0} \end{array}\right\} \tag{8-4}$$

Since I_g is essentially zero for all FET's, both y_{11} and y_{12} are zero and the y-parameter equivalent circuit for the FET is the rather simple circuit shown in Fig. 8-18. The input lead is left open to signify that $I_g = 0$. As

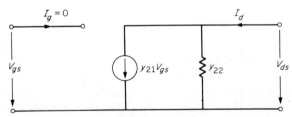

FIG. 8-18. The y-parameter equivalent circuit for the FET (both JFET and MOSFET).

an example of how to determine both y_{21} and y_{22} refer to Fig. 8-19 where it is assumed that the Q point of a 2N2498 JFET is located at the intersection of $v_{GS} = V_{GS} = 1$ volt and $v_{DS} = V_{DS} = -10$ volts. Remember now that the I_d, V_{gs}, and V_{ds} symbols used to define y_{21} and y_{22} in Eq. set 8-4 are the effective or rms values of the incremental or alternating (a-c) variations about a given point. To determine y_{21} first note that $V_{ds} = 0$. This means that there is to be no variation of v_{DS} about the chosen Q point, i.e., draw a vertical line at $v_{DS} = V_{DS} = -10$ volts. For convenience in reading the desired values of V_{gs} and I_d, extend this vertical line to adjacent given values for v_{GS}. Now calculate y_{21} by recognizing

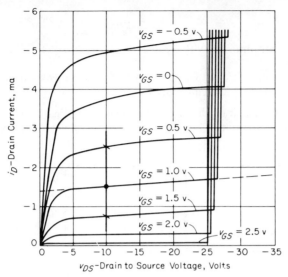

FIG. 8-19. Determining y_{21} and y_{22} from the 2N2498 JFET common-source drain characteristics.

that the a-c voltages and currents are directly related to the incremental variations or

$$y_{21} = \frac{I_d}{V_{gs}}\bigg|_{V_{ds}\,=\,0} = \frac{\Delta i_D}{\Delta v_{GS}}\bigg|_{\Delta v_{DS}\,=\,0}$$

$$= \frac{[-2.6 - (-0.7)] \times 10^{-3}}{0.5 - (1.5)} = \frac{-1.9 \times 10^{-3}}{-1} \text{ mhos}$$

$$= 1900 \ \mu\text{mhos}$$

To evaluate y_{22} draw a line tangent to the $v_{GS} = 1$-volt curve at the Q point. This assures that $V_{gs} = 0$ as is required in the equation for y_{22} in Eq. set 8-4. For greater accuracy in reading the slope of this tangent line, extend the line as far as possible on either side of the Q point and evaluate y_{22} as follows

$$y_{22} = \frac{I_d}{V_{ds}}\bigg|_{V_{gs}\,=\,0} = \frac{\Delta i_D}{\Delta v_{DS}}\bigg|_{\Delta v_{GS}\,=\,0}$$

$$= \frac{[-1.8 - (-1.4)]10^{-3}}{-35 - 0} = \frac{-0.40 \times 10^{-3}}{-35} \text{ mhos}$$

$$= 11.4 \ \mu\text{mhos}$$

Usually y_{21} and y_{22} are redesignated as

$$y_{21} \triangleq g_m \tag{8-5}$$

$$y_{22} \triangleq \frac{1}{r_d} \qquad (8\text{-}6)$$

when used in conjunction with FET's.

EXAMPLE 8-4. Find the voltage gain of the amplifier used in Example 8-1 by using an a-c equivalent circuit analysis.

Solution: First redraw the circuit given in either Fig. 8-5a or Fig. 8-5b in terms of its a-c equivalent. This is shown in Fig. 8-20 where it is recognized that

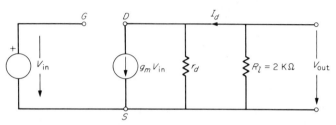

FIG. 8-20. A-c equivalent circuit of Fig. 8-5.

$V_{in} = V_{gs}$, that the capacitors and batteries are replaced by short circuits, and that g_m and r_d are to be determined at the approximate Q point shown on Fig. 8-6. It is left as an exercise for the reader to confirm that

$$g_m \simeq 3100 \ \mu\text{mhos}$$
$$r_d \simeq 60 \ \text{K ohms}$$

Because r_d, the drain resistance, is so much larger than the parallel load resistance R_l, r_d can be neglected in the analysis with little error resulting. The simplified a-c or small-signal equivalent circuit with all parameters identified and evaluated is shown in Fig. 8-21.

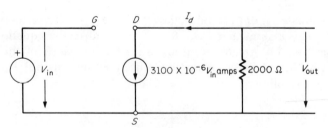

FIG. 8-21. Circuit of Fig. 8-20 completed and simplified for final analysis.

To complete the solution note that

$$
\begin{aligned}
V_{out} &= -I_d \times 2000 \\
&= -3100 \times 10^{-6} V_{in} \times 2000 \\
&= -6.2 V_{in}
\end{aligned}
$$

thus

$$A_v = \frac{V_{out}}{V_{in}} = -6.2$$

which compares quite favorably with the A_v of -7 determined by graphical analysis in Example 8-1.

Additional Remarks. For either the JFET or MOSFET biased properly in the linear region of operation it is normally assumed that

$$R_{in} \triangleq \frac{V_{gs}}{I_g} = \infty$$

r_d The drain resistance is usually between 50 KΩ and 500 KΩ and thus is usually negligible when connected in parallel with the external drain-circuit load resistors.

R_{out} The drain-circuit output resistance must be calculated just as R_{out} was calculated for the BJT circuits. In Example 8-4 $R_{out} = R_l = 2$ K ohms.

The reader is referred to Chapters 9 and 10 for possible additional applications of the FET. In many cases the FET can replace the vacuum tube in a given circuit application.

8-5. Integrated Circuits

Microcircuits, integrated circuits, small-scale integration (SSI), medium-scale integration (MSI), large-scale integration (LSI), silicon-integrated circuits (SIC), and hybrid-integrated circuits (HIC) are some of the terms currently used to describe what promises to be the most exciting of the first three eras of electronics. The first era (1895–1950) was dominated by the development and utilization of the vacuum tube. The second era (1950–1970) was dominated by the use of the transistor as a discrete device along with other discrete circuit elements, such as resistors, capacitors and inductors. The third and present era (1970–?) is and will be dominated by integrated circuits with particular emphasis on large-scale integration (LSI).

Although integrated circuits are miniaturized circuits, it must be stressed that in general a circuit designed for discrete components can not be converted into a miniaturized integrated circuit. The two main reasons for this are costs and manufacturing methods. It turns out that as well as being used as a transistor, the MOSFET can be substituted for or eliminates the use of resistors, capacitors, and diodes in electronic circuits.

New Design Concepts. It was almost a decade after the invention of the transistor before college and university faculties and industrial personnel, involved in the design of electronics, fully accepted and appreciated the impact of transistors on electronic circuit design. The impact of integrated-circuit technology is placing severe demands on the teachers and practitioners of electronic circuit design. The stress is even more severe than that placed on the vacuum tube designer who had to adapt to solid state semiconductor technology. Vacuum tube circuits were and still are designed on the discrete component level. The changeover from discrete vacuum tube circuits to discrete semiconductor circuits was trying but was really not too severe because there was a great deal of similarity between the two circuits; i.e., the transistor replaced the tube but the resistors, capacitors, and inductors continued to play much the same role in either circuit. This is not so in integrated circuits.

In addition to using the MOSFET as a transistor, a diode, and a resistor (the use of MOS devices often enables one to eliminate capacitors but where this is not possible thin film techniques allow capacitors to be fabricated), the designer must be prepared to accept rather large component tolerances or parameter variations. For instance discrete resistors with tolerances of ± 5 percent or ± 10 percent are commonplace and if the designer needs a resistor of ± 1 percent tolerance this can be provided at little extra cost. Such tolerances are not possible in integrated circuit technology unless the bulkier thin-film IC's are used. Using present day SIC technology resistance tolerances of ± 25 percent must be accepted by the designer. Although transistor parameters can vary by a factor of 2 or more from one integrated-circuit device to another, transistors fabricated on one silicon chip tend to be remarkably alike in their characteristics.

Lest the aspiring designer of electronic circuits becomes discouraged and wonder of what use the material in Chapters 6 and 7 is to him in this age of integrated circuits, the authors hasten to add that not only are many circuits still being designed and built using discrete components but that the design methods and ideas presented in these earlier chapters should at least be instructive in understanding integrated-circuit design. One can be sure that Ohm's and Kirchhoff's laws are still valid and that equivalent circuit techniques similar to those studied in Chapter 7 can be used in designing integrated circuits.

Perhaps one of the most difficult aspects of integrated-circuit design is learning how to employ feedback to make the circuit behave properly. The subject of feedback is touched upon in Chapter 21. One of the desirable features of feedback is that it makes a circuit relatively insensitive to parameter changes; thus, resistor tolerances of ± 25 percent can be accommodated. In order to use feedback successfully it is generally necessary that amplification be present in the circuit. Since it is easy to fabricate hundreds or thousands of transistors on a single silicon chip, lavish use of transistors can be made to achieve the amplification necessary to make feedback a useful design procedure.

Although the bipolar integrated circuit is utilized in some electronic designs, the MOS integrated circuit is expected to dominate the IC market in the near future because unipolar MOS devices can be packed much more densely on a silicon chip or substrate than can the bipolar devices. MOS devices lend themselves very nicely to batch production methods which means that cost per device is minimized.

Large-Scale Integration. Although SSI, MSI, thin-film, and hybrid assemblies are and will continue to be designed and fabricated, major attention will probably be focused on LSI circuits and systems for the next decade or so. To be considered large-scale integration a circuit should contain about 1000 or so transistors in a volume of about 2×10^{-3} cubic centimeters. If the circuit contains less than 1000 transistors then it is considered either a medium- or small-scale integrated circuit.

The fabrication of any integrated circuit is no simple matter. The circuit must be designed and a scaled diagram of the circuit produced with great precision on a large piece of paper. This diagram is then photographed and reduced to 1/400 or less of its original size. Approximately 100 closely controlled steps involving photolithographic and chemical processes are then used to make the original schematic or circuit diagram into a practical integrated circuit. These steps include chemical polishing and cleaning of the silicon (or other semiconductor material), oxidation, masking, etching, deposition, alloying, and other processes, some of which are repeated several times during the manufacturing process.

Because of mass production methods which allow many LSI circuits to be processed simultaneously, the unit cost is quite economical. In fact it is the assembling, connecting, and packaging of the LSI chips that accounts for about half of the cost of each unit.

Although LSI has opened the door to a new era in electronics, no one seems to be sure of the impact LSI will actually have on the electronics industry. There is certainly a pressing need and a wonderful opportunity for imaginative designers of electronic circuits and systems. For instance, a present day transistor radio or television may contain from 1 to 100 transistors, but in the near future it will be possible to design electronic systems that contain 10,000 or more transistors at a cost of say $500 to $3000 per system. If these new systems are as attractive or necessary as a television set or automobile to the consumer, then the new system would have a tremendous impact on our economy. In other words, it should be possible to design a very sophisticated and useful device (such as a small digital computer for home use) using 10,000 or more transistors. Only LSI technology can make the device inexpensive enough to be afforded by the average consumer.

Problems

8-1. Plot i_D versus v_{GS} for $v_{DS} = -10$ volts for the 2N2498 JFET. Comment on the linearity of the curve.

8-2. Repeat Prob. 8-1 for the 2N3819 JFET.

8-3. Repeat Example 8-1 with $R_1 = 4 \text{ K}\Omega$ and $V_{GS} = -2$ volts.

8-4. Plot i_D versus v_{GS} for $v_{DS} = -12$ volts for the 3N160 MOSFET. Comment on the linearity of the resulting curve.

8-5. Repeat Example 8-2 if a) $V_{GS} = -12$ volts, and b) $V_{GS} = -14$ volts.

8-6. Plot v_{out} versus ωt for Example 8-2 and for the two values of V_{GS} given in Prob. 8-5. Note that $v_{in} = 2 \sin \omega t$ volts remains a pure sine wave for all three plots of v_{out}, but v_{out} deviates from a pure sine wave. It will be seen in Chapter 13 how any repetitive waveshape can be expressed as a series of harmonically related sine waves by means of a Fourier series. This means that a pure tone of ω radians/second at the input is converted into an output containing $\omega, 2\omega, 3\omega, 4\omega, -n\omega$ frequency components. In a high fidelity circuit the introduction of these harmonics is unacceptable.

8-7. Determine A_i and A_p for Example 8-3 that accompanies Fig. 8-14. (*Hint:* Assume that the input resistance to Q_2 as seen by v_{in} is 10^9 ohms.)

8-8. Repeat Example 8-3 accompanying Fig. 8-14, if Q_1 and Q_2 are 2N3819 JFET's.

8-9. Determine R_{in} (as seen by v_{in}) and R_{out} for the circuit given in Example 8-1.

8-10. Determine R_{in} (as seen by v_{in}) and R_{out} (as seen looking into the terminal across which v_{out} is taken) for the circuit of Example 8-2.

8-11. Calculate A_v for Example 8-2 using equivalent circuit techniques and compare to the value obtained graphically.

8-12. Calculate A_v for the example problem accompanying Fig. 8-14 using equivalent circuit techniques and compare to the value obtained graphically.

9 / *The Vacuum Tube*

There seems to be a widespread belief that the transistor has almost completely replaced the vacuum tube in electronic circuits. This is not the case. Not only are vacuum tubes able to deliver greater output power at higher frequencies than the best transistors available, but they also are much more stable at differing temperatures and are less susceptible to damage when electrically overloaded or subjected to high energy electromagnetic radiation (such as X-rays or gamma rays). For the foreseeable future, both transistors and vacuum tubes will be important components in electronic devices.

Although the transistor was developed comparatively recently (by J. Bardeen and W. H. Brattain of the Bell Telephone Laboratories in 1948), the vacuum tube has enjoyed almost 70 years of development and refinement. In 1905 Lee De Forest added a third electrode to the already existing diode, and this new tube opened the door to the electronic age. This third electrode, called a grid, is usually constructed of a length of fine wire in the form of a spiral and is situated much closer to the cathode than to the plate. The addition of the grid permits a tube to amplify, and the amplification ability of an electronic device is the very heart of almost all electronic equipment. The operation of this three-element tube (called a triode) is similar to the control of water in a pipe by the opening and closing of a valve. In fact, the British word for a vacuum tube is *valve*, which is perhaps a more apt designation for the device.

The analytic methods used in the study of the vacuum-tube characteristics and of its behavior in a circuit are identical to those used in the study of the transistor. The major difficulty that will be encountered is in learning the new symbolism or "language" of vacuum tubes. One might rightly argue that the symbols for voltage, current, and other parameters should be uniform for tubes and transistors, but, unfortunately, this is wishful thinking. The terminology for vacuum tubes is more standardized and more uniformly adhered to by authors than is the case with transistors. The reader is urged to learn this new terminology immediately, for it will be very helpful in understanding this and the following chapters.

The study and application of the vacuum triode will be stressed in this text, but it should be noted that the external characteristics of the multigrid tubes (such as the tetrode or pentode) are used basically in the same way as are the triode characteristics. Although it is assumed that the reader is familiar with the material in the previous eight chapters, knowledge of the material in the first five chapters is all that is required to understand this chapter and the following chapter.

9-1. Triode Construction and Behavior

Figure 9-1 shows the structure of a typical triode. The grids of older tubes were constructed of wire mesh, but most grid structures in modern

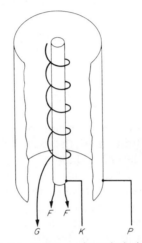

FIG. 9-1. Structure of a typical triode.

tubes are in the form of a fine-wire spiral. The grid is usually placed at the potential minimum (called the *virtual cathode*, as described in Chapter 5) of the tube, and the grid-wire diameter and spiral spacings determine, to a large extent, the tube characteristics. A qualitative explanation of how the triode operates and why it can amplify a signal is as follows: 1) The grid is situated much closer to the cathode than to the anode. 2) Since the force on an electron is $\mathbf{F} = e\mathcal{E}$, a small voltage difference between the grid and the cathode creates a much greater electric-field intensity $\mathcal{E}$ than would be the case if the same voltage were applied between the anode and the cathode (recall that $\mathcal{E}$ = voltage/distance). 3) The net result of statements 1 and 2 implies that a small voltage applied between the grid and the cathode can effectively control the flow of emitted electrons between the cathode and the anode.

9-2. Vacuum Tube Terminology

Since several different currents and voltages are used in discussing vacuum-tube circuits, it is necessary to define clearly the symbols for these terms. In vacuum-tube circuits the symbols E and e are used for voltage, instead of V and v as in transistor circuits. As with transistor symbols, capital I's and E's represent rms or d-c values, while lowercase i's and e's represent instantaneous values. Uppercase subscripts are not used in vacuum-tube circuits; instead, different lower case letters are used to distinguish between total values and the a-c component only. The subscript letters b and p refer to plate-circuit quantities, c and g to grid-circuit quantities, and k to the cathode circuit. A partial list of these symbols, as used in this text, follows:

e_c = instantaneous total voltage rise from cathode to grid

e_g = instantaneous value of the a-c component of the voltage rise from cathode to grid

E_c = average or quiescent value of the voltage rise from cathode to grid

E_{cc} = grid-circuit d-c supply voltage (magnitude only)

E_s = effective or rms value of the a-c input voltage rise

E_g = effective or rms value of the a-c component of the cathode to grid voltage rise

i_b = instantaneous total anode current

i_p = instantaneous value of the a-c component of the anode current

I_b = average or quiescent value of the anode current

I_p = effective or rms value of the a-c component of the anode current

e_b = instantaneous total voltage rise from cathode to anode

e_p = instantaneous value of the a-c component of the cathode to anode voltage rise

E_b = average or quiescent value of the cathode to anode voltage rise

E_p = effective or rms value of the a-c component of the cathode to anode voltage rise

E_{bb} = anode circuit d-c supply voltage, sometimes referred to as the B$^+$ supply (magnitude only)

e_{out} = instantaneous value of the a-c component of the output voltage rise

Notice that the cathode is the reference point for almost all the voltages listed above. It is also well to remember that the word *ground*, when applied to electronic circuits, usually means the negative terminal of the B$^+$ (or E_{bb}) supply voltage.

Figure 9-2 illustrates how some of these voltages and currents are used on a circuit diagram. The grid-circuit d-c supply voltage E_{cc}, which is also known as the grid-biasing voltage, helps to "set up" the tube so that it can operate as a linear device. One very important fact should now be noted: No grid current is shown on the circuit diagram of Fig. 9-2, nor is a definition of a grid current included in the list of symbols just given.

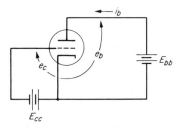

Fig. 9-2. A basic triode circuit used for obtaining the triode characteristics.

The grid current of a vacuum tube, for most tube applications, is assumed to be equal to zero (the grid current is usually less than 1 nanoamp). If the grid current is always zero, the vacuum tube cannot have a meaningful current or power gain but can only have a voltage gain A_v. Furthermore, if the grid current is zero but the grid voltage is some finite value, the input resistance must be infinite $R_{in} = E_g/I_g = \infty$. Because there is no grid current, the grid-cathode and plate-cathode circuits can be considered to be isolated from each other. This results in a much simplified a-c equivalent circuit for the vacuum tube, as will be seen in the next chapter.

From a qualitative standpoint it is easy to see why there is almost no grid current in the usual vacuum-tube circuit. To control properly the flow of electrons, the grid-cathode-bias supply E_{cc} is always negative. Since the grid is negative with respect to the cathode, almost all the electrons emitted by the cathode (the grid is cold and thus emits no electrons, except perhaps by secondary emission) are repelled by the grid. Because there is no transfer of electrons between the cathode and the grid, there is no current flow. Actually, some of the cathode-emitted electrons strike the grid and produce some small current flow, but, as mentioned earlier, this grid current is usually so small as to be negligible. The vacuum tube, just like the field-effect transistor, is thus a voltage controlled device (as opposed to the bipolar junction transistor, which is a current controlled device).

9-3. Triode Characteristics

The behavior of the transistor is frequently described by sets of graphical input and output characteristics. Since the grid current is assumed to be zero for the vacuum tube, input characteristics are not necessary and only output characteristics need be given. The grounded- (or common) cathode connection is the most commonly used circuit configuration for the vacuum tube, and it will be studied first. Figure 9-2 is the basic grounded-cathode circuit from which the output characteristics can be obtained experimentally. As indicated on the diagram, only three instead of four variables are used in vacuum-tube circuits—the grid voltage e_c, the plate voltage e_b, and the plate current i_b. Since there are three variables that can be controlled, there are three possible sets of experimental curves which can graphically describe the output characteristics. These three sets are called the plate, the mutual or transfer, and the constant-current characteristics and are illustrated in Fig. 9-3a, b, and

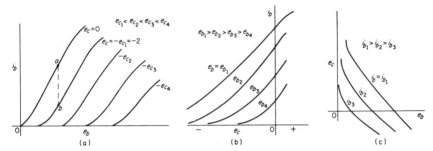

FIG. 9-3. a) Typical triode plate characteristics. b) Typical triode mutual or transfer characteristics. c) Typical triode constant-current characteristics.

c, respectively. The plate characteristics are a family of curves of i_b versus e_b for various constant values of e_c; the transfer-characteristics plot i_b versus e_c for various constant values of e_b; and the constant-current characteristics show e_c versus e_b for constant values of i_b. Probably the most useful of these are the plate characteristics, for they can be used in the same way as can the collector characteristics of the common-emitter transistor; that is, to obtain the vacuum-tube parameters for use in the a-c equivalent circuits and to be used directly in graphical circuit-analysis techniques. Since the three sets of characteristic curves are obtained by varying d-c voltages and currents, they are frequently referred to as *static characteristics*. It is also of interest that the last two sets of characteristics can be obtained from the first set, so that no information is really lost by using only the plate characteristics and ignoring the other two sets.

Referring to the plate characteristics of Fig. 9-3a, the plot of i_b versus e_b for $e_c = 0$ is identical to the vacuum-diode characteristic studied in Chapter 5. This is because the grid is at the same potential as the cathode for $e_c = 0$ (we can consider the grid to be physically connected to the cathode). To argue qualitatively as to why the family of curves shifts to the right as e_c becomes progressively more negative, let us consider point a on the curve for $e_c = 0$ in Fig. 9-3a. If the plate voltage e_b is held constant and e_c is decreased to -2 volts, the grid tends to force many of the cathode-emitted electrons back to the cathode, thereby reducing the plate current to the value indicated by point b on the curve for $e_c = -2$ volts.

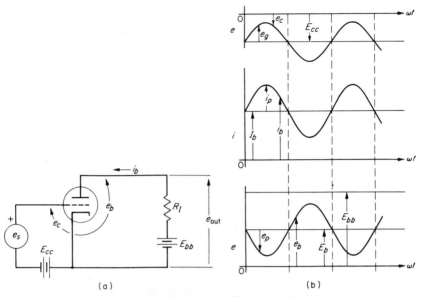

FIG. 9-4. a) The basic grounded-cathode amplifier circuit. b) Voltage-current relations for the circuit of part a, with $e_s = E_m \sin \omega t$.

9-4. The Grounded-Cathode Amplifier

Figure 9-4a shows a grounded-cathode amplifier circuit, which is similar to Fig. 9-2, except that a signal source e_s is added in series with the grid-bias supply E_{cc}, and a plate load resistance R_l is added in series with the plate supply E_{bb}. In this basic circuit diagram it is evident that $e_s = e_g$ and $e_{out} = e_p$, but in general this is not true, as will be seen later.

Several voltage-current equations can be written for the circuit of Fig. 9-4a. For the grid-cathode circuit,

$$e_c = e_s - E_{cc}$$
$$= e_g - E_{cc} \qquad (9\text{-}1)$$

For the plate-to-cathode circuit, some of the voltage-current equations that can be written are as follows:

The plate current i_b:

$$i_b = I_b + i_p \qquad (9\text{-}2)$$

The plate voltage e_b:

$$e_b = E_{bb} - i_b R_l = E_{bb} - (I_b + i_p) R_l$$
$$= (E_{bb} - I_b R_l) - i_p R_l$$
$$= E_b - i_p R_l \qquad (9\text{-}3)$$

where

$$E_b = E_{bb} - I_b R_l \qquad (9\text{-}4)$$

Finally, by definition,

$$e_{\text{out}} = -i_p R_l \qquad (9\text{-}5)$$

For the grounded-cathode circuit the grid voltage and plate current are in phase, while the grid voltage and plate voltage are 180° out of phase. These phase relations change if the load becomes a complex impedance instead of just resistive in nature.

The explanation for the curves of Fig. 9-4b is straightforward. If the grid voltage is assumed to increase in the positive direction (become less negative), the grid allows more electrons to pass to the plate. This causes the plate current to increase and thereby increases the voltage drop across the load resistance R_l. Since the plate-supply voltage E_{bb} is fixed, an increase in the plate-load-resistance voltage drop means that the plate voltage has to decrease.

9-5. Triode Parameters

Before proceeding with the analysis of vacuum-tube circuits, it is well to define the parameters that can be used to describe the external behavior of the tube so that the tube can be replaced by an equivalent circuit. The vacuum-tube dynamic or a-c parameters are obtained in exactly the same manner as were the transistor dynamic h parameters—either directly from the graphical characteristics or from experimental measurements. Figure 9-5 shows the basic two-port network configuration for a vacuum tube.

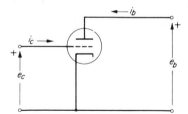

FIG. 9-5. Basic two-port configuration for a vacuum tube.

In analyzing transistors, the hybrid h parameters were used in defining the transistor parameters, but for vacuum tubes as with field-effect transistors it is better to use the admittance y parameters. (The reason for changing from the h to the y parameters is perhaps not obvious, but it is necessitated by the fact that the tube grid current i_c is assumed to be zero.) Repeating Eq. set 3-22 in terms of instantaneous quantities and substituting $e_c = v_1$, $i_c = i_1$, $e_b = v_2$, and $i_b = i_2$,

$$\left. \begin{aligned} i_c &= y_{11}e_c + y_{12}e_b \\ i_b &= y_{21}e_c + y_{22}e_b \end{aligned} \right\} \qquad (9\text{-}6)$$

In Chapter 6 the separation of the transistor d-c and a-c voltages and currents was discussed. The instantaneous voltages and currents shown in Eq. set 9-6 are also a combination of a-c and d-c components, and so, before the admittance parameters can be defined, the incremental linear a-c and the nonlinear d-c voltages and currents must be separated. This can be accomplished by taking the total derivative of both sides of Eq. set 9-6, as the d-c terms will drop out when this is done. The y parameters can best be defined by rewriting Eq. set 9-6 in functional form, or

$$\left. \begin{aligned} i_c &= i_c(e_c, e_b) \\ i_b &= i_b(e_c, e_b) \end{aligned} \right\} \tag{9-7}$$

and then taking the total derivative of each functional equation, or

$$\left. \begin{aligned} di_c &= \frac{\partial i_c}{\partial e_c} de_c + \frac{\partial i_c}{\partial e_b} de_b \\ di_b &= \frac{\partial i_b}{\partial e_c} de_c + \frac{\partial i_b}{\partial e_b} de_b \end{aligned} \right\} \tag{9-8}$$

The dynamic y parameters can now be defined from Eq. set 9-8 as

$$\left. \begin{aligned} y_{11} &= \frac{\partial i_c}{\partial e_c} = \frac{\Delta i_c}{\Delta e_c}\bigg|_{e_b = \text{ a constant}} \\[2mm] y_{12} &= \frac{\partial i_c}{\partial e_b} = \frac{\Delta i_c}{\Delta e_b}\bigg|_{e_c = \text{ a constant}} \\[2mm] y_{21} &= \frac{\partial i_b}{\partial e_c} = \frac{\Delta i_b}{\Delta e_c}\bigg|_{e_b = \text{ a constant}} \\[2mm] y_{22} &= \frac{\partial i_b}{\partial e_b} = \frac{\Delta i_b}{\Delta e_b}\bigg|_{e_c = \text{ a constant}} \end{aligned} \right\} \tag{9-9}$$

The constraints e_b and e_c equal a constant means that $e_b = E_b$ and $e_c = E_c$.

As discussed in Chapter 7, the Δ variations of the total voltages and currents (Δe_b, Δi_b, Δe_c, Δi_c) are identically equal to the a-c components only (e_p, i_p, e_g, i_g), or

$$\left. \begin{aligned} \Delta e_b &\equiv e_p \\ \Delta i_b &\equiv i_p \\ \Delta e_c &\equiv e_g \\ \Delta i_c &\equiv i_g \end{aligned} \right\} \tag{9-10}$$

Since i_c can usually be assumed to be equal to zero for vacuum tubes, both y_{11} and y_{12} equal zero and in addition y_{21} and y_{22} are given new symbols and definitions:

$$y_{21} = g_m \tag{9-11}$$

the mutual or transconductance of the tube and

$$y_{22} = \frac{1}{r_p} \qquad (9\text{-}12)$$

where r_p is defined as the plate resistance of the tube, in ohms. In addition, another term μ is defined as

$$\mu = -\frac{y_{21}}{y_{22}} = -\frac{\Delta i_b / \Delta e_c}{\Delta i_b / \Delta e_b} = -\frac{\Delta e_b}{\Delta e_c}\bigg|_{i_b \,=\, \text{a constant}} \qquad (9\text{-}13)$$

the amplification factor; or

$$\mu = g_m r_p \qquad (9\text{-}14)$$

where the negative $(-)$ sign in Eq. 9-13 is used so that μ will be a positive quantity.

The g_m, r_p, and μ parameters, as defined in Eqs. 9-11 through 9-14, are well recognized as standard symbols for vacuum tubes. In the next section it will be shown how μ, g_m, and r_p can be obtained graphically from the plate characteristics, and in Chapter 10 the a-c equivalent circuit using these three parameters will be presented.

9-6. Graphical Analysis

The definitions of the three basic vacuum-tube parameters are repeated here for convenience.

$$\left.\begin{aligned} \mu &= -\frac{\Delta e_b}{\Delta e_c}\bigg|_{i_b \,=\, I_b} \\[2mm] r_p &= \frac{\Delta e_b}{\Delta i_b}\bigg|_{e_c \,=\, E_c} \\[2mm] g_m &= \frac{\Delta i_b}{\Delta e_c}\bigg|_{e_b \,=\, E_b} \end{aligned}\right\} \qquad (9\text{-}15)$$

Although these parameters can be defined at any point on the characteristic curves, they are usually defined only in the linear region of the characteristics. This means that μ, r_p, and g_m can be considered constant for a given tube. Unlike the transistor parameters, vacuum-tube parameters remain remarkably constant over a wide temperature variation of the tube.

Figure 9-6 shows the plate characteristics of a typical triode with an operating or Q point arbitrarily placed within the linear region of the characteristics (the portion of the curves enclosed by the dotted line can be considered the linear region). To calculate μ, draw a constant i_b line through the Q point. One such line is aa', and it is general practice to draw equal excursions on either side of this Q point. The reason for stopping at the adjacent e_c curves is simply one of expediency and ease of

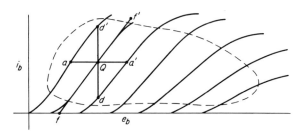

FIG. 9-6. Triode plate characteristics, showing how
the triode parameters can be determined.

calculation. μ then becomes

$$\mu = -\frac{e_{b \text{ at } a'} - e_{b \text{ at } a}}{e_{c \text{ at } a'} - e_{c \text{ at } a}} \tag{9-16}$$

To calculate r_p, draw a straight line tangent to the e_c curve at the Q point. Again for ease of calculation, it is convenient to extend this line until it reaches the X axis such as the line ff', and

$$r_p = \frac{e_{b \text{ at } f'} - e_{b \text{ at } f}}{i_{b \text{ at } f'} - i_{b \text{ at } f}} \tag{9-17}$$

To find g_m, draw a constant e_b line of convenient length, such as dd', through the Q point. Then g_m is

$$g_m = \frac{i_{b \text{ at } d'} - i_{b \text{ at } d}}{e_{c \text{ at } d'} - e_{c \text{ at } d}} \tag{9-18}$$

A convenient check on the accuracy of these results is to use the relation

$$\mu = g_m r_p \tag{9-19}$$

These parameters could now be used in the equivalent-circuit approach to vacuum-tube circuits, but this discussion will be reserved for the next chapter.

Because the graphical circuit-analysis techniques for the transistor and the vacuum tube are so similar, a numerical example will best illustrate not only how to calculate μ, g_m, and r_p, but also how to carry out a graphical circuit analysis on a grounded-cathode amplifier.

EXAMPLE 9-1. Using the circuit of Fig. 9-7, a) find the quiescent operating point of the tube; b) find μ, r_p, and g_m around the Q point found in part a; c) draw in the a-c load line; d) if the input voltage is $e_s = 2 \sin \omega t$, find I_p and E_p; e) find the voltage gain A_v, where $A_v \triangleq E_{\text{out}}/E_s$. The frequency ω is of such a value that $1/(\omega C_c) \ll R_i$.

Solution: The plate characteristics for the 6J5 triode are given in Fig. 9-8. The first step is to draw the d-c load line on the characteristics, as was explained in both Chapters 3 and 6. The d-c load line has a slope of $-1/20,000$ and passes through the two points $e_b = 240, i_b = 0$ and $e_b = 0, i_b = 12$ ma. The second step

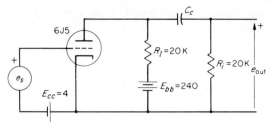

FIG. 9-7. Circuit for Example 9-1.

is to find the Q point. When $e_s = 0$ (the quiescent condition), $e_c = E_c = -E_{CC} = -4$ volts. Thus the Q point is located at the intersection of the d-c load line and the plate-characteristic curve for $e_c = -4$ volts, as indicated in Fig. 9-8.

The third step is to draw in the a-c load line if it differs from the d-c load line. The a-c load resistance R_a is simply the parallel equivalent of R_l and R_i, or

$$R_a = \frac{R_l R_i}{R_l + R_i} = \frac{20 \times 10^3 \times 20 \times 10^3}{20 \times 10^3 + 20 \times 10^3} = 10,000 \text{ ohms} \qquad (9\text{-}20)$$

The slope of the a-c load line can best be determined by letting e_b equal some convenient value—say 100 volts. Then, dividing this value by R_a, we have $i_b = 100/10 \times 10^3 = 10$ ma. By plotting a straight line through the two points $e_b = 0$, $i_b = 10$ ma and $e_b = 100$, $i_b = 0$ on the plate characteristics, and then drawing a line parallel to this through the Q point, the proper a-c load line is determined. (This construction is shown in Fig. 9-8.)

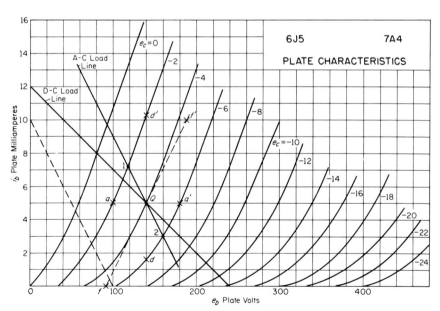

FIG. 9-8. Diagram for Example 9-1.

Finally, we let e_s vary about this Q point from $+2$ to -2 volts (the sinusoidal voltage e_s has a positive maximum of $+2$ volts and a negative minimum of -2 volts). Thus

$$e_c = -E_{cc} + e_s$$

or

$$e_{c_{max}} = -4 + 2 = -2 \text{ volts}$$

and

$$e_{c_{min}} = -4 - 2 = -6 \text{ volts}$$

The two points corresponding to an $e_{c_{max}}$ of -2 and an $e_{c_{min}}$ of -6 volts are indicated as points *1* and *2*, respectively, on the a-c load line. The solutions that were asked for in the statement of the problem can now be obtained.

a) At the Q point

$$I_b = 5 \text{ ma}$$
$$E_b = 140 \text{ volts}$$

b)
$$\mu = -\left.\frac{\Delta e_b}{\Delta e_c}\right|_{I_b = 5 \text{ ma}}$$

$$= -\frac{e_{b \text{ at } a'} - e_{b \text{ at } a}}{e_{c \text{ at } a'} - e_{c \text{ at } a}} = -\frac{180 - 100}{(-6) - (-2)}$$

$$= 20$$

where the points a and a' were chosen on the adjacent e_c curves for convenience.

$$r_p = \left.\frac{\Delta e_b}{\Delta i_b}\right|_{E_c = -4 \text{ volts}}$$

$$= \frac{e_{b \text{ at } f'} - e_{b \text{ at } f}}{i_{b \text{ at } f'} - i_{b \text{ at } f}} = \frac{187 - 95}{(10 - 0) 10^{-3}}$$

$$= 9200 \text{ ohms}$$

where the line ff' is tangent to the curve $e_c = -4$ volts at the Q point.

$$g_m = \left.\frac{\Delta i_b}{\Delta e_c}\right|_{E_b = 140} = \frac{i_{b \text{ at } d'} - i_{b \text{ at } d}}{e_{c \text{ at } d'} - e_{c \text{ at } d}} = \frac{(10.6 - 1.7) 10^{-3}}{-2 - (-6)} = \frac{8900 \times 10^{-6}}{4}$$

$$= 2225 \text{ } \mu\text{mhos}$$

As a check, we calculate g_m by using Eq. 9-19 or

$$g_m = \frac{\mu}{r_p} = \frac{20}{9200}$$

$$g_m = 2170 \text{ } \mu\text{mhos} \quad (check)$$

The reason that the two values do not check more closely is because point d is situated in the nonlinear region of the plate characteristics.

c) The a-c load line is shown in Fig. 9-8.

d) I_p is the rms value of the a-c component of i_b, or

$$I_p = \frac{i_{b_1} - i_{b_2}}{2\sqrt{2}}$$

$$= \frac{7.3 - 3.0}{2\sqrt{2}}$$

$$= 1.52 \text{ ma}$$

where the factor of 2 changes $(i_{b_1} - i_{b_2})$ from a peak-to-peak value of i_p to a maximum value of i_p, and the factor of $\sqrt{2}$ changes the maximum value of i_p to an rms value (I_p). All a-c components of the voltages and currents are assumed to be purely sinusoidal in nature, since the tube is being operated in the linear portion of its characteristics.

E_p is the rms value of the a-c component of e_b, which is e_p, or

$$E_p = \frac{e_{b_2} - e_{b_1}}{2\sqrt{2}} = \frac{160 - 117}{2\sqrt{2}} = 15.2 \text{ volts}$$

As a check, $E_p = I_p R_a$, or $1.52 \times 10^{-3} \times 10^4 = 15.2 \text{ volts}$.   (*check*)

e) The voltage gain A_v is

$$A_v = \frac{E_{\text{out}}}{E_s} = \frac{E_p}{E_s}$$

$$= -\frac{15.2}{2/\sqrt{2}}$$

$$= -10.8$$

where the negative sign is attached because E_{out} and E_s are known to be 180° out of phase (see the arguments used in the preceding section).

9-7. Self-Biasing of a Vacuum Tube

The grounded-cathode amplifier is shown in Fig. 9-9a and is the same as the amplifier discussed in Sec. 9-4, except for the biasing arrangement. Few vacuum-tube circuits now employ a separate d-c supply for

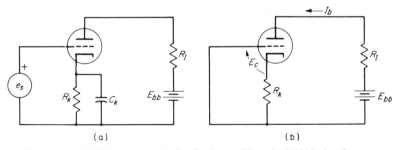

(a)   (b)

FIG. 9-9. a) Self-biased grounded-cathode amplifier. b) Self-biasing for an amplifier.

the grid-biasing voltage of the tubes. Instead, some type of self-biasing arrangement is provided, as is shown in Fig. 9-9b. Under static operating conditions the tube voltage must settle down to some equilibrium, or Q point. There are two ways to find the Q point. In the first method we draw the d-c load line (the d-c resistance is the sum of R_l and R_k, but if $R_l \gg R_k$ the d-c load resistance is usually considered to be just R_l) and then, by trial and error, move up and down the load line until the equation $-I_b R_k = E_c$ is satisfied. In the second method we draw the d-c load line as before, and then, for various values of E_c, calculate I_b where $I_b = -E_c/R_k$. We plot points corresponding to the intersection of I_b and the particular E_c curve for which the value of I_b was calculated. Finally, we draw a curve between these points. The intersection of this curve with the d-c load line is the Q point.

EXAMPLE 9-2. In Fig. 9-9b, let the tube be a 6J5, $E_{bb} = 300$ volts, $R_k = 1000$ ohms, and $R_l = 20,000$ ohms. Find the Q point, using the two previously mentioned methods.

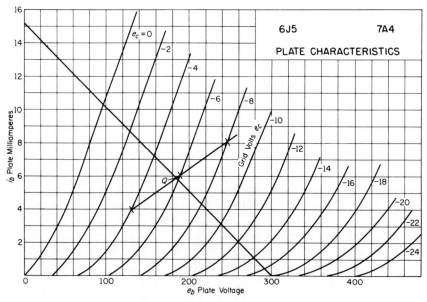

FIG. 9-10. Determination of the Q point when using self-biasing.

Solution: To a good approximation, R_l can be used for the d-c load-line resistance. We draw the load line as indicated in Fig. 9-10.

For $E_c = -8$ volts $I_b = 8/1000 = 8$ ma
For $E_c = -6$ volts $I_b = 6$ ma
For $E_c = -4$ volts $I_b = 4$ ma

We then plot the following points: $I_b = 8$, $E_c = -8$; $I_b = 6$, $E_c = -6$; $I_b = 4$, $E_c = -4$; and draw a curve passing through these points. The intersection of this curve with the d-c load line is the Q point, as indicated in Fig. 9-10.

The trial-and-error method involves choosing points on the d-c load line until the equation $-I_b R_k = E_c$ is satisfied. As a check on the above method, we can see if $-I_b R_k = E_c$ at the Q point:

$$-I_b R_k = -5.9 \times 10^{-3} \times 1000 = -5.9 \text{ volts} \quad (check)$$

As will be seen later, the addition of the self-biasing resistor R_k reduces the overall gain of the amplifier circuit. To restore the higher gain, it is necessary to bypass the cathode resistor with a shunt capacitance, as shown in Fig. 9-9a. Remember that the gain of an a-c amplifier is dependent on the a-c impedances in the circuit. If $X_{C_k} \ll R_k$, then, for all practical purposes, the cathode is restored to ground potential for the a-c signals. It is thus evident that the parallel combination of the resistance and capacitance effectively replaces the d-c grid bias battery.

9-8. Power Relations in a Vacuum-Tube Circuit

As pointed out in Chapter 6, the instantaneous power in any portion of a circuit is defined as the product of the appropriate instantaneous voltage and instantaneous current. Using Fig. 9-11, the instantaneous

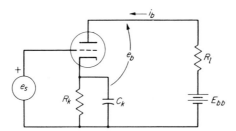

FIG. 9-11. Grounded-cathode amplifier circuit used in defining power relations.

tube plate power loss is defined as

$$p_{\text{tube}} = e_b i_b \tag{9-21}$$

and the average tube power is

$$P_{\text{tube}} = \frac{1}{2\pi} \int_0^{2\pi} e_b i_b \, d\omega t \tag{9-22}$$

If the input signal e_s is sinusoidal, that is, $e_s = \sqrt{2} E_s \sin \omega t$, then

$$i_b = I_b + \sqrt{2} I_p \sin \omega t$$
$$e_b = E_b - \sqrt{2} E_p \sin \omega t$$

where the signs indicate that I_p and E_p are 180° out of phase with each other if the plate load impedance is purely resistive in nature. The average tube power is

$$P_{tube} = \frac{1}{2\pi} \int_0^{2\pi} (I_b + \sqrt{2}I_p \sin \omega t)(E_b - \sqrt{2}E_p \sin \omega t)\, d\omega t \quad (9\text{-}23)$$

and, upon integrating and substituting in limits,

$$P_{tube} = E_b I_b - E_p I_p \quad (9\text{-}24)$$

The reader should verify the correctness of Eq. 9-24 as this final equation has a very important interpretation. The $E_b I_b$ term represents the d-c or quiescent power loss in the tube, whereas the $-E_p I_p$ term represents the a-c power returned. Note that for the purely resistive plate load impedance $E_p = I_p R_l$ and $E_p I_p = I_p^2 R_l$. Thus, *when an a-c signal is applied to the tube, the tube power is less than when no signal is applied.* To obtain maximum tube utilization, many vacuum-tube circuits designed for continuous a-c operation can take advantage of this fact, but it must be remembered that rated tube plate dissipations can be exceeded if the circuit is operated under quiescent conditions.

The power source E_{bb} supplies all the energy for the d-c and a-c energy losses in the circuit (not including the filament supply). The instantaneous input power from the power supply is

$$p_{bb} = E_{bb} i_b \quad (9\text{-}25)$$

and the average power supplied is

$$P_{bb} = \frac{1}{2\pi} \int_0^{2\pi} E_{bb}(I_b + \sqrt{2}I_p \sin \omega t)\, d\omega t \quad (9\text{-}26)$$

Upon integrating and substituting in limits, we obtain

$$P_{bb} = E_{bb} I_b \quad (9\text{-}27)$$

An example will help to clarify the power relations in a vacuum-tube circuit.

EXAMPLE 9-3. Using Fig. 9-12, assume that $i_b = 5 + 3 \sin \omega t$ ma, as obtained from a graphical analysis of the circuit. Find all of the power relations.

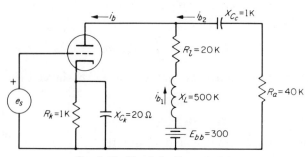

FIG. 9-12. Circuit for Example 9-3.

Solution: Perhaps the first step in a problem such as this is to find the a-c and d-c currents that flow in each path. In this particular problem the capacitive and inductive reactances are of such a magnitude that the plate load impedance is almost purely resistive for the a-c as well as the d-c voltage and currents. The plate current i_b is comprised of the a-c i_p and d-c I_b currents where

$$I_p = \frac{3}{\sqrt{2}} \text{ ma}$$

$$I_b = 5 \text{ ma}$$

In Fig. 9-12 the total plate current is shown as being the sum of two currents i_{b_1} and i_{b_2}, where

$$i_b = i_{b_1} + i_{b_2}$$

Since a capacitor blocks all direct current, all of I_b flows through X_L and the 20-K plate resistance and through the 1-K cathode-biasing resistance. For all practical purposes, the large inductive reactance in comparison to the 40-K resistance and 1-K capacitive reactance forces almost all the a-c current to flow through the 40-K resistance. Since very little a-c current flows through the inductive reactance, i_{b_2} is essentially equal to i_p, or

$$I_{b_2} \simeq I_p = \frac{3}{\sqrt{2}} \text{ ma}$$

Also note that essentially all of the a-c plate current is shunted around R_k and through C_k.

With this information at hand, we can proceed with the calculation for the various power relations in the circuit of Fig. 9-12.

Power supplied to the circuit (less filament power):

$$P_{bb} = E_{bb} I_b = 300 \times 5 \times 10^{-3} = 1.5 \text{ watts}$$

Power in the 20-K plate load resistance:

$$P_{20K} = I_b^2 \times R_l = (5 \times 10^{-3})^2 \times 20 \times 10^3 = 0.5 \text{ watt}$$

Power in the 40-K output resistance:

$$P_{40K} = I_p^2 \times R_a$$
$$= \left(\frac{3}{\sqrt{2}} \times 10^{-3}\right)^2 \times 40 \times 10^3 = 0.18 \text{ watt}$$

Power in the 1-K cathode-bias resistance:

$$P_{1K} = I_b^2 \times R_k = (5 \times 10^{-3})^2 \times 10^3 = 0.025 \text{ watt}$$

Power in the tube:

$$P_{tube} = E_b I_b - I_p^2 R_a$$

where

$$E_b = E_{bb} - I_b R_{d\text{-}c} \qquad \text{and} \qquad R_{d\text{-}c} = R_l + R_k$$

thus

$$E_b = 300 - 5 \times 10^{-3} \times (20 + 1) 10^3 = 195 \text{ volts}$$
$$P_{tube} = 195 \times 5 \times 10^{-3} - 0.18$$
$$= 0.795 \text{ watt}$$

As a check, the total input power must equal the sum of all other powers in the circuit:

$$P_{bb} = P_{20K} + P_{40K} + P_{1K} + P_{tube}$$
$$1.5 = 0.5 + 0.18 + 0.025 + 0.795$$
$$1.5 = 1.5 \quad (check)$$

EXAMPLE 9-4. If the reader objects to the integral relations of Eqs. 9-22, 9-23, and 9-26, let us solve for all of the power losses in the circuit accompanying Example 9-3 using a nonintegral calculus approach.

Solution: Using the same physical reasoning as in Example 9-3 regarding the currents obtains

$$P_{1K} = 0.025 \text{ watt}$$
$$P_{20K} = 0.5 \text{ watt}$$
$$P_{40K} = 0.18 \text{ watt}$$

by use of the $P = I^2 R$ relationship.

By inspection of Fig. 9-12 the E_{bb} power supply can supply only d-c energy or power, thus

$$P_{E_{bb}} = E_{bb}I_b = 1.5 \text{ watts}$$

The power lost in the tube must be the power supplied minus the powers lost in the resistors (capacitors and inductors cannot dissipate energy)

$$P_{tube} = P_{E_{bb}} - P_{1K} - P_{20K} - P_{40K}$$
$$= 1.5 - 0.025 - 0.5 - 0.18$$
$$= 0.795 \text{ watt}$$

which checks the result given by application of Eq. 9-24.

The reader will recognize that this same technique was used to develop the power relations in the transistor circuits of Chapter 6.

9-9. Multigrid Vacuum Tubes

Soon after the triode was developed, it was found that the operation of the tube was limited to a rather narrow range of frequencies. The inherent capacitance that exists between any two pieces of metal caused the difficulty. Even though the grid is a fine spiral of wire, the capacitance between the grid and the plate C_{gp} is about 5 picofarads (pico = 10^{-12}), and similar capacitances exist between the grid and cathode C_{gk}, and the plate and cathode C_{pk}. The capacitive reactance of a 5-picofarad capacitor at a frequency of 1 MHz is $1/\omega C = 1/(2\pi \times 10^6 \times 5 \times 10^{-12}) = 31,700$ ohms. This means that an a-c signal at a frequency of 1 MHz or so has a rather low impedance path from the plate to the grid. Recall that if a positive-going voltage is applied to the grid, then a negative-going voltage is coupled to the grid because of C_{gp}, and the net result is a reduction of the voltage applied to the grid and a much-reduced amplifier gain.

The Tetrode. Various circuit schemes were employed to overcome the serious frequency limitation of the triode, but none was as effective as the tetrode tube which was introduced in 1927. Since the principal frequency limitation of the triode was caused by the grid-to-plate capacitance, a second grid was placed between the first grid (now called the *control grid*) and the plate. The addition of this second grid (aptly called the *screen grid*) reduced the control grid-to-plate capacitance C_{gp} by a factor of 1000 or better. One of the first questions that arises is: At what potential should the screen grid be operated? For the most effective shielding the screen grid should consist of a fine wire mesh, although often a spiral of wire is used. The wire-mesh construction causes most of the plate electric-field-intensity lines to terminate on the screen grid rather than on the cathode. Since the plate voltage no longer attracts many electrons from the cathode, it is necessary to apply a positive voltage to the screen grid. In other words, the screen grid acts as a secondary plate to attract the cathode electrons, but, since the screen grid is perforated most of the electrons pass on through to the actual plate. Figure 9-13a shows a tetrode connected in a circuit suitable for measuring the plate characteristics, and Fig. 9-13b shows a typical family of characteristics for a tetrode.

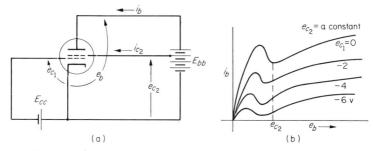

FIG. 9-13. a) A tetrode circuit. b) Typical tetrode plate characteristics.

The reason for the shape of the curves is rather interesting. As already mentioned, the screen-grid potential attracts the cathode electrons that the control grid allows to escape. For small values of e_b, the screen attracts practically all the electrons, since even those electrons that pass through the screen openings find the screen potential more attractive than the plate potential. However, some of the higher-energy electrons do reach the plate, and a small plate current begins to flow. As e_b is increased further, the number of electrons attracted to the plate increases, as might be expected. As e_b increases even more, a decrease in i_b is encountered. This phenomenon is explained by recalling that high-energy bombarding electrons are capable of releasing many other electrons from a metal surface. The decrease in plate current is due to these secondary electrons moving to the screen grid instead of returning to the plate, as they do in a triode. In some tubes the plate current can even go negative, indicating that more electrons are leaving the plate than are striking it. As the plate voltage increases beyond the screen-grid voltage, these secondary electrons return to the plate, and the curves again have a positive slope.

The use of the tetrode in amplifier circuits is seriously hampered by the extreme nonlinearity of the curves. The graphical and equivalent-circuit analyses are the same for the tetrode as for the triode, but, to operate the tetrode in its linear region, a high plate-supply voltage must be used, as can be seen by observation of the plate characteristics. Notice that there is a region of negative slope in the plate curves. This is a region of negative resistance, and, in some circuit applications, a negative resistance can be of value.

The Pentode. It was quickly recognized that secondary emission was the undesirable feature of the tetrode, and that, if the secondary electrons could be forced to return to the plate, the undesirable "dip" in the tetrode plate characteristics could be eliminated.

This was done by adding a third grid between the plate and the screen grid. Since most of the secondary electrons have only a few electron volts of energy, a small repelling field is sufficient to cause them to return to the plate. To ensure that the third grid, appropriately called the *suppressor grid*, has a repelling effect on the electrons, it is necessary that the potential of this grid always be negative when compared to the plate. The most obvious source of negative potential is the cathode, hence many pentodes have the suppressor grid internally connected to the cathode (see Fig. 9-14a). This figure shows a circuit that can be used to determine experi-

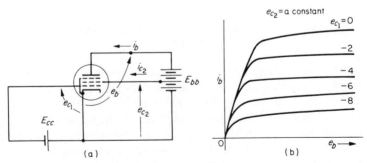

FIG. 9-14. a) A pentode circuit. b) Typical pentode plate characteristics.

mentally the plate characteristics of the pentode, and Fig. 9-14b shows a set of typical characteristics for a pentode.

For both the tetrode and the pentode, the plate resistance r_p is 1 megohm or so, the amplification factor μ is several thousand, and the mutual conductance g_m is several thousand mhos.

The same techniques of graphical or a-c equivalent-circuit analysis are applicable whether a triode, tetrode, or pentode is used in the circuit. The major differences between a triode and a tetrode or pentode are the relative magnitudes of μ, g_m, r_p, the interelectrode capacitances, and the power that is consumed by the screen grid.

PROBLEMS

9-1. A grounded-cathode amplifier using a 6J5 triode has an E_{bb} of 250 volts and a load resistance of 25,000 ohms. Plot the plate voltage e_b as a function of the grid voltage e_c. Over what range of grid voltage can the plate voltage be considered linearly related to the grid voltage?

9-2. Determine μ, r_p, and g_m of a 6J5 triode when operating with 220 volts on the plate and a grid bias of -10 volts.

9-3. Determine μ, r_p, and g_m of a 6J5 triode when operating with 160 volts on the plate and a grid bias of -4 volts.

9-4. If a 6J5 triode is used in the circuit of the accompanying figure, a) draw the d-c load line, b) draw the a-c load line, and c) graphically determine μ, g_m, and r_p. d) If $e_s = 2 \sin \omega t$, graphically determine the a-c gain of the amplifier and e) determine the output voltage E_{out}.

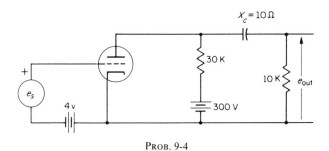

PROB. 9-4

9-5. Replace the bias battery in the circuit of the figure for Prob. 9-4 with a self-biasing circuit. Estimate the size of the capacitor used in the biasing circuit if the lowest value of ω is known to be 5000 radians/sec.

9-6. A 6J5 triode is used in the circuit of the accompanying figure. If $R_l = 8000$ ohms and the Q point is to be located at $E_b = 180$ volts, $I_b = 4$ ma, find the necessary values of E_{bb} and E_{cc}.

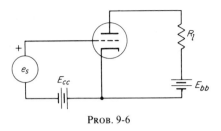

PROB. 9-6

9-7. In the circuit shown in the accompanying figure, a) find the Q point. b) If E_s = 4 volts peak to peak, what will be the peak-to-peak value of E_{out}? c) Find μ, r_p, and g_m at the operating point in part a.

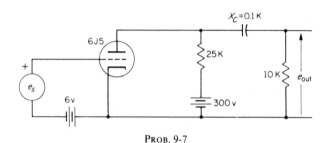

PROB. 9-7

9-8. Repeat Prob. 9-4 if E_{bb} is changed to 180 volts and E_{cc} to 2 volts.

9-9. Repeat Prob. 9-4 if E_{bb} and E_{cc} are reduced to 90 and 2 volts, respectively. Is this a satisfactory operating condition if the amplifier is used in a high-fidelity audio amplifier?

9-10. Compare the values for μ, g_m, and r_p, and the gain for Prob. 9-4, 9-8, and 9-9.

9-11. Replace the bias battery in the circuit of the figure in Prob. 9-6 with a self-biasing circuit. If $\omega \geq 900$ radians/sec, estimate the proper value for the shunt capacitance.

9-12. Find the power in the tube and in each of the resistors in Prob. 9-4. Also find the power supplied by E_{bb}.

9-13. Find the power in the tube, the power supplied by E_{bb}, and the a-c power in R_l in Prob. 9-6.

10 / *The Vacuum Tube Used as a Linear Device*

In Chapter 9 the graphical vacuum-tube characteristics were studied. The vacuum-tube parameters were obtained by using the graphical plate or output characteristics, but the vacuum-tube circuit was analyzed by graphical methods. In this chapter the triode parameters (μ, g_m, and r_p), as obtained either graphically or experimentally, will be used in the linear equivalent-circuit approach to triode-circuit analysis. In Chapters 3 and 7 the a-c equivalent circuit using the hybrid h parameters was presented. The hybrid parameters cannot conveniently be used in vacuum-tube analysis (because the grid current is assumed to be zero), but the admittance y parameters do yield an acceptable equivalent circuit just as they do for the FET.

It is the purpose of this chapter to develop the analytic techniques necessary to handle three basic vacuum-tube circuits, and to show how these basic circuits can be cascaded to give a desired effect.

The three circuits to be studied are the grounded-cathode amplifier, the cathode-follower circuit, and the grounded-grid amplifier. For each circuit it is desired to obtain 1) the a-c equivalent circuit, 2) the circuit gain, and 3) the input and output resistances.

10-1. The Grounded-Cathode Amplifier

A-C Equivalent Circuit. In studying either transistors or vacuum-tube circuits, one must be constantly aware of the frequency-response characteristics or limitations of the transistors and tubes being used, as well as other elements in the circuit. This text deals primarily with the low-frequency applications of tubes and transistors, but the term *low frequency* needs additional defining. For ordinary vacuum tubes, low frequency usually means frequencies under 500,000 Hz, but, for many transistors, low frequencies are defined as 50,000 Hz or less. In Chapter 13 an investigation of the frequency-response characteristics of tubes and transistors will be presented.

Although this chapter deals primarily with the low-frequency characteristics and equivalent circuits of the triode, the higher-frequency equivalent circuit is considered when defining the input resistance of the triode grounded-cathode amplifier. The first step is to develop the a-c equivalent circuit for the vacuum tube. As pointed out earlier, the admittance y parameters will be used in the equivalent circuit for the tube just as they were used for the FET. Since the a-c equivalent circuit using the admittance parameters was not shown in Chapter 3, it will be presented here. The development of the a-c equivalent circuit given here is identical to that presented in Chapter 8 except for the changes in the parameter and variable symbols. Repeating Eq. set 3-21 with the voltage rise symbol, E, substituted for V

$$I_1 = y_{11}E_1 + y_{12}E_2 \tag{10-1a}$$

$$I_2 = y_{21}E_1 + y_{22}E_2 \tag{10-1b}$$

It is now possible to develop an equivalent circuit using Eq. set 10-1. Both Eqs. 10-1a and 10-1b have the dimensions of current, and thus the right-hand sides of each equation suggest the summing of currents enter-

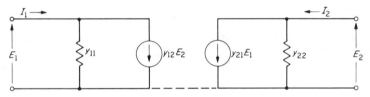

FIG. 10-1. Two-current-generator equivalent circuit using the two-port y parameters.

ing a junction (KCL or Kirchhoff's current law). Figure 10-1 illustrates the simplest and most obvious two-current-generator equivalent circuit than can be developed from Eq. set 10-1. The left-hand portion of the circuit satisfies Eq. 10-1a, and the right-hand portion satisfies Eq. 10-1b.

The a-c equivalent circuit of Fig. 10-1 can be used to describe the incremental linear behavior of a vacuum tube by redefining the symbols so that

$$
\left.
\begin{aligned}
E_1 &= E_g & y_{11} &= \left.\frac{I_g}{E_g}\right|_{E_p=0} & \text{mhos} \\[2mm]
I_1 &= I_g & y_{12} &= \left.\frac{I_g}{E_p}\right|_{E_g=0} & \text{mhos} \\[2mm]
E_2 &= E_p & y_{21} &= g_m & \text{mhos} \\[2mm]
I_2 &= I_p & y_{22} &= \frac{1}{r_p} & \text{mhos}
\end{aligned}
\right\} \tag{10-2}
$$

but, since I_g is assumed to be negligible (let $I_g = 0$), both y_{11} and y_{12} are equal to zero. After substituting the values from Eq. set 10-2 into Eq. set 10-1,

$$0 = 0E_g + 0E_p \tag{10-3a}$$

$$I_p = g_m E_g + \frac{1}{r_p} E_p \tag{10-3b}$$

Thus Eq. 10-3b is the only equation of any importance in vacuum-tube a-c equivalent-circuit analysis. The physical vacuum tube and its a-c equivalent circuit are shown in Figs. 10-2a and b, respectively.

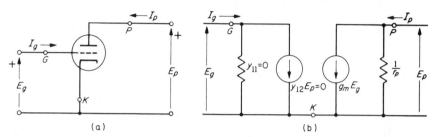

FIG. 10-2. a) Physical vacuum triode-circuit symbol. b) Low-frequency a-c equivalent circuit of part a, using the admittance y parameters with $I_g = 0$.

The assumption that $I_g = 0$ greatly simplifies the equivalent circuit for the vacuum tube. This assumption also emphasizes that the vacuum tube is not a current amplifier (like the bipolar junction transistor) but a voltage amplifier (like the field-effect transistor). The equivalent circuit shown in Fig. 10-2b involves a current source and a shunt resistance, but this can easily be transformed into another equivalent circuit by using a voltage source and a series resistance (a Norton generator to a Thévenin generator). The two commonly used a-c equivalent circuits for the vacuum tube are shown in Fig. 10-3a and b.

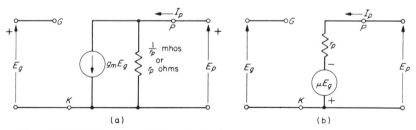

FIG. 10-3. a) Norton equivalent circuit for the vacuum tube of Fig. 10-2a. b) Thévenin equivalent circuit for the vacuum tube of Fig. 10-2a (recalling that $\mu = g_m r_p$).

The a-c equivalent circuit for the R-C-coupled grounded-cathode amplifier can now be presented. Let us consider the circuit of Fig. 9-7, repeated here as Fig. 10-4a (note that a self-bias circuit replaces E_{CC}). Figure 10-4b shows the low-frequency a-c equivalent circuit for the circuit

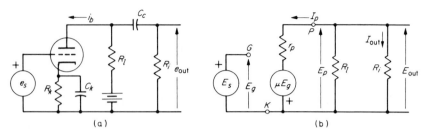

FIG. 10-4. a) *R-C* coupled grounded cathode amplifier. b) A-c equivalent circuit for part a.

shown in Fig. 10-4a, where the tube itself has been replaced by the Thévenin equivalent circuit of Fig. 10-3b, and $1/(\omega C_k)$ is assumed small in comparison to R_k; that is, $1/(\omega C_k) \leq 0.1 R_k$. Also $1/(\omega C_c) \leq 0.1 R_i$.

The derivation of the voltage gain, the input resistance, and the output resistance of the *R-C*-coupled grounded-cathode amplifier will now be presented.

Gain. The voltage gain A_v is defined as

$$A_v = \frac{E_{\text{out}}}{E_s} \tag{10-4}$$

and, in Fig. 10-4b, it is evident that $E_g = E_s$ and that the output voltage E_{out} is

$$E_{\text{out}} = I_{\text{out}} R_i \tag{10-5}$$

The output current I_{out} in terms of the input current I_p (using the current splitting theorem) is

$$I_{\text{out}} = - \frac{R_l}{R_l + R_i} I_p \tag{10-6}$$

where the minus sign is necessary because of the arbitrary way in which I_{out} is chosen to flow.

The total or plate current I_p is the ratio of the Thévenin voltage source μE_g and the series resistance, or

$$I_p = \frac{\mu E_g}{r_p + R_{ac}} \tag{10-7}$$

where R_{ac} is the parallel equivalent of R_l and R_i, or

$$R_{ac} = \frac{R_l \times R_i}{R_l + R_i} \tag{10-8}$$

which can be considered the a-c plate load resistance just as R_l is considered the d-c plate load resistance.

To solve for the voltage gain A_v, substitute I_p from Eq. 10-7 into Eq. 10-6, or

$$I_{\text{out}} = - \frac{R_l}{R_l + R_i} \frac{\mu E_g}{r_p + R_{ac}} \tag{10-9}$$

Next, we substitute I_{out} from Eq. 10-9 into Eq. 10-5 and also substitute $E_s = E_g$.

$$E_{out} = -\frac{R_l}{R_l + R_i}\frac{\mu E_s}{r_p + R_{ac}}R_i$$

$$= -\frac{\mu R_{ac}}{r_p + R_{ac}}E_s \tag{10-10}$$

Upon substituting R_{ac} from Eq. 10-8 into Eq. 10-10 and solving for the ratio E_{out}/E_s, and rearranging

$$A_v = \frac{-\mu R_{ac}}{r_p + R_{ac}} = \frac{-\mu R_l R_i}{r_p R_l + r_p R_i + R_l R_i} \tag{10-11}$$

This equation can be expressed as

$$A_v = M_v \underline{/\theta_v} \tag{10-12}$$

where M_v is the magnitude of the voltage gain and θ_v is the phase relation between the input and output voltages. The study of phase shift and its importance in electronic circuits is deferred to Chapter 12 and the chapters that follow. Comparing Eqs. 10-11 and 10-12 obtains

$$M_v = \frac{\mu R_l R_i}{r_p R_l + r_p R_i + R_l R_i} \tag{10-13}$$

$$\theta_v = 180° \tag{10-14}$$

where the minus $(-)$ sign indicates a phase shift of $180°$ [a plus $(+)$ sign indicates a phase shift of $0°$]. Thus, if the frequency ω is high enough, the grounded-cathode amplifier has a constant gain magnitude M_v and a constant phase shift of $180°$ between the input and output voltages. Notice that as $R_l R_i \rightarrow \infty$ the gain of the amplifier can approach but not exceed the amplification factor μ of the tube.

Input Resistance. The input resistance for the low-frequency application of the grounded-cathode amplifier is particularly easy to obtain, since the grid current is assumed to be zero, or

$$R_{in} = \frac{E_g}{I_g} = \frac{E_g}{0}$$

$$= \infty \tag{10-15}$$

Output Resistance. The output resistance can be determined either by placing an assumed voltage source across the output terminals and calculating the resulting voltage-current ratio or by calculating the open-circuit voltage to short-circuit current ratio using the original circuit (Fig. 10-4b). Using the former method Fig. 10-5a shows the circuit that can be used to obtain the output resistance, which is defined as $R_{out} \triangleq V_1/I_1$. Figure 10-5a is the same as Fig. 10-4b except that a voltage source V_1 is applied to the output terminals, and Fig. 10-5b is a simplification of Fig. 10-5a in that the independent voltage sources E_s and μE_g have been

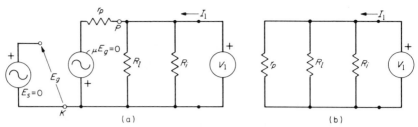

FIG. 10-5. a) Circuit used to determine the output resistance of Fig. 10-4b.
b) Part a simplified.

properly removed by replacing them by a short circuit. Figure 10-5 shows
that the output resistance is simply the parallel combination of r_p, R_l,
and R_i, or

$$\frac{1}{R_{out}} = \frac{1}{r_p} + \frac{1}{R_l} + \frac{1}{R_i}$$

and, rearranging,

$$R_{out} = \frac{r_p R_l R_i}{r_p R_l + r_p R_i + R_l R_i} = \frac{r_p R_{ac}}{r_p + R_{ac}} \qquad (10\text{-}16)$$

The simplifying assumption has again been made that the capacitive
reactances $1/\omega C$ are quite small so that the reactive terms are small compared to the real terms (0.1 or less). The reader should derive Eq. 10-16
using the open and short circuit test since R_{out} is the same as the Thévenin
equivalent resistance of the circuit.

EXAMPLE 10-1. The R-C-coupled amplifier shown in Fig. 10-4a has the following circuit values:

$$\mu = 20 \qquad\qquad C_c = 1 \ \mu f$$
$$r_p = 10{,}000 \ \Omega \qquad C_k = 20 \ \mu f$$
$$R_l = \ \ 5{,}000 \ \Omega \qquad R_k = 1000 \ \Omega$$
$$R_i = 20{,}000 \ \Omega$$

If $e_s = 3 \sin 10{,}000t$, find a) A_v, b) R_{in}, c) R_{out}, and d) E_{out}.

Solution: We first check to see if the capacitive reactances $1/\omega C_k$ and $1/\omega C_c$
are small enough to be neglected.

$$\frac{1}{\omega C_k} = \frac{1}{10{,}000 \times 20 \times 10^{-6}}$$
$$= 5 \text{ ohms}$$

which is small in comparison to R_k; that is, $1/\omega C_k = 1/200 \ R_k$ and so can be
neglected with little error being introduced in the calculation. The reactance of the
coupling capacitor is

$$\frac{1}{\omega C_c} = \frac{1}{10{,}000 \times 1 \times 10^{-6}}$$
$$= 100 \text{ ohms}$$

which is small in comparison to R_i, R_l, or r_p and so can be neglected. A check should always be made to see if the reactance terms can be neglected for both the gain and the output-impedance calculations, for, although the reactances might be negligible as far as the gain calculation is concerned, they might not be negligible for the output-impedance calculation.

Since both capacitive reactances are negligible, Eqs. 10-12, 10-15, and 10-16 can be used, or

$$A_v = -\frac{20 \times 5000 \times 20,000}{10,000 \times 5000 + 10,000 \times 20,000 + 5000 \times 20,000} = 5.7\underline{/180°}$$

$$R_{in} = \infty$$

$$R_{out} = \frac{10,000 \times 5000 \times 20,000}{350 \times 10^6} = 2860 \text{ ohms}$$

$$E_{out} = |A_v| E_s = 5.7 \times \frac{3}{\sqrt{2}} = 12.1 \text{ volts}$$

See Chapter 13 for examples where the coupling capacitors cannot be replaced by short circuits and thus where the voltage-current ratios are termed impedances rather than resistances for quantities such as R_{out}.

To summarize the grounded cathode relationships

$$R_{in} = \infty$$

$$R_{out} = \frac{r_p \cdot R_{ac}}{r_p + R_{ac}} \qquad (10\text{-}17)$$

$$A_{GC} = \frac{E_{out}}{E_s} = \frac{-\mu R_{ac}}{r_p + R_{ac}} \qquad (10\text{-}18)$$

10-2. The Cathode Follower

As will soon be seen, the cathode follower is really not an amplifier, since the overall gain of the circuit is always less than 1; however, the device is very useful in isolating one circuit from another and in matching circuit impedances. Figure 10-6a shows a diagram of a cathode follower, and Fig. 10-6b is the corresponding a-c equivalent circuit. The interelectrode capacitances are neglected, since only the low-frequency application will be considered here.

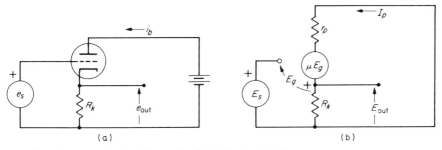

(a)   (b)

FIG. 10-6. a) Cathode-follower circuit. b) Cathode-follower a-c equivalent circuit.

Gain. In this circuit R_k is both the bias and the load resistance. The equations necessary to describe the circuit behavior can be written using either instantaneous or rms quantities. For the grid-cathode circuit using the assumed voltage polarities and current directions shown in Fig. 10-6b,

$$E_s = E_g + I_p R_k \qquad (10\text{-}19)$$

By definition

$$E_{out} = I_p R_k \qquad (10\text{-}20)$$

where the reference for the output is the ground node (the negative side of E_{bb}) instead of the cathode. The plate current is

$$I_p = \frac{\mu E_g}{r_p + R_k} \qquad (10\text{-}21)$$

Let us calculate $A_{CF} = E_{out}/E_s$ from Eqs. 10-19, 10-20, and 10-21 by eliminating I_p and E_g.

$$E_s = \frac{r_p + R_k}{\mu} I_p + R_k I_p = \left[\frac{r_p + R_k(\mu + 1)}{\mu R_k} \right] E_{out}$$

or

$$A_{CF} = \frac{E_{out}}{E_s} = \frac{\mu R_k}{r_p + R_k(\mu + 1)} \qquad (10\text{-}22)$$

It is instructive to compare Eq. 10-22 to the gain equation for the grounded-cathode amplifier. To do so, we rewrite Eq. 10-22 by dividing both numerator and denominator by the factor $(\mu + 1)$.

$$A_{CF} = \frac{E_{out}}{E_s} = \frac{\dfrac{\mu}{\mu + 1} R_k}{\dfrac{r_p}{\mu + 1} + R_k} \qquad (10\text{-}23)$$

Output and Input Resistance. The reason for writing the cathode-follower gain equation in the form of Eq. 10-23 will be clear after the output resistance has been calculated. The output resistance is obtained by applying a source voltage V_1 at the output terminals, as shown in Fig. 10-7, and then calculating $V_1/I_1 = R_{out}$ (the reader is encouraged to use the open-circuit voltage/short-circuit current calculation as a check on the following evaluation for R_{out}). The necessary voltage-current equations using I_p and I_1 as loop currents are

$$(r_p + R_k)I_p + R_k I_1 = \mu E_g \qquad (10\text{-}24)$$

$$R_k I_p + R_k I_1 = V_1 \qquad (10\text{-}25)$$

$$V_1 = -E_g \qquad (10\text{-}26)$$

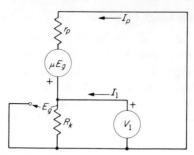

FIG. 10-7. Circuit used in calculating the output resistance of the cathode follower.

Using determinants to solve for I_1 after substituting Eq. 10-26 into Eq. 10-24, we obtain

$$I_1 = \frac{\begin{vmatrix} r_p + R_k & -\mu V_1 \\ R_k & V_1 \end{vmatrix}}{\begin{vmatrix} r_p + R_k & R_k \\ R_k & R_k \end{vmatrix}} = \frac{(r_p + R_k + \mu R_k)V_1}{r_p R_k + R_k^2 - R_k^2}$$

or

$$R_{out} = \frac{V_1}{I_1} = \frac{r_p R_k}{r_p + (\mu + 1)R_k} \qquad (10\text{-}27a)$$

and, dividing both numerator and denominator by $(\mu + 1)$,

$$R_{out} = \frac{\dfrac{r_p}{\mu + 1} R_k}{\dfrac{r_p}{\mu + 1} + R_k} \qquad (10\text{-}27b)$$

Notice that Eqs. 10-23 and 10-27b have the same denominator and that the output resistance is simply the parallel sum of the resistances shown in the denominator of the gain equation. The vacuum tube has effectively transferred the plate resistance r_p to the cathode side of the vacuum tube by dividing the plate resistance by the factor $(\mu + 1)$.

Since the input signal is applied to the grid of the tube, the input resistance is again considered infinite. To sum up the pertinent cathode follower relationships at low frequencies:

$$R_{in} = \infty \qquad (10\text{-}28)$$

$$R_{out} = \frac{\dfrac{r_p}{\mu + 1} R_k}{\dfrac{r_p}{\mu + 1} + R_k} \qquad (10\text{-}29)$$

$$A_{CF} = \frac{E_{out}}{E_s} = \frac{\dfrac{\mu}{\mu + 1} R_k}{\dfrac{r_p}{\mu + 1} + R_k} \qquad (10\text{-}30)$$

10-3. The Grounded-Grid Amplifier

The grounded-grid-circuit configuration is particularly useful in high-frequency applications, since the grid acts as an electrostatic shield or screen between the input and output circuits, thereby reducing the undesirable effects of feedback. Figure 10-8a is a schematic diagram of a grounded-grid amplifier, and Fig. 10-8b is the usual low-frequency a-c equivalent circuit with the interelectrode capacitances neglected.

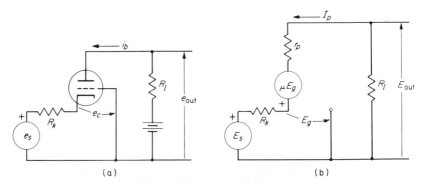

Fig. 10-8. a) Grounded-grid amplifier. b) A-c equivalent circuit of part a.

If we assume that plus-to-minus cathode-to-ground signal voltages are applied and that the other circuit voltages and currents assume their standard polarities and directions, the following relations can be written:

$$E_g = -E_s - I_p R_k \tag{10-31}$$

$$E_{out} = -I_p R_l \tag{10-32}$$

$$I_p = \frac{\mu E_g - E_s}{r_p + R_l + R_k} \tag{10-33}$$

Gain. To solve for the grounded-grid amplifier voltage gain A_{GG}, we eliminate E_g and I_p from Eqs. 10-31, 10-32, and 10-33.

$$-\frac{E_{out}}{R_l} = \frac{-\mu[E_s + (-E_{out}/R_l)R_k] - E_s}{r_p + R_l + R_k}$$

$$\left(\frac{r_p + R_l + R_k}{R_l} + \mu \frac{R_k}{R_l}\right)E_{out} = (\mu + 1)E_s$$

or

$$A_{GG} = \frac{E_{out}}{E_s} = \frac{(\mu + 1)R_l}{r_p + (\mu + 1)R_k + R_l} \tag{10-34}$$

Output and Input Resistance. Before continuing with the calculation of R_{in} and R_{out} of the grounded-grid amplifier, it is instructive to understand the behavior of the $(\mu + 1)$ factor in regard to the resistances. In the grounded-cathode type[1] of gain equation for the cathode follower, the plate resistance was transferred to the cathode side by dividing r_p by $(\mu + 1)$. In the cathode follower it was desired to calculate the output resistance on the cathode side of the vacuum tube. However, in the grounded-cathode type of gain equation for the grounded-grid amplifier, the cathode resistance is transferred to the plate side by multiplying it by the $(\mu + 1)$ factor. The tube thus behaves much like a transformer in regard to resistances, with the square of the transformer turns ratio equaling $(\mu + 1)$. To illustrate the proper use of the $(\mu + 1)$ factor, the output and input resistances of the grounded-grid amplifier will now be calculated.

The gain equation (Eq. 10-34) can be used in calculating the plate-side output resistance, since all resistances are already referred to the plate side of the tube. From Fig. 10-8 it can be seen that r_p and R_k are in series, and, in turn, these two series resistances are in parallel with R_l as far as the calculation for the output resistance is concerned.

$$R_{\text{out}} = \frac{R_l[r_p + R_k(\mu + 1)]}{R_l + r_p + R_k(\mu + 1)} \tag{10-35}$$

Figure 10-8 can also be used in calculating the input resistance, which is defined as $R_{\text{in}} = E_s/I_s = E_s/-I_p$, and it is evident that we are "looking into" the cathode-side of the tube circuit for this situation. The plate resistance r_p and the load resistance R_l must be "transformed" to the cathode side of the tube. The input resistance is the series sum of R_k and the two transformed plate resistances, or

$$R_{\text{in}} = \frac{E_s}{I_s} = \frac{E_s}{-I_p} = R_k + \frac{r_p}{\mu + 1} + \frac{R_l}{\mu + 1} \tag{10-36}$$

The reader should check Eqs. 10-35 and 10-36 by using conventional analysis methods such as loop current or node voltage equations.

To sum up for the grounded-grid amplifier:

$$R_{\text{in}} = R_k + \frac{r_p + R_l}{\mu + 1} \tag{10-37}$$

$$R_{\text{out}} = \frac{R_l[r_p + R_k(\mu + 1)]}{R_l + r_p + (\mu + 1)R_k} \tag{10-38}$$

$$A_{GG} = \frac{E_{\text{out}}}{E_s} = \frac{(\mu + 1)R_l}{r_p + (\mu + 1)R_k + R_l} \tag{10-39}$$

[1]The grounded-cathode type of gain equation simply refers to the form of Eqs. 10-30 and 10-34 when compared to the grounded-cathode gain equation (Eq. 10-18). In each of these equations, the resistance across which the output voltage E_{out} is taken appears by itself in the denominator of the gain equations.

10-4. Cascading of Amplifiers

A complete electronic circuit usually consists of many elementary transistor and/or tube circuits tied together. As mentioned in Chapter 7, the overall circuit behavior can be very complex, and the solution can require advanced techniques in analysis. In this section only a simple cascading[2] of circuits will be considered.

The problem to be solved is to find the overall circuit gain, and Thévenin's theorem will be the primary tool of analysis. In developing the analysis we usually start from the circuit input and proceed in an orderly manner to the circuit output. Let us consider Fig. 10-9. The first

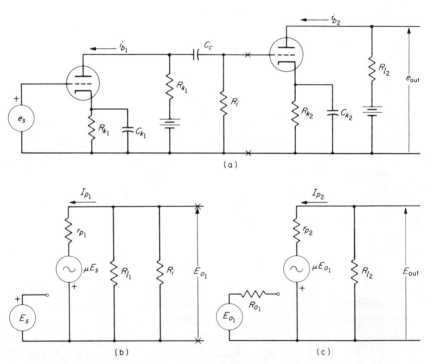

Fig. 10-9. Circuit reduction of two cascaded grounded-cathode amplifiers.

step is to recognize the individual units that comprise the entire amplifier. It should be apparent that there are two grounded cathode amplifiers in the circuit and they are cascaded.

If the frequency is high enough so that X_{C_c}, X_{Ck_1} and X_{Ck_2} are very small, the a-c equivalent circuit is given in Figs. 10-9b and c. Figure 10-9b shows the equivalent circuit of the first stage only. According to Thévenin's theorem, any circuit consisting of linear bilateral elements can

[2]Cascading means connecting the output of one stage to the input of the next stage.

be replaced at any given pair of terminals by a voltage source and an equivalent resistance. The voltage source is the voltage appearing across the open-circuited terminals, and the equivalent resistance is the resistance seen looking into the open-circuited terminals. The terminals marked XX represent such a pair. The open-circuit voltage E_{o_1} is seen to be

$$E_{o_1} = A_{GC_1} E_s$$

or

$$E_{o_1} = \frac{-\mu_1 \dfrac{R_{l_1} R_i}{R_{l_1} + R_i}}{r_{p_1} + \dfrac{R_{l_1} R_i}{R_{l_1} + R_i}} E_s \qquad (10\text{-}40)$$

where A_{GC_1} is the previously derived gain expression for the grounded-cathode amplifier (see Eq. 10-18).

The Thévenin equivalent resistance of the first stage is

$$R_{o_1} = \frac{1}{\dfrac{1}{r_{p_1}} + \dfrac{1}{R_{l_1}} + \dfrac{1}{R_i}} \qquad (10\text{-}41)$$

If it is assumed that no grid current flows, the total value of E_{o_1} is applied to the grid of the second stage, regardless of the value of R_{o_1}. Thus the gain of the second stage is

$$A_{GC_2} = \frac{E_{\text{out}}}{E_{o_1}} = \frac{-\mu_2 R_{l_2}}{R_{l_2} + r_{p_2}} \qquad (10\text{-}42)$$

The overall or total gain is defined to be

$$A_T = \frac{E_{\text{out}}}{E_s} = \frac{E_{o_1}}{E_s} \frac{E_{\text{out}}}{E_{o_1}} = A_{GC_1} A_{GC_2}$$

or

$$A_T = \frac{\mu_1 \mu_2 R_{l_1} R_i R_{l_2}}{[r_{p_1}(R_{l_1} + R_i) + R_{l_1} R_i](R_{l_2} + r_{p_2})} \qquad (10\text{-}43)$$

and, as expected, the phase shift for two cascaded grounded amplifiers is $2 \times 180°$ or a net $0°$ phase shift.

As another example, let us consider Fig. 10-10, in which a cathode follower and a grounded-grid amplifier are cascaded. Again employing Thévenin's theorem, we first separate or open the circuit at the terminals marked XX. We then calculate the open-circuit terminal voltage and the Thévenin equivalent resistance (which is also the output resistance of this first stage):

$$E_{o_1} = A_{CF} E_s$$

or

$$E_{o_1} = \frac{\dfrac{\mu_1}{\mu_1 + 1} R_{k_1}}{\dfrac{r_{p_1}}{\mu_1 + 1} + R_{k_1}} E_s \tag{10-44}$$

$$R_{\text{out}} = R_{o_1} = \frac{\dfrac{r_{p_1}}{\mu_1 + 1} R_{k_1}}{\dfrac{r_{p_1}}{\mu_1 + 1} + R_{k_1}} \tag{10-45}$$

This Thévenin generator is then connected to the XX terminals, as shown in Fig. 10-10c. In this case, R_{o_1} is definitely not to be neglected and

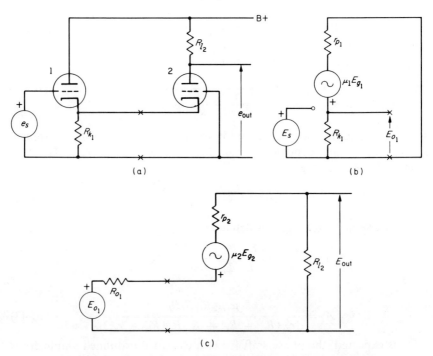

FIG. 10-10. Cascaded cathode follower and grounded-grid amplifier.

must be included in the calculation for the gain of the grounded-grid amplifier (refer to Eq. 10-39).

$$A_{GG} = \frac{E_{\text{out}}}{E_{o_1}}$$

$$= \frac{(\mu_2 + 1) R_{l_2}}{r_{p_2} + R_{l_2} + (\mu_2 + 1) R_{o_1}} \tag{10-46}$$

The overall gain is

$$A_T = \frac{E_{out}}{E_s} = \frac{E_{o_1}}{E_s} \frac{E_{out}}{E_{o_1}} \qquad (10\text{-}47)$$

or

$$A_T = A_{CF} A_{GG} \qquad (10\text{-}48)$$

and, upon substituting Eqs. 10-44, 10-45, and 10-46 into Eq. 10-48, we obtain

$$A_T = \frac{\mu_1 (\mu_2 + 1) R_{k_1} R_{l_2}}{(r_{p_2} + R_{l_2})[r_{p_1} + (\mu_1 + 1) R_{k_1}] + (\mu_2 + 1) r_{p_1} R_{k_1}} \qquad (10\text{-}49)$$

The application of Thévenin's theorem makes possible an orderly, step-by-step reduction of the circuit. Other methods employing loop or nodal analysis, or even more sophisticated techniques, provide a more comprehensive and elegant means of analyzing circuits, but, for the types of problems encountered in this text, Thévenin's theorem is quite adequate.

Problems

10-1. In the circuit of the accompanying illustration, $E_s = 3$ volts. a) Draw the a-c equivalent circuit, b) calculate the circuit gain, and c) find E_{out}.

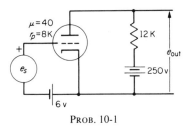

PROB. 10-1

10-2. If the tube used in the accompanying illustration has a μ of 25 and an r_p of 9000 ohms, what must R_l be in order for $E_{out}/E_s = -16$?

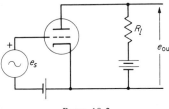

PROB. 10-2

10-3. If $e_s = \sin 10,000t$ in the circuit shown in the accompanying illustration, find a) the plate-circuit resistance, b) the output resistance, and c) the circuit gain. Justify all simplifying assumptions.

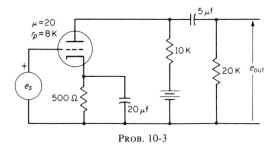

PROB. 10-3

10-4. Using the circuit shown in the accompanying illustration, find a) the circuit gain, if the output voltage is taken from plate to ground, and b) the output resistance.

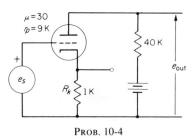

PROB. 10-4

10-5. Repeat Prob. 10-4 if the output voltage is taken across the cathode resistance R_k.

10-6. In the circuit shown in the accompanying illustration, $R_k = 5000$ ohms, $\mu = 40$, $r_p = 12,000$ ohms and $E_{bb} = 250$ volts. What must the value of R_1 be to give maximum a-c power output to this load resistance? (Recall the maximum-power-transfer theorem).

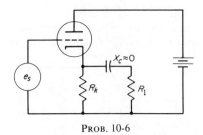

PROB. 10-6

10-7. If $e_s = 2.5 \sin 50,000t$ volts in the circuit shown in the accompanying illustration and $\mu = 30$, $g_m = 1000$ μmhos, find a) the gain, b) E_{out}, and c) the output resistance. Justify all simplifying assumptions. (*Hint:* $X_C = 1/\omega C$, $X_L = \omega L$.)

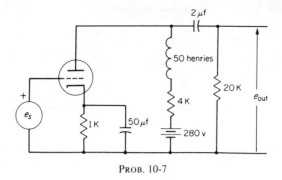

PROB. 10-7

10-8. If all the capacitive reactances are negligibly small in the accompanying illustration, find the overall circuit gain E_{out}/E_s.

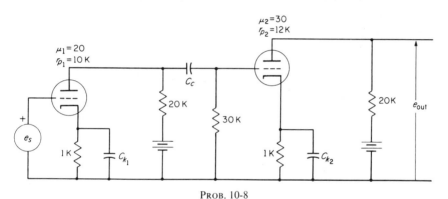

PROB. 10-8

10-9. The two triodes in the accompanying illustration are identical with $\mu = 20$ and $r_p = 10$ K. The quiescent plate currents are known to be 2 ma, and the desired value of E_c for each tube is -6 volts. a) Find R_{k_1} and R_{k_2} and b) find the overall circuit gain E_{out}/E_s.

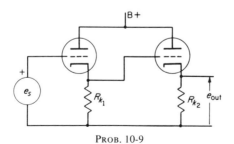

PROB. 10-9

10-10. The two triodes in this accompanying illustration are identical with $\mu = 18$ and $r_p = 9000$ ohms. Find the overall circuit gain E_{out}/E_s.

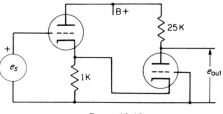

PROB. 10-10

10-11. All the triodes in the accompanying illustration are identical with $\mu = 25$ and $r_p = 8$ K. If all the capacitive reactances are negligibly small, find the overall circuit voltage gain.

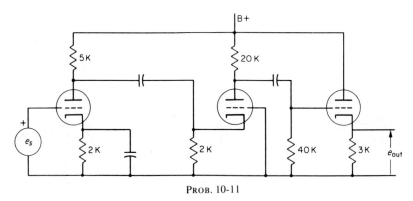

PROB. 10-11

11 / *Digital Computers*

One of the greatest impacts of the advent of the transistor and its progeny has been in the fantastic growth of the digital computer industry. In slightly more than a decade, the fourth generation of computers is making its appearance, with each generation being faster, larger and lower in cost per unit computational element than the preceding generation. Two characteristics of the electronic digital computer distinguish it from its mechanical predecessor, namely, speed and memory. Operational times are measured in microseconds and nanoseconds as compared to seconds for mechanical calculators. The ability to store millions of separate pieces of information in the memory of a large computer and to recall any particular piece of information upon demand simply has no analog in a mechanical calculator. To illustrate further that speed and memory constitute the computer's main advantage, it should be pointed out that most computer's mathematical abilities are restricted to being able to add, to subtract, to multiply, to divide and to compare two numbers for relative size. This illustrates the fact that the computer is not using some mysterious or sophisticated method of solving problems. As an example, suppose one wanted the sum of the first 1000 positive integers. It might take a computer about 6 milliseconds to produce the result of 500,500. Man on the other hand would look for some other way of doing the problem. Recognizing that the sum of a simple arithmetic progression is desired, looking up the formulation which is $n(n + 1)/2$, and performing the arithmetic rewards the intelligent individual with the result without actually having to add 1000 numbers. This simpler calculation, however, still requires some seconds so that the computer is many times faster even though it does not possess man's cleverness. And of course, if a sum cannot be put in closed form or must be done many times, the computer then gives a real advantage, since it neither tires nor becomes bored. Even though a computer may have upwards of millions of components, the heart of the computational unit is relatively simple to understand in terms of its fundamental operations. Before examining the internal organization and operation, a brief account of how this computational speed and enormous memory can help man might be useful.

11-1. Applications

There is scarcely an area of human endeavor that is not touched, knowingly or unknowingly, by this age of computers. Thus only a few broad areas can be mentioned with the intent of showing how widespread are the uses of the digital computer.

Data Reduction. Man's capacity to absorb data and make a decision is limited by his speed to assimilate the data fast enough to be useful. As an example, all would agree that the more temperature, pressure, wind velocity, cloud formation, and other pieces of data a weather observer had, the more intelligent weather prediction he could make. By the use of weather satellites, widespread communication networks, etc., he now has access to so much data available at any instant, that to assimilate it all would take days at which time his prediction would be worthless since the data he used is then old or worthless. He is hampered by his lack of speed. If he can transfer his data evaluation criteria to a computer, it can give him the evaluation in a matter of seconds so that his prediction based on this evaluation is now timely. Other areas where time would not be so important but where the sheer volume of data to be reduced would be prohibitive if done by man would be census data reduction, income tax processing, inventory status of products of national and international corporations, deployment of military equipment and replacement parts, air-line reservation service, to name a few.

Monitoring. Advantage can be taken of the computer's ability to perform the same task continuously without becoming bored. An example might be the intensive care section of a large hospital. Each patient must be closely monitored which previously meant that an individual attendant was assigned to each patient. Devices are now available for continuously measuring blood pressure, heart activity, temperature, etc. These devices produce electrical signals which are appropriate inputs to a digital computer (its signals are electrical). It will monitor many patients simultaneously and call for help whenever it is needed. Meanwhile the attendant is released for more productive work until he is actually needed by a patient. In this same manner, the computer can figuratively watch continuously gauges which may represent temperature, water level, pressure, voltage, power, energy and a host of other quantities that are measurable via electrical instrumentation.

Control. In the same manner that the digital computer can be made to accept and to properly interpret electrical signals representing some information, it can be made to provide electrical signals in accordance with a preplanned course of action which is stored in the computer's memory. This course of action or decision is provided by the individual responsible for the control of the system being monitored by the computer. An excellent example of this is the control of a space rocket. Man can, in an analytic manner, determine all of the factors that must be taken into account in order to make a thrust adjustment upon launching a space vehicle. However, all these factors must be considered in a fraction of a second and the necessary action taken immediately to keep the rocket under control and to achieve the proper orbit. This analytic procedure is programmed into a computer and the computer then provides the proper control signals based on the input signals. Many industrial processes are now automated in the same manner whereby a computer can issue corrective action or control signals whenever some disturbance takes place in the process. In numerical-controlled cutting machines, a digital computer can analyze a mathematical description of the cutting contour and manipulate the machine accordingly in a much more precise manner than previously, thereby holding much closer tolerances.

Simulation. By simulation, the operation of a system or piece of equipment can be predicted by setting up the necessary descriptive equations on the computer. The computer can, in many instances, determine the ultimate success or failure of a particular system without the equipment actually having to be built, thereby saving much time and expense. Also, once a given system has been chosen or designed, the computer can aid in seeking the best possible value or values for the many parameters that are usually present in the equations.

Library. The computer's ability to maintain vast amounts of information in its memory makes it useful as a library. However, it can do much more through use of stored search programs. As an example, a medical library on diseases provides a physician with all recorded information (symptoms, treatment, further references, etc.) when he types the name of the disease on his teletypewriter in his office connected to a computer via a telephone line. The actual computer may be anywhere as long as a telephone link can be established. Information resulting from days of library research are at his disposal in a matter of seconds. The physician may need the reverse procedure, i.e., he types all the symptoms from his diagnosis and the search program gives him all the diseases exhibiting these symptoms. He then can concentrate on this information until he has isolated the disease. At the same time, he can provide daily patient information to the system and the computer can detect the incidence of an epidemic in a given area and alert the proper authorities.

A library on civil and criminal court cases is providing a similar service for the legal profession.

Education. Computers are being used at all levels to help in the educational process. Coupled with an audio and visual display device, a tutor-student relationship can be emulated. Again, the computer's insensitivity to boredom lets it exhibit the utmost patience in proceeding at the learner's pace, however slow. The instructional material can also be structured according to the level of each individual.

Arts. The frequency characteristics of musical instruments can be stored in the memory of a digital computer. A composer can then take his score for each instrument and input it to the computer whereby the computer processes this information and provides the proper electrical signals to play the composition through an amplifier and speaker. In this way a composer gets some idea as to the symphonic sound of his work without the need and expense of the orchestra in the early stages of the development of the work.

Some electronic visual artwork has been performed on digital computers with the results then printed on paper or used on television to create special visual effects.

11-2. Computer Organization

It is traditional to break the digital computer into five sections based on a functional point of view. These are shown in Fig. 11-1. Information

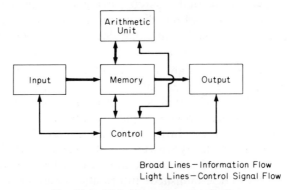

Broad Lines — Information Flow
Light Lines — Control Signal Flow

FIG. 11-1. Digital-computer organization.

is brought into the computer's memory section by way of the input section. The input section converts the information from the user's form or language to that required by the machine. Information is processed arithmetically or logically in the arithmetic section. Eventually when all processing is completed the results are sent to the output section where they are converted back to the form desired by the user. All of these operations are under the guidance of the control section which sends out a sequence of signals to perform a given task in one of the other sections. In many computers, different tasks may be performed simultaneously with a "finished" signal being sent back to the control section upon completion of the task. A further discussion of each section follows.

Input Section. The user must provide his information to the computer on punched cards, punched tape or magnetic tape. This information must be in a so-called programming language; e.g., FORTRAN, so as to be unambiguous when electrically "read" and interpreted by the computer. The information may also be input by means of a teletypewriter when the user is remotely located from the computer. Specialists or "programmers" preparing programs for the computer are generally highly trained in the use of several programming languages but even the most sophisticated computer can be used by anyone with some knowledge of the more universal programming languages which are "understood" by most machines. The major task of this input section is to convert the user information into *binary* form which is the language of the memory and arithmetic sections that are to be looked at shortly.

It is not to be construed that the input section does this conversion independent of the other sections. Reference to Fig. 11-1 shows that control signals for all tasks emanate from the control section. The arithmetic and memory sections would both have roles to play in this conversion as well.

Output Section. The task of the output section is the reverse of the input section. It must convert the results from binary form into a form readable by the user; i.e., text and numbers. It may output this information via any number of different devices. A typewriter can be used but this is slow in comparison to the speed at which the computer can provide this information. The user may request that this information be punched on cards for use as input back into the machine at a later time. A magnetic-tape recording can be made for the same reason. If he wants to view the information but not retain it, he may direct that it be placed on a visual display device (CRT) or if the output information is amenable, output it through an audio device. If he wants so-called "hard-copy" he may request that a photograph of the visual display be made, or if applicable, an x-y plot of the information be made. By far, the most common type of output device of reasonable speed is an inpact printer which can print several hundred to a thousand lines per minute.

Memory Section. Before discussing the memory section of the computer, some aspects of the machine language must be given. A binary language is used which is best explained in terms of numbers. Recall that the decimal number system uses a base or radix of 10 and the number 4038 means

$$4038_{10} = 4 \times 10^3 + 0 \times 10^2 + 3 \times 10^1 + 8 \times 10^0$$

The decimal system has 10 allowable digits (0 to 9) and any number system has as many digits (0 to b-1) as given by the base (b). Hence the binary system (base = 2) has two digits, 0 and 1. Any integer is expressible in any number system. Hence

$$21_{10} = \underbrace{1 \times 2^4}_{16_{10}} + \underbrace{0 \times 2^3}_{0} + \underbrace{1 \times 2^2}_{4_{10}} + \underbrace{0 \times 2^1}_{0} + \underbrace{1 \times 2^0}_{1_{10}} = 10101_2$$

and

$$4038_{10} = 111111000110_2$$

The utility of the binary system comes from the fact that the system has only two digits and therefore any bistable device (on-off, high-low) can be used to store binary numbers with each state of the device representing one of the two digits (on $\rightarrow$ 0, off $\rightarrow$ 1). If the 26 letters of the English alphabet plus any other desired symbols are also "encoded" in binary, i.e., each letter has its own unique binary code and a way is developed of handling literals and digits together in the machine, a machine "language" results.

What are possible devices for storing these binary digits (bits) in an electrical system? The transistor could be used with its conducting and nonconducting switch-like property. A relay with its energized and de-energized two-state property is suitable. A magnetic material fashioned into a loop (core) with its bidirectional magnetic property is also a candidate. The first requirement on the two-state element is that it be fast in operation so as not to limit the speed of the arithmetic operation. If millions of these bits are to be stored then the device must be relatively inexpensive. The relay suffers from both speed and cost deficiencies. The transistor, while exceptionally fast, is expensive if millions are required for a single digital computer. The impact of microcircuits and large scale integration (LSI) might change this economic factor (see Chapter 8) but the magnetic core is the predominant device now in use for the computer's memory section. This memory device is also nonvolatile; i.e., in the event of a power failure the information in a core memory is not lost since the cores retain their residual magnetic property.

A brief description of the principle of operation of a magnetic-core memory will now be given. Magnetic cores are quite small, and the ferrite material of which they are made has an almost rectangular *B-H* hysteresis loop, as shown in Fig. 11-2a. Recall that *B* is the magnetic flux density (see Chapter 4) and *H* is the magnetic field intensity which, in turn, is proportional to current. The *B* remaining, when $H = 0$, is called the residual flux; that is, the core is in a magnetized state with no current being applied. Figure 11-2b shows how six of these cores might be utilized in a magnetic-core matrix array. Lines *a, b, c,* 1, and 2 represent insulated electrical conductors that pass through the centers of these cores. When a pulse of current is sent through any one of these wires, nothing happens if the current produces only 1 unit of magnetic field intensity *H* since

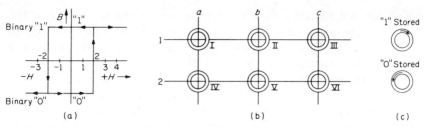

FIG. 11-2. a) *B-H* curve of ferritic material used in magnet cores. b) 2 × 3 magnetic-core matrix. c) Binary-bit storage by residual magnetic flux direction.

Fig. 11-2a shows that 2 units are required to force the core to go through the transition region, "0" → "1" or "1" → "0." If a pulse of current corresponding to 1 unit of *H* is simultaneously sent through line *b* and line 2, then core *V* is magnetized in a given direction. If the two current pulses are reversed, corresponding to −2 units of *H*, the residual flux in the core reverses direction. If the residual flux is clockwise or counterclockwise in the core, a binary bit 1 or 0 is stored, respectively as indicated in Figs. 11-2c. This illustrates how a core matrix can serve as a memory, but of course much additional circuitry is necessary to make this a usable storage unit.

Other memory elements have also been used but to a much lesser extent than magnetic cores. A thin film memory element has the advantage that it is faster operating than a core element but also is more expensive. It consists essentially of small spots of magnetic material vaporized onto a thin sheet of dielectric material. The same advantages apply to the plated wire memory element. These and any other elements utilizing magnetism of a ferrite material essentially have the same principle of operation as the core and therefore will not be discussed further. Other memory elements have been proposed using deposition of charges, laser techniques, crystal technology, etc. They offer advantages in speed, volume, i.e., bit packing density, etc., but none has been used outside of the laboratory to any great extent.

The single most descriptive measure of the memory of a digital computer is its *memory cycle time.* This is the time required to retrieve (read from core) a piece of information from memory after its address has been specified. This time is generally measured in microseconds and nanoseconds for most computers. A magnetic core memory is also classified as a *random-access* memory since each location is accessible in the same amount of time. Most large digital computers have a second level of memory in addition to the magnetic core first level. The arithmetic unit utilizes individual pieces of data from the first level while blocks of information are generally transferred between the first and second levels of memory. The second level can then afford to be somewhat slower since it does not communicate directly with the fast arithmetic unit. This loss in speed is compensated by decreased cost per unit bit and increased bit density per unit volume. Here again magnetic devices predominate with the disc, drum and tape cited in order of increasing memory cycle time. These devices are nonrandom access since the time to retrieve a piece of information depends on the location of the information relative to the read-head positions at the time the address is specified. Naturally the read operation must wait until the information is passing below the read-heads. The magnetic disc and magnetic drum are rotating continuously at a fixed speed while the magnetic tape is generally idling until an address is received that is to be accessed for reading or writing (storing) a piece of information. The magnetic tape consists of a thin layer of magnetic material deposited on a flexible tape, such as acetate, polyester or Mylar. A magnetic drum consists of a cylinder whose surface is coated with a magnetic material. Read/write heads are located along tracks around the cylinder. A disc pack consists of a series of discs with both surfaces coated with a magnetic material and arranged with their axis coincident and a separation space into which the read/write heads are inserted. A binary bit is recorded on or read from the tape, drum or disc by means of a magnetic read/write head. Many drums and disc packs will contain more than one set of read/write heads. This decreases the average cycle time since it is proportional to the average relative location of a piece of information with respect to the nearest read/write head.

Control Section. The function of the control unit is to see that the operations are carried out in the sequence ordered by the original program. The instructions are transferred from the memory to the control unit and are held there until the instruction is carried out. Thus the control units contain registers that provide for temporary storage. These registers or memory elements are always constructed from transistors, microcircuits or integrated circuits. They are few in number compared to the memory section and therefore speed rather than cost is of the utmost importance.

The control unit also generates the timing pulses that cause the computer to perform the required operations in an orderly manner. A *synchronous* computer proceeds with the calculations at a predetermined rate in accordance with an internal clock. An *asynchronous* computer starts each operation immediately upon receiving a signal indicating that the preceding operation is completed.

Arithmetic Section. The primary function of this section is to perform the arithmetic and logical operations when provided with the proper signals from the control section. Here again, transistors, microcircuits or integrated circuits are used almost exclusively because of their speed advantage over other devices that could perform the same functions. The design of the arithmetic section is based on binary arithmetic. A later section will present the essentials of binary arithmetic in order to understand the electronic implementation of these arithmetic operations. However, some material on transistor logic circuits to complement the diode logic circuits explained in Chapter 5 will now be presented.

11-3. Transistor Logic Circuits

In Chapter 5, diode logic gates were discussed and the AND and OR logical function electronic implementations were carried out in detail. Diode logic gates cannot provide any energy to the signals so that the output signal is degenerated after passing through a diode gate. An active device such as a transistor can be used to maintain the integrity of the signal as it passes through the gate and also provide inverting or complementing of a logical variable where required.

Using the transistor as a gate or a switch requires operation at the two extremes of the linear region. Consider the circuit in Fig. 11-3a and the

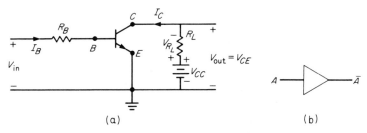

(a) (b)

Fig. 11-3. a) Common-emitter NPN circuit. b) Standard inverter symbol.

typical transistor characteristics in Fig. 11-4. The two states of this device are indicated by the regions labeled saturation and cutoff. The active or linear region separates the ON-OFF states of the transistor switch. The switch is off when I_B is zero or analogously the collector to emitter path is open circuited and V_{CE} is approximately equal to V_{CC} since I_C is essentially zero. If a positive voltage V_{in} is applied and R_B is adjusted so that I_B drives the transistor into the saturation region, the switch is on. The collector to emitter path approaches a short circuit condition ($V_{CE} = V_{CE_{sat}}$) and V_{CE} is essentially zero. Note the inversion, i.e., a low input gives a high output and vice-versa. If low and high stand for the two states of some logical variable A, then this transistor provides the complement function, i.e., A on the input gives $\bar{A}$ on the output and conversely, $\bar{A}$ on the input gives A on the output. This is also referred to as the negation or NOT logical function and the circuit symbol for the NOT logic circuit is given in Fig. 11-3b.

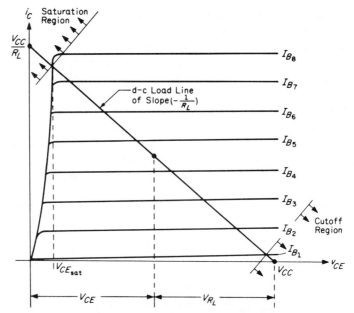

FIG. 11-4. Typical NPN transistor characteristics.

If the load resistor is placed in the emitter leg as in Fig. 11-5 and the output taken across R_L, then no inversion occurs. The output voltage V_{R_L} will be high when the input voltage is high (switch on) and low when the input voltage is low (switch off). No logical function is performed by this circuit but it can be used in combination with the diode logic circuits discussed in Chapter 5 to restore the signals to their correct voltage levels thereby compensating for nonideal diodes and resistive losses. The circuit

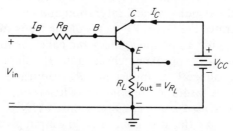

FIG. 11-5. Emitter-follower NPN circuit.

in Fig. 11-6a shows a diode transistor logical AND gate. As was the case in Chapter 5 a true state of the logical variable is electrically represented by +5 volts and a false state is electrically represented by 0 volts or ground potential (positive logic). The truth table for this circuit is given in Fig. 11-6b. The proper base current flow is regulated by the resistor R_B so

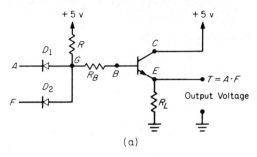

(a)

Logical Inputs		Diode Connection		Base		Transistor	Output
A	F	D_1	D_2	G	B		T
0 v	0 v	FB	FB	0 v	0 v	OFF	0 v
0 v	5 v	FB	BB	0 v	0 v	OFF	0 v
5 v	0 v	BB	FB	0 v	0 v	OFF	0 v
5 v	5 v	BB	BB	5v	+	ON	5 v

FB; Forward–Biased

BB; Back–Biased

(b)

FIG. 11-6. a) Diode-transistor positive-logic AND gate. b) Biasing and output conditions for all possible input conditions.

that with a positive voltage at point *G*, the transistor is driven into saturation reducing the collector-to-emitter voltage to approximately zero giving 5 volts potential at output point *T*. Relating the output voltage
column to the two input voltage columns shows that the output is high if
AND only if both inputs are high. Although for simplicity only two inputs are provided in this circuit, additional inputs could be added without
changing the behavior of the circuit. Diode-transistor logic gates are
referred to as DTL type logic implementation. The collector supply voltage can be chosen such that the output voltage is of the proper level.
Unlike diode logic gates, this capability allows DTL gates to be used without concern for levels (number of logic gates a signal has traversed) since
the signal is always restored to its proper level before leaving a gate. A
DTL OR gate is given in Fig. 11-7. The construction of the truth table is
left as an exercise (see Prob. 11-3). Other implementations for logic
gates consist of resistor-transistor combinations (RTL), transistor-
transistor combinations (TTL), and other combinations with each having
some advantages and some disadvantages. All of these circuit combinations electronically implement various logical functions. The transistor
inverter or logical NOT circuit can be combined with the diode logical
AND circuit to give an AND followed by a NOT, or NAND. The logical
NOT circuit can also be combined with the diode logical OR circuit to

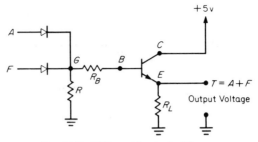

FIG. 11-7. DTL positive-logic OR gate.

give an OR followed by a NOT, or NOR. It can be shown that any logical expression can be implemented using either NAND or NOR logic circuits alone. For large digital systems, this can result in economies through volume purchasing discounts, minimum replacement inventory and simpler maintenance. The NAND and NOR logic circuits are shown in Fig. 11-8a and Fig. 11-9a respectively.

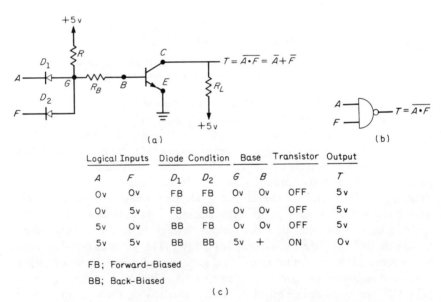

Logical Inputs		Diode Condition		Base		Transistor	Output
A	F	D_1	D_2	G	B		T
0v	0v	FB	FB	0v	0v	OFF	5v
0v	5v	FB	BB	0v	0v	OFF	5v
5v	0v	BB	FB	0v	0v	OFF	5v
5v	5v	BB	BB	5v	+	ON	0v

FB; Forward–Biased

BB; Back–Biased

(c)

FIG. 11-8. DTL positive-logic NAND gate. b) Standard logical NAND gate symbol. c) Biasing and output conditions for all possible input conditions.

The truth table for the NAND circuit is given in Fig. 11-8c. Comparing the output voltage column to the two input voltage columns confirms the implementation of the AND-NOT or NAND logical function. Of course, additional inputs could be added without changing the circuit behavior. Note the expression

$$T = \overline{A \cdot F} = \overline{A} + \overline{F}$$

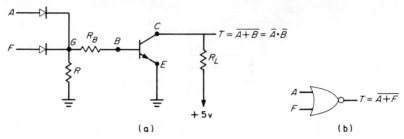

This is an illustration of De Morgan's theorem which states that the complement of any Boolean logical expression may be found by

1. replacing all OR symbols connecting terms with AND symbols and

    replacing all AND symbols connecting terms with OR symbols, and

2. complementing each term in the expression.

    Some examples are:

$$\overline{(A + B) \cdot C} = \overline{(A + B)} + \overline{C} = (\overline{A} \cdot \overline{B}) + \overline{C}$$

$$\overline{(A \cdot \overline{B} \cdot C) + (D + \overline{E})} = \overline{(A \cdot \overline{B} \cdot C)} \cdot \overline{(D + \overline{E})} = (\overline{A} + B + \overline{C}) \cdot (\overline{D} \cdot E)$$

The construction of the truth table for the NOR gate is left as an exercise. (See Prob. 11-5.) The standard circuit symbols are shown in Fig. 11-8b and Fig. 11-9b respectively. Note that the symbols are almost identical to the AND and OR gate symbols with a circle added to the output point to indicate the negation and converting the symbol to represent a NAND and NOR gate respectively.

A simple but effective memory element, called an *RS* flip-flop can be constructed by cross-coupling two NOR gates as shown in Fig. 11-10a. Other than during a transition, the two output terminals are the logical complements of each other. Applying a high voltage pulse to the *S* Input will force the flip-flop into the set state if it is in the reset state prior to the pulse. It will then remain in the set state until a positive pulse is applied to the *R* Input. Consider both the *R* and *S* Inputs at zero with the Set Output high and therefore the Reset Output low. A check of the gates will show that this is a stable condition. A positive pulse on the *S* Input will not affect the lower gate since the other lower gate input is already high from the cross-connection. A positive pulse on the *R* Input of the upper gate will drive the Set Output low which in turn through the cross-coupling drives the Reset Output high which through the cross-coupling puts a high voltage on the upper gate to hold the Reset Output low after

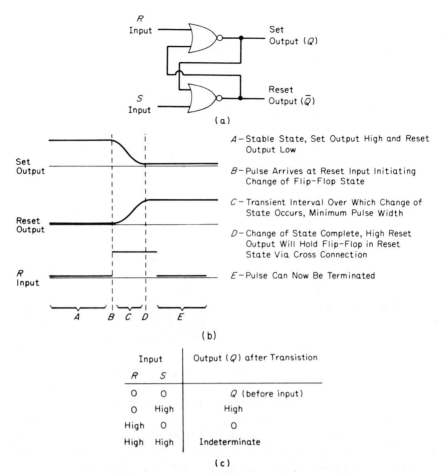

A — Stable State, Set Output High and Reset Output Low

B — Pulse Arrives at Reset Input Initiating Change of Flip–Flop State

C — Transient Interval Over Which Change of State Occurs, Minimum Pulse Width

D — Change of State Complete, High Reset Output Will Hold Flip–Flop in Reset State Via Cross Connection

E — Pulse Can Now Be Terminated

(b)

Input		Output (Q) after Transistion
R	S	
O	O	Q (before input)
O	High	High
High	O	O
High	High	Indeterminate

(c)

FIG. 11-10. a) *RS* flip-flop. b) Timing diagram for changing state of flip-flop. c) Characteristic table of *RS* flip-flop.

the pulse period has terminated. The *R* Input pulse width must be long enough for this transition to occur. See the timing diagram in Fig. 11-10b.

The characteristic table of the *RS* flip–flop is given in Fig. 11-10c. The flip-flop will maintain its state as long as no input pulses occur. The last row shows that both inputs cannot simultaneously be pulsed since the resultant state of the flip-flop would be indeterminate. The resultant state would be dependent on which of the two pulses terminated first rather than which had arrived first. Flip-flops of this sort can be used as memory elements since they maintain indefinitely whatever state they are forced to assume. They are most commonly used as memory elements in the arithmetic section of the computer because of the speed advantage over magnetic cores.

11-4. Binary Arithmetic

It has already been pointed out that the binary system lends itself to representation by two-state electronic devices such as afforded by the transistor. For these devices to be useful in an arithmetic section, they are going to have to perform the arithmetic operations of addition $(+)$, subtraction $(-)$, multiplication $(\times)$ and division $(\div)$. The increased complexity, and hence cost, of the implementation of each operation is in the above order $(+ - \times \div)$. Through the use of complementation, it will be demonstrated that all these operations can be done by addition $(+)$ alone although the speed of the nonaddition operations $(- \times \div)$ will generally be slow in comparison to other methods. As in many engineering designs, a trade-off between cost and speed has to be made.

Boolean Algebra. A few rules relating to Boolean algebra are necessary to understand the relationship of binary logical variables to the binary arithmetic operations to be discussed. A binary number system contains only two elements, zero and one. A binary logical variable can assume one of two states so that it can be used to represent a variable that is to be a binary digit. However, there is one slight difference in the connective OR operation when trying to compare it to the binary addition operation. If the false state of a logical variable is indicated by a binary zero and the true state of a binary variable is indicated by a binary one, then the parallels are seen in Table 11-1.

TABLE 11-1
COMPARISON OF LOGICAL OR AND
BINARY ADD OPERATIONS

Logical OR(+)	Binary ADD(+)
$0 + 0 \Rightarrow 0$	$0 + 0 = 0$
$0 + 1 \Rightarrow 1$	$0 + 1 = 1$
$1 + 0 \Rightarrow 1$	$1 + 0 = 1$
$1 + 1 \Rightarrow 1$	$1 + 1 = 0, \text{carry } 1$

There is an exact parallel between the logical AND operation and the binary multiply operation as seen in Table 11-2. With respect to the binary digits, 0 and 1, the logical OR and AND operations in Tables 11-1

TABLE 11-2
COMPARISON OF LOGICAL AND AND
BINARY MULTIPLY OPERATIONS

Logical AND($\cdot$)	Binary MULTIPLY($\cdot$)
$0 \cdot 0 \Rightarrow 0$	$0 \cdot 0 = 0$
$0 \cdot 1 \Rightarrow 0$	$0 \cdot 1 = 0$
$1 \cdot 0 \Rightarrow 0$	$1 \cdot 0 = 0$
$1 \cdot 1 \Rightarrow 1$	$1 \cdot 1 = 1$

and 11-2 are termed logical addition and logical multiplication respectively. The complementation operation is given in Table 11-3. The logical addition, multiplication and complementation rules form the basis of Boolean algebra operations. In the ensuing material, it will be clear from context when a Boolean algebra operation is being discussed as opposed to a binary arithmetic operation. The key will generally be that a Boolean variable, that can take on either a 0 or 1 value, will be specified as a symbol or literal while binary arithmetic operations will be discussed explicitly in terms of the binary digits 0 and 1.

TABLE 11-3
LOGICAL COMPLEMENTATION OF
BOOLEAN ALGEBRA

Logical NOT($\bar{}$)
$\bar{0} \Rightarrow 1$
$\bar{1} \Rightarrow 0$

Number Forms. The first question is how is a number represented in a digital computer? Two types of numbers are generally accommodated, so-called fixed point numbers or integers and floating point numbers. The maximum integer is determined by the computer word size, i.e., the number of bits making up one piece of information that can be stored in memory. Typical word sizes are 8, 16, 24, 32, 48 and 64 bits. One bit will generally be used to identify the sign of the number so that the range of integers is usually from approximately -2^{N-1} to $+2^{N-1}$ where N is the number of bits in the computer word. This range is insufficient for many problems so that a second form, floating point, is used where the computer word is structured into two parts. One part contains the most significant digits (in binary form) of the number while the other part contains a power of 10 (in binary form) that is applied to the number. This greatly increases the magnitude of numbers that can be handled with the limit being imposed by the number of binary bits used for representing the power. Since the arithmetic operations are almost the same for the two types of numbers but require different handling within the arithmetic unit, only integers will be considered to simplify the analysis.

Complements. Binary numbers may be represented in three different ways relating to the form of the representation of negative numbers. Signed-magnitude is the most straightforward representation where the leftmost bit is reserved for indicating the sign ($0 \rightarrow +$, $1 \rightarrow -$) of the number. For example

$$8_{10} \rightarrow 0 \,|\, 1000 \qquad 11_{10} \rightarrow 0 \,|\, 1011 \qquad 0_{10} \rightarrow 0 \,|\, 0000$$
$$-8_{10} \rightarrow 1 \,|\, 1000 \qquad -11_{10} \rightarrow 1 \,|\, 1011 \qquad -0_{10} \rightarrow 1 \,|\, 0000$$

are some decimal numbers and their signed-magnitude representation in binary form.

The 2's complement of a positive number N given by

$$N^{**} = \text{2's complement of } N = 2^n - N$$

may be used to represent the negative numbers where the leftmost bit will still indicate whether a number is positive or negative and n is the number of bits available for representing the number. Of course, if it is a negative number it is in 2's complement form. For example, with $n = 4$,

	$8_{10} \rightarrow 0 \mid 1000$	$11_{10} \rightarrow 0 \mid 1011$	$0_{10} \rightarrow 0 \mid 0000$
Invert	$1 \mid 0111$	$1 \mid 0100$	$1 \mid 1111$
Add "1"	1	1	1
	$-8_{10} \rightarrow 1 \mid 1000$	$-11_{10} \rightarrow 1 \mid 0101$	$-0_{10} \rightarrow 0 \mid 0000$

are some decimal numbers and their representation in a computer using 2's complement representation. Note the positive numbers are essentially signed magnitude. An easy algorithm for obtaining the 2's complement is to invert (complement) each bit of the number and then add "1" to the result. A feature of the 2's complement system is that "zero" has only one representation unlike the next system to be discussed.

The 1's complement of a positive number N given by

$$N^{*} = \text{1's complement of } N = 2^n - N - 1$$

may be used to represent the negative numbers where the left most bit will again indicate whether a number is positive or negative. For example, with $n = 4$,

$8_{10} \rightarrow 0 \mid 1000$	$11_{10} \rightarrow 0 \mid 1011$	$0_{10} \rightarrow 0 \mid 0000$
$-8_{10} \rightarrow 1 \mid 0111$	$-11_{10} \rightarrow 1 \mid 0100$	$-0_{10} \rightarrow 1 \mid 1111$

are some decimal numbers and their representation in a computer using 1's complement representation. In this system "zero" has two different representations, but obtaining the 1's complement is easier than obtaining the 2's complement since each bit is simply inverted. Again, it is noted that the positive number representation is the same in all three systems. No attempt will be made to compare the different systems in detail since there are advantages and disadvantages associated with each and each system has been used in one or another computer. However, examples will be given which will show the utility of the different systems for different applications.

Conversion. The input section must convert the numbers brought into the machine into binary form. Since the numbers are coded on a card by way of punched holes, this is done by a code conversion which will not be discussed here since attention is directed toward the arithmetic of numbers in binary form. An algorithm useful for hand conversion of numbers from decimal to binary form is the so-called "successive division by 2" method. For example

$$113_{10} \rightarrow 2 \underline{|113}$$
$$\phantom{113_{10} \rightarrow} 2 \underline{|56} \quad + 1$$
$$\phantom{113_{10} \rightarrow} 2 \underline{|28} \quad + 0$$
$$\phantom{113_{10} \rightarrow} 2 \underline{|14} \quad + 0$$
$$\phantom{113_{10} \rightarrow} 2 \underline{|7} \quad + 0$$
$$\phantom{113_{10} \rightarrow} 2 \underline{|3} \quad + 1$$
$$\phantom{113_{10} \rightarrow} 2 \underline{|1} + 1$$
$$\phantom{113_{10} \rightarrow} 0 + 1 \rightarrow 0 \vdots 1110001_2$$

$$75_{10} \rightarrow 2 \underline{|75}$$
$$\underline{|37} + 1$$
$$\underline{|18} + 1$$
$$\underline{|9} + 0$$
$$\underline{|4} + 1$$
$$\underline{|2} + 0$$
$$\underline{|1} + 0$$
$$\underline{|0} + 1 \rightarrow 0 \vdots 1001011_2$$

demonstrate such conversions. Once the information is coded in binary form, it remains in this form for all subsequent processing within the machine. The output section converts information from binary form back to the numbers and literals familiar to the user. An algorithm useful for hand conversion of numbers from binary to decimal form is the so-called "successive multiplication by 2" method which is the inverse of the decimal to binary conversion. For example

$$1110001_2 \rightarrow \qquad 1$$
$$1 \times 2 + 1 = 3$$
$$3 \times 2 + 1 = 7$$
$$7 \times 2 + 0 = 14$$
$$14 \times 2 + 0 = 28$$
$$28 \times 2 + 0 = 56$$
$$56 \times 2 + 1 = 113_{10}$$

$$1001011_2 \rightarrow 1$$
$$ 0 \quad 2$$
$$ 0 \quad 4$$
$$ 1 \quad 9$$
$$ 0 \quad 18$$
$$ 1 \quad 37$$
$$ 1 \quad 75_{10}$$

illustrate such conversions. The reader can prove the validity of these conversions by using the general representation of a binary number given by

$$N_2 = a_n 2^n + a_{n-1} 2^{n-1} + a_{n-2} 2^{n-2} + \cdots + a_2 2^2 + a_1 2^1 + a_0 2^0$$

where the a_i's are the binary digits (0 or 1).

Arithmetic Rules. Certain rules must be followed in the binary arithmetic operations as in the arithmetic of decimal numbers. These can best be given in tabular form since there are only four possible combinations of the two binary digits. Table 11-4 demonstrates that logic gates are

TABLE 11-4
RULES FOR BINARY ARITHMETIC

Addition					Subtraction					Multiplication				
Augend	0	0	1	1	Minuend	0	0	1	1	Multiplicand	0	0	1	1
Addend	0	1	0	1	Subtrahend	0	1	0	1	Multiplier	0	1	0	1
Sum	0	1	1	0	Difference	0	1	1	0	Product	0	0	0	1
Carry	0	0	0	1	Borrow	0	1	0	0					

candidates for implementing these binary arithmetic operations since each binary digit must be one of two possible values, 0 or 1, and therefore each value can be represented by one of the two states of a two-state device. This will be discussed in the next section.

11-5. Electronic Binary Adder

For the purposes of this presentation, it is assumed that the literals used represent a binary digit with a high voltage representing the bit "1" and a low voltage representing the bit "0." Digital computers are divided into two classes according to how they perform the arithmetic. A *serial* machine adds two binary numbers by operating on one pair of bits at a time as illustrated in Fig. 11-11a where the numbers five (00101) and six (00110) are added to give eleven (01011). The same adding unit can then be used to process each pair of bits but the result is not completely available until all the addition operations have been performed sequentially in *n* bit times where *n* is the length of the binary numbers to be added. This

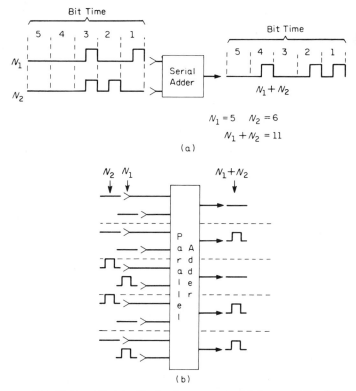

(a)

(b)

FIG. 11-11. a) Serial adder for five successive bit times. b) Parallel adder.

adder sacrifices speed for hardward economy and simplicity. Many early computers used such serial modes of operation. A *parallel* machine adds two binary numbers by working on all of the bits of the two numbers simultaneously. (See Fig. 11-11b.) The speed advantage over the serial adder increases as n (number of bits in computer word) increases. However, with some exceptions as to how the carry is accommodated both type adders use essentially the same algorithm (step-by-step procedure in performing a given task) in implementing binary addition. The addition circuits can therefore be discussed without regard as to whether a serial or parallel mode is used.

Half Adder. The half adder is best understood by looking at Table 11-5 which exhausts the combinations for the addition of a pair of binary digits. Note that Table 11-5 is a systematic tabulation of the rules

TABLE 11-5
LOGIC TABLE FOR A BINARY HALF ADDER

Input		Output	
Augend Bit A_i	Addend Bit B_i	Sum Bit S_i	Carry Bit C_i
0	0	0	0
0	1	1	0
1	0	1	0
1	1	0	1
		$S_i = (\bar{A_i} \cdot B_i) + (A_i \cdot \bar{B_i})$	$C_i = A_i \cdot B_i$

for the binary addition of two binary digits as given in Table 11-4. However, logical variables are now going to represent the augend, addend, sum and carry bits. Rather than stating the truth or falseness of the logical variable, it is discussed directly in terms of the two binary values it may assume, i.e., either a 0 or a 1. The expressions for the output logical variables are obtained by utilizing the rules of Boolean algebra. The reader should check the expressions for the sum bit and carry bit to verify that they satisfy the Boolean algebra rules while generating the proper binary arithmetic result for all possible input combinations. Inspection of this logic, or truth table, shows the logical expressions for the sum and carry which can be implemented by the logic circuitry given in Fig. 11-12. In a practical binary adder where the numbers are several bits in length, the addition of a pair of bits from the two numbers must also consider the carry from the one less significant pair of bits. The half adder would only suffice for the addition of the pair of least significant bits from the two numbers. Of course the results from the addition of the pair of bits from the two numbers could be looked upon as an intermediate result and operated on by a second half adder as indicated in Fig. 11-12b. This total configuration gives the proper sum considering the carry from the sum of the one less significant pair of bits.

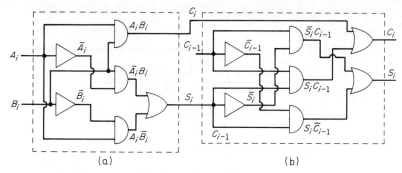

FIG. 11-12. a) Logic circuit implementation of half adder. b) Second half adder to consider carry bit from addition of the one less significant pair of bits and give full adder.

Full Adder. The logic table for a full adder that considers the carry from the sum of the one less significant pair of bits is given in Table 11-6. The logical expressions for the sum and carry are

$$S_i = \bar{A}_i \cdot \bar{B}_i \cdot C_{i-1} + \bar{A}_i \cdot B_i \cdot \bar{C}_{i-1} + A_i \cdot \bar{B}_i \cdot \bar{C}_{i-1} + A_i \cdot B_i \cdot C_{i-1}$$

and

$$C_i = \bar{A}_i \cdot B_i \cdot C_{i-1} + A_i \cdot \bar{B}_i \cdot C_{i-1} + A_i \cdot B_i \cdot \bar{C}_{i-1} + A_i \cdot B_i \cdot C_{i-1}$$

The expression for the carry can be simplified to

$$C_i = A_i \cdot B_i + B_i \cdot C_{i-1} + A_i \cdot C_{i-1}$$

since a carry results when *any* two of the input binary bits are "1." The implementation of the full adder is given in Fig. 11-13. In comparing Fig. 11-13 with Fig. 11-12, note that the maximum number of gates that is traversed by any one input signal is three in Fig. 11-13 as compared to six in Fig. 11-12. Since there is a propagation time of the signal associated with each gate, the circuit of Fig. 11-13 can be considerably faster than that of Fig. 11-12. This is so even though there is only a difference of one gate between the two circuits. These adder circuits can also be imple-

TABLE 11-6
LOGIC TABLE FOR A BINARY FULL ADDER

Input			Output	
Augend Bit A_i	Addend Bit B_i	Carry Bit C_{i-1}	Sum Bit S_i	Carry Bit C_i
0	0	0	0	0
0	0	1	1	0
0	1	0	1	0
0	1	1	0	1
1	0	0	1	0
1	0	1	0	1
1	1	0	0	1
1	1	1	1	1

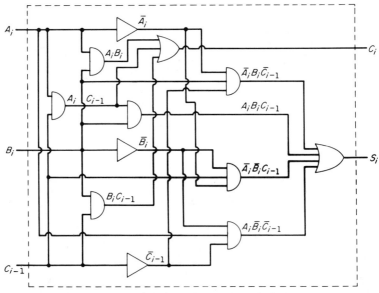

FIG. 11-13. Logic circuit implementation of a full adder.

mented with either NAND or NOR gates but the necessary analysis goes beyond the scope of the treatment given here.

11-6. Electronic Binary Subtracter

Digital computers that represent negative numbers in signed magnitude form could use electronic binary subtracters in a straight-forward manner according to the rules of binary subtraction. Digital computers that represent negative numbers in a complement form generally do not perform straight-forward subtraction but take advantage of complement arithmetic. Both types will be discussed.

Half Subtracter. An electronic half subtracter can be implemented in much the same way as the half adder. The logic table for a binary half subtracter is given in Table 11-7. Note that the logical expression for the

TABLE 11-7
LOGIC TABLE FOR A BINARY HALF SUBTRACTER

| Input | | Output | |
Minuend Bit A_i	Subtrahend Bit E_i	Difference Bit D_i	Borrow Bit B_i
0	0	0	0
0	1	1	1
1	0	1	0
1	1	0	0
		$D_i = \bar{A}_i \cdot E_i + A_i \cdot \bar{E}_i$	$B_i = \bar{A}_i \cdot E_i$

difference is identical to that for the sum in a half adder. The logic circuit implementation of the half subtracter is given in Fig. 11-14. The logic circuit implementation of a full subtracter is left as an exercise. Subtraction can also be implemented by use of a complementer with subsequent addition. Either 1's or 2's complement may be used but the algorithms differ slightly depending on the complement used.

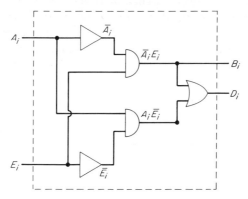

Fig. 11-14. Logic circuit implementation of a half subtracter.

Two's Complement Subtracter. Consider that the difference of two binary numbers is desired, say $A-E$. Note that adding A to the two's complement of E gives

$$A + E^{**} = A + (2^n - E) = 2^n + (A - E) \qquad (11\text{-}1)$$

The term in parenthesis is the difference being sought. The term 2^n is represented by a "1" followed by n "0's." The "1" flows or carries out of the computer word and is ignored while the remaining "0's" do not change the difference term. The digital computer would also represent all negative numbers in 2's complement form. Table 11-8 demonstrates that the proper result is obtained for all possible signs of the two numbers. To read a negative number represented in 2's complement form take 2's

TABLE 11-8
EXAMPLE OF SUBTRACTION THROUGH 2's COMPLEMENT ADDITION

A(decimal)		5		5		-5		-5
A(binary)	0	0101	0	0101	1	1011	1	1011
E(decimal)		3		-3		3		-3
E(binary)	0	0011	1	1101	0	0011	1	1101
E**	1	1101	0	0011	1	1101	0	0011
A$-$E = A + E**	0	0010	0	1000	1	1000	1	1110
(A+E**)**					0	1000	0	0010
A$-$E(decimal)		2		8		-8		-2

complement of the number which will give its positive counterpart or absolute value.

Since the adder has already been electronically implemented, 2's complement subtraction needs only a complementer in addition to the adder circuits. Note that to obtain the complement of the subtrahend, each bit is first inverted and "1" is added to the result. The inversion is easily accomplished by inverting each bit of the subtrahend. The addition of the "1" can be incorporated into the addition of the minuend with the complemented subtrahend by always considering a carry, i.e., a "1," being input into the addition of the least significant pair of bits where normally there is no carry. This is indicated in Fig. 11-15 for the first two

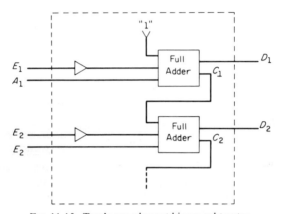

FIG. 11-15. Two's complement binary subtracter.

stages (subtraction of first two least significant bit pairs) of a 2's complement binary subtracter. Note that by bypassing the inverters and eliminating the "1" input on the first stage, i.e., replacing the "1" with a "0," the numbers will be added. This could be done with a few additional gates controlled by "ADD" and "SUBTRACT" control signals which would allow the same unit to be used for both addition and subtraction. Many digital computers utilize this technique for adding and subtracting. A similar dual adder-subtracter unit can be implemented for a computer in which negative numbers are represented in 1's complement form. This implementation is left as an exercise.

11-7. Electronic Binary Multiplication and Division

There are several different algorithms that can be implemented electronically for performing multiplication and division. Because of the relatively higher expense of these implementations, many computers use a software (programmed algorithm) implementation of multiplication and division. However, a penalty is paid in time since software implementation consists of a sequence of instructions involving only addition and subtraction mathematical operations along with other necessary instructions rather than a single multiply or division instruction or command. Only the simplest algorithms will be given for the multiplication and division of binary numbers in signed magnitude form. No detailed circuitry will be given since it is beyond the scope of the present treatment.

Electronic Binary Multiplication. Consider a "pencil and paper" multiplication of two binary numbers, e.g.,

$$
\begin{array}{rl}
3_2 \rightarrow & 0011 \\
5_2 \rightarrow & 0101 \\
\hline
& 0011 \\
& 0000 \\
& 0011 \\
& 0000 \\
\hline
15_2 \rightarrow & 0001111
\end{array}
\qquad
\begin{array}{rl}
7_2 \rightarrow & 0111 \\
6_2 \rightarrow & 0110 \\
\hline
& 0000 \\
& 0111 \\
& 0111 \\
& 0000 \\
\hline
42_2 \rightarrow & 0101010
\end{array}
$$

Note that an equivalent operation would be the following:

1. Start with a zero *partial product*.
2. Add multiplicand multiplied by least significant bit (l.s.b.) of multiplier to partial product. This has the effect of adding multiplicand or zero to the partial product depending upon whether l.s.b. of the multiplier is "1" or "0" respectively.
3. Shift partial product right one bit.
4. Repeat step 2 for next to least significant bit of multiplier.
5. Repeat step 3.
6. Continue this procedure until all bits of the multiplier have played their role.

In essence, this technique simply keeps a running sum of the rows of digits produced by hand calculation as soon as they are generated. The electronic implementation of the add operation has already been shown. The technique of shifting is explained by referring to Fig. 11-16. The partial product is formed in an accumulator or register made up of

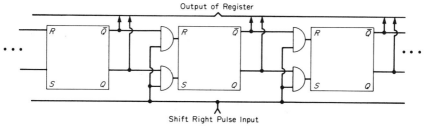

Fig. 11-16. Shift right circuitry for RS flip-flop register.

memory elements, say, RS flip-flops (see Fig. 11-10). A "0" is stored in the flip-flop by pulsing the "R" input, i.e., making the "$\bar{Q}$" output terminal high for a positive logic system. Conversely, a "1" is stored in the flip-flop by pulsing the "S" input, i.e., making the "Q" terminal high for a positive logic system. Suppose a partial product has been formed and is stored in the register. A pulse is now applied to the line labeled "Shift Right Pulse Input." The output of each flip-flop is transferred to the flip-flop to the right through the appropriate AND gate. The only restriction is that the shift pulse width should be narrower than the time required for the flip-flop to change state. This will prevent shifting more than one bit to the right for each shift pulse. The shifted partial product is now available at the register output for use in the next addition. To increase speed, a "0" multiplier bit can be routed to give a shift pulse and bypass the addition of zero to the partial product since addition of zero has no effect on the partial product. Upon shifting, the system is ready to act in accordance with the next multiplier bit. The next section will examine the division implementation.

Electronic Binary Division. The division operation is the most difficult of the four arithmetic operations to implement electronically and will almost always be the slowest of the four operations. Some computers have broken the division operation into two parts. The reciprocal of the divisor is first calculated and then this reciprocal is multiplied by the dividend to produce the quotient. Another technique simulates the division process as accomplished by hand computation where one quotient bit is determined with each operation, starting with the most significant bit and continually finding each successive bit until the desired accuracy in terms of the remainder is achieved if the division is not exact. As an example consider

$$\frac{27_{10}}{3_{10}} \rightarrow 11 \overline{\left) 11011. \right.} \rightarrow$$

	1001.	11011
	11	11 Subtract divisor if possible
	0011	00011 Yes → 1
	11	0 0011 Shift left one bit
	0000	11 Subtract divisor if possible
		No → 0
		0 011 Shift left one bit
		11 Subtract divisor if possible
		No → 0
		0 11 Shift left one bit
		11 Subtract divisor if possible
		00 Yes → 1
		$1001_2 = 9_{10}$

or

$$44_{10} \over 4_{10}$$ $\rightarrow$ $100 \overline{)\ 1011 \atop 101100.}$ $\rightarrow$

```
          1011
100 | 101100.
      100
      110
      100
      100
      000
```

```
101100
100
001100  →  1

0 01100
  100
        →  0

0 1100
  100
  0100   →  1

0 100
  100
  000   →  1
```

$$1011_2 = 11_{10}$$

where the divisions are exact for simplicity. The subtraction operation and the shifting operation have already been covered so they will not be repeated here. The only new operation is a comparison of the divisor with the partial remainder to determine if a subtraction should be made prior to the shift (resulting in a quotient bit of "1") or omission of subtraction if divisor is greater than quotient (resulting in a quotient bit of "0"). This can be accomplished by shifting bit pairs of the partial quotient and divisor into a pair of flip-flops for comparison. The bit pairs are compared successively with the most significant pair first and then the next most significant pair, etc. The comparison circuit is given in Fig. 11-17a along with the flip-flop output implication table in Fig. 11-17b. As long as the bits in the successive pairs are identical, the output of the NOR gate will be high (positive logic) and allow the shift pulses to pass through the AND gate and continually shift successive bit pairs into the comparison flip-flops. Whenever the first mismatched bit pair reaches the comparison flip-flops, one of the two AND gates becomes high, forcing the NOR gate low and thereby opening the AND gate so no more shift pulses enter the system. The mismatched bit pair are therefore residing in the comparison flip-flops and provide the proper signal, i.e., perform the subtraction if the partial quotient is greater than the divisor or skip the subtraction if the divisor is greater than the partial quotient. The line labeled counter goes to a binary counter that keeps track of the number of shifts of the two binary numbers so they may be restored to their original positions in the accumulator registers after the comparison is made. Of course if all n bits of the two binary numbers are shifted through the comparator without an inequality indication, the divisor and partial quotient are equal and a

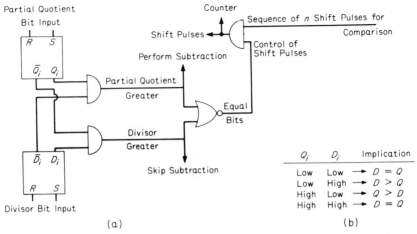

FIG. 11-17. a) Partial quotient and divisor comparison circuit. b) Truth table for comparison flip-flop outputs.

subtraction is performed leaving a zero remainder, i.e., giving an exact quotient.

This concludes the design aspects of the arithmetic unit of a digital computer. The next chapter turns attention to electronic circuits that are concerned primarily with a-c excitation.

Problems

Transistor Logic Circuits

11-1. The 2N3704 NPN transistor, whose output characteristics are given below, is to be used as a switch in the circuit of Fig. 11-3. The output voltage is to

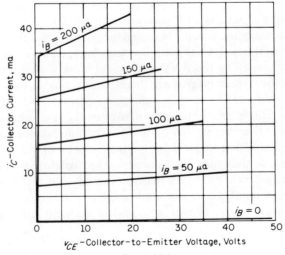

PROB. 11-1

be 40 volts in the "open" switch mode and the collector current is to be 25 ma in the "closed" switch mode. Determine values of V_{CC}, R_L and R_B that will meet these specifications if V_{in} takes on the values 0 volts or 5 volts for open and close conditions respectively.

11-2. Draw the schematic diagram of a DTL negative-logic AND gate and OR gate. Use 0 and 5 volts as the two logic voltage levels.

11-3. Construct the truth or logic table for the DTL OR gate given in Fig. 11-7.

11-4. Draw the schematic diagram of a DTL negative-logic NAND and NOR gate using 0 and 5 volts as the two-logic voltage levels.

11-5. Construct the truth table for the DTL NOR gate given in Fig. 11-9a.

Binary Arithmetic

The next few problems will refer to the following numbers:

$$N_1 = 150_{10} \qquad N_2 = 93_{10} \qquad N_3 = -112_{10}$$
$$N_4 = -87_{10} \qquad N_5 = 58_{10} \qquad N_6 = -128_{10}$$

11-6. Express N_1 through N_6 in signed-magnitude binary form.

11-7. Express N_1 through N_6 in one's complement binary form.

11-8. Express N_1 through N_6 in two's complement binary form.

11-9. Using one's complement binary form, add $N_1 + N_3$, $N_3 + N_4$, and $N_5 + N_6$.

11-10. Using two's complement binary form, add $N_2 + N_3$, $N_3 + N_6$, and $N_1 + N_4$.

Electronic Arithmetic Operations

11-11. Develop an algorithm for handling the sign bits associated with each of two numbers to be multiplied so as to give the correct sign to the product. Draw the schematic diagram of a logic circuit that will implement the algorithm. Would this algorithm suffice for obtaining the correct sign of the quotient of two numbers as well?

11-12. Give the logic table and logic circuit implementation of a full subtracter.

11-13. Give the logic circuit implementation of a dual adder-subtracter unit in a digital system in which negative numbers are represented in one's complement form.

12 / A-C Circuit Analysis Methods

In the previous chapters only the solution of circuits containing resistors was considered. Although capacitors and sometimes inductors appeared in the original circuit, they were replaced by either short circuits or open circuits before a solution to the circuit was attempted. The reader probably questioned the assumption that allowed the capacitive reactance $1/\omega C$ (called "equivalent resistance" in the earlier chapters) to be set equal to zero or infinity. Obviously there are many circuit applications for which the assumption that $1/\omega C = 0$ or $1/\omega C = \infty$ is not warranted. In fact there are many, many important and interesting electronic circuits where it is desired that $1/\omega C$ or ωL not have the extreme values of zero or infinity. Before discussing some of these applications it is necessary that one know how to analyze circuits containing the capacitance C, the inductance L, as well as the resistance R parameter. A discussion of the general voltage and current relationships for the inductance and capacitance parameters is found in Appendix G.

The material in this chapter can be studied or skipped (except perhaps Sec. 12-11) depending on whether or not the reader is familiar with steady state a-c circuit analysis methods.

In this chapter a technique will be developed for application to networks meeting the following specifications:

1. Excitation to the network is provided by sources being sinusoidal in form and of a single frequency.
2. If more than one source is present, all are of the same frequency.
3. The steady-state (in a periodic sense) portion of the solution is desired, with the transient portion being neglected. A network that does not meet these specifications could not be analyzed with the technique to be developed, but one could turn to the Laplace-transform technique (see Chapter 19) to analyze such a system, or one could approach it from the classical technique of solving the differential equations that describe the system.
4. In view of specification 3, all initial conditions are arbitrary and can be taken as zero if desired.

12-1. Excitation Functions

The networks to be analyzed will be excited or driven by sinusoidal voltage and/or current sources of the form

$$e = E_m \sin \omega t = E_m \sin 2\pi f t = E_m \sin \frac{2\pi}{T} t \qquad (12\text{-}1)$$

or

$$i = I_m \sin \omega t = I_m \sin 2\pi f t = I_m \sin \frac{2\pi}{T} t \qquad (12\text{-}2)$$

A plot of Eq. 12-1, called the waveform of e, is given in Fig. 12-1, where e is the instantaneous value of the source voltage; E_m is the maximum value attained by the source voltage; f is the cyclic frequency, in hertz or cycles per second; $\omega = 2\pi f$ is the radian frequency, in radians per second; and $T = 1/f$ is the period of time necessary to complete 1 cycle of the repetitive waveform. Since only single-frequency sources are to be allowed, one could plot e against (ωt) rather than (t), and the waveform

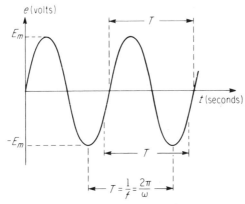

FIG. 12-1. Plot of $e = E_m \sin \omega t$.

would be the same. However, the units of (ωt) would be electrical radians or electrical degrees (to distinguish from angular spatial measurement) where

$$\theta \text{ (electrical degrees)} = \frac{360 \text{ (electrical degrees/cycle)}}{2\pi \text{ (electrical radians/cycle)}} \omega t \text{ (electrical radians)}$$

$$(12\text{-}3)$$

Thus, for a current source of the form

$$i = I_m \sin [2\pi (100) t] \qquad (12\text{-}4)$$

the frequency is 100 hertz, and 1 cycle of the waveform is given in Fig.

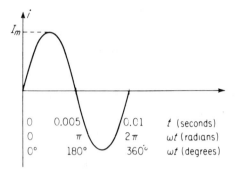

FIG. 12-2. Plot of $i = I_m \sin [2\pi(100)t]$.

12-2. *It is henceforth understood that the radian and degree measures are electrical.* One cycle of another sinusoidal current of the form

$$i = I_m \sin [2\pi(25)t] \qquad (12\text{-}5)$$

is given in Fig. 12-3. Comparing Fig. 12-2 with Fig. 12-3 shows that 1 cycle of either waveform represents 2π radians or $360°$ independent of the

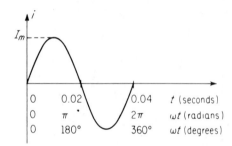

FIG. 12-3. Plot of $i = I_m \sin [2\pi(25)t]$.

frequency, thus making radian or degree measure a useful variable for plotting sinusoidal waveforms. The two numbers which completely describe the sinusoidal waveform are the magnitude, or maximum value, attained by the waveform and the frequency, or period.

12-2. Complex Algebra

Complex algebra serves as the mathematical basis for the steady-stage analysis of sinusoidally excited networks. This section will serve as a quick refresher guide to the rules governing the manipulations of complex numbers; as such, it may be omitted by those already familiar with this branch of mathematics. A complex number is composed of two parts—a real part and a so-called imaginary part. When expressed in this form, it is given by

$$a + jb \qquad (12\text{-}6)$$

where a is termed the real component and is a real number (positive or negative), $j \overset{\Delta}{=} \sqrt{-1}$ is called the imaginary operator, b is termed the coefficient of the imaginary component and is a real number (positive or negative), and jb is termed the imaginary component.

Since the number is composed of two parts, it can be represented quite nicely on a plane (two-coordinate representation). Such a plane is termed the complex plane, where the abscissa is called the axis of the reals (the real component serving as this coordinate) and the ordinate is called the axis of the imaginaries (the coefficient of the imaginary component serving as this coordinate). The number given by Eq. 12-6 is plotted as a point with coordinates (a, b) on the complex plane in Fig. 12-4. Another

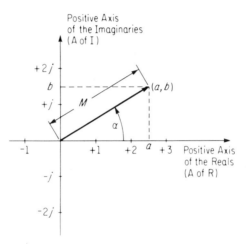

FIG. 12-4. Complex-plane representation
of $a + jb$.

set of two parameters can also be used to locate a point on the complex plane. *By convention, let the positive axis of the reals be the reference line for positive angular measurement in the counterclockwise direction.* Now with M (termed the modulus, or magnitude, of the complex number) being the distance from the origin to the point representation of the complex number, and α being the angle, in radians, that the modulus M makes with the positive real axis, then, from plane geometry, it is apparent that

and
$$\left.\begin{aligned} a &= M \cos \alpha \\ b &= M \sin \alpha \end{aligned}\right\} \qquad (12\text{-}7)$$

Therefore,
$$a + jb = M \cos \alpha + jM \sin \alpha = M (\cos \alpha + j \sin \alpha) \qquad (12\text{-}8)$$

Now, through Euler's identity relating the exponential functions to the trigonometric functions ($e^{jx} = \cos x + j \sin x$).

$$a + jb = Me^{j\alpha} \tag{12-9}$$

If the components a and b are considered to be known, then, from Fig. 12-4 or Eq. set 12-7, it is obvious that

$$\left. \begin{aligned} M &= +\sqrt{a^2 + b^2} \\[2ex] \alpha &= \tan^{-1}\!\left(\frac{b}{a}\right) \end{aligned} \right\} \tag{12-10}$$

and

with due regard to the signs of the components in the second expression of Eq. set 12-10, since the signs of the components determine the quadrant in which the complex number is located. A complex number expressed in the form given by the left side of Eq. 12-9 is said to be in *rectangular* form; if it is expressed in the form given by the right side of Eq. 12-9, it is said to be in *polar or exponential* form. Analogously, (a, b) and (M, α) are referred to as the rectangular and polar coordinates, respectively. Either pair of coordinates completely specifies a complex number, and Eq. sets 12-7 and 12-10 can be used to transform from one set of coordinates to the other set.

Given a complex number N, where $N = a + jb = Me^{j\alpha}$, then N^* is called the conjugate of N and is given by $N^* = a - jb = Me^{-j\alpha}$. The conjugate of a complex number differs from the complex number itself by

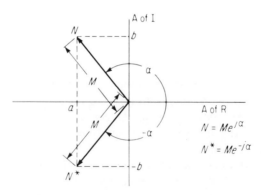

FIG. 12-5. Complex-plane representation of a complex number N and its conjugate.

having an *oppositely signed* imaginary component. A complex number and its conjugate are shown in Fig. 12-5.

The addition and subtraction of two complex numbers are accomplished with the numbers in rectangular form. Given

$$N_1 = a + jb \qquad \text{and} \qquad N_2 = c + jd$$

then

$$N_1 + N_2 = (a + jb) + (c + jd) \triangleq (a + c) + j(b + d)$$

and

$$N_1 - N_2 = (a + jb) - (c + jd) \triangleq (a - c) + j(b - d)$$

or

$$N_1 - N_2 = (N_1) + (-N_2)$$
$$= (a + jb) + (-c - jd) \triangleq [a + (-c)] + j[b + (-d)]$$

It is noted that addition or subtraction is performed in the same way as for real numbers—by treating the real components and the imaginary components as separate entities. In this sense, the operation is analogous to that performed on two-dimensional vector quantities, and the sum can be obtained by the so-called parallelogram law; that is, the sum is the diagonal of the parallelogram whose sides are made up of the two complex numbers located on the complex plane. This is illustrated in Fig. 12-6. By

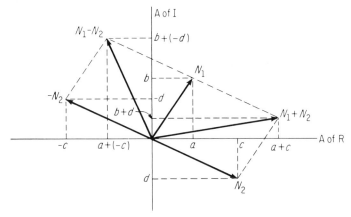

FIG. 12-6. Sum of two complex numbers.

changing the signs of the components of the subtrahend, the process of subtraction of complex numbers becomes addition, as far as the interpretation is concerned. The definition of addition or subtraction of two complex numbers can be extended readily to the addition and/or subtraction of several complex numbers by treating a pair at a time, taking the resultant, and combining it with the next complex number in the addition and/or subtraction process.

Two complex numbers are said to be equal to each other *if and only if* both their real components are equal to each other and the coefficients of their imaginary components are equal to each other. Hence, if

$$N_1 = a + jb \quad \text{and} \quad N_2 = c + jd$$

then $N_1 = N_2$ implies that

$$a = c \quad \text{and} \quad b = d$$

The multiplication or division of two complex numbers may be done in either rectangular form or polar form, the latter method being preferable because of its relative simplicity. Consider the two complex numbers

$$N_1 = a + jb = M_1 e^{j\alpha_1}$$

and

$$N_2 = c + jd = M_2 e^{j\alpha_2}$$

Now, since $(j)^2 = (\sqrt{-1})^2 = -1,$

$$\begin{aligned}
(N_1)(N_2) = (a + jb)(c + jd) &= ac + ajd + jbc + jbjd \\
&= ac + (jj)bd + j(ad + bc) \\
&= (ac - bd) + j(ad + bc) \quad \text{(12-11)}
\end{aligned}$$

or

$$(N_1)(N_2) = M_1 e^{j\alpha_1} M_2 e^{j\alpha_2} = M_1 M_2 e^{j\alpha_1} e^{j\alpha_2} = M_1 M_2 e^{j(\alpha_1 + \alpha_2)}$$
$$\text{(12-12)}$$

Note that, in polar form, the product of two complex numbers can readily be seen to be a resultant complex number whose modulus is the product of the moduli of the two complex numbers and whose angle is the algebraic sum of the angles of the two complex numbers. It is left to the reader to show, by using either Eq. set 12-7 or Eq. set 12-10, that the right-hand sides of Eqs. 12-11 and 12-12 are identical.

Now,

$$\begin{aligned}
\frac{N_1}{N_2} = \frac{N_1}{N_2} \cdot \underbrace{\left[\frac{N_2{}^*}{N_2{}^*}\right]} = \frac{a + jb}{c + jd} \cdot \frac{c - jd}{c - jd} &= \frac{ac - ajd + jbc - jbjd}{c^2 - j^2 d^2} \\
&= \frac{(ac + bd) + j(bc - ad)}{c^2 + d^2} \quad \text{(12-13)}
\end{aligned}$$

The process of multiplying the quotient of two complex numbers by one, where one is expressed as the ratio of the conjugate of the complex number forming the denominator of the quotient, is called *rationalization*. It clears the denominator of the imaginary operator j. Now,

$$\frac{N_1}{N_2} = \frac{M_1 e^{j\alpha_1}}{M_2 e^{j\alpha_2}} = \left(\frac{M_1}{M_2}\right) e^{j\alpha_1} e^{-j\alpha_2} = \left(\frac{M_1}{M_2}\right) e^{j(\alpha_1 - \alpha_2)} \quad \text{(12-14)}$$

In polar form, it is evident that division of two complex numbers results in a complex number whose modulus is the ratio of the moduli of the two complex numbers and whose angle is the angle of the complex number serving as the dividend minus the angle of the complex number serving as the divisor. Again, the reader may want to assure himself that the right-hand sides of Eqs. 12-13 and 12-14 are identical by utilizing Eq. set 12-7 or Eq. set 12-10. An illustration of the product and quotient of two complex numbers is given in Fig. 12-7.

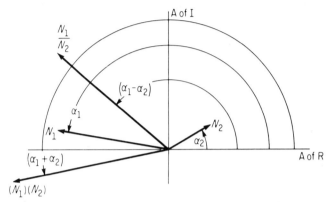

FIG. 12-7. Product and quotient of two complex numbers.

It is left to the reader to prove that

$$(N_1)(N_2)(N_3)\cdots(N_P) = (M_1M_2M_3\cdots M_P)e^{j(\alpha_1 + \alpha_2 + \alpha_3 + \cdots + \alpha_P)}$$

and

$$\frac{(N_1)(N_2)\cdots(N_P)}{(N_a)(N_b)\cdots(N_R)} = \frac{(M_1M_2\cdots M_P)}{(M_aM_b\cdots M_R)}e^{j[(\alpha_1 + \alpha_2 + \cdots + \alpha_P) - (\alpha_a + \alpha_b + \cdots + \alpha_R)]}$$

Another useful relation is given by

$$\frac{1}{N_1} = \frac{1}{N_1}\cdot\frac{N_1{}^*}{N_1{}^*} = \frac{M_1e^{-j\alpha_1}}{M_1e^{j\alpha_1}M_1e^{-j\alpha_1}} = \frac{M_1e^{-j\alpha_1}}{(M_1M_1)e^{j0}} = \left(\frac{1}{M_1}\right)e^{-j\alpha_1}$$

A geometric interpretation of the effect of multiplication or division by the imaginary operator j will be most useful in the work that is to follow. The imaginary operator is a complex number itself, given in rectangular form by

$$0 + j1$$

and in polar form by

$$1\,e^{j\pi/2}$$

A complex number with a zero real component is sometimes called a *pure imaginary*.

Multiplication of a complex number N_1 by the imaginary operator j gives

$$jN_1 = 1e^{j\pi/2} M_1 e^{j\alpha_1} = (1 \cdot M_1) e^{j(\pi/2 + \alpha_1)} = M_1 e^{j(\alpha_1 + \pi/2)}$$

The magnitude of the resultant complex number is the same as the magnitude of the original complex number, and the angle of the resultant complex number is equal to the angle of the original complex number plus $\pi/2$ radians or 90°. The net effect of multiplication of the complex number by the imaginary operator j is to advance the complex number by $\pi/2$ radians or 90° in the positive angular direction in the complex plane. In a similar manner,

$$\frac{N_1}{j} = \frac{M_1 e^{j\alpha_1}}{1 e^{j\pi/2}} = \left(\frac{M_1}{1}\right) e^{j(\alpha_1 - \pi/2)}$$

or

$$\frac{N_1}{j} \cdot \frac{j}{j} = \frac{jN_1}{j^2} = -jN_1$$

Division of a complex number results in a retardation of the complex number by $\pi/2$ radians or 90° in the negative angular direction in the complex plane. Refer to Fig. 12-8 for an illustration of such operations.

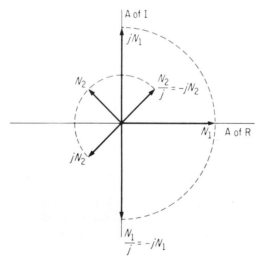

FIG. 12-8. Multiplication and division of complex numbers by an imaginary operator j.

One final remark on notation: In the polar or exponential form, α is measured in radians, since the exponent must be dimensionless. However, for convenience in plotting, angles are expressed in degrees. There is also a greater familiarity with angular positions given in degrees rather than in radians. The following notation will be used later on, but, for completeness, it is defined here. A complex number N_1 given by $M_1 e^{j\alpha_1}$ can be represented by $M_1 \underline{/\alpha_1}$ where

$$\alpha_1 \text{ (degrees)} = \alpha_1 \text{ (radians)} \cdot \left(\frac{360°}{2\pi \text{ radians}}\right)$$

Thus, in terms of this simpler notation

$$(N_1)(N_2) = (M_1 M_2)\underline{/\alpha_1 + \alpha_2} \qquad \text{and} \qquad \frac{N_1}{N_2} = \frac{M_1}{M_2} \underline{/\alpha_1 - \alpha_2}$$

12-3. Complex-Number Representation of Sinusoidal Functions

Suppose one poses the following situation: Take a complex number of modulus M and allow it to rotate about the origin in the complex plane, in the positive angular direction with a constant angular velocity of ω radians/sec. At $t = 0$, let the complex number be in some position such that the angle that the modulus makes with the positive real axis is θ degrees. Determine the magnitude of the coefficient of the imaginary component of this complex number for all positive time. See Fig. 12-9.

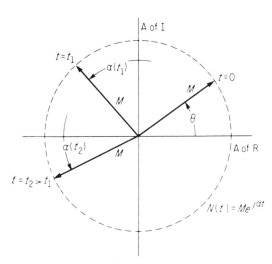

FIG. 12-9. Rotating complex number.

Let the angle that the modulus makes with the positive real axis at any time t be denoted by $\alpha(t)$. Then $\alpha(t) = \theta + \omega t$, where $\alpha(0) = \theta$ and $N(t) = M e^{j\alpha(t)}$. But, from Eq. set 12-7, the coefficient of the imaginary component is given by

$$M \sin [\alpha(t)] = M \sin (\omega t + \theta) \qquad (12\text{-}15)$$

Thus, from Eq. 12-15, it is evident that the coefficient of the imaginary component of the rotating complex number can represent a sinusoidal function. If it is understood that the complex number is rotating with an angular velocity of ω radians/sec in the counterclockwise direction, in what fixed position in the complex plane should the complex number be depicted in order to impart additional information? It is certainly not necessary that a sinusoidal function be zero at the arbitrary time picked to start recording time, that is, $t = 0$. A sinusoidal function that is not zero at $t = 0$ is of the form

$$M \sin (\omega t + \theta) \tag{12-16}$$

Two sinusoidal functions of the form given by Eq. 12-16, and with both functions having the same frequency and the same maximum value, are plotted in Fig. 12-10. The function f_1 is said to have an angle displacement

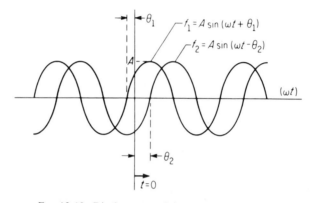

FIG. 12-10. Displacement of sinusoidal wave forms.

of lead (positive angle) of θ_1 degrees, while f_2 is said to have an angle displacement of lag (negative angle) of θ_2 degrees. This angle displacement is measured from the $t = 0$ *axis* to the point where the sinusoidal function is *zero* and has a *positive slope*. By being able to measure in either direction from the $t = 0$ *axis*, the angle displacement θ will always be

$$|\theta| \quad (\text{angle displacement}) \leq 180°$$

The words "lead" and "lag" can also be interpreted, from the same figure, by noting that one observes f_1 passing through zero with a positive slope *prior* (lead) to the $t = 0$ axis, and f_2 passing through zero with a positive slope *after* (lag) the $t = 0$ axis. It should now be apparent that one logical placement of the complex number (representing a sinusoidal function) on the complex plane would be in the position it occupies at the time $t = 0$. The complex-plane representation of the two sinusoidal functions depicted in Fig. 12-10 is shown in Fig. 12-11. The coefficients of the imaginary components of these complex numbers for their depicted posi-

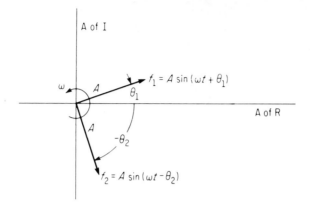

FIG. 12-11. Complex-plane representation of sinusoidal functions.

tions represent the value of the functions at $t = 0$. Previously, it has been
stated that only the steady-state (in a periodic sense) behavior of the electrical quantities is to be obtained, with the transient behavior neglected.
In this respect, the time $t = 0$ is of little consequence, since one will have
to wait sufficiently long enough for all transients to die out, after which
only the steady-state sinusoidal functions will be present. The question,
then, is not what are the values of the functions at $t = 0$, the answer to
which would require both the steady-state and the transient solutions, but,
in the steady-state (in a periodic sense), what is the relationship of one
sinusoidal function to another? Analysis of Fig. 12-12 should provide the

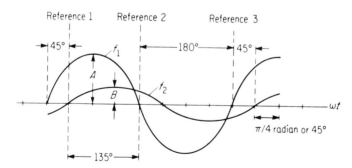

FIG. 12-12. Plot of two sinusoidal functions.

information called for in this question: First, since $t = 0$ is of no special
significance in the steady-state solution, one could just as well refer to the
$t = 0$ axis as a nominal reference axis which might be arbitrarily chosen.
Three such reference axes are shown in Fig. 12-12. Now, with respect to

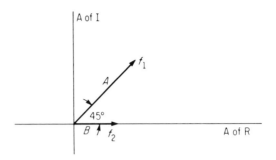

FIG. 12-13. Complex-plane representation of
Eq. set 12-17.

reference axis 1, the waveforms are

$$f_1 = A \sin (\omega t + 45°)$$
and
$$f_2 = B \sin (\omega t)$$

$$(12\text{-}17)$$

Their complex-plane representation is given in Fig. 12-13. If reference
axis 2 is used, then

$$f_1 = A \sin (\omega t + 180°)$$
and
$$f_2 = B \sin (\omega t + 135°)$$

$$(12\text{-}18)$$

Their complex-plane representation is given in Fig. 12-14. Finally, if

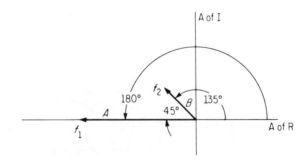

FIG. 12-14. Complex-plane representation of Eq.
set 12-18.

reference axis 3 is used, then

$$f_1 = A \sin (\omega t)$$
and
$$f_2 = B \sin (\omega t - 45°)$$

$$(12\text{-}19)$$

Their complex-plane representation is given in Fig. 12-15.

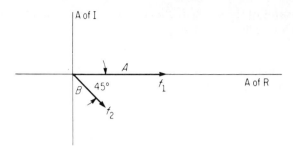

FIG. 12-15. Complex-plane representation of Eq. set 12-19.

Upon examining Figs. 12-13 through 12-15, it is apparent that the complex numbers, representing the sinusoidal functions, maintain their *same positions with respect to each other*, but the orientation of the system of complex numbers in the complex plane will depend upon the choice of the reference axis. In this sense, for any of the representations, one would speak of f_1 leading f_2 by 45°, or, analogously, f_2 lagging f_1 by 45°, since both are rotating with the same angular velocity in a counterclockwise direction. Since the choice of the reference axis is arbitrary, it will generally be convenient to choose the reference axis in such a manner that one of the sinusoidal waveforms is passing through zero with a positive slope, so that its complex-number representation on the complex plane is in the so-called "reference position," that is, along the positive real axis. All other complex numbers (representing sinusoidal waveforms of the same frequency) will then assume their relative positions to the reference complex number. Since it is understood that all the complex numbers are rotating counterclockwise with an angular velocity of ω radians/sec, the graphical representation can be visualized as a photograph of this system taken when one of the complex numbers is oriented along the positive real axis. To distinguish this photograph of rotating complex numbers from a vector diagram, which depicts space orientation of vector quantities (quantities having magnitude and direction), the name *phasor diagram* is used, and the complex numbers representing sinusoidal waveforms are called *phasors* or *sinors*, the former nomenclature being used in this text. This distinction is necessary, since a vector (space-oriented quantity) could also be sinusoidal; for example, its magnitude could be a sinusoidal function of time while its space orientation remained fixed.

In summation, it has been demonstrated how a sinusoidally varying quantity can be represented by a phasor. This phasor can be plotted on the complex plane, with its magnitude indicating the maximum value attained by the sinusoid and the position of the phasor indicating its orientation with respect to the reference axis and to any other phasors representing sinusoids of the same frequency. In order to distinguish a phasor quantity, boldface letters will be used; hence

$$\mathbf{I} \triangleq I\underline{/\theta} \tag{12-20}$$

is a phasor of magnitude I making an angle of θ degrees with the reference axis when plotted on the complex plane.

12-4. Addition of Sinusoidal Quantities

Frequently, it will be necessary, when analyzing sinusoidally excited networks, to add several sinusoidal functions. This section will demonstrate that the addition of two sinusoidal functions of the same frequency can be accomplished by performing a geometrical or vector addition of their phasor representations. This will eliminate the necessity of having to consider explicitly the time behavior of the two functions. Consider

and
$$
\left.
\begin{aligned}
A &= M_1 \sin(\omega t + \theta) \\
B &= M_2 \sin(\omega t + \alpha)
\end{aligned}
\right\} \tag{12-21}
$$

First the phasor addition. The phasors **A** and **B** are shown in Fig. 12-16.

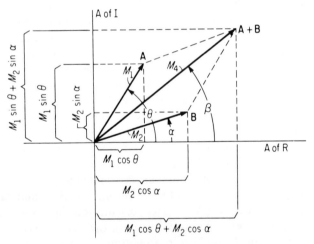

FIG. 12-16. Complex-plane representation of Eq. set 12-21.

From plane geometry, it is apparent that the sum is a phasor $M_4\underline{/\beta}$, where

$$M_4 = +[(M_1 \sin\theta + M_2 \sin\alpha)^2 + (M_1 \cos\theta + M_2 \cos\alpha)^2]^{1/2}$$

and

$$\beta\text{ (degrees)} = \tan^{-1}\left(\frac{M_1 \sin\theta + M_2 \sin\alpha}{M_1 \cos\theta + M_2 \cos\alpha}\right)$$

Second the sum of the functions. Using the trigonometric identity

$$\sin (x + y) = \sin x \cos y + \cos x \sin y \qquad (12\text{-}22)$$

the sum of A and B can be expressed in the form

$$A + B = \underbrace{\{M_1 \cos \theta + M_2 \cos \alpha\}}_{\text{constant with time}} \sin \omega t + \underbrace{\{M_1 \sin \theta + M_2 \sin \alpha\}}_{\text{constant with time}} \cos \omega t$$

This is recognized as being in the form of

$$A + B = M_3 \sin (\omega t + \gamma) = M_3 \cos \gamma \sin \omega t + M_3 \sin \gamma \cos \omega t$$

where

$$M_3 \cos \gamma = M_1 \cos \theta + M_2 \cos \alpha \qquad (12\text{-}23)$$
$$M_3 \sin \gamma = M_1 \sin \theta + M_2 \sin \alpha \qquad (12\text{-}24)$$

Now, if both sides of Eqs. 12-23 and 12-24 are squared and then summed, there results

$$M_3^2(\sin^2 \gamma + \cos^2 \gamma) = (M_1 \cos \theta + M_2 \cos \alpha)^2 + (M_1 \sin \theta + M_2 \sin \alpha)^2$$

from which

$$M_3 = +[(M_1 \sin \theta + M_2 \sin \alpha)^2 + (M_1 \cos \theta + M_2 \cos \alpha)^2]^{1/2} \quad (12\text{-}25)$$

By comparison, it is evident that M_3 is equal to M_4. If Eq. 12-24 is divided by Eq. 12-23, there results

$$\frac{M_3 \sin \gamma}{M_3 \cos \gamma} = \tan \gamma = \frac{M_1 \sin \theta + M_2 \sin \alpha}{M_1 \cos \theta + M_2 \cos \alpha}$$

from which

$$\gamma \text{ (degrees)} = \tan^{-1} \left(\frac{M_1 \sin \theta + M_2 \sin \alpha}{M_1 \cos \theta + M_2 \cos \alpha} \right) \qquad (12\text{-}26)$$

It is further evident that γ is equal to β. This shows that the addition of two *phasors* in the complex plane results in a *phasor* that is the complex-number representation of the resulting sinusoid from the addition of two sinusoids in the time domain. It is left to the reader to extend this same notion to the difference of two sinusoidal functions. (*Hint:* View subtraction as the sum of one sinusoid with the negative of a second sinusoid.)

12-5. Resistive Element

Consider a sinusoidal voltage impressed across a resistor, such as that depicted in Fig. 12-17. Now, from Ohm's law

$$v_R = iR = V_m \sin \omega t \qquad (12\text{-}27)$$

from which

$$i = \frac{V_m}{R} \sin \omega t = I_m \sin \omega t \qquad (12\text{-}28)$$

where $I_m = V_m / R$.

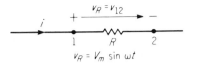

FIG. 12-17. Resistor with sinusoidal voltage.

Note that the current flowing as a result of the sinusoidal voltage is also sinusoidal in form. Furthermore, note that *the current flowing through the resistor is in phase with the voltage drop across the resistor in the direction of the assumed positive current.* A phasor diagram, showing the relationships of the phasors representing the sinusoidal voltage and current, appears in Fig. 12-18. The magnitude of the current is determined by the resistance for a given magnitude of the sinusoidal voltage impressed across the resistor. Since the resistor is a linear element, if a sinusoidal current had been postulated, it would have resulted in a sinusoidal voltage being developed across the resistor.

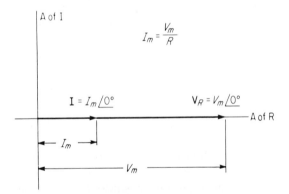

FIG. 12-18. Phasor diagram of voltage and current relationships for a resistor.

It was pointed out, in Chapter 1, that the resistor dissipates energy any time a voltage difference exists across its terminals or, analogously, any time a current flows through it. Referring to Fig. 12-17, the power dissipated, representing the rate at which the energy is being absorbed from the network, is given by any of the following expressions:

$$p_{\text{dissipated}} = \underbrace{v_R^2 \cdot \frac{1}{R}}_{1^{\text{st}}} = \underbrace{i^2 \cdot R}_{2^{\text{nd}}} = \underbrace{v_R \cdot i}_{3^{\text{rd}}} \qquad (12\text{-}29)$$

From Eq. 12-29, it is apparent that the instantaneous power varies with time, that is,

$$p = v_R \cdot i = (V_m \sin \omega t) \left(\frac{V_m}{R} \sin \omega t\right)$$

$$= \frac{V_m^2}{R} \sin^2 \omega t = \frac{V_m^2}{2R} (1 - \cos 2\omega t) \qquad (12\text{-}30)$$

A plot of the instantaneous power, as well as the voltage across the resistor and the current through the resistor, is shown in Fig. 12-19. Note

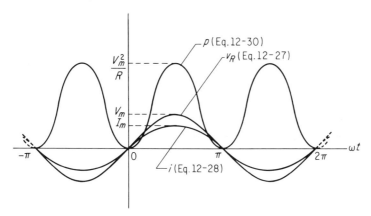

FIG. 12-19. Instantaneous power in a resistor.

that the power is also periodic or cyclic; that is, it repeats itself in time. With respect to the sinusoidal voltage and current, the power is a sinusoid displaced upward, *always positive*, and it has a frequency twice that of the voltage or current waveform. It would be most convenient to have some measure of power that would be independent of time. Since the power is periodic, the *average power dissipated per cycle* would be a constant. We shall assign the uppercase letter P to indicate the average power dissipated per cycle. Furthermore, it would be desirable to have some measure or index of the voltage across the resistor such that, when this voltage was squared and divided by the resistance, the result would be the average power dissipated per cycle. This would then give an expression for the average power that would be identical in form to the dc power dissipated in a resistor. This voltage is called the *effective or rms value*. Thus, by definition,

$$[(V_R)_{\text{eff}}]^2 \cdot \frac{1}{R} \triangleq P \qquad (12\text{-}31)$$

which is an equation analogous to the instantaneous relationship given by the 1st term in Eq. 12-29. The average power per cycle is also given by

$$P \triangleq \frac{1}{T} \int_0^T p \, dt = \frac{1}{T} \int_0^T \frac{v_R^2}{R} \, dt \qquad (12\text{-}32)$$

Equating Eqs. 12-31 and 12-32 gives

$$(V_R)_{\text{eff}} = \left(\frac{1}{T} \int_0^T v_R^2 \, dt \right)^{1/2} \tag{12-33}$$

which indicates that, to determine the effective value, one *squares* the function, then finds the *mean* of the resultant, and then takes the square *root*; hence, *rms* value. Now $v_R = V_m \sin \omega t$, so that

$$(V_R)_{\text{eff}} = \left(\frac{1}{T} \int_0^T V_m^2 \sin^2 \omega t \, dt \right)^{1/2} = \left\{ \frac{V_m^2}{T} \left[\int_0^T \left(\frac{1 - \cos 2\omega t}{2} \right) dt \right] \right\}^{1/2}$$

$$= \left[\frac{V_m^2}{T} \left(\frac{t}{2} - \frac{1}{4\omega} \sin 2 \omega t \right)_0^T \right]^{1/2} = \left(\frac{V_m^2}{2} \right)^{1/2} = \frac{V_m}{\sqrt{2}} \tag{12-34}$$

since $\sin(2\omega T) = \sin(2 \cdot 2\pi f T) = \sin[2 \cdot 2\pi(1/T)T] = \sin 4\pi = 0$. Equation 12-34 indicates that the effective value is obtainable directly from the maximum value. The effective value of the current is defined by the equation

$$(I_{\text{eff}})^2 \cdot R \triangleq P \tag{12-35}$$

which is analogous to the instantaneous relationship given by the 2nd term in Eq. 12-29. The average power per cycle can also be expressed as

$$P = \frac{1}{T} \int_0^T i^2 \cdot R \, dt \tag{12-36}$$

and equating Eqs. 12-35 and 12-36 gives

$$I_{\text{eff}} = \left(\frac{1}{T} \int_0^T i^2 \, dt \right)^{1/2} = \left(\frac{1}{T} \int_0^T I_m^2 \sin^2 \omega t \right)^{1/2}$$

which, by comparison with Eq. 12-34, can be evaluated similarly to give

$$I_{\text{eff}} = \frac{I_m}{\sqrt{2}} \tag{12-37}$$

Now, utilizing Eq. 12-30,

$$P = \frac{1}{T} \int_0^T p \, dt = \frac{1}{T} \int_0^T \frac{V_m^2}{2R} (1 - \cos 2 \omega t) \, dt = \frac{V_m^2}{2RT} \left(t - \frac{\sin 2\omega t}{2\omega} \right)_0^T$$

$$= \frac{V_m^2}{2R} = \frac{V_m}{\sqrt{2}} \frac{V_m}{\sqrt{2}R} = \frac{V_m}{\sqrt{2}} \cdot \frac{I_m}{\sqrt{2}} = V_{\text{eff}} \cdot I_{\text{eff}} \tag{12-38}$$

which is analogous to the instantaneous relationship given by the 3rd term in Eq. 12-29.

The use of effective values is a great convenience in the steady-state analysis of sinusoidally excited networks. Comparing Eq. 12-31 with Eq. 12-35 shows that

$$(V_R)_{\text{eff}} = I_{\text{eff}} \cdot R \qquad (12\text{-}39)$$

so that one may, for convenience, express the phasor magnitude in terms of the effective value rather than in terms of the maximum value, since they are related by a constant, namely, $\sqrt{2}$. The resistance R may also be plotted on the complex plane along the positive real axis, with the understanding that it is fixed, that is, does not rotate with time. In this respect, Eq. 12-39 may be considered *the describing phasor equation* for the relationship of the sinusoidal voltage across and current through a resistor. Thus, the phasor current and the phasor voltage drop, in the direction of the assumed positive current flow in a resistor, are related by

or

$$\left.\begin{aligned}
\mathbf{V}_R &= \mathbf{I} \cdot \mathbf{R} = (I\,\underline{/\alpha}) \cdot R\,\underline{/0^\circ} = I \cdot R\,\underline{/\alpha + 0^\circ} = I \cdot R\,\underline{/\alpha} \\
\mathbf{I} &= \frac{\mathbf{V}_R}{\mathbf{R}} = \frac{V_R\,\underline{/\beta}}{R\,\underline{/0^\circ}} = \frac{V_R}{R}\,\underline{/\beta - 0^\circ} = \frac{V_R}{R}\,\underline{/\beta}
\end{aligned}\right\} \qquad (12\text{-}40)$$

where

$$\mathbf{V}_R = V_R\,\underline{/\beta} \qquad \text{in the phasor domain}$$

represents

$$v_R = \sqrt{2}\,V_R \sin(\omega t + \beta) \qquad \text{in the time domain}$$

and

$$\mathbf{I} = I\,\underline{/\alpha} \qquad \text{in the phasor domain}$$

represents

$$i = \sqrt{2}\,I \sin(\omega t + \alpha) \qquad \text{in the time domain}$$

The boldface letter indicates a phasor quantity. The lightface print, or a stated value of voltage or current, indicates the effective value of the phasor quantity, unless otherwise designated; for example, I_m refers to the maximum value, as before. A common example is the designation of 125 volts for residential voltage, meaning that the effective value is 125 volts.

A question that immediately comes to mind is whether the effective value of a voltage or current is dependent upon the choice of a reference axis. Hopefully not, if it is to be of any use in circuit analysis. Suppose that the voltage across the resistor is given by

$$v_R = V_m \sin(\omega t + \theta) \qquad (12\text{-}41)$$

It is quite evident that the average power per cycle will not depend on θ, since the power itself is cyclic. From the definition, the effective value of the voltage is given by

$$(V_R)_{\text{eff}} = \left(\frac{1}{T} \int_0^T v_R^2\, dt\right)^{1/2} = \left\{\frac{1}{T} \int_0^T [V_m^2 \sin^2(\omega t + \theta)]\, dt\right\}^{1/2}$$

$$= \left\{\frac{V_m^2}{T \cdot 2} \int_0^T [1 - \cos 2(\omega t + \theta)]\, dt\right\}^{1/2}$$

$$= \left\{\frac{V_m^2}{T \cdot 2} \left[t - \frac{\sin 2(\omega t + \theta)}{2\omega}\right]_0^T\right\}^{1/2} = \left(\frac{V_m^2}{2}\right)^{1/2}$$

$$= \frac{V_m}{\sqrt{2}} \tag{12-42}$$

The effective value of a sinusoidal voltage or current is independent of the choice of reference axis or, analogously, it is independent of its position on the phasor diagram.

EXAMPLE 12-1. A sinusoidal current of the form $i = 3 \sin(400t + \pi/6)$ is flowing through a resistor of 2 ohms resistance. What is the resultant phasor voltage drop *in* the direction of current flow? What is the average power dissipated? Draw the resultant phasor diagram. The circuit diagram involving voltages and currents for the depicted resistive element is given in Fig. 12-20.

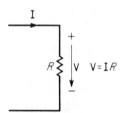

FIG. 12-20. Circuit diagram for Example 12-1.

Solution: From the expression for the current,

$$I_m = 3 \text{ amps} \quad \text{and} \quad \frac{\pi}{6} \text{ radian} = 30°$$

so that

$$I = \frac{I_m}{\sqrt{2}} = 2.12 \text{ amps}$$

In phasor notation,

$$\mathbf{I} = 2.12 \underline{/30°} \text{ amps} \quad \text{and} \quad R = 2 \text{ ohms}$$

so that

$$\mathbf{V} = \mathbf{I} \cdot R = (2.12 \underline{/30°})(2) = (2.12)(2) \underline{/30°} = 4.24 \underline{/30°} \text{ volts}$$

Now

$$P = I^2 R = (2.12)^2 \cdot (2) = 9.00 \text{ watts}$$

or

$$P = \frac{V^2}{R} = \frac{(4.24)^2}{2} = 9.00 \text{ watts}$$

or

$$P = VI = (4.24)(2.12) = 9.00 \text{ watts}$$

The resultant phasor diagram is given in Fig. 12-21.

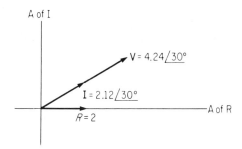

FIG. 12-21. Phasor diagram for Example 12-1.

12-6. Inductive Element

The general equation (Eq. G-4) for the voltage and current relationship in an inductor is developed in Appendix G. It is used here under

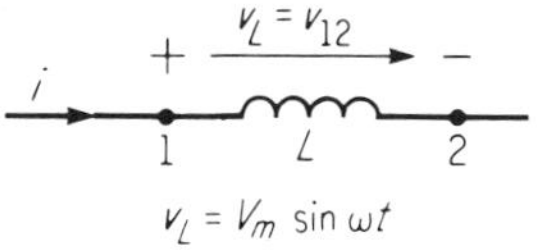

FIG. 12-22. Inductor with sinusoidal voltage.

the assumption that a sinusoidal voltage is impressed upon the inductor, such as that depicted in Fig. 12-22.

Now,

$$v_L = L\frac{di}{dt} = V_m \sin \omega t \tag{12-43}$$

and, by integrating both sides of Eq. 12-43,

$$i = \frac{1}{L} \int_{-\infty}^{t} V_m \sin \omega t \, dt = \frac{1}{L} \int_{-\infty}^{0} V_m \sin \omega t \, dt + \frac{1}{L} \int_{0}^{t} V_m \sin \omega t \, dt$$

$$= i(0^+) + \frac{1}{L} \int_{0}^{t} V_m \sin \omega t \, dt$$

$$= i(0^+) + \frac{V_m}{L} \left(\frac{-\cos \omega t}{\omega} + \frac{1}{\omega} \right)$$

For convenience, choose the initial condition for the current as $i(0^+) = -V_m/\omega L$. Then

$$i = \frac{-V_m}{\omega L} \cos \omega t = \frac{V_m}{\omega L} \sin \left(\omega t - \frac{\pi}{2} \right) = I_m \sin \left(\omega t - \frac{\pi}{2} \right) \quad (12\text{-}44)$$

where $I_m = V_m/\omega L$.[1]

A number of observations can be made. It is seen that, as a result of a sinusoidal voltage being impressed across an inductor, a sinusoidal current flow results. This current lags the voltage drop, in the direction of the assumed positive current flow, by $\pi/2$ radians or 90°. A phasor diagram, showing the relationships of the phasors representing the sinusoidal voltage and current is given in Fig. 12-23. The magnitude of the current is determined not only by the magnitude of the voltage and the inductance of the element itself but also by the radian frequency of the sinusoidal voltage. For a fixed amplitude of voltage and a constant inductance, the current decreases with increasing frequency. The quantity ωL that determines the magnitude of the current for a given magnitude of the voltage is called the *inductive reactance*. It is denoted by the symbol X_L; that is,

$$X_L = \omega L = 2\pi f L \quad (12\text{-}45)$$

The same relationship exists between the effective values of the voltage

[1] The reader may be alarmed by what appears to be some sleight-of-hand manipulations made with the initial conditions. In general, as is evident from the equations, a d-c (non-varying) component of current would be established in a *pure* inductor if a sinusoidal voltage were placed across its terminals. This is necessary since the current cannot change instantaneously in an inductor; that is, the energy stored in the magnetic field is a continuous function of time. If one supposes that the negative of this d-c component of current was flowing prior to impressing the sinusoidal voltage, then only a sinusoidal component of current would be present after impressing the sinusoidal voltage. Many authors avoid this problem by assuming that a sinusoidal current is flowing in the inductor, and then differentiating the current expression to find the resultant sinusoidal voltage. However, the same voltage results if in addition to the sinusoidal current, a d-c component of current is also assumed to be flowing in the inductor, since the derivative of a constant is zero. A final word without belaboring the point: A practical inductor would always have some resistance associated with it, and the d-c component of current would decay out, since the resistance would afford a way for this energy to be dissipated. This is not possible for a pure inductor.

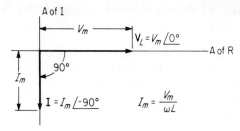

FIG. 12-23. Phasor diagram of the voltage
and current relationships for an inductor.

and current. Since

$$I_m = \frac{V_m}{\omega L}$$

then

$$I_{eff} = \frac{I_m}{\sqrt{2}} = \frac{V_m}{\omega L \sqrt{2}} = \frac{V_{eff}}{\omega L}$$

Now, since

$$\omega L = \frac{V_m}{I_m} = \frac{V_{eff}}{I_{eff}}$$

the inductive reactance has the dimensions of volts per ampere, or *ohms.*

It should be mentioned that one could start by assuming a sinusoidal current flowing through the inductor and then finding the resultant voltage drop due to this current flow. However, the relationship between the voltage and current obtained in this manner would be the same as in the analysis just completed.

In the preceding section it was pointed out that the resistance could be plotted on the phasor diagram, along the positive axis of the reals, even though it does not vary with time. With this artifice, an equation relating the phasor voltage and current for the resistor was obtained (see Eq. 12-40). With a similar goal in mind, let the inductive reactance be plotted along the positive axis of the imaginaries, so that the pseudophasor representation of the inductive reactance is given by

$$\mathbf{X}_L = j\omega L$$

With this complex-reactance representation, *the phasor current and the phasor voltage drop in the assumed positive direction of current flow in an inductor* are related by

$$\mathbf{V}_L = \mathbf{I} \cdot \mathbf{X}_L = (I\,\underline{/\alpha}) \cdot (X_L\,\underline{/90°}) = (IX_L)\,\underline{/\alpha + 90°}$$

or

$$\mathbf{I} = \frac{\mathbf{V}_L}{\mathbf{X}_L} = \frac{V_L\,\underline{/\beta}}{X_L\,\underline{/90°}} = \frac{V_L}{X_L}\,\underline{/\beta - 90°}$$

(12-46)

The pseudophasor $\mathbf{X}_L$ provides the 90° displacement between the appropriate voltage and current phasors for the inductive element.

EXAMPLE 12.2. A sinusoidal-voltage source of the form $v = 120\sqrt{2}\sin(2\pi60t + \pi/4)$ is applied to the terminals of an inductor having an inductance of 0.1 henry. What is the resultant phasor current? Draw the resultant phasor diagram. The circuit diagram involving phasor voltages and currents for the depicted inductive element is given in Fig. 12-24.

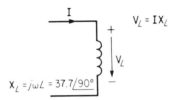

FIG. 12-24. Circuit diagram for Example 12-2.

Solution: Note, from the expression of the voltage source, $\omega = 2\cdot\pi\cdot60$ radians/sec, $V_m = 120\sqrt{2}$ volts, and $\pi/4$ radians = 45°, so that $X_L = \omega L = 2\cdot\pi\cdot60\cdot(0.1) = 37.7$ ohms, $V = V_m/\sqrt{2} = 120$ volts and therefore, $V = 120\underline{/45°}$ volts and $\mathbf{X}_L = 37.7\underline{/90°}$ ohms. From Eq. 12-46,

$$\mathbf{I} = \frac{\mathbf{V}}{\mathbf{X}_L} = \frac{120\underline{/45°}}{37.7\underline{/90°}} = \frac{120}{37.7}\underline{/45° - 90°} = 3.18\underline{/-45°}\ \text{amps}$$

The resultant phasor diagram is given in Fig. 12-25.

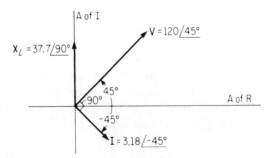

FIG. 12-25. Phasor diagram for Example 12-2.

12-7. Capacitive Element

The general equation (Eq. G-8) for the voltage and current relationship in a capacitor is developed in Appendix G. It is used here under the assumption that a sinusoidal voltage is impressed upon a capacitor, such as that depicted in Fig. 12-26.

$$v_C = \frac{1}{C} \int_{-\infty}^{t} i \, dt = \underbrace{\frac{1}{C} \int_{-\infty}^{0} i \, dt}_{v_C(0^+)} + \frac{1}{C} \int_{0}^{t} i \, dt = V_m \sin \omega t \qquad (12\text{-}47)$$

By taking the derivatives of both sides of Eq. 12-47, that is,

$$\frac{d}{dt} \left(v_C(0^+) + \frac{1}{C} \int_{0}^{t} i \, dt \right) = \frac{d}{dt} \left(V_m \sin \omega t \right)$$

one obtains

$$\frac{1}{C} i = V_m \omega \cos \omega t$$

or

$$i = \frac{V_m}{1/(\omega C)} \cos \omega t = \frac{V_m}{1/(\omega C)} \sin \left(\omega t + \frac{\pi}{2} \right)$$

$$= I_m \sin \left(\omega t + \frac{\pi}{2} \right) \qquad (12\text{-}48)$$

where $I_m = [V_m \div 1/(\omega C)]$.

Here it is seen that, as a result of a sinusoidal voltage being impressed across a capacitor, a sinusoidal-current flow results. This current leads the voltage drop in the direction of the assumed positive current flow by $\pi/2$ radians or 90°. Note that the word "leads" is significant in dif-

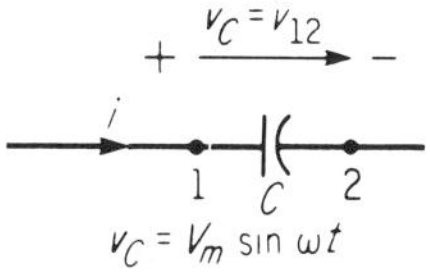

FIG. 12-26. Capacitor with sinusoidal voltage.

ferentiating the voltage-current relationship in a capacitor from that in an inductor, where the displacement is also 90° but the current lags the voltage drop in the direction of the assumed positive current flow. A phasor diagram, showing the relationships of the phasors representing the sinusoidal voltage and current is given in Fig. 12-27.

The magnitude of the current is a function not only of the magnitude of the voltage and the capacitance of the element but also of the radian frequency of the sinusoidal voltage. Given a fixed amplitude of voltage and a constant capacitance, the current increases with increasing frequency. It might be well to note that this is just the opposite variation

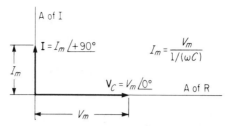

FIG. 12-27. Phasor diagram of the voltage
and current relationships for a capacitor.

with frequency as compared to an inductor, where the current decreases
with increasing frequency for a fixed voltage magnitude and a fixed value
of inductance. The quantity $1/(\omega C)$, which determines the magnitude of
the current for a given magnitude of the voltage, is called the *capacitive
reactance.* It is denoted by the symbol X_C, that is,

$$X_C = \frac{1}{\omega C} = \frac{1}{2\pi f C} \tag{12-49}$$

Since

$$I_m = \frac{V_m}{1/(\omega C)}$$

then

$$I_{\text{eff}} = \frac{I_m}{\sqrt{2}} = \frac{V_m/\sqrt{2}}{1/(\omega C)} = \frac{V_{\text{eff}}}{1/(\omega C)}$$

which shows that the same relation exists between the effective values of
the voltage and current as between the maximum values of the voltage and
current. Therefore,

$$\frac{1}{\omega C} = \frac{V_m}{I_m} = \frac{V_{\text{eff}}}{I_{\text{eff}}}$$

indicating that the capacitive reactance has the dimensions of volts per
ampere, or *ohms.*

The utility of the phasor representation of resistance and inductive
reactance has been demonstrated in preceding sections and will now be
extended to the capacitive reactance. Let the capacitive reactance be
plotted along the negative axis of the imaginaries, so that the pseudophasor representation is given by

$$\mathbf{X}_C = -j\frac{1}{\omega C} = \frac{1}{j\omega C}$$

Using this complex-reactance representation, *the phasor current and
the phasor voltage drop in the assumed positive direction of current flow in*

the capacitor are related by

$$\mathbf{V}_C = \mathbf{I} \cdot \mathbf{X}_C = (I\underline{/\alpha}) \cdot (X_C\underline{/-90°}) = (IX_C)\underline{/\alpha - 90°} \left.\begin{matrix} \\ \\ \\ \\ \\ \end{matrix}\right\}$$

or

$$\mathbf{I} = \frac{\mathbf{V}_C}{\mathbf{X}_C} = \frac{V_C\underline{/\beta°}}{X_C\underline{/-90°}} = \frac{V_C}{X_C}\underline{/\beta + 90°}$$

(12-50)

The pseudophasor $\mathbf{X}_C$ provides the 90° displacement between the appropriate voltage and current phasors for the capacitive element.

EXAMPLE 12-3. A sinusoidal current of the form $i = 14.14 \sin (500t + \pi/3)$ is flowing through a capacitor having a capacitance of 2 μf. What is the resultant phasor voltage? Draw the resultant phasor diagram. The circuit diagram involving phasor voltages and currents for the depicted capacitive element is given in Fig. 12-28.

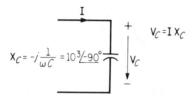

FIG. 12-28. Circuit diagram for
Example 12-3.

Solution: From the expression for the current, one deduces that $\omega = 500$ radians/sec, $I_m = 14.14$ amps, and $\pi/3$ radians $= 60°$, so that $X_C = 1/(\omega C) = 1/[(500)(2 \times 10^{-6})] = 10^3$ ohms, $I = I_m/\sqrt{2} = 10$ amps, and, therefore, $\mathbf{I} = 10\underline{/60°}$ amps and $\mathbf{X}_C = 10^3\underline{/-90°}$ ohms. From Eq. set 12-50,

$$\mathbf{V}_C = \mathbf{I} \cdot \mathbf{X}_C = (10\underline{/60°})(10^3\underline{/-90°}) = 10^4\underline{/-30°} \text{ volts}$$

The resultant phasor diagram is given in Fig. 12-29.

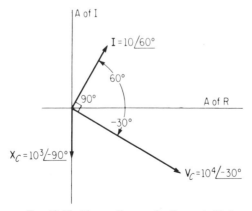

FIG. 12-29. Phasor diagram for Example 12-3.

12-8. Combined Elements

In general, networks are formed by series and parallel combinations of single elements, be they resistors, inductors, and/or capacitors. The preceding sections have afforded the phasor relations for the voltage and current for each element (Eqs. 12-40, 12-46, and 12-50). Before proceeding, one other concept will be introduced. It has been noted that R, X_L, and X_C each has the unit of ohms, and, since resistance has traditionally been associated with the symbol R, the term *impedance* is used for the dimension of the three quantities taken collectively. As has already been discussed in several of the earlier chapters, the term impedance follows naturally from the idea that each quantity impedes current flow when a sinusoidal voltage is impressed across the particular element. Through the use of a complex reactance, the equations describing the voltage and current relationships for each element are essentially the same; that is, *in the assumed positive direction of current flow, the voltage drop across an element is the product of the current and the impedance of the particular element under consideration.* The question now to be answered is: How does one handle a network where more than one kind of element is present? The results of Sec. 12-4 will be used to provide the answer. Recall that it was demonstrated that the addition of two sinusoidal quantities could be accomplished by adding their phasor representations, which greatly facilitates the addition.

Consider a resistor and an inductor, connected in series, with a sinusoidal current flowing through the series combination. Such a combination is depicted in Fig. 12-30, with appropriate phasors indicated on

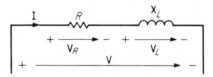

FIG. 12-30. Series combination of
a resistor and an inductor.

the diagram. From Kirchhoff's voltage law, the voltage across the series combination is equal to the sum of the voltages across each of the elements which, in terms of the phasors, is given by $\mathbf{V} = \mathbf{V}_R + \mathbf{V}_L$. But $\mathbf{V}_R = \mathbf{I} \cdot R$ and $\mathbf{V}_L = \mathbf{I} \cdot X_L$, and, therefore.

$$\mathbf{V} = \mathbf{I} \cdot R + \mathbf{I} \cdot \mathbf{X}_L = \mathbf{I} \cdot (R + \mathbf{X}_L) \tag{12-51}$$

The first observation from Eq. 12-51 is that the total impedance of the 2 elements in series is the sum of the individual impedances. Note that this is the same relationship as 2 resistors in series, with one exception. The sum is a phasor sum rather than an algebraic sum, as in the case of resistors. Let the symbol $\mathbf{Z}$ represent a general impedance so that, for the above case of the resistor in series with the inductor,

$$\mathbf{Z} = Z\,\underline{/\theta} = R + \mathbf{X}_L = R + jX_L = R + j\omega L$$

where

$$Z = (R^2 + X_L^2)^{1/2} \quad \text{and} \quad \theta = \tan^{-1}(X_L/R)$$

The impedance $\mathbf{Z}$ then determines the proper relationship between the phasors $\mathbf{V}$ and $\mathbf{I}$.

EXAMPLE 12-4. Consider a sinusoidal current of 4 amps flowing through a series combination of a resistor and an inductor as shown in Fig. 12-31. The re-

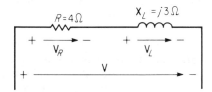

FIG. 12-31. Circuit diagram for Example 12-4.

sistor has a resistance of 4 ohms, and the inductor has an inductive reactance of 3 ohms. What is the equivalent impedance of the series combination? What is the voltage across the series combination? Draw a phasor diagram.

Solution: The equivalent impedance is given by $\mathbf{Z} = R + \mathbf{X}_L = 4 + j3 = 5\,\underline{/36.9°}$ ohms. The current is chosen as the reference phasor and is given by $\mathbf{I} = 4\,\underline{/0°}$ amps. The voltage drops $\mathbf{V}_R$ and $\mathbf{V}_L$ are given by $\mathbf{V}_R = \mathbf{I} \cdot R = (4\,\underline{/0°})(4\,\underline{/0°}) = 16\,\underline{/0°}$ volts and $\mathbf{V}_L = \mathbf{I} \cdot \mathbf{X}_L = (4\,\underline{/0°})(3\,\underline{/90°}) = 12\,\underline{/90°}$ volts. The voltage across the series combination is $\mathbf{V} = \mathbf{V}_R + \mathbf{V}_L = 16\,\underline{/0°} + 12\,\underline{/90°} = 16 + j12 = 20\,\underline{/36.9°}$ volts, or $\mathbf{V} = \mathbf{I} \cdot \mathbf{Z} = (4\,\underline{/0°})(5\,\underline{/36.9°}) = 20\,\underline{/36.9°}$ volts.

This approach can be extended to several elements in series whereby the total impedance of the elements is the sum of the individual impedances. The phasor diagram for Example 12-4 is given in Fig. 12-32.

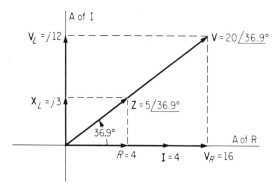

FIG. 12-32. Phasor diagram for Example 12-4.

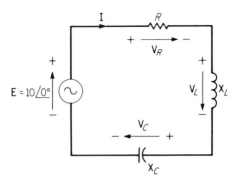

FIG. 12-33. Circuit diagram for Example
12-5.

EXAMPLE 12-5. A resistor of 2 ohms, an inductor of 18 millihenries (mh), and a capacitor of 50 μf are connected in series, as shown in Fig. 12-33, and are to be supplied by a sinusoidal voltage source of 10 volts rms having a radian frequency of 1000 radians/sec. Determine the impedance of each element and the total impedance as presented to the source. Determine the voltage across each element. Draw a phasor diagram.

Solution: Now R = 2 ohms, L = 0.018 henry, C = 50 $\times$ 10^{-6} farad, and ω = 1000 radians/sec. The individual impedances are R = 2 ohms, $\mathbf{X}_L = j\omega L$ = $j(1000)(0.018)$ = $j18$ ohms, $\mathbf{X}_C = -j\,1/\omega C$ = $-j\,1/(1000)(50 \times 10^{-6})$ = $-j20$ ohms, and $\mathbf{Z} = R + \mathbf{X}_L + \mathbf{X}_C$, $\mathbf{Z}$ = $2 + j18 - j20$ = $2 - j2$ = $2.828 \underline{/-45°}$ ohms.

Note that the total impedance is *less* than the individual impedance of two of the elements. This is a very useful characteristic of such circuits, and it will be exploited later. This can be further illustrated by drawing the impedances alone on the complex plane. This type of diagram is referred to as an impedance diagram and is shown in Fig. 12-34a. Since the inductive reactance increases with increasing frequency while the capacitive reactance decreases with increasing frequency, one can find a frequency at which the two quantities will exactly cancel each other, and the total impedance will then be that of the resistance alone. This frequency is called the series resonance frequency. It is obtained from $X_L = X_C$; that is, $2\pi f L = 1/(2\pi f C)$, whereby $f = 1/(2\pi\sqrt{LC})$.

Now, the current in the network of this example is given by $\mathbf{I} = \mathbf{E}/\mathbf{Z}$ = $10 \underline{/0°}/2.828 \underline{/-45°}$ = $3.54 \underline{/45°}$ amps. The voltage across the resistor is $\mathbf{V}_R = \mathbf{I}R = (3.54 \underline{/45°})(2)$ = $7.07 \underline{/45°}$ volts. The voltage across the inductor is $\mathbf{V}_L = \mathbf{I}\mathbf{X}_L = (3.54 \underline{/45°})(18 \underline{/90°})$ = $63.6 \underline{/135°}$ volts. The voltage across the capacitor is $\mathbf{V}_C = \mathbf{I}\mathbf{X}_C = (3.54 \underline{/45°})(20 \underline{/-90°})$ = $70.7 \underline{/-45°}$ volts.

A phasor diagram of the voltages and currents is given in Fig. 12-34b. It is to be observed that the voltages across the inductor and the capacitor are several times greater than the magnitude of the source voltage. This is a result of the mutual canceling effects of the reactances when the impedance of the network presented to the source is determined. From the viewpoint of the source, a capacitive reactance of 2 ohms in series with a resistance of 2 ohms determines the current

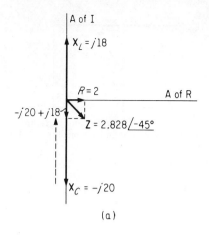

(a)

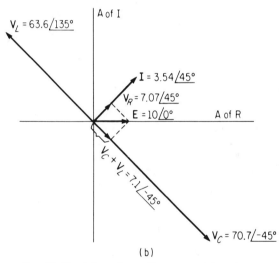

(b)

FIG. 12-34. a) Impedance diagram and b) phasor diagram for Example 12-5.

flow. This current, in turn, flows through the high individual reactances to give the relatively large voltage drops. Kirchhoff's voltage law is naturally satisfied, but, as a check

$$\mathbf{V}_R + \mathbf{V}_L + \mathbf{V}_C = (5 + j5) + (-45 + j45) + (50 - j50)$$
$$= 10 + j0 = 10 \underline{/0^\circ} = \mathbf{E}$$

Now consider a resistor and a capacitor in parallel, with a sinusoidal voltage impressed across the pair. The combination, with appropriate phasors, is shown in Fig. 12-35. Using Kirchhoff's current law, one obtains $\mathbf{I} = \mathbf{I}_R + \mathbf{I}_C$ as the relation describing the phasor currents. But, for

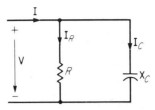

FIG. 12-35. Parallel combination
of a resistor and a capacitor.

the individual elements, $I_R = V/R$ and $I_C = V/X_C$, so that

$$I = \frac{V}{R} + \frac{V}{X_C} = V\left(\frac{X_C + R}{R \cdot X_C}\right)$$

or

$$\frac{V}{I} = \frac{R \cdot X_C}{R + X_C}$$

But the ratio V/I is the equivalent impedance of the 2 elements in parallel; that is,

$$Z = \frac{R \cdot X_C}{R + X_C} \qquad (12\text{-}52)$$

Note that the product-over-sum relationship (Eq. 12-52) is the same as that for 2 resistors in parallel, but, in this case, the product and sum involve phasors and must be treated accordingly.

EXAMPLE 12-6. Consider a sinusoidal voltage of 120 volts impressed on a resistor and a capacitor in parallel as depicted in Fig. 12-36. The resistor has a resistance of 40 ohms, and the capacitor has a capacitive reactance of 30 ohms. What is the equivalent impedance of the parallel combination? What is the total current supplied to the 2 elements? Draw a phasor diagram.

Solution: The equivalent impedance is given by

$$\begin{aligned}
Z &= \frac{R \cdot X_C}{R + X_C} \\
&= \frac{(40)(-j30)}{40 - j30} = \frac{-j1200}{50\,\underline{/-36.9°}} \\
&= \frac{1200\,\underline{/-90°}}{50\,\underline{/-36.9°}} = 24\,\underline{/-53:1°}\text{ ohms}
\end{aligned}$$

The voltage is chosen as the reference phasor and is given by $V = 120\,\underline{/0°}$ volts. The currents I_R and I_C are given by $I_R = V/R = 120\,\underline{/0°}/40\,\underline{/0°} = 3\,\underline{/0°}$ amps and $I_C = V/X_C = 120\,\underline{/0°}/30\,\underline{/-90°} = 4\,\underline{/90°}$ amps. The total current is $I = I_R + I_C = 3\,\underline{/0°} + 4\,\underline{/90°} = 3 + j4 = 5\,\underline{/53.1°}$ amps, or $I = V/Z = 120\,\underline{/0°}/24\,\underline{/-53.1°} = 5\,\underline{/53.1°}$ amps.

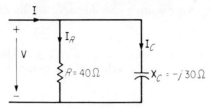

FIG. 12-36. Circuit diagram for
Example 12-6.

It should be recognized that the total current delivered by the voltage source can be determined by the sum of the individual currents flowing to each branch. Alternatively, the equivalent impedance of the branches, with respect to the voltage source, can be calculated and the total current determined on the basis of this equivalent impedance. The notion of the equivalent impedance of a group of elements can be used quite effectively in calculating various circuit quantities. The phasor diagram is given in Fig. 12-37.

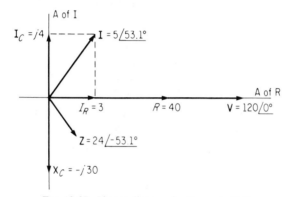

FIG. 12-37. Phasor diagram for Example 12-6.

It is left to the reader to show that the equivalent impedance of several impedances in parallel is given by

$$Z_{eq} = \frac{1}{\dfrac{1}{Z_1} + \dfrac{1}{Z_2} + \cdots + \dfrac{1}{Z_n}}$$

(*Hint:* This can be demonstrated in the same manner as for resistances in parallel with the exception that all operations involve complex quantities.)

12-9. Network Solutions

The steady-state solutions of networks with sinusoidal-voltage and/or -current sources is exactly the same as in the d-c case, except that the equations now involve phasors and must be treated accordingly. This makes the calculations somewhat more complicated, since complex algebra must be utilized. An example should serve to illustrate the treatment of such networks.

EXAMPLE 12-7.   Solve the network, given in Fig. 12-38, by the method of loop currents. Note that if only the values of the inductors and capacitors are

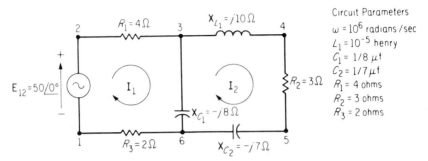

Circuit Parameters

$\omega = 10^6$ radians/sec
$L_1 = 10^{-5}$ henry
$C_1 = 1/8 \mu f$
$C_2 = 1/7 \mu f$
$R_1 = 4$ ohms
$R_2 = 3$ ohms
$R_3 = 2$ ohms

FIG. 12-38. Circuit diagram for Example 12-7.

given, then the frequency must be specified explicitly in order that the reactances can be calculated. If the reactances are given, then there is no need to state explicitly the frequency in order to solve the problem. However, without knowing the frequency, one cannot determine the values of the inductive and capacitive components from their corresponding reactances.

Solution: One can begin by writing an expression for Kirchhoff's voltage law around each loop, in terms of the phasors, giving

$$\left. \begin{aligned} \mathbf{V}_{12} + \mathbf{V}_{23} + \mathbf{V}_{36} + \mathbf{V}_{61} = 0 \\ \\ \mathbf{V}_{34} + \mathbf{V}_{45} + \mathbf{V}_{56} + \mathbf{V}_{63} = 0 \end{aligned} \right\} \qquad (12\text{-}53)$$

and

The next step is to substitute appropriate expressions for each of the voltage drops, in terms of the loop currents and circuit impedances. This gives

$$\left. \begin{aligned} -\mathbf{E}_{12} + \mathbf{I}_1 \cdot R_1 + \mathbf{I}_1 \cdot \mathbf{X}_{C_1} - \mathbf{I}_2 \cdot \mathbf{X}_{C_1} + \mathbf{I}_1 \cdot R_3 = 0 \\ \\ \mathbf{I}_2 \cdot \mathbf{X}_{L_1} + \mathbf{I}_2 \cdot R_2 + \mathbf{I}_2 \cdot \mathbf{X}_{C_2} + \mathbf{I}_2 \cdot \mathbf{X}_{C_1} - \mathbf{I}_1 \cdot \mathbf{X}_{C_1} = 0 \end{aligned} \right\} \qquad (12\text{-}54)$$

and

Equation set 12-54 can be rearranged to group the coefficients of each of the unknown loop currents, to give

$$\left. \begin{aligned} (R_1 + \mathbf{X}_{C_1} + R_3) \cdot \mathbf{I}_1 + (-\mathbf{X}_{C_1}) \cdot \mathbf{I}_2 = \mathbf{E}_{12} \\ (-\mathbf{X}_{C_1}) \cdot \mathbf{I}_1 + (\mathbf{X}_{L_1} + R_2 + \mathbf{X}_{C_2} + \mathbf{X}_{C_1}) \mathbf{I}_2 = 0 \end{aligned} \right\} \qquad (12\text{-}55)$$

It should be quite evident that the form of the equations is the same as the equations describing d-c networks, the only difference being the use of phasors and the impedance concept.

A generalized notation could be used to indicate the impedance summations so that the describing equations could be written by inspection as in the case of the resistive networks treated in Chapter 2. This gives

$$\left.\begin{array}{c} \mathbf{Z}_{11}\mathbf{I}_1 - \mathbf{Z}_{12}\mathbf{I}_2 = \mathbf{E}_{12} \\ -\mathbf{Z}_{21}\mathbf{I}_1 + \mathbf{Z}_{22}\mathbf{I}_2 = 0 \end{array}\right\} \qquad (12\text{-}56)$$

where

$$\left.\begin{array}{l} \mathbf{Z}_{11} = (R_1 + \mathbf{X}_{C_1} + R_3) = 4 - j8 + 2 \\ \mathbf{Z}_{12} = \mathbf{Z}_{21} = \mathbf{X}_{C_1} = -j8 \\[1em] \mathbf{Z}_{22} = (\mathbf{X}_{L_1} + R_2 + \mathbf{X}_{C_2} + \mathbf{X}_{C_1}) = j10 + 3 - j7 - j8 \end{array}\right\} \qquad (12\text{-}57)$$

and

Substituting the values of the several impedances into Eq. set 12-55 gives

$$(4 - j8 + 2)\cdot\mathbf{I}_1 + (+j8)\mathbf{I}_2 = 50\underline{/0°}$$
$$(+j8)\cdot\mathbf{I}_1 + (+j10 + 3 - j7 - j8)\mathbf{I}_2 = 0$$

or

$$\left.\begin{array}{c} (6 - j8)\cdot\mathbf{I}_1 \qquad + j8\cdot\mathbf{I}_2 = 50\underline{/0°} \\ j8\cdot\mathbf{I}_1 + (3 - j5)\cdot\mathbf{I}_2 = 0 \end{array}\right\} \qquad (12\text{-}58)$$

A set of simultaneous linear algebraic equations results. The only difference from those previously obtained for d-c circuits is the nature of the coefficients which are complex numbers. It now remains to practice solving various networks, in order to become familiar with the complex-algebra manipulations necessary to obtain the solutions.

The determinant solution can be obtained in a straightforward manner. From Eq. set 12-58.

$$\mathbf{I}_1 = \frac{\begin{vmatrix} 50\underline{/0°} & j8 \\ 0 & 3 - j5 \end{vmatrix}}{\begin{vmatrix} 6 - j8 & j8 \\ j8 & 3 - j5 \end{vmatrix}} = \frac{\begin{vmatrix} 50\underline{/0°} & 8\underline{/90°} \\ 0 & 5.83\underline{/-59°} \end{vmatrix}}{\begin{vmatrix} 10\underline{/-53.1°} & 8\underline{/90°} \\ 8\underline{/90°} & 5.83\underline{/-59°} \end{vmatrix}}$$

$$= \frac{292\underline{/-59°}}{58.3\underline{/-112.1°} - 64\underline{/180°}}$$

$$= \frac{292\underline{/-59°}}{68.5\underline{/-52°}} = 4.26\underline{/-7°}\text{ amps}$$

and

$$\mathbf{I}_2 = \frac{\begin{vmatrix} 6 - j8 & 50\underline{/0°} \\ j8 & 0 \end{vmatrix}}{68.5\underline{/-52°}} = \frac{-(8\underline{/90°})(50)}{68.5\underline{/-52°}} = 5.84\underline{/-38°}\text{ amps}$$

EXAMPLE 12-8. Solve the network, given in Fig. 12-39, by the method of node voltages.

Solution: Since there are 3 nodes and no voltage sources, then, by Eq. 2-53 of Chapter 2,

$$N_v = n - 1 - N_{vs} = 2$$

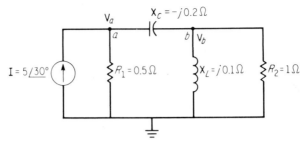

FIG. 12-39. Circuit diagram for Example 12-8.

That is, two equations are to be written, and the nodes at which Kirchhoff's current law is to be applied are labeled *a* and *b*. The equation resulting at node *a* is

$$\frac{\mathbf{V}_a}{R_1} + \frac{\mathbf{V}_a - \mathbf{V}_b}{\mathbf{X}_C} = \mathbf{I} \qquad (12\text{-}59)$$

and the equation resulting at node *b* is

$$\frac{\mathbf{V}_b - \mathbf{V}_a}{\mathbf{X}_C} + \frac{\mathbf{V}_b}{\mathbf{X}_L} + \frac{\mathbf{V}_b}{R_2} = 0 \qquad (12\text{-}60)$$

Putting the equations in standard form gives

$$\left. \begin{aligned} \left(\frac{1}{R_1} + \frac{1}{\mathbf{X}_C}\right)\mathbf{V}_a - \left(\frac{1}{\mathbf{X}_C}\right)\mathbf{V}_b &= \mathbf{I} \\[2mm] -\left(\frac{1}{\mathbf{X}_C}\right)\mathbf{V}_a + \left(\frac{1}{\mathbf{X}_C} + \frac{1}{\mathbf{X}_L} + \frac{1}{R_2}\right)\mathbf{V}_b &= 0 \end{aligned} \right\} \qquad (12\text{-}61)$$

Again, the form of the equations is identical to those developed in Chapter 2, with the voltages and currents being phasor quantities and with complex impedances being included rather than resistors alone. Substituting the values of the impedances into the set of equations gives

$$\left(\frac{1}{0.5} + \frac{1}{-j0.2}\right)\mathbf{V}_a - \left(\frac{1}{-j0.2}\right)\mathbf{V}_b = 5\underline{/30°}$$

and

$$-\left(\frac{1}{-j0.2}\right)\mathbf{V}_a + \left(\frac{1}{-j0.2} + \frac{1}{j0.1} + \frac{1}{1}\right)\mathbf{V}_b = 0$$

or

$$\left. \begin{aligned} (2 + j5)\,\mathbf{V}_a + (-j5)\,\mathbf{V}_b &= 5\underline{/30°} \\[2mm] (-j5)\,\mathbf{V}_a + (1 - j5)\,\mathbf{V}_b &= 0 \end{aligned} \right\} \qquad (12\text{-}62)$$

The solution to Eq. set 12-62 is

$$\mathbf{V}_a = \frac{\begin{vmatrix} 5\underline{/30°} & -j5 \\ 0 & 1-j5 \end{vmatrix}}{\begin{vmatrix} 2+j5 & -j5 \\ -j5 & 1-j5 \end{vmatrix}} = \frac{25.5\underline{/-48.7°}}{52-j5} = \frac{25.5\underline{/-48.7°}}{52\underline{/-5.5°}} = 0.49\underline{/-43.2°} \text{ volt}$$

and

$$\mathbf{V}_b = \frac{\begin{vmatrix} 2+j5 & 5\underline{/30°} \\ -j5 & 0 \end{vmatrix}}{52\underline{/-5.5°}} = \frac{25\underline{/120°}}{52\underline{/-5.5°}} = 0.48\underline{/125.5°} \text{ volt}$$

12-10. Power Relationships

Of the 3 circuit elements, only the resistor dissipates energy. The inductor and capacitor have the property of being able to store energy in a magnetic and an electric field, respectively. In a circuit with sinusoidal excitation, energy transfer can occur between the inductive and capacitive elements as well as between the source and all the reactive elements taken collectively. Since this cyclic transfer of energy between the source and the circuit demands an energy capacity of the source in excess of that being fully utilized by the circuit, it has very important economic consequences in large electric-power systems. Also, this cyclic transfer of energy always has some losses associated with its transmission (i^2R loss in the copper transmission conductors) which are of economic importance. Although the economic questions will not be considered here, they point to the importance of the energy and/or power relationships that exist in sinusoidally excited power systems, which will now be examined.

It has already been shown (Sec. 12-5) that the average power dissipated per cycle for a resistor can be calculated through the use of the effective voltage across and/or the effective current through the resistor. It has also been demonstrated that the inductive and capacitive reactances have the dimensions of ohms. Consider a sinusoidal current i flowing through an inductor with a resultant sinusoidal voltage v_L developed across the inductor, such as that depicted in Fig. 12-40. The expression

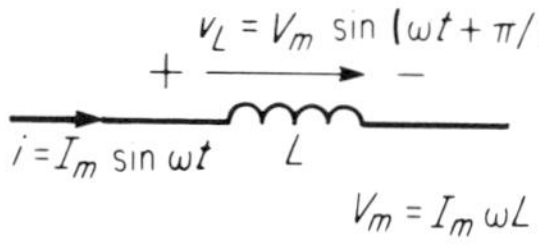

$$v_L = V_m \sin(\omega t + \pi/2)$$

$$i = I_m \sin \omega t \qquad L$$

$$V_m = I_m \omega L$$

FIG. 12-40. Steady-state sinusoidal voltage and current in an inductor.

for the power dissipated is

$$p = v_L \cdot i = \left[V_m \sin \left(\omega t + \frac{\pi}{2} \right) \right] \cdot (I_m \sin \omega t) = (I_m \omega L \cos \omega t) \cdot (I_m \sin \omega t)$$

$$= \frac{I_m^2 \omega L}{2} \sin 2\omega t = \left(\frac{I_m}{\sqrt{2}} \right)^2 (\omega L) \sin 2\omega t$$

$$= I^2 X_L \sin 2\omega t \tag{12-63}$$

Thus, Eq. 12-63, which gives the instantaneous power relation for the inductor, shows:

1. The average power is zero, since the variation is sinusoidal.
2. The dimension of $I^2 X_L$ is power.
3. The maximum power attained is $I^2 X_L$.
4. Energy is absorbed from the circuit and stored in the magnetic field when p is positive.
5. Energy is returned to the circuit from the magnetic field when p is negative.

This relationship is shown in Fig. 12-41. The term $I^2 X_L$ provides a measure of the maximum power that represents a transfer of energy from

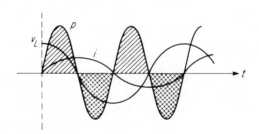

▨ Energy Absorbed from Circuit

▩ Energy Returned to Circuit

FIG. 12-41. Voltage, current, and power
relations for circuit of Fig. 12-40.

the network to the element, and vice versa. We designate this power by the symbol Q and call it the *reactive* power or *reactive volt-amperes* to distinguish it from the *real* power P that represents energy actually dissipated and lost from the circuit in resistive elements. The unit of reactive power Q is the *var*, that is, *volt-ampere reactive*. Thus,

$$Q_L = I^2 X_L = (I) \cdot \left(\frac{V_L}{X_L} \right) X_L = (I) \cdot V_L = \left(\frac{V_L}{X_L} \right) \cdot V_L = \frac{V_L}{X_L^2}$$

or

$$Q_L = I^2 X_L = I \cdot V_L = \frac{V_L^2}{X_L} \qquad (12\text{-}64)$$

so that the reactive power can be found from expressions identical to those for the real power in a resistor simply by replacing R with X_L. The subscript L is used to identify the inductive reactive power from the capacitive reactive power.

In a similar manner, if i is a sinusoidal current flowing through a capacitor with a resultant sinusoidal voltage (v_C) developed across the capacitor so that

$$i = I_m \sin \omega t$$

$$v_C = V_m \sin \left(\omega t - \frac{\pi}{2} \right) = -V_m \cos \omega t$$

and

$$p = v_C \cdot i = (-V_m \cos \omega t) \cdot (I_m \sin \omega t)$$

$$= \left(\frac{-I_m}{\omega C} \cos \omega t \right) (I_m \sin \omega t)$$

$$= \frac{-I_m^2}{2} \frac{1}{\omega C} \sin 2\omega t = - \left(\frac{I_m}{\sqrt{2}} \right)^2 \cdot \left(\frac{1}{\omega C} \right) \sin 2\omega t$$

$$= -I^2 X_C \sin 2\omega t \qquad (12\text{-}65)$$

Equation 12-65 is identical to Eq. 12-63, with the exception of the negative sign. What is the significance of the negative sign? With the *same* sinusoidal current in the inductor and the capacitor, the power as given by Eq. 12-63 will be positive during one half-cycle while the power as given by Eq. 12-65 will be negative during the same half-cycle. This means that during the time interval in which the inductor is absorbing energy from the network, the capacitor will be returning energy to the network. Thus, the energy that must be supplied from the network and, ultimately, from the source is the difference of these two energies. It must be emphasized that this condition is based on the phase displacement of the current flowing in the inductor being equal to the phase displacement of the current flowing in the capacitor. Therefore, for the capacitor, in a manner similar to the inductor, expressions accounting for this transfer of energy are

$$Q_C = I^2 X_C = I \cdot V_C = \frac{V_C^2}{X_C} \qquad (12\text{-}66)$$

The next question that might be asked is: What about the situation when both an energy-dissipating and an energy-storage element are present in a network which is being subjected to sinusoidal excitation?

Through the use of the technique for combining impedances, any network or portion of a network not containing an energy source can always be reduced to, at most, a resistor in series with a reactive element. This means that from the viewpoint of the source there are two general cases to be examined, namely, a resistor in series with an inductor and a resistor in series with a capacitor. Before considering these cases, it will be convenient to talk about still another power, namely, the *apparent* power that is indicated by the product of the effective voltage and effective current for any network or portion of a network. The symbol (V-A) is used to indicate the apparent power, and the unit (*va*) is quite naturally called the volt-ampere.

This apparent power is an important quantity, since it specifies the power capacity of an energy source by including both the real- and the reactive-power components, as will be demonstrated in the following case, where it is desired that the power relations, for the network given in Fig. 12-42, be determined. The impedance of the network of Fig. 12-42 is

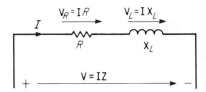

FIG. 12-42. An *R-L* network for illustrating power relations.

given by $\mathbf{Z} = R + \mathbf{X}_L = R + jX_L = Z \underline{/\theta}$, and the voltage $\mathbf{V}$ is equal to

$$\mathbf{V} = \mathbf{IZ} = \mathbf{I}R + \mathbf{I}\mathbf{X}_L = \mathbf{I}R + j\mathbf{I}X_L = \mathbf{I}[R + jX_L] = \mathbf{I}Z\underline{/\theta}$$

The impedance diagram is given in Fig. 12-43a, with θ being called the impedance angle. If the current phasor is chosen as the reference phasor, then it is a real number, since it is directed along the positive real axis. Multiplication of any other phasor by this reference current phasor will simply be a scalar multiplication, since it will have no effect on the angle. With this in mind, one can multiply each of the components of the impedance diagram by the reference current phasor, obtaining the phasor diagram shown in Fig. 12-43b. Now, if the magnitudes of the components of the phasor diagram are again multiplied, in turn, by the reference current, a so-called power diagram results, as shown in Fig. 12-43c. This power diagram readily provides a measure of the real, reactive, and apparent power. Even though the quantities are not strictly phasors, they are plotted on the complex plane, and, from the diagram, it can be seen that the complex relationship

$$\mathbf{V}\text{-}\mathbf{A} = P + jQ_L \qquad (12\text{-}67)$$

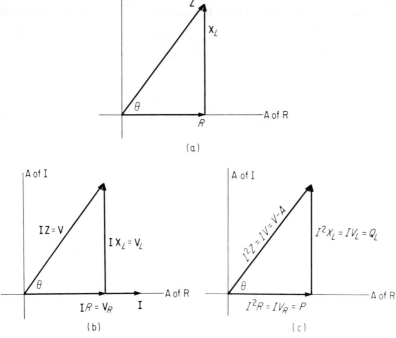

FIG. 12-43. a) Impedance diagram, b) phasor diagram, and c) power diagram for Fig. 12-42.

properly describes the relationship between the three components. In much the same way that the arguments were presented for the utility of considering R and X_L always to be directed along the positive real and positive imaginary axes, respectively, for determining the equivalent impedance Z, it will be advantageous to direct P along the positive real axis and Q_L along the positive imaginary axis, in order to determine the apparent power V-A through use of Eq. 12-67. It can also be seen that the relationships

$$P = V\text{-}A \cos \theta \tag{12-68}$$

and

$$Q_L = V\text{-}A \sin \theta \tag{12-69}$$

can be used to calculate P and Q_L, if V-A and θ are known quantities. In this respect, the angle θ is referred to as the power-factor angle when observed on the power diagram, in the same sense that one refers to θ as the phase angle or the impedance angle when observing the phasor and impedance diagrams, respectively. From Eq. 12-68,

$$\cos \theta = \frac{P}{V\text{-}A} \triangleq \text{power factor}$$

The power factor of a network or portion of a network affords a measure of the real to the apparent power for the network or portion of the network under consideration.

EXAMPLE 12-9. Determine the power relations for the network of Fig. 12-44.

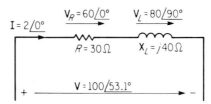

FIG. 12-44. Network for Example 12-9.

Solution: Several expressions have been developed for calculating each of the different powers. The choice of an expression will generally be determined by what information is given or calculated for the particular network in question. Consider the expressions

$$\mathbf{Z} = Z\underline{/\theta} = 30 + j40 = 50\underline{/53.1^\circ}$$

$$P = I^2R = 2^2 \cdot 30 = 120 \text{ watts}$$

$$= \frac{(V_R)^2}{R} = \frac{(60)^2}{30} = 120 \text{ watts}$$

$$= V_R I = (60)(2) = 120 \text{ watts}$$

$$= VI \cos\theta = (100)(2) \cos 53.1^\circ = 120 \text{ watts}$$

$$Q_L = I^2 X_L = 2^2 \cdot 40 = 160 \text{ vars}$$

$$= \frac{(V_L)^2}{X_L} = \frac{(80)^2}{40} = 160 \text{ vars}$$

$$= V_L I = (80)(2) = 160 \text{ vars}$$

$$= VI \sin\theta = (100)(2) \sin 53.1^\circ = 160 \text{ vars}$$

$$V\text{-}A = V \cdot I = (100)(2) = 200 \text{ va}$$

$$= I^2 Z = 2^2 \cdot 50 = 200 \text{ va}$$

and

$$V\text{-}A = \sqrt{P^2 + Q_L^2} = \sqrt{(120)^2 + (160)^2} = 200 \text{ va}$$

EXAMPLE 12-10. Attention is now directed to the case where the portion of the network under study is composed of a resistor in series with a capacitor, as shown in Fig. 12-45.

Solution: The impedance of the network is given by $\mathbf{Z} = \mathbf{R} + \mathbf{X}_C = R - jX_C = Z\underline{/\theta}$, and the voltage $\mathbf{V}$ is given by $\mathbf{V} = \mathbf{IZ} = \mathbf{IR} + \mathbf{IX}_C = IR - jIX_C = I(R - jX_C) = \mathbf{I} \cdot Z\underline{/\theta}$. With the current phasor being the refer-

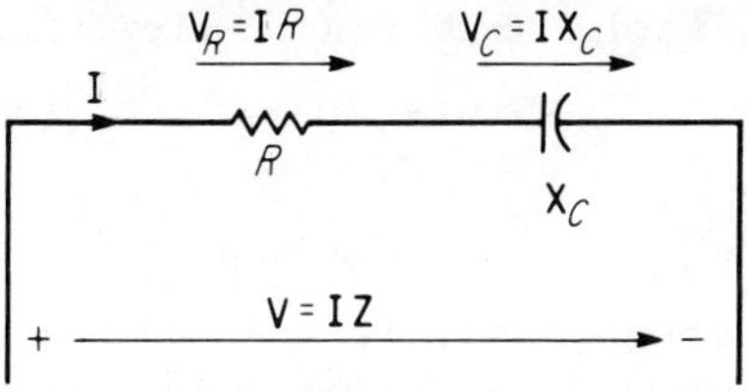

$V_R = IR$ $V_C = IX_C$

$V = IZ$

FIG. 12-45.   *R-C* network for Example 12-10.

ence phasor, the impedance diagram, phasor diagram, and power diagram are as shown in Fig. 12-46a, b, and c, respectively. Note that the complex-plane relationship of the components on the power diagram is given by

$$\mathbf{V\text{-}A} = P - jQ_C \tag{12-70}$$

Comparison with Eq. 12-67 shows that whereas the reactive-power component associated with an inductor is directed along the positive imaginary axis, the reactive-power component associated with a capacitor is directed along the negative imaginary axis. The power-factor angle, in this latter case, is negative, whereas it was positive in the case of the resistor-inductor series combination. In determining the components P and Q by Eqs. 12-68 and 12-69, respectively, the

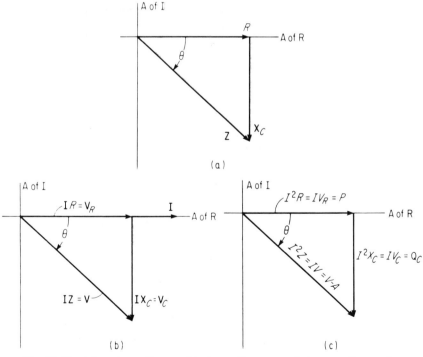

FIG. 12-46. a) Impedance diagram, b) phasor diagram, and c) power diagram for Example 12-10.

calculation for P is unaffected, since the cosine is an even function; that is,

$$\cos \alpha = \cos (-\alpha)$$

However,

$$Q_? = V\text{-}A \sin \theta \qquad (12\text{-}71)$$

will be positive when θ is positive and negative when θ is negative. Advantage can be taken of this fact by using the sign of the right-hand side of Eq. 12-71 to determine the subscript to be affixed to Q. A plus sign indicates that the reactive power is inductive, whereas a minus sign indicates that the reactive power is capacitive. Furthermore, the sign ($+$ or $-$) indicates that the component of reactive power is directed along the (positive or negative) imaginary axis. It goes without saying that the angle θ in Eq. 12-71 is between the voltage and the current, with the current being considered the reference phasor.

A final example will be presented to demonstrate how these concepts apply to any arbitrary network.

EXAMPLE 12-11. Determine the power relations for the network of Fig. 12-47, where the currents have already been determined.

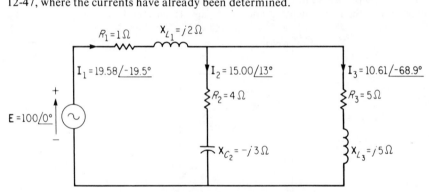

FIG. 12-47. Figure to accompany Example 12-11.

Solution: Since the currents are given, one can make use of expressions for real and reactive power involving the magnitude of the currents. Thus,

$$P_{R_1} = I_1^2 R_1 = (19.58)^2 \cdot 1 = 383 \text{ watts}$$

$$P_{R_2} = I_2^2 R_2 = (15.00)^2 \cdot 4 = 900 \text{ watts}$$

$$P_{R_3} = I_3^2 R_3 = (10.61)^2 \cdot 5 = 565 \text{ watts}$$

$$Q_{L_1} = I_1^2 X_{L_1} = (19.58)^2 \cdot 2 = 766 \text{ vars}$$

$$Q_{C_2} = I_2^2 X_{C_2} = (15.00)^2 \cdot 3 = 675 \text{ vars}$$

$$Q_{L_3} = I_3^2 X_{L_3} = (10.61)^2 \cdot 5 = 565 \text{ vars}$$

$$P_{\text{total}} = P_{R_1} + P_{R_2} + P_{R_3} = 1848 \text{ watts}$$

and

$$Q_{\text{total}} = Q_{L_1} - Q_{C_2} + Q_{L_3} = 656 \text{ vars}$$

The total volt-amperes is given by $V\text{-}A_{total} = \sqrt{(P_{total})^2 + (Q_{total})^2} = 1960$ va. This can be compared to the volt-amperes being delivered by the source; that is, $E \cdot I_1 = 100(19.58) = 1958$ va, which is equal to that being demanded by the circuit elements and is well within slide-rule accuracy. It is well to note that even though the individual real and reactive powers can be algebraically added to obtain the total real and reactive power demands, this is not the case for the total volt-amperes. Why? (HINT: Observe Eq. 12-67 and show that a direct sum can be made if the rules for adding complex numbers are followed.)

With the techniques that have been presented, one ought now to be able to analyze the steady-state response of circuits excited by sinusoidal voltage and/or current sources. The application of these methods to electronic circuits will now be considered in the following section as well as in the succeeding chapters.

12-11. Complex Load Impedances in Electronic Circuits

In all the preceding graphical and equivalent-circuit-analysis applications the collector load impedance was assumed to be purely resistive in nature. This is not always a valid assumption over certain frequency ranges. Figure 12-48a shows the a-c equivalent for a common-emitter amplifier with an $R\text{-}L$ (resistive-inductive) load, and Fig. 12-48b shows the a-c equivalent circuit for the typical $R\text{-}C$-coupled amplifier which, of course, has an $R\text{-}C$ load impedance. Graphical-analysis techniques for

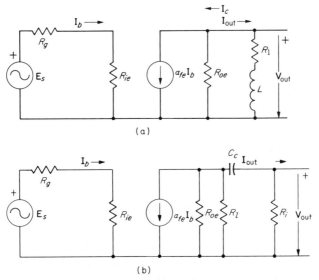

FIG. 12-48. A-c equivalent circuits, for a common-emitter amplifier with an a) $R\text{-}L$ load impedance and b) an $R\text{-}C$ load impedance.

circuits with complex load impedances can be utilized, but, since the resultant load line is elliptical in nature, the graphical method is rather tedious and thus not too popular.

Complex load impedances present no particular problem when equivalent-circuit methods are used. In fact, it is necessary only to replace the collector-circuit resistance, through which I_{out} flows with the complex load impedance Z in many (but not all) of the previously derived equations in Chapters 7, 8, or 10. For the series R-L load impedance in Fig. 12-48a the substitution of $Z_1 = R_1 + j\omega L$ for R_i can be made in many of the equations derived for the amplifier circuits of Chapters 7, 8, or 10, but, for the parallel impedance combination shown in Fig. 12-48b, caution must be exercised when calculating the voltage or power gain. The reason for this cautionary statement for the parallel impedance is because the output voltage V_{out} is not taken as the total a-c collector-emitter voltage V_{ce}.

For the circuit shown in Fig. 12-48a, the current gain A_i is obtained from

$$A_i = \frac{I_{out}}{I_b} = -\frac{I_c}{I_b} = -\frac{\alpha_{fe} R_{oe}}{R_{oe} + Z_1}$$

where $Z_1 = R_1 + j\omega L$.

$$A_i = \frac{-\alpha_{fe} R_{oe}}{R_{oe} + R_1 + j\omega L} \tag{12-72a}$$

$$A_i = \alpha_{fe} \frac{R_{oe}}{[(R_{oe} + R_1)^2 + (\omega L)^2]^{1/2}} \left/ 180° - \tan^{-1} \frac{\omega L}{R_{oe} + R_1} \right. \tag{12-72b}$$

$$A_i = M \left/ \theta \right. \tag{12-72c}$$

where

$$M = \alpha_{fe} \frac{R_{oe}}{[(R_{oe} + R_1)^2 + (\omega L)^2]^{1/2}} \tag{12-72d}$$

$$\theta = 180° - \tan^{-1} \frac{\omega L}{R_{oe} + R_1} \tag{12-72e}$$

and where the additional phase angle $[-\tan^{-1} \omega L/(R_{oe} + R_1)]$ means that I_b and I_{out} are no longer simply 180° out of phase when the frequency ω is high enough to cause the inductive reactance ωL to satisfy the inequality $\omega L > 0.1(R_{oe} + R_1)$. Figure 12-49a is a plot of the magnitude M of A_i versus ω and Fig. 12-49b is a plot of the phase angle θ of A_i versus ω. The shape of the M versus ω curve can be qualitatively explained by noting that as $\omega \to 0$ the inductive reactance $\omega L \to 0$, and thus the collector load impedance is purely resistive again. For large values of frequency ω the inductive reactance $\omega L \to \infty$, causing $I_{out} \to 0$ and thus $M \to 0$. At the same time, the collector load impedance appears purely inductive reactive, and so a phase shift of +90° is introduced in the denominator of Eq.

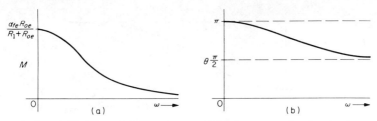

FIG. 12-49. Plots of a) M versus ω and b) θ versus ω for the circuit shown in Fig. 12-48a.

12-72a which becomes $-90°$ when transferred to the numerator, thus making $\theta \rightarrow 90°$ (see Eq. 12-72e).

The voltage and power gains for Fig. 12-48a can be obtained similarly. The current gain for the circuit of Fig. 12-48b can be calculated from the current splitting theorem and with R_b the parallel sum of R_l and R_{oe} as

$$\mathbf{A}_i = -\alpha_{fe} \frac{R_b}{R_b + R_i - j\dfrac{1}{\omega C_c}} \tag{12-73a}$$

$$\mathbf{A}_i = \alpha_{fe} \frac{R_b}{\left[(R_b + R_i)^2 + \left(\dfrac{1}{\omega C_c}\right)^2\right]^{1/2}} \Big/\!\underline{180° - \tan^{-1}\dfrac{-1}{(R_b + R_i)\omega C_c}} \tag{12-73b}$$

$$\mathbf{A}_i = M \underline{/\theta} \tag{12-73c}$$

where

$$M = \alpha_{fe} \frac{R_b}{\left[(R_b + R_i)^2 + \left(\dfrac{1}{\omega C_c}\right)^2\right]^{1/2}} \tag{12-73d}$$

$$\theta = 180° - \tan^{-1}\frac{-1}{(R_b + R_i)\omega C_c} \tag{12-73e}$$

Figure 12-50a is a plot of M versus ω and Fig. 12-50b is a plot of θ versus ω for the R-C load impedance of Fig. 12-48b.

The shape of the magnitude M versus ω curve can be explained qualitatively by noting that as $\omega \rightarrow 0$, $1/\omega C_c \rightarrow \infty$, and thus $\mathbf{I}_{out} \rightarrow 0$. As $\omega \rightarrow \infty$, $1/\omega C_c \rightarrow 0$, and the collector load impedance becomes purely resistive again. The phase angle $\theta \rightarrow 270°$ as $\omega \rightarrow 0$, since the capacitive reactance introduces a $-90°$ phase shift in the denominator of Eq. 12-73a. The circuit of Fig. 12-48b and the curves of Fig. 12-50a and b will be discussed further in Chapter 13.

EXAMPLE 12-12. A 2N334 NPN transistor is used in the circuit of Fig. 12-48b. If $C_c = 0.01$ μf, $R_l = 10,000$ ohms, $R_i = 5000$ ohms, $R_g = 1000$ ohms,

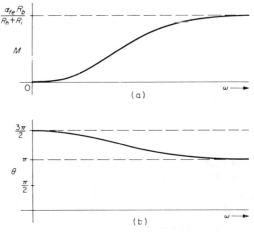

FIG. 12-50. Plots of a) M versus ω and b) θ versus ω for the circuit shown in Fig. 12-48b.

$R_{ie} = 200$ ohms, $\alpha_{fe} = 46$, and $R_{oe} = 115,000$ ohms, find the magnitude of the current gain M and the necessary frequency ω to obtain a desired phase-shift angle θ of 225°.

Solution: From Eq. 12-73e, the phase angle θ of 225° is obtained when

$$180° - \tan^{-1} \frac{-1}{(R_b + R_i)\omega C_c} = 225°$$

or

$$\tan^{-1} \frac{-1}{(R_b + R_i)\omega C_c} = -45°$$

thus

$$\frac{1}{(R_b + R_i)\omega C_c} = 1$$

Substituting in values and noting that $R_b \simeq R_l$ obtains

$$\omega = \frac{1}{(10,000 + 5000)0.01 \times 10^{-6}}$$

$$= 6667 \text{ radians/sec}$$

Therefore

$$M = \frac{\alpha_{fe} R_b}{\left[(R_b + R_i)^2 + \left(\frac{1}{\omega C_c}\right)^2\right]^{1/2}} = \frac{46 \times 10,000}{[(15 \times 10^3)^2 + (15 \times 10^3)^2]^{1/2}}$$

$$= 21.6$$

or

$$\mathbf{A}_i = 21.6 \underline{/225°}$$

The frequency that causes a phase shift of 180° ± 45° between the output and input currents is used as an important figure of merit for the transistor amplifier. In Chapter 13 this particular frequency will be called the −3-db cutoff frequency and will be used in defining the bandwidth of the amplifier.

Although the transistor has some interesting internal energy-storage effects that limit its usefulness at high frequencies (high frequency for one transistor may be 50 KHz and for another 100 MHz), these effects are not emphasized in this text. Refer to Chapter 13 and other texts on solid state electronics for further information about transistor equivalent cir- behavior characteristics of vacuum tubes will be presented next.

Vacuum Triode A-C Equivalent Circuit. Thus far, only the low-frequency a-c equivalent circuit for the grounded-cathode amplifier has been presented, and the input impedance was assumed to be infinite. This is not realistic at higher frequencies, for the tube interelectrode capaci- tances have a pronounced effect on the tube operation at the higher fre- quencies. In Chapter 9 it was pointed out that the triode has inter- electrode capacitances of a few picofarads, and for frequencies above 500 KHz, these capacitive reactances are comparable in magnitude to the other circuit impedances. Figure 12-51a shows the interelectrode ca-

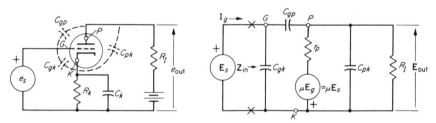

Fig. 12-51. a) A self-biased grounded-cathode amplifier showing the interelectrode capacitances. b) The a-c equivalent circuit for part a.

pacitances of the tube sketched in. The exact a-c equivalent circuit of the grounded-cathode amplifier which includes these interelectrode capac- itances is shown in Fig. 12-51b. The input impedance is the impedance "looking in" at grid terminals (XX). It should be noted that when a voltage E_s is applied to the input terminals, a voltage μE_s appears in the internal portion of the circuit. This is a dependent voltage source and so cannot be removed from the circuit before calculating the input impedance.

Input Impedance. Figure 12-52 illustrates *one way* to solve for the input impedance by a successive application of Thévenin's and Norton's theorems. Loop or nodal analysis would yield equally valid results, but it

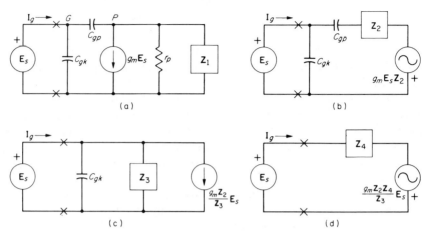

FIG. 12-52. Reduction of Fig. 12-51b to find $\mathbf{Z}_{in}$ by using Thévenin's and Norton's theorems.

is well to practice the usage of the other network-analysis methods. Figure 12-52a is identical to Fig. 12-51b except that a Norton generator has replaced the Thévenin generator, and

$$\mathbf{Z}_1 = \frac{R_l \left(-j \dfrac{1}{\omega C_{pk}}\right)}{R_l - j \dfrac{1}{\omega C_{pk}}}$$

Figure 12-52b involves a Thévenin generator and a new impedance. Figures 12-52c and d should be self-explanatory. The impedances indicated in these figures are defined as

$$\mathbf{Z}_2 = \frac{r_p \mathbf{Z}_1}{r_p + \mathbf{Z}_1} \tag{12-74}$$

$$\mathbf{Z}_3 = \mathbf{Z}_2 - j \frac{1}{\omega C_{gp}} \tag{12-75}$$

$$\mathbf{Z}_4 = \frac{\mathbf{Z}_3 \left(-j \dfrac{1}{\omega C_{gk}}\right)}{\mathbf{Z}_3 - j \dfrac{1}{\omega C_{gk}}} \tag{12-76}$$

From Fig. 12-52d solve for $\mathbf{I}_g$.

$$\mathbf{I}_g = \frac{\mathbf{E}_s + \dfrac{g_m \mathbf{Z}_2 \mathbf{Z}_4 \mathbf{E}_s}{\mathbf{Z}_3}}{\mathbf{Z}_4} = \frac{(\mathbf{Z}_3 + g_m \mathbf{Z}_2 \mathbf{Z}_4) \mathbf{E}_s}{\mathbf{Z}_3 \mathbf{Z}_4} \tag{12-77}$$

and

$$\mathbf{Z}_{in} = \frac{\mathbf{E}_s}{\mathbf{I}_g} = \frac{\mathbf{Z}_3 \mathbf{Z}_4}{\mathbf{Z}_3 + g_m \mathbf{Z}_2 \mathbf{Z}_4} \tag{12-78}$$

Since the expansion of Eq. 12-78 involves a considerable number of terms, no further expression for $\mathbf{Z}_{in}$ will be developed. Suffice it to say that $\mathbf{Z}_{in}$ is resistive and capacitive in nature if the plate load impedance is purely resistive.

Let us now consider the limiting cases of both high and low frequencies. For high frequencies, all the capacitive reactances approach zero and $\mathbf{Z}_{in} \rightarrow 0$. Thus a vacuum tube very definitely has frequency limitations that are usually dictated by the magnitude of the interelectrode capacitances. At frequencies where all the capacitive reactances are much greater than the resistances, it can be shown that Eq. 12-78 becomes

$$\mathbf{Z}_{in} = -j \frac{r_p + R_l}{[(r_p + R_l)(C_{gk} + C_{gp}) + \mu R_l C_{gp}]\omega}$$

or

$$\mathbf{Z}_{in} = -j \frac{1}{\omega C_{in}} \tag{12-79}$$

where

$$C_{in} = \frac{(r_p + R_l)(C_{gk} + C_{gp}) + \mu R_l C_{gp}}{r_p + R_l}$$

or

$$C_{in} = C_{gk} + \left(1 + \frac{\mu R_l}{r_p + R_l}\right) C_{gp}$$

$$= C_{gk} + (1 - \mathbf{A}_{GC}) C_{gp} \tag{12-80}$$

where $\mathbf{A}_{GC}$ is the low-frequency voltage gain $\mathbf{E}_{out}/\mathbf{E}_s$ of the grounded-cathode amplifier shown in Fig. 12-51a. Thus it is evident that the input impedance is wholly capacitive in nature at low frequencies. For frequencies in the audio range (up to 20,000 Hz), $\mathbf{Z}_{in}$ is well over 1 megohm, so it is sometimes said that the circuit input impedance "looking in" at the grid of a vacuum tube is infinity.

Output Impedance. By definition, the output impedance is obtained by applying a source voltage $\mathbf{V}_1$ to the output terminals and making the grid signal voltage $\mathbf{E}_s$ zero. Figure 12-53a shows the resulting circuit (from Fig. 12-51a) for frequencies low enough so that the interelectrode capacitances can be neglected. The output impedance is simply the parallel sum of r_p and R_l, or

$$\mathbf{Z}_{out} = \frac{r_p R_l}{r_p + R_l} \tag{12-81}$$

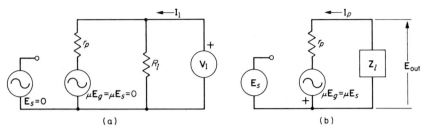

FIG. 12-53. a) The low-frequency a-c equivalent circuit used to obtain the output impedance of the grounded-cathode amplifier. b) Basic grounded-cathode a-c equivalent circuit.

If a more general expression for the output impedance is desired, then Fig. 12-51b should be used instead of Fig. 12-53.

Problems

Excitation Functions

12-1. a) Sketch 2 cycles of a sinusoidal-voltage waveform having an amplitude of 10 volts and a frequency of 100 hertz. On the abscissa, indicate both the values of time and of the electrical radians. b) What is the period of the voltage waveform? c) What is the radian frequency?

12-2. The radian frequency of a sinusoidal-current waveform is 5500 radians/min. a) Calculate the first three instances of time at which the current has attained a positive maximum. b) Can you deduce the period from these values? How? Compare this value of the period with that calculated directly from the radian frequency.

Complex Algebra

Each of Probs. 12-3 through 12-8 has eight parts. Thus working any one part of all these problems will give a representative sample of the different operations.

12-3. Locate graphically the following complex numbers:

a) $4 + j3$ b) $5 - j5$ c) $j7$ d) $-3 + j6$
e) $-8 - j11$ f) $4 + j0$ g) $3 - j2$ h) $-6 + j0$

12-4. Express each number, in Prob. 12-3, in exponential form.

12-5. Convert each rectangular complex number in Prob. 12-3, to polar form.

12-6. Multiply each complex number, in Prob. 12-3, by its complex conjugate and give the result. Comment on the form of the result.

12-7. Add the following complex numbers, and express the result in rectangular form.

a) $(6 + j2) + (-9 + j7)$ b) $(-1 + j4) + (1 - j14)$
c) $8\underline{/45°} + 4e^{-j60°}$ d) $(2 + j8) + (-5e^{j127°})$
e) $(9\underline{/-120°}) + 17\underline{/90°}$ f) $22e^{j80°} + (-10 - j6) + 14\underline{/160°}$
g) $e^{j90°} + (0 + j)$ h) $(3 + j6) + (3 + j6)^*$

12-8. Perform the following indicated multiplication or division, and express the answer in both polar and exponential forms.

a) $(5 \underline{/20°}) \times (6 \underline{/-80°})$

b) $3e^{j18°} \times 31 \underline{/-143°}$

c) $(6 + j2) \times (-9 - j7)$

d) $(2 + j8) \times (-5e^{j127°})$

e) $(9 \underline{/-128°})/(17 \underline{/90°})$

f) $(-10 - j6)/(14 \underline{/160°})$

g) $(-1 + j4)/(1 - j14)$

h) $(8 \underline{/56°})/(4 \underline{/-226°})$

12-9. a) If $a + jb = 8 - j6$, find a and b. b) If $a + j10 = 3 - jb$, find a and b.

Complex-Number Representation of Sinusoidal Functions

12-10. If $V_0 = 2 + j2$, show graphically the original phasor V_0 and the following phasor expressions:

a) jV_0

b) $-jV_0$

c) j^2V_0

d) $-j^2V_0$

e) j^3V_0

f) $-j^3V_0$

g) j^4V_0

h) $-j^4V_0$

12-11. Find V_m and θ for the wave $v = V_m \sin(\omega t + \theta)$, represented by the following complex expressions at $\omega t = 0$:

a) $V = 100 + j0$

b) $V = 60 + j80$

c) $V = 0 + j100$

d) $V = -100 + j0$

12-12. Express the following voltages and currents in phasor form:

a) $e = 141.4 \sin \omega t$

b) $i = 100 \sin(\omega t - 30°)$

c) $v = 50 \cos(\omega t + 20°)$

d) $i = 7.07 \cos(\omega t - 45°)$

Resistive Element

12-13. An a-c voltage of the form $v = 28.28 \sin(377t + 30°)$ volts is impressed across a resistor of 20 ohms' resistance. a) What is the resultant phasor current? b) Draw the phasor diagram, showing the voltage and current.

12-14. An a-c current of the form $i = 14.14 \sin(100t - 60°)$ amp is passing through a resistor of 4 ohms' resistance. What is the phasor voltage drop a) in the direction of the assumed positive current flow and b) in the opposite direction? c) Draw the phasor diagram, showing the voltage and current.

Inductive Element

12-15. An a-c voltage of the form $v = 170 \sin(377t)$ is impressed across an inductor having an inductance of 0.6 henry. a) What is the impressed frequency, in hertz? b) What is the magnitude of the inductive reactance? c) What is the current, expressed in phasor form? d) Draw a phasor diagram, showing the voltage and current.

12-16. An a-c current of the form $i = 20 \sin(1000t - 45°)$ is passing through an inductor of 250 mh. a) What is the inductive reactance? b) What is the phasor voltage drop in the direction of the assumed positive current? c) Draw a phasor diagram, showing the voltage and current.

Capacitive Element

12-17. An a-c voltage of the form $V = 100 \underline{/0°}$, with a frequency of 60 hertz, is impressed across a capacitor having a capacitance of 10 μf. a) What is the magnitude of the capacitive reactance? b) What is the current, expressed in phasor form? c) Draw a phasor diagram, showing the voltage and current.

12-18. An a-c current having an effective value of 5 amp and a frequency of 1000 hertz is passing through a variable capacitor. a) What should be the value of capacitance to limit the maximum voltage across the capacitor to 100 volts? b) For this value of capacitance, draw a phasor diagram, showing the voltage and current.

Combined Elements

12-19. An inductor of 0.4 henry and a resistor of 200 ohms are connected in parallel to a voltage source of the form $v = 250 \sin (200 \pi t)$. a) What is the total impedance presented to the source? What is the phasor current b) in the inductor and c) in the resistor? d) Draw a phasor diagram, showing all voltages and currents.

12-20. A current of the form $i = 12 \sin (500t)$ passes through a series combination of a resistor of 2 ohms and a capacitor of 400 μf. What is the value of the voltage developed across a) the capacitor and b) the resistor? c) What is the total impedance of the series combination? d) Draw a phasor diagram, showing all voltages and currents.

Network Solutions

12-21. The circuit in the accompanying figure has the following values: $R = 10$ ohms, $L = 10/6280$ henry, $C = 1/(6280 \times 10)$ farad, $\mathbf{E}_{g1} = 100\underline{/0°}$, $\mathbf{E}_{g2} = 100\underline{/180°}$, and the frequency $= 1000$ hertz. Find $\mathbf{I}_R$, $\mathbf{I}_L$, and $\mathbf{I}_C$, using loop currents $\mathbf{I}_1$ and $\mathbf{I}_2$.

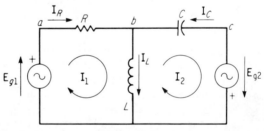

PROB. 12-21

12-22. Consider the network in the accompanying figure. Determine the phasor expressions for the node voltages $\mathbf{V}_a$ and $\mathbf{V}_b$.

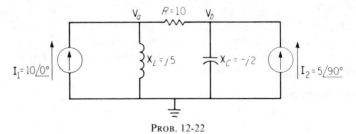

PROB. 12-22

12-23. The circuit in the accompanying figure is excited by a voltage source having a frequency of 1000 hertz. Find the indicated current and voltage.

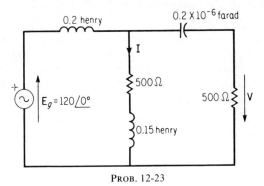

PROB. 12-23

12-24. Find the branch currents I_1 and I_2 for the circuit in the accompanying figure. Show that if the impedance of branch 1 is Z_1 and of branch 2 is Z_2 then

$$I_1 = \frac{Z_2}{Z_1 + Z_2} I \quad \text{and} \quad I_2 = \frac{Z_1}{Z_1 + Z_2} I$$

(*Hint:* What can be said about the voltage across each branch since they are in parallel?)

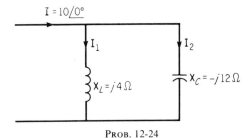

PROB. 12-24

Power Relationships

12-25. The black box Z, in the accompanying figure, contains a series circuit of 2 elements. The polar expressions for the voltage and current are $V_0 = 100\underline{/-30°}$ and $I = 10\underline{/30°}$, respectively. a) Draw the 2 phasors in a complex

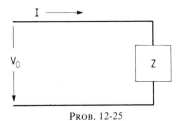

PROB. 12-25

plane to represent the phasor diagram. b) Is **Z** an *R-L* or an *R-C* circuit? c) What is the power-factor angle of the circuit? d) Is the power factor leading or lagging? e) What are the rectangular expressions for **V₀** and **I**? f) Draw the power diagram.

12-26. An *R-L* circuit, consisting of R = 3 ohms and X_L = 4 ohms, is connected across the terminals of a 1000-hertz generator. The voltage drop across the resistance is $V_R = 60\underline{/36.8°}$, and the voltage drop across the inductance is $V_L = 80\underline{/126.8°}$. a) Determine the polar expression for the total voltage drop. b) Draw phasors that represent the 3 voltage drops. c) Draw the power diagram.

12-27. The accompanying figure shows an a-c bridge circuit. a) Find the phasor expressions for the loop currents. b) Determine the power losses in R_1, R_2, and R_5.

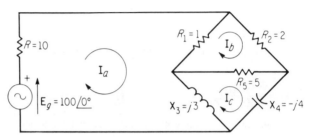

PROB. 12-27

12-28. The circuit in the accompanying figure has the following values: R = 10 ohms, $E_{g1} = 100\underline{/0°}$, and $E_{g2} = E_{g3} = 50\underline{/90°}$. a) Find the power supplied by the generator E_{g1}. b) Draw the power diagram.

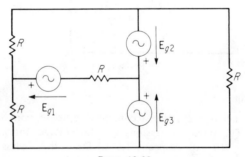

PROB. 12-28

Electronic Circuits

12-29. A 2N3704 NPN transistor is used in the circuit of the accompanying figure. If R_{ie} = 200 ohms, α_{fe} = 140, R_{oe} = 80 K, and $e_s = 0.2 \sin (10^5 \, t)$, find a) the current gain, b) the voltage gain, and c) the output voltage V_{out}.

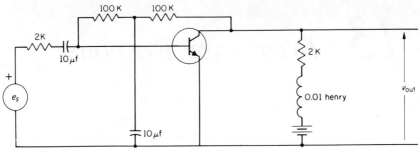

PROB. 12-29

12-30. Repeat Prob. 12-29 if $e_s = 0.2 \sin 4 \times 10^5 t$.

12-31. A 2N3704 NPN transistor is used in the circuit of the accompanying figure. If $e_s = 0.35 \cos 10^5 t$, $R_{ie} = 200$ ohms, $\alpha_{fe} = 140$, and $R_{oe} = 80$ K, find a) the current gain, b) the voltage gain, c) the output voltage $\mathbf{V}_{out}$, and d) the power gain.

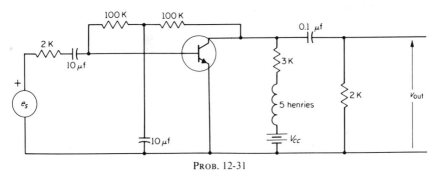

PROB. 12-31

12-32. Repeat Prob. 12-31 if $e_s = 0.35 \cos 10^4 t$.

12-33. Repeat Prob. 12-31 if $e_s = 0.35 \cos 10^3 t$.

12-34. Find the -3-db cutoff frequency in Prob. 12-29.

12-35. Find the -3-db cutoff frequency in Prob. 12-31.

12-36. Find the input and output impedance of the circuit used in a) Prob. 12-29 and b) Prob. 12-30.

12-37. Find the input and output impedance of the circuit used in a) Prob. 12-31, b) Prob. 12-32, and c) Prob. 12-33.

13 / *Frequency Response*

The frequency-response characteristics of an electrical (or mechanical) system constitute a very important portion of the study of overall system performance. In electrical circuits the expression *frequency response* usually means a comparison of an output to an input signal over a selected range of frequencies.

Before an investigation of a particular transistor or vacuum-tube circuit is undertaken, the frequency-response characteristics of a simple series R-C network will be given. Then the transistor and vacuum-tube amplifier will be compared to the R-C circuit.

13-1. Basic Low-Pass R-C Filter

Transfer Function. The ratio $\mathbf{E}_{out}/\mathbf{E}_{in}$ in the two-port network of Fig. 13-1 is called a *transfer function.* A transfer function is, in general, a

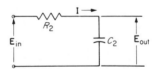

FIG. 13-1. Low-pass filter.

ratio of *any* output to any input quantity in a given system and can have any dimension. The concept of the transfer function will be further developed in Chapter 21. The transfer function $\mathbf{E}_{out}/\mathbf{E}_{in}$ for sinusoidal excitations follows from a circuit analysis of Fig. 13-1. First, we solve for the phasor current $\mathbf{I}$, or

$$\mathbf{I} = \frac{\mathbf{E}_{in}}{R_2 - j\,\dfrac{1}{\omega C_2}} \tag{13-1}$$

Next, we solve for the output voltage which is the product of X_C and $\mathbf{I}$, or

$$\mathbf{E}_{out} = \left(-j\,\frac{1}{\omega C_2}\right)\mathbf{I}$$

$$= \left(-j\,\frac{1}{\omega C_2}\right)\frac{\mathbf{E}_{in}}{R_2 - j\,\dfrac{1}{\omega C_2}}$$

$$\frac{E_{out}}{E_{in}} = \frac{-j\,\dfrac{1}{\omega C_2}}{R_2 - j\,\dfrac{1}{\omega C_2}} \qquad (13\text{-}2)$$

and multiplying numerator and denominator of Eq. 13-2 by $j\omega C_2$,

$$\frac{E_{out}}{E_{in}} = \frac{1}{1 + j\omega R_2 C_2} \qquad (13\text{-}3a)$$

$$\frac{E_{out}}{E_{in}} = \frac{1}{[1 + (\omega R_2 C_2)^2]^{1/2}} \underline{/-\tan^{-1} \omega R_2 C_2} \qquad (13\text{-}3b)$$

or, in another form,

$$\frac{E_{out}}{E_{in}} = M_L \underline{/\phi_L} \qquad (13\text{-}4)$$

where

$$M_L = \left|\frac{E_{out}}{E_{in}}\right| = \frac{1}{[1 + (\omega R_2 C_2)^2]^{1/2}} \qquad (13\text{-}5)$$

$$\phi_L = -\tan^{-1} \omega R_2 C_2 \qquad (13\text{-}6)$$

where the subscript L means low pass.

Gain and Phase Plots. A plot of M_L versus radian frequency ω is shown in Fig. 13-2a, and a plot of the phase angle ϕ_L versus radian frequency ω, is shown in Fig. 13-2b.

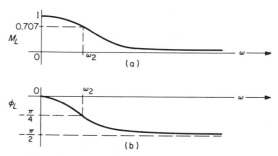

FIG. 13-2. a) M versus ω from Eq. 13-5. b) ϕ_L versus ω from Eq. 13-6.

It is generally accepted practice to choose a specific frequency in order to have some figure of merit with which to judge the circuit performance. Quite arbitrarily, the frequency is defined such that $\omega_2 R_2 C_2 = 1$, where ω_2 is denoted by a variety of names: the half-power-point frequency, the -3-db-point frequency, the breakpoint frequency, the corner frequency, or the frequency at which the voltage gain is reduced by a factor of

$1/\sqrt{2}$ or 0.707. That all of these terms are indeed equivalent will now be shown.

Decibel Gain. The plot of M versus frequency can be presented in a much more satisfactory manner if M is converted to decibels and plotted versus $\log_{10}\omega$. By definition a decibel is

$$\text{db} \triangleq 10 \log_{10}\frac{P_1}{P_2} \qquad (13\text{-}7)$$

where P_1/P_2 is the ratio of any two powers in a circuit. If the two power measurements are considered to involve the same magnitude of resistance, then,

$$P_1 = \frac{E_1^2}{R} \qquad P_2 = \frac{E_2^2}{R}$$

and

$$\text{db} = 10 \log_{10}\frac{E_1^2/R}{E_2^2/R} = 20 \log_{10}\frac{E_1}{E_2} \qquad (13\text{-}8)$$

The voltages E_1 and E_2 are magnitude quantities only and can be compared to the voltages E_{in} and E_{out} in Fig. 13-1 if we consider that a voltmeter, whose internal resistance is R_v, is permanently connected across the output terminals. We let E_1 be the output voltage E_{out} at any frequency and let E_2 be the output voltage at E_{out} only at very low frequencies (low frequencies are defined such that $X_C \ggg R_2$). At very low frequencies the capacitive reactance X_{C_2} is so much greater than R_2 that E_{out} is almost equal to E_{in}. Under these conditions the ratio E_1/E_2 in Eq. 13-8 is identically equal to E_{out}/E_{in}, as defined in Eq. 13-3, or

$$\left.\frac{E_{out}}{E_{in}}\right|_{db} = M_{L_{db}} = 20 \log_{10}\left(\frac{1}{1 + (\omega R_2 C_2)^2}\right)^{1/2}$$

or

$$M_L = -10 \log [1 + (\omega R_2 C_2)^2] \text{ db} \qquad (13\text{-}9)$$

where $\log \equiv \log_{10}$. A plot of $M_{L_{db}}$ versus $\log \omega$ is shown in Fig. 13-3.

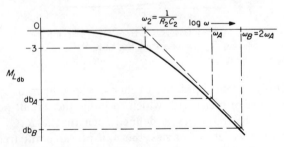

FIG. 13-3. Plot of $M_{L_{db}}$ versus $\log \omega$ as given in Eq. 13-9.

The asymptotes of Eq. 13-9 are of particular interest, since they aid in sketching the overall curve. The first asymptote is the abscissa, for, as $\omega \to 0$, $M_{L_{db}} \to 0$. The second asymptote is found by allowing ω to approach high frequencies such that $\omega RC \gg 1$. Equation 13-9 then reduces to

$$M_{L_{db}} = -10 \log (\omega R_2 C_2)^2 = -20 \log \omega R_2 C_2 \tag{13-10}$$

By inspection, it is evident that Eq. 13-10 defines a straight line which passes through $\omega = 1/R_2 C_2 \triangleq \omega_2$. It can be shown that the straight line has a slope of -6 db/octave or -20 db/decade. To prove this, we choose any high frequency ω_A, and then double it, $\omega_B = 2\omega_A$; that is, a frequency span of 1 octave. As Fig. 13-3 shows,

$$db_A = -20 \log \omega_A R_2 C_2 \tag{13-11}$$

$$db_B = -20 \log \omega_B R_2 C_2 \tag{13-12}$$

and, upon subtracting Eq. 13-11 from Eq. 13-12,

$$db_B - db_A = -20 \log \omega_B R_2 C_2 + 20 \log \omega_A R_2 C_2$$

$$= 20 \log \frac{\omega_A R_2 C_2}{\omega_B R_2 C_2} = 20 \log \frac{\omega_A R_2 C_2}{2\omega_A R_2 C_2}$$

$$= 20 \log \frac{1}{2} = -20 \log 2$$

$$\simeq -6 \, db \tag{13-13}$$

Thus the slope of the straight line defined by Eq. 13-10 is indeed -6 db/octave. If we consider that $\omega_B = 10\omega_A$ (a frequency span of 1 decade), it can easily be shown that the slope of the line defined by Eq. 13-10 is -20 db/decade.

Note that at the point $\omega = 1/R_2 C_2 = \omega_2$, the actual decibel drop is not 0, as indicated by Eq. 13-10, but is $-10 \log (1 + 1) = -10 \log 2 \simeq -3$ db, as given by Eq. 13-9, which is the exact expression for the $M_{L_{db}}$ versus ω relation. The actual frequency-response curve can be sketched in (and quite accurately) by knowing this corner or -3-db frequency ω_2. At the point defined by the corner frequency, the asymptotes of Eq. 13-9 intersect, and the value of -3 db at this frequency gives one point on the actual response curve.

13-2. Basic High-Pass R-C Filter

Transfer Function. A similar log-modulus versus log-frequency curve can be sketched for the transfer function E_{out}/E_{in}, for the circuit shown in Fig. 13-4. The transfer function for this circuit is

$$\frac{E_{out}}{E_{in}} = \frac{R_1}{R_1 - j\dfrac{1}{\omega C_1}}$$

$$= \frac{1}{1 - j \dfrac{1}{\omega R_1 C_1}} \tag{13-14a}$$

$$\frac{\mathbf{E}_{out}}{\mathbf{E}_{in}} = \frac{1}{\left[1 + \left(\dfrac{1}{\omega R_1 C_1}\right)^2\right]^{1/2}} \left/ \tan^{-1} \dfrac{1}{\omega R_1 C_1}\right. \tag{13-14b}$$

or

$$\frac{\mathbf{E}_{out}}{\mathbf{E}_{in}} = M_H \underline{/\phi_H} \tag{13-15}$$

where

$$M_H = \left[\frac{1}{1 + \left(\dfrac{1}{\omega R_1 C_1}\right)^2}\right]^{1/2} \tag{13-16}$$

$$\phi_H = \tan^{-1} \frac{1}{\omega R_1 C_1} \tag{13-17}$$

The subscript *H* means high pass.

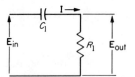

FIG. 13-4. High-pass filter.

Gain and Phase Plots. Figure 13-5a and b show the M_H versus ω, and the ϕ_H versus ω plots, respectively, for this transfer function.

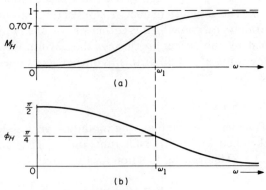

FIG. 13-5. Plots of M_H and ϕ_H versus ω from Eqs. 13-16 and 13-17.

Decibel Gain. By arguments very similar to those used in obtaining the log-modulus versus log-frequency plot for the low-pass filter, we have for the high-pass filter

$$M_{H_{db}} = 20 \log_{10} \left[\frac{1}{1 + \left(\frac{1}{\omega R_1 C_1}\right)^2} \right]^{1/2}$$

$$M_{H_{db}} = -10 \log_{10} \left[1 + \left(\frac{1}{\omega R_1 C_1}\right)^2 \right] \qquad (13\text{-}18)$$

and the plot of this equation is shown in Fig. 13-6.

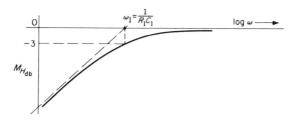

FIG. 13-6. Plot of $M_{H_{db}}$ versus log ω from Eq. 13-18.

13-3. The R-C-Coupled Grounded-Cathode Amplifier

The previous work with R-C networks will prove to be very useful in describing the frequency response of the R-C-coupled amplifier. An R-C-coupled, grounded-cathode amplifier is shown in Fig. 13-7a, with the exact a-c equivalent circuit given in Fig. 13-7b. R_g is the impedance of the source voltage, and C_{in} represents the input capacitance of the next amplifier. R'_{in} in Fig. 13-7b is the parallel sum of R_i and R_{in}, as shown in Fig. 13-7a.

If E_s is considered to be a rather low-impedance source voltage, the voltage-dividing effect of the input impedance of the first amplifier stage is negligible ($R_g \ll R_{in}$). This means that E_g is, for all practical purposes, equal to E_s.

As shown in Fig. 13-8a and b, the input circuit of the actual triode can be represented by a shunt impedance and ideal triode. (An ideal triode is a triode with no interelectrode capacitances.) The triode input impedance $Z_{in} = V_{in}/I_{in}$ of Fig. 13-8 corresponds to $Z_{in} = E_s/I_g$ given in Eq. 12-78. (Figures 13-8 and 12-51 are intended to convey the same concept even though drawn a bit differently.) Usually Z_{in} of Eq. 12-78 can be expressed in terms of an equivalent resistance, R_{in}, and capacitance, C_{in}, connected in series or in parallel. As depicted in Fig. 13-8b, a practical triode (a triode with significant interelectrode capacitances) is replaced by

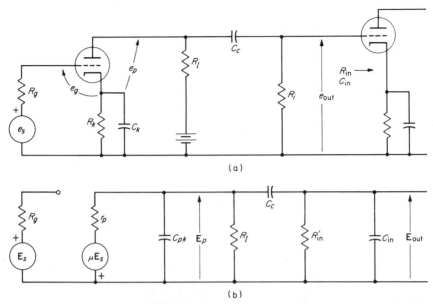

FIG. 13-7. a) A schematic diagram of an *R-C*-coupled grounded-cathode amplifier. b) An exact a-c equivalent circuit of part a.

an input impedance consisting of a parallel combination of R_{in} and C_{in} and an ideal triode (a triode with no interelectrode capacitances).

When considering the output circuit, the plate-to-cathode capacitance C_{pk} is included in the equivalent circuit, but the plate-to-grid capacitance C_{pg} and the grid-to-cathode capacitance C_{gk} are usually neglected. This introduces some error into the equivalent circuit, but, for frequencies below 1 MHz, this is a good approximation.

The equivalent circuit of Fig. 13-7b is a somewhat involved array of resistances and capacitances, but, under normal operation, C_{pk} and C_{in} are

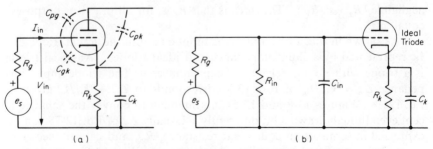

FIG. 13-8. a) Input section of a grounded-cathode circuit. b) Actual triode of part a is shown replaced by a shunt impedance and an ideal triode.

$\ll C_C$. This means that it is possible to describe the behavior of the circuit over various discrete frequency ranges—low frequencies, mid-frequencies and high-frequencies.

Low-Frequency Operation. Low frequencies are defined such that $X_{C_{pk}}$, X_{C_c}, and $X_{C_{in}} \gg R_i$ or R_l. Thus it is possible to neglect the effect of the shunt capacitive reactances ($X_{C_{pk}}$ and $X_{C_{in}}$) but not the series reactance X_{C_c}. This gives a low frequency equivalent circuit as shown in Fig. 13-9.

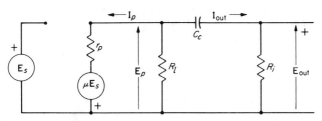

FIG. 13-9. Low-frequency a-c equivalent circuit of Fig. 13-7.

From previous work (see Chapter 10) the gain, as defined by the ratio E_p/E_s, is

$$\mathbf{A}'_L = \frac{\mathbf{E}_p}{\mathbf{E}_s} = -\frac{\mu \mathbf{Z}_{l_L}}{r_p + \mathbf{Z}_{l_L}} \tag{13-19}$$

where a complex plate load impedance, $\mathbf{Z}_{l_L}$, replaces the purely resistive load R_l used in Chapter 10.

$$\mathbf{Z}_{l_L} = \frac{R_l \left(R_i - j \dfrac{1}{\omega C_c} \right)}{R_l + R_i - j \dfrac{1}{\omega C_C}} \tag{13-20}$$

But the desired low-frequency gain is $\mathbf{A}_L = \mathbf{E}_{out}/\mathbf{E}_s$, so a relation involving $\mathbf{E}_{out}$ and $\mathbf{E}_p$ is needed. This is readily achieved by noting that the current through R_i is

$$\mathbf{I}_{out} = \frac{\mathbf{E}_p}{R_i - j \dfrac{1}{\omega C_c}} \tag{13-21}$$

and

$$\mathbf{E}_{out} = \mathbf{I}_{out} R_i$$
$$= \frac{\mathbf{E}_p}{R_i - j \dfrac{1}{\omega C_C}} R_i \tag{13-22}$$

or

$$E_p = E_{out} \left(\frac{R_i - j \dfrac{1}{\omega C_C}}{R_i} \right) \tag{13-23}$$

Then A'_L can be rewritten as

$$A'_L = \frac{E_p}{E_s} = \frac{E_{out}}{E_s} \left(\frac{R_i - j \dfrac{1}{\omega C_C}}{R_i} \right) \tag{13-24}$$

But

$$\frac{E_{out}}{E_s} \triangleq A_L \tag{13-25}$$

or

$$A_L = A'_L \frac{R_i}{R_i - j \dfrac{1}{\omega C_C}}$$

$$= -\frac{\mu Z_{l_L}}{r_p + Z_{l_L}} \left(\frac{R_i}{R_i - j \dfrac{1}{\omega C_C}} \right) \tag{13-26}$$

and, upon substituting the value of Z_{l_L} from Eq. 13-20 into Eq. 13-26 and simplifying,

$$A_L = -\frac{\mu R_l R_i}{r_p R_l + r_p R_i + R_l R_i - j \dfrac{1}{\omega C_C} (r_p + R_l)} \tag{13-27}$$

Mid-Frequency Operation. As the input frequency is increased, there is a range of frequencies where $X_{C_{pk}}$ and $X_{C_{in}} \ggg R_l$ or R_i but where $X_{C_C} \gg R_l$ or R_i. This is because C_C is usually 100 to 10,000 times larger than C_{in} or C_{pk}. In this frequency band a new equivalent circuit can be evolved, since not only can the shunt capacitive reactances $X_{C_{pk}}$ and $X_{C_{in}}$ still be considered quite large and hence neglected, but the series capacitive reactance X_{C_C} can now be considered a short circuit.

In the circuit of Fig. 13-10, $E_p = E_{out}$, and the mid-frequency gain A_M

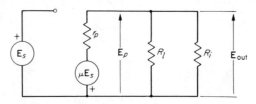

FIG. 13-10. Mid-frequency equivalent circuit of Fig. 13-7.

is recognized as

$$\mathbf{A}_M = \frac{\mathbf{E}_{out}}{\mathbf{E}_s} = -\frac{\mu \mathbf{Z}_{l_M}}{r_p + \mathbf{Z}_{l_M}} \tag{13-28}$$

where

$$\mathbf{Z}_{l_M} = \frac{R_l R_i}{R_l + R_i} \tag{13-29}$$

and, substituting Eq. 13-29 into Eq. 13-28,

$$\mathbf{A}_M = \frac{-\mu \dfrac{R_l R_i}{R_l + R_i}}{r_p + \dfrac{R_l R_i}{R_l + R_i}}$$

$$= -\frac{\mu R_l R_i}{r_p R_l + r_p R_i + R_l R_i} \tag{13-30}$$

Note that Eq. 13-30 can be derived from Eq. 13-27 simply by letting $\omega C_C \rightarrow \infty$.

High-Frequency Operation. As the frequency is increased still further, the capacitive reactances $X_{C_{pk}}$ and $X_{C_{in}}$ approach or become less than the values of the circuit resistances; therefore, the shunt effect of these reactances can no longer be neglected. Of course X_{C_C}, the series reactance, becomes even smaller and is still negligible. Figure 13-11 shows this higher-frequency equivalent circuit, where $C_T = C_{pk} + C_{in}$.

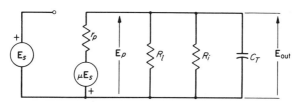

Fig. 13-11. High-frequency equivalent circuit of Fig. 13-7.

The high-frequency gain $\mathbf{A}_H$ is

$$\mathbf{A}_H = \frac{\mathbf{E}_{out}}{\mathbf{E}_s} = \frac{-\mu \mathbf{Z}_{l_H}}{r_p + \mathbf{Z}_{l_H}} \tag{13-31}$$

where

$$\frac{1}{\mathbf{Z}_{l_H}} = \frac{1}{R_l} + \frac{1}{R_i} + \frac{1}{-j\dfrac{1}{\omega C_T}}$$

or

$$\frac{1}{Z_{l_H}} = \frac{R_i + R_l + jR_lR_i\omega C_T}{R_lR_i} \tag{13-32}$$

After substituting from Eq. 13-32 into Eq. 13-31, the expression for the high-frequency gain becomes

$$\mathbf{A}_H = \frac{-\mu R_l R_i}{r_p(R_l + R_i + jR_lR_i\omega C_T) + R_lR_i}$$

$$= -\frac{\mu R_l R_i}{r_p R_l + r_p R_i + R_l R_i + jR_lR_ir_p\omega C_T} \tag{13-33}$$

Comparison of the *R-C*-Coupled Amplifier to the Basic *R-C* Circuit.
By observation of Eqs. 13-27, 13-30, and 13-33, it should be apparent that the maximum gain of the circuit occurs in the mid-frequency band. For a better comparison of the gain characteristics, it is convenient to normalize or compare each of the gain expressions to that of the mid-frequency gain.

First, let us divide $\mathbf{A}_L$ by $\mathbf{A}_M$, or

$$\frac{\mathbf{A}_L}{\mathbf{A}_M} = \frac{\dfrac{-\mu R_l R_i}{r_p R_l + r_p R_i + R_l R_i - j\dfrac{1}{\omega C_C}(r_p + R_l)}}{\dfrac{-\mu R_l R_i}{r_p R_l + r_p R_i + R_l R_i}}$$

$$= \frac{1}{1 - j\dfrac{1}{\omega C_C}\left(\dfrac{r_p + R_l}{r_p R_l + r_p R_i + R_l R_i}\right)} \tag{13-34}$$

Next, we divide $\mathbf{A}_M$ by $\mathbf{A}_M$, or

$$\frac{\mathbf{A}_M}{\mathbf{A}_M} = 1 \tag{13-35}$$

Finally, if we divide $\mathbf{A}_H$ by $\mathbf{A}_M$,

$$\frac{\mathbf{A}_H}{\mathbf{A}_M} = \frac{\dfrac{-\mu R_l R_i}{r_p R_l + r_p R_i + R_l R_i + jR_lR_ir_p\omega C_T}}{\dfrac{-\mu R_l R_i}{r_p R_l + r_p R_i + R_l R_i}}$$

$$= \frac{1}{1 + j\omega C_T\left(\dfrac{R_l R_i r_p}{r_p R_l + r_p R_i + R_l R_i}\right)} \tag{13-36}$$

Now let us compare Eq. 13-34 with Eq. 13-14a for the high-pass filter. The form of the two equations is seen to be the same when a term-by-term comparison is made, or, if

$$C_1 = C_C \tag{13-37}$$

$$R_1 = \frac{r_p R_l + r_p R_i + R_l R_i}{r_p + R_l} \tag{13-38}$$

Similarly, Eq. 13-36 is identical in form to that of Eq. 13-3a for the low-pass filter, and, upon comparing the two equations, the following relations are obtained:

$$C_2 = C_T \tag{13-39}$$

$$R_2 = \frac{R_l R_i r_p}{r_p R_l + r_p R_i + R_l R_i} \tag{13-40}$$

The corner or -3-db frequencies in terms of vacuum-tube circuit parameters are thus

$$\omega_1 = \frac{1}{R_1 C_1} = \frac{1}{C_C}\left(\frac{r_p + R_l}{r_p R_l + r_p R_i + R_l R_i}\right) \tag{13-41}$$

$$\omega_2 = \frac{1}{R_2 C_2} = \frac{1}{C_T}\left(\frac{r_p R_l + r_p R_i + R_l R_i}{r_p R_l R_i}\right) \tag{13-42}$$

Equations 13-34, 13-35, and 13-36 can be written as

$$\frac{\mathbf{A}_L}{\mathbf{A}_M} = \frac{1}{1 - j\,\dfrac{\omega_1}{\omega}} = \frac{1}{\sqrt{1 + \left(\dfrac{\omega_1}{\omega}\right)^2}}\bigg/\!\tan^{-1}\frac{\omega_1}{\omega} \tag{13-43}$$

$$\frac{\mathbf{A}_M}{\mathbf{A}_M} = 1 \tag{13-44}$$

$$\frac{\mathbf{A}_H}{\mathbf{A}_M} = \frac{1}{1 + j\,\dfrac{\omega}{\omega_2}} = \frac{1}{\sqrt{1 + \left(\dfrac{\omega}{\omega_2}\right)^2}}\bigg/\!-\tan^{-1}\frac{\omega}{\omega_2} \tag{13-45}$$

Keep in mind that ω_1 and ω_2 are particular or constant frequencies while ω is the variable frequency.

Depending on the order of magnitude of the separation of C_C and C_T, which determines the mid-frequency bandwidth, a typical frequency-response curve for an R-C-coupled amplifier is shown in Fig. 13-12. The abscissa is log ω and the ordinate is 20 log $\left|\dfrac{\mathbf{A}}{\mathbf{A}_M}\right|$, where $\mathbf{A}$ is either $\mathbf{A}_L$,

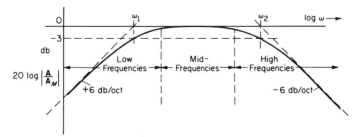

FIG. 13-12. Frequency-response characteristics of the R-C-coupled grounded-cathode amplifier.

A_M, or A_H. It is evident that this curve is similar to a superposition of the two curves given in Figs. 13-3 and 13-6 for the high- and low-pass *R-C* filter circuits, respectively.

Cascading Identical Stages. A moment's reflection should indicate that the low- and high-frequency response curves are mirror images, and that, since ω_1 and ω_2 represent any fixed frequencies, it is possible to plot a universal curve for all *R-C*-coupled amplifiers. This is shown in Fig. 13-13

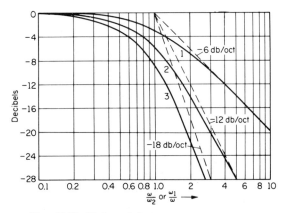

FIG. 13-13. Universal frequency-response curve for *N*-identical cascaded *R-C*-coupled stages (N = 1, 2, or 3).

where the response curves for one, two, or three identical, cascaded, *R-C*-coupled amplifiers are given. An example will illustrate how Fig. 13-13 can be used.

EXAMPLE 13-1. The overall f_2 of four identical, cascaded, *R-C*-coupled amplifier stages is 80 KHz. What is the f_2 of one stage?

Solution: When the term f_2 is seen, it should immediately be identified with the -3-db point on the proper frequency-response curve. Thus f_2 for *four* stages of amplification means that the gain of *all four* amplifier stages is decreased 3 db as compared to the mid-frequency gain, which, in turn, means that *each single* stage is down only $\frac{3}{4}$ db. However, the f_2 for *one stage* is the frequency at which *one* stage is down 3 db. The ω_1 and ω_2 frequencies noted on Fig. 13-13 are the -3-db frequencies for *one* stage of amplification.

Armed with this information, the solution to the problem is straightforward. Using Fig. 13-13, the frequency ratio ω/ω_2 at which one stage is down $\frac{3}{4}$ db, or two stages are down $\frac{6}{4}$ db, or three stages are down $\frac{9}{4}$ db, is 0.38. (Had a four-stage curve been plotted on Fig. 13-13, we could have used -3 db and obtained the same ratio.)

Thus

$$\frac{\omega}{\omega_2} = \frac{f}{f_2} = 0.38$$

and f is given as 80 KHz. Therefore,

$$f_2 = \frac{f}{0.38} = \frac{80}{0.38} = 210 \text{ KHz}$$

13-4. The Common-Emitter Transistor Amplifier

The low- and mid-frequency equivalent circuits of the common-emitter transistor R-C-coupled amplifier are easy to explain and are identical in form to the grounded-cathode vacuum-tube equivalent circuits. The high-frequency response of the vacuum-tube amplifier is largely determined by the plate-cathode interelectrode capacitance and other miscellaneous external capacitances such as stray wiring capacitances. Unfortunately, the transistor does not lend itself to such a simple explanation. The high-frequency response of the transistor is complicated by hole-and-electron diffusion times and internal reactances. The simplest procedure, in developing the high-frequency equivalent circuit, is to show a shunt collector-emitter capacitance and to use this capacitance in the same way that the plate-cathode capacitance is used for a vacuum tube. Stray wiring capacitance must also be considered to be present in transistors as in tube circuits. Figure 13-14a shows the actual diagram for a transistor R-C-coupled amplifier in the common-emitter configuration.

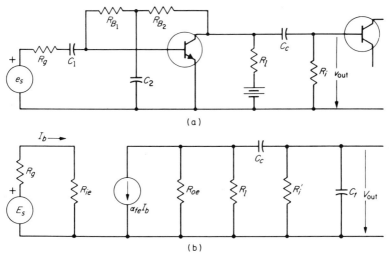

FIG. 13-14. a) Transistor R-C-coupled amplifier. b) Exact a-c equivalent circuit for part a with $\mu_{re} = 0$. Also $X_{C_1} \ll R_g$ for all frequencies of interest, and $R_i' = R_i R_{ie}/(R_i + R_{ie})$. See Prob. 13-13 for a more accurate high frequency model.

Figure 13-14b shows the exact a-c equivalent circuit for Fig. 13-14a, assuming that a simple collector-emitter shunt capacitance adequately determines the high-frequency response characteristics. The transistor a-c equivalent circuit of Fig. 13-14b is so similar to the vacuum-tube equivalent circuit of Fig. 13-7b that it is easy to compare the low-, mid-, and high-frequency gains of the transistor to the tube amplifier. (The transistor current gain is compared to the vacuum tube voltage gain.)

The low-frequency equivalent circuit for Fig. 13-14a is shown in Fig.

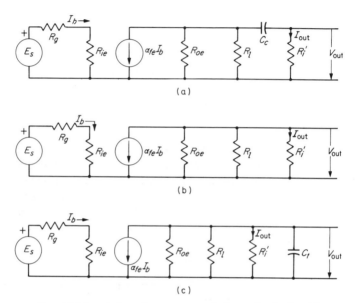

FIG. 13-15. a) Low-frequency equivalent circuit for Fig. 13-14a. This figure is similar to Fig. 13-7b. b) Mid-frequency equivalent circuit for Fig. 13-14a. c) High-frequency equivalent circuit for Fig. 13-14a.

13-15a. It can be shown[1] that the current gain is:

$$\mathbf{A}_{i_L} = -\alpha_{fe} \frac{R_{oe} R_l}{R_{oe} R_l + (R_{oe} + R_l)\left(R_i' - j\,\dfrac{1}{\omega C_C}\right)}$$

$$= -\alpha_{fe} \frac{\dfrac{R_{oe} R_l}{R_{oe} R_l + R_{oe} R_i' + R_l R_i'}}{1 - j\,\dfrac{R_{oe} + R_l}{\omega C_C(R_{oe} R_l + R_{oe} R_i' + R_l R_i')}} \tag{13-46}$$

[1]This can best be accomplished by replacing R_i in Eq. 7-20 with $R_i - j1/\omega C_C$ and R_b with Eq. 7-19. Also refer to Fig. 12-48b and Eqs. 12-73a and 12-73b.

The mid-frequency equivalent is shown in Fig. 13-15b. This circuit is similar to Fig. 7-1b, and the mid-frequency-gain expression is identical to the combination of Eqs. 7-19 and 7-20, repeated here as Eq. 13-47.

$$\mathbf{A}_{i_M} = -\alpha_{fe}\frac{R_{oe}R_l}{R_{oe}R_l + R_{oe}R_i' + R_lR_i'} \tag{13-47}$$

The high-frequency equivalent circuit is shown in Fig. 13-15c.

The high-frequency current-gain expression can also be obtained from Eq. 7-20 by replacing R_b with the parallel sum of R_b and X_{C_t} where

$$C_t = C_{ce} + C_{stray}$$

C_{ce} is the effective internal collector-emitter capacitance and C_{stray} accounts for miscellaneous capacitances such as wiring capacitance. The high-frequency current gain is

$$\mathbf{A}_{i_H} = \frac{-\alpha_{fe}\dfrac{R_{oe}R_l}{R_{oe}+R_l}\left(-j\dfrac{1}{\omega C_t}\right)}{\dfrac{R_{oe}R_l}{R_{oe}+R_l}\left(-j\dfrac{1}{\omega C_t}\right)+R_i'\left(\dfrac{R_{oe}R_l}{R_{oe}+R_l}-j\dfrac{1}{\omega C_t}\right)}$$

Upon simplifying,

$$\mathbf{A}_{i_H} = -\alpha_{fe}\frac{\dfrac{R_{oe}R_l}{R_{oe}R_l + R_i'R_{oe} + R_i'R_l}}{1+j\omega C_t\dfrac{R_i'R_{oe}R_l}{R_{oe}R_l + R_i'R_{oe} + R_i'R_l}} \tag{13-48}$$

Upon normalizing each of the gain expressions (Eqs. 13-46, 13-47, and 13-48) by dividing by the mid-frequency gain, we obtain

$$\frac{\mathbf{A}_{i_L}}{\mathbf{A}_{i_M}} = \frac{1}{1-j\dfrac{R_{oe}+R_l}{\omega C_c(R_{oe}R_l + R_{oe}R_i' + R_lR_i')}} \tag{13-49}$$

$$\frac{\mathbf{A}_{i_M}}{\mathbf{A}_{i_M}} = 1 \tag{13-50}$$

$$\frac{\mathbf{A}_{i_H}}{\mathbf{A}_{i_M}} = \frac{1}{1+j\omega C_t\dfrac{R_i'R_{oe}R_l}{R_{oe}R_l + R_i'R_{oe} + R_i'R_l}} \tag{13-51}$$

As with the vacuum-tube circuit, we compare Eq. 13-49 with Eq. 13-14a for the high-pass filter. A term-by-term comparison shows that

$$C_1 = C_C \tag{13-52}$$

$$R_1 = \frac{R_{oe}R_l + R_{oe}R_i' + R_lR_i'}{R_{oe}+R_l} \tag{13-53}$$

Similarly, Eq. 13-51 is identical in form to Eq. 13-3a for the low-pass filter, and, upon comparing terms,

$$C_2 = C_t \tag{13-54}$$

$$R_2 = \frac{R_i' R_{oe} R_l}{R_{oe} R_l + R_i' R_{oe} + R_i' R_l} \tag{13-55}$$

The corner or -3-db frequencies are

$$\omega_1 = \frac{1}{R_1 C_1} = \frac{1}{C_C} \frac{R_{oe} + R_l}{R_{oe} R_l + R_{oe} R_i' + R_l R_i'} \tag{13-56}$$

$$\omega_2 = \frac{1}{R_2 C_2} = \frac{1}{C_t} \frac{R_{oe} R_l + R_i' R_{oe} + R_i' R_l}{R_i' R_{oe} R_l} \tag{13-57}$$

Equations 13-49, 13-50, and 13-51 are identical to Eqs. 13-43, 13-44, and 13-45, and thus the curves of Fig. 13-13 can just as well be used for the transistor as for the vacuum-tube R-C-coupled amplifier.

However, it must be stressed that the correctness of the high-frequency equivalent circuit of Fig. 13-15c is open to question. Some transistor high-frequency equivalent circuits are quite complicated (involving inductances as well as capacitances), and it is beyond the scope of this text to discuss the more complex (and, it is hoped, more accurate) equivalent circuits. One other approach used in describing the high-frequency operation of the transistor is to assume that the short-circuit current gain α_{fe} (or β) varies with frequency, or

$$\alpha_{fe_H} = \frac{1}{1 + j \dfrac{\omega}{\omega_{2\alpha}}} \alpha_{fe} \tag{13-58}$$

where α_{fe} is the low-frequency α_{fe} which is used throughout this text, and $\omega_{2\alpha}$ is the corner frequency for α_{fe_H}. (This ω_2 can be calculated in terms of the transistor parameters, such as the type of material out of which the transistor is constructed, the base width, the temperature, and so on, but this derivation will not be given here.) If ω_2 is known (by calculation or measurement), we can use the mid-frequency-gain expression to obtain the high-frequency-gain expression by simply using α_{fe_H} instead of α_{fe}. Substituting Eq. 13-58 into Eq. 13-47, we obtain

$$\mathbf{A}_{i_H} = -\alpha_{fe} \frac{R_{oe} R_l}{R_{oe} R_l + R_{oe} R_i' + R_l R_i'} \frac{1}{1 + j \dfrac{\omega}{\omega_{2\alpha}}} \tag{13-59}$$

and, upon normalizing,

$$\frac{\mathbf{A}_{i_H}}{\mathbf{A}_{i_M}} = \frac{1}{1 + j \dfrac{\omega}{\omega_{2\alpha}}} \tag{13-60}$$

which is identical to Eq. 13-45. Although the derivation and explanation for the value of ω_2 in Eq. 13-47 and for the value of $\omega_{2\alpha}$ in Eq. 13-60 differ, we can use the curves of Fig. 13-13 to evaluate R-C-amplifier behavior once ω_2 or $\omega_{2\alpha}$ is known. See Prob. 13-13 for a model that describes more accurately the high frequency behavior of the transistor.

Problems

13-1. In Fig. 13-7a, $R_l = 50$ K, $C_C = 0.01$ μf, $R_i = 200$ K, $C_{in} = 60$ picofarads, $R_{in} = 2$ megohms, $C_{pk} = 10$ picofarads, $\mu = 25$, and $r_p = 12$ K. a) Calculate the -3-db frequencies ω_1 and ω_2. b) Calculate the mid-frequency gain A_M. c) Sketch the frequency-response curve $|A|$ versus log ω.

13-2. If C_C is changed to 1 μf in Prob. 13-1, with all other values remaining the same, a) calculate the amplifier-frequency bandwidth $\omega_2 - \omega_1$ and b) compare the amplifier bandwidth of Probs. 13-1 and 13-2. Which amplifier could best be used to amplify audio frequencies?

13-3. The transistor R-C-coupled amplifier shown in Fig. 13-14a has the following circuit parameters: $R_g = 10$ K, $C_1 = 10$ μf, $C_C = 0.01$ μf, $R_{B_1} = R_{B_2} = 150$ K, $R_l = 5$ K, $R_i = 10$ K, $R_{ie} = 1.5$ K, $\mu_{re} = 2.5 \times 10^{-4}$, $\alpha_{fe} = 46$, $R_{oe} = 122$ K, and $C_t = 70$ picofarads. a) Determine the -3-db frequencies ω_1 and ω_2. b) Calculate the mid-frequency gain A_{iM}. c) Sketch the frequency-response curve $|A|$ versus log ω.

13-4. Compare and comment on the frequency-response characteristics of the respective tube and transistor amplifiers of Probs. 13-1 and 13-3.

13-5. If an R-C-coupled amplifier has an $\left|\dfrac{A_H}{A_M}\right|$ of -12 db at 300 KHz, at what frequency is $\left|\dfrac{A_H}{A_M}\right|$ -2 db?

13-6. The overall gain of a three-identical-stage R-C-coupled amplifier is -6 db at a frequency of 150 KHz. a) Find the -3-db or f_2 frequency for a single stage. b) What is $\left|\dfrac{A_H}{A_M}\right|$ for the three-stage amplifier at a frequency of 600 KHz?

13-7. If three R-C-coupled stages are cascaded and $A_1 = 25$ db, $A_2 = -18$ db, and $A_3 = 61$ db at a given frequency, find the overall gain of the three-stage amplifier a) in decibels and b) as a numerical gain.

13-8. If the input voltage of a cascaded two-stage R-C-coupled tube amplifier is 0.05 volt, find the output voltage if the respective stage gains are $A_1 = 15$ db and $A_2 = 20$ db.

13-9. If four identical R-C-coupled tube stages are cascaded in an amplifier and the ratio of the overall amplifier output to input is 3800, find the gain of each amplifier, in decibels.

13-10. Four identical R-C-coupled transistor amplifiers are cascaded. If all the amplifiers are down -18 db from the mid-frequency gain at a frequency of 650 KHz, what is the decibel gain of all four amplifiers at a frequency of 325 KHz?

13-11. A three-identical-stage transistor, R-C-coupled amplifier has an $\left|\dfrac{A_{iH}}{A_{iM}}\right|$

of -28 db at a frequency of 1200 KHz. At what frequency is the gain of one of the stages -20 db below the mid-frequency current gain?

13-12. Two R-C-coupled transistor stages are cascaded to form an amplifier. One stage has an f_2 frequency of 300 KHz and the other has an f_2 of 550 KHz. What is the $\dfrac{A_{i_H}}{A_{i_M}}$ for the two-stage amplifier at a frequency of a) 500 KHz and b) 800 KHz?

13-13. A more exact high frequency equivalent circuit for the transistor includes a capacitor in parallel with R_{ie}. This capacitance (termed the diffusion capacitance C_D) appears because electric charges are capable of being stored in the basic-emitter region of the transistor. The modified equivalent circuit for Fig. 13-15c is shown in the accompanying figure. Derive the expression for the high frequency current gain $\mathbf{A}_{i_H}$ and the high frequency voltage gain $\mathbf{A}_{v_H}$. Identify the -3 db frequencies.

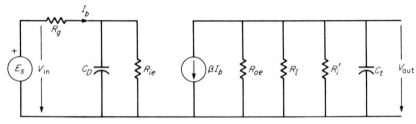

PROB. 13-13

13-14. If in Prob. 13-13 $R_g = 2\text{K}\,\Omega$, $R_{ie} = 0.5\text{K}\,\Omega$, $R_{oe} = 20\text{K}\,\Omega$, $R_l = 4\text{K}\,\Omega$, $R_i' = 3\text{K}$, $C_D = 1$ nf, $C_t = 50$ pf, $\beta = 150$, determine the -3 db frequency.

14 / Nonlinear Graphical Analysis

In the preceding chapters it was assumed that transistors and vacuum tubes behave as linear devices. This is a good assumption if the transistor or tube is properly biased and if the input and output signal voltages are relatively small. If large input and output signals are considered, the tube or transistor probably operates as a nonlinear device. Figure 14-1a shows how the plate voltage can become a distorted sine wave (assuming a sinusoidal input) owing to the nonlinear or nonuniform effect of the triode plate characteristics. Figure 14-1b shows the same effect for a transistor.

When large signals are applied to a tube or transistor amplifier, the amplifier is usually termed a *power amplifier*. A power amplifier can supply a considerable amount of power to drive such devices as loudspeakers, motors, relays, antenna arrays, and so on.

14-1. Amplifier Classification

Power amplifiers can be operated as class A, B, AB, or C. These classifications are applicable for transistor or tube circuits and are defined as follows:

Class A operation means that the tube plate or transistor collector current flows for a full 360° of the input cycle. In addition, class A operation usually implies that the tube or transistor is operated only in the linear region of the plate or collector characteristics. The theoretical power efficiency of a class A amplifier is 50 percent, but, in practice, efficiencies of only 1 to 25 percent are obtained.

Class B operation means that the tube plate or transistor collector current flows for 180° of the input cycle. Class B power amplifiers have a theoretical plate or collector efficiency of 78.5 percent, with 50 to 60 percent being the practical limit of attainment.

Class AB operation means that the tube or transistor is operated so that the plate or collector current flows for more than 180° but less than 360° of the input cycle.

In Class C operation the plate or collector current flows for less than 180° of the input cycle. Both the distortion and the efficiency are very high for this class of operation. In theory, 100 percent plate or collector effi-

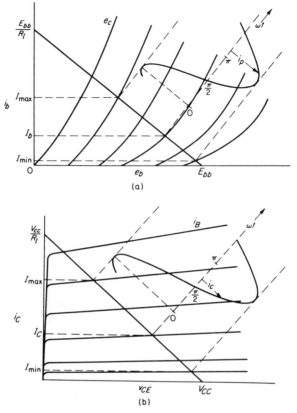

FIG. 14-1. a) A distorted sinusoidal output caused by the nonuniformity of the plate characteristics $e_g = E_m \sin \omega t$. b) The same as part a, except that a transistor is used in place of a tube.

ciency can be obtained, but, in practice, values of 60 to 80 percent are common.

Efficiency is an important consideration when studying power amplifiers. It will be considered in the last section of this chapter.

In this chapter only low-frequency (up to 20,000 Hz or so) class A or AB power amplifiers will be considered. Class C amplifiers are a separate study in themselves and will not be taken up in this text.

14-2. Output Distortion in Tubes and Transistors

If the static collector characteristics of a transistor or the static plate characteristics of a tube are examined, it is evident that the family of curves are not straight lines, nor are they separated equal distances from each other. This results in a different shape between the output and input wave forms, which is termed *nonlinear* or *amplitude distortion*.

Fourier Cosine Series. Since the input and output wave forms are not linearly related, it is important to find a way to describe mathematically the relation between the input and output wave forms. One well-known mathematical technique is to use a power-series expansion.[1] For a tube, the power series is written in terms of the total plate current i_b as a function of the a-c grid voltage e_g, or

$$i_b = I_b + a_1 e_g + a_2 e_g^2 + a_3 e_g^3 + \cdots + a_n e_g^n \qquad (14\text{-}1)$$

and, for the common-emitter transistor, the power series is written in terms of the collector current i_C as a function of the a-c base current i_b, or

$$i_C = I_C + d_1 i_b + d_2 i_b^2 + d_3 i_b^3 + \cdots + d_n i_b^n \qquad (14\text{-}2)$$

where a_n and d_n are constants that remain to be determined.

Many a-c signal inputs are sinusoidal in nature, and so, if it is assumed that e_g and i_b are sinusoids,

$$e_g = E_m \cos \omega t \qquad (14\text{-}3)$$

$$i_b = I_m \cos \omega t \qquad (14\text{-}4)$$

where the cosine form for the input is chosen because it is mathematically the simplest and easiest sinusoidal form to handle.

Substituting Eqs. 14-3 and 14-4 into Eqs. 14-1 and 14-2, respectively,

$$i_b = I_b + a_1 E_m \cos \omega t + a_2 (E_m \cos \omega t)^2 + \cdots + a_n (E_m \cos \omega t)^n \qquad (14\text{-}5)$$

and

$$i_C = I_C + d_1 I_m \cos \omega t + d_2 (I_m \cos \omega t)^2 + \cdots + d_n (I_m \cos \omega t)^n \qquad (14\text{-}6)$$

By expanding Eqs. 14-5 and 14-6 and using the trigonometric identities for $\cos n\omega t$, we can rewrite Eqs. 14-5 and 14-6 as

$$i_b = I_b + A_0 + A_1 \cos \omega t + A_2 \cos 2\omega t$$
$$+ A_3 \cos 3\omega t + \cdots + A_n \cos n\omega t \qquad (14\text{-}7)$$

and

$$i_C = I_C + D_0 + D_1 \cos \omega t + D_2 \cos 2\omega t$$
$$+ D_3 \cos 3\omega t + \cdots + D_n \cos n\omega t \qquad (14\text{-}8)$$

Since Eqs. 14-7 and 14-8 contain only cosine terms, the tube plate and transistor collector currents are considered to be even functions of time. Had a sinusoidal input signal of the form $e_g = E_m \sin \omega t$ or $i_b = I_m \sin \omega t$ been assumed instead of Eqs. 14-3 and 14-4, Eqs. 14-7 and 14-8 would have contained odd sine and even cosine components. To anyone unfamiliar with Fourier series analysis techniques,[2] this may seem confusing, since a phase shift of 90° in the input signal apparently produces two different output signals. Actually, a given wave shape can be described

---

[1] The power series approximates the voltage-current relations near the Q point of the tube or transistor. This particular form of the power-series expansion is known as Taylor's series. The method is valid and valuable for the small distortions found in class A and AB operation, but it is not valid for the large distortions present in class C operation.

[2] A brief discussion of Fourier series will be found in Appendix D.

in an infinite number of ways, depending on how the coordinate axes are chosen. Wherever possible, one should take advantage of symmetry in order to simplify the Fourier series expression. Equations 14-7 and 14-8 represent one of the simplest possible ways to describe the distorted output signal for a sinusoidal input signal.

Graphical Determination of Coefficients. The next step in the analysis is to determine the values for the A and D constants in Eqs. 14-7 and 14-8. This can be done graphically, and it greatly simplifies matters to know beforehand that harmonics above the fourth can be omitted with little error (this becomes evident if several higher order analyses are performed). If only harmonics up to the fourth are considered, Eqs. 14-7 and 14-8 become

$$i_b = I_b + A_0 + A_1 \cos \omega t + A_2 \cos 2\omega t$$
$$+ A_3 \cos 3\omega t + A_4 \cos 4\omega t \qquad (14\text{-}9)$$
$$i_C = I_C + D_0 + D_1 \cos \omega t + D_2 \cos 2\omega t$$
$$+ D_3 \cos 3\omega t + D_4 \cos 4\omega t \qquad (14\text{-}10)$$

To determine graphically the five A or D constants in Eq. 14-9 or Eq. 14-10, it is necessary to obtain five simultaneous equations from the graphical plot of i_b or i_C. The output signals i_b and i_C for a sinusoidal input signal are shown in Fig. 14-2a and b, respectively. (Compare Fig. 14-2 and 14-1.) Since the A and D constants are determined in exactly the same manner, let us consider Eq. 14-10 and the transistor characteristics of Fig. 14-2b to illustrate the procedure for obtaining the D constants. The five different values of ωt (which is the independent variable) and the corresponding values of i_C must be suitably chosen in order to obtain five independent equations involving the D constants. It is usually best to divide the ωt axis into equally spaced increments, but this is really not necessary. For ease in performing the calculations, it is expedient to choose the values $\omega t = 0, \pi/3, \pi/2, 2\pi/3$, and π. (Since the wave shape of i_C has half-wave symmetry, there is no need to choose values of ωt from π to 2π, as no new information can be obtained by this procedure.) The values for i_C corresponding to the chosen values for ωt are indicated in Figs. 14-1 and 14-2 and are designated I_{max} at $\omega t = 0$, I_2 at $\omega t = \pi/3$, I_C at $\omega t = \pi/2$, I_4 at $\omega t = 2\pi/3$ and I_{min} at $\omega t = \pi$. If these five sets of values for i_C and ωt are substituted one at a time into Eq. 14-10, the result is

$$\left. \begin{array}{l} I_{max} = I_C + D_0 + D_1 \quad + D_2 \quad + D_3 + D_4 \\ I_2 = I_C + D_0 + D_1/2 - D_2/2 - D_3 - D_4/2 \\ I_C = I_C + D_0 \qquad\quad - D_2 \quad + \qquad D_4 \\ I_4 = I_C + D_0 - D_1/2 - D_2/2 + D_3 - D_4/2 \\ I_{min} = I_C + D_0 - D_1 \quad + D_2 \quad - D_3 + D_4 \end{array} \right\} \quad (14\text{-}11)$$

The solution of the five simultaneous equations in Eq. set 14-11 yields the values for D_0, D_1, D_2, D_3, and D_4, or

$$D_0 = \frac{I_{max} + I_{min} + 2(I_2 + I_4) - 6I_C}{6} \tag{14-12}$$

$$D_1 = \frac{I_{max} - I_{min} + I_2 - I_4}{3} \tag{14-13}$$

$$D_2 = \frac{I_{max} + I_{min} - 2I_C}{4} \tag{14-14}$$

$$D_3 = \frac{I_{max} - I_{min} - 2(I_2 - I_4)}{6} \tag{14-15}$$

$$D_4 = \frac{I_{max} + I_{min} - 4(I_2 + I_4) + 6I_C}{12} \tag{14-16}$$

If harmonics higher than the fourth are desired, we simply divide the ωt axis into more segments and substitute the sets of ωt and the corresponding values of i_C into Eq. 14-8. Then we solve the resulting simultaneous equations for the D constants.

Needless to say, the same analysis used in obtaining the transistor D constants can be used in obtaining the vacuum-tube A constants.

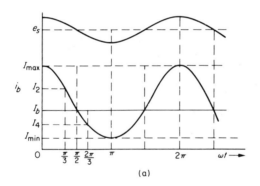

(a)

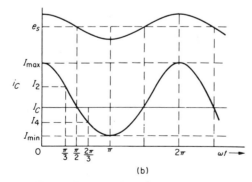

(b)

FIG. 14-2. a) A plot of i_b versus time.
b) A plot of i_C versus time.

Note that D_0 and A_0 add a new d-c term to the quiescent value of the respective collector I_C and plate I_b currents. This means that the Q-point location and the value of the collector- and plate-circuit input powers, as supplied by V_{CC} and E_{bb}, are changed materially when distortion is present. The new collector and plate input powers, when distortion is present, are

$$P_C = V_{CC}(I_C + D_0) \qquad (14\text{-}17)$$

$$P_b = E_{bb}(I_b + A_0) \qquad (14\text{-}18)$$

Harmonic Distortion. The harmonic contents of the plate or collector currents can be described as the ratio of the amplitude of the various harmonics to the fundamental component. Second-harmonic distortion is defined as

$$\Delta_2 \triangleq \left(\frac{A_2}{A_1} \text{ or } \frac{D_2}{D_1}\right) \times 100\% \qquad (14\text{-}19)$$

Third-harmonic distortion is

$$\Delta_3 \triangleq \left(\frac{A_3}{A_1} \text{ or } \frac{D_3}{D_1}\right) \times 100\% \qquad (14\text{-}20)$$

and so on for the higher harmonics.

The total harmonic distortion is defined as

$$\Delta \triangleq \sqrt{\Delta_2^2 + \Delta_3^2 + \Delta_4^2 + \cdots + \Delta_n^2} \qquad (14\text{-}21)$$

A-C Power Output. The average a-c power delivered to the a-c load resistance R_a for the transistor circuit is

$$P_{R_a} = \frac{1}{2\pi} \int_0^{2\pi} (D_1 \cos \omega t + D_2 \cos 2\omega t + D_3 \cos 3\omega t + \cdots + D_n \cos n\omega t)^2 R_a \, d\omega t \qquad (14\text{-}22)$$

and upon integrating and substituting in limits, we obtain

$$P_{R_a} = \left(\frac{D_1}{\sqrt{2}}\right)^2 R_a + \left(\frac{D_2}{\sqrt{2}}\right)^2 R_a + \left(\frac{D_3}{\sqrt{2}}\right)^2 R_a + \cdots + \left(\frac{D_n}{\sqrt{2}}\right)^2 R_a \qquad (14\text{-}23)$$

where the first term on the right-hand side of Eq. 14-23 is the a-c power output at the fundamental frequency ω, the second term is the a-c power output of the second harmonic 2ω, and so on.

EXAMPLE 14-1. A 2N334 NPN transistor is used in the circuit of Fig. 14-3a. If $e_s = 1.5 \cos \omega t$ volts, a) find i_C and b) calculate the individual and total harmonic distortion.

Solution: First, we draw the d-c and a-c load lines (which, in this case, are identical) on the graphical collector characteristics for the 2N334 transistor which are shown in Fig. 14-3b. Next, we locate the Q point. By trial and error (or by drawing a base-current bias curve satisfying $V_{BE} = I_C \times 180,000$), we locate the Q point at $V_{CE} = 18$ volts and $I_B = 100 \ \mu a$.

From $e_s = 1.5 \cos \omega t$, we calculate i_b as

$$i_b = \frac{e_s}{20,000} = 75 \cos \omega t \ \mu a$$

Next, we identify the points on the a-c load line corresponding to I_{max}, I_{min}, I_C, I_2, and I_4, as indicated on Fig. 14-3b. Note that the points corresponding to

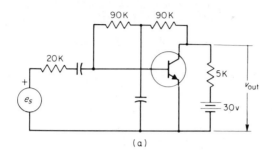

(a)

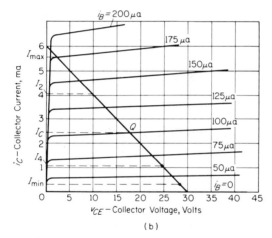

(b)

FIG. 14-3. a) Transistor circuit used in Example 14-1. b) Common-emitter collector characteristics for the 2N334 transistor.

I_2 and I_4 *are not* the bisectors of the line between the Q points and the I_{max} or I_{min} points but lie halfway between the Q point and the maximum or minimum values for i_B. The values of I_{max}, I_2, I_C, I_4, and I_{min}, from Fig. 14-3b, are

$$I_{max} = 5.55 \ ma$$
$$I_2 = 4.1 \ ma$$
$$I_C = 2.5 \ ma$$
$$I_4 = 1.1 \ ma$$
$$I_{min} = 0.3 \ ma$$

The D constants, as defined in Eqs. 14-12, 14-13, 14-14, 14-15, and 14-16, can now be evaluated:

$$D_0 = \frac{5.55 + 0.3 + 2(4.1 + 1.1) - (6 \times 2.5)}{6} = \frac{1.25}{6} \text{ ma}$$

$$D_1 = \frac{5.55 - 0.3 + 4.1 - 1.1}{3} = \frac{8.25}{3} \text{ ma}$$

$$D_2 = \frac{5.55 + 0.3 - (2 \times 2.5)}{4} = \frac{0.85}{4} \text{ ma}$$

$$D_3 = \frac{5.55 - 0.3 - 2(4.1 - 1.1)}{6} = \frac{-0.75}{6} \text{ ma}$$

$$D_4 = \frac{5.55 + 0.3 - 4(4.1 + 1.1) + (6 \times 2.5)}{12} = \frac{0.05}{12} \text{ ma}$$

a) The total collector current can now be written

$$i_C = 2.5 + \frac{1.25}{6} + \frac{8.25}{3} \cos \omega t + \frac{0.85}{4} \cos 2\omega t$$
$$- \frac{0.75}{6} \cos 3\omega t + \frac{0.05}{12} \cos 4\omega t \text{ ma}$$

b) The distortion contents are

$$\Delta_2 = \frac{0.85/4}{8.25/3} \times 100 = 7.63\%$$

$$\Delta_3 = \frac{0.75/6}{8.25/3} \times 100 = 4.55\%$$

$$\Delta_4 = \frac{0.05/12}{8.25/3} \times 100 = 0.152\%$$

$$\Delta = \sqrt{\Delta_2^2 + \Delta_3^2 + \Delta_4^2} = \sqrt{(7.63)^2 + (4.55)^2 + (0.152)^2} = 8.88\%$$

14-3. Class A Amplifier Efficiency

If the tube plate or transistor collector characteristics are considered to be ideal and linear, the typical characteristics would appear as in Fig. 14-4a and b.

Shunt-Fed Amplifier. Let us consider Fig. 14-4a for the ideal triode. If the triode is used in a shunt-fed circuit (see Fig. 14-5a) such that the a-c load resistance R_i is much greater than the d-c load resistance R_l (in this analysis we will consider that $R_l = 0$), then the Q point is located at $e_b = E_b = E_{bb}$ and $i_b = I_b$, as shown in Fig. 14-4a. If we let a sinusoidal input signal e_s be chosen such that the grid voltage just causes the minimum i_b to equal zero, the a-c current I_p becomes

$$I_p \triangleq \frac{I_{max} - I_{min}}{2\sqrt{2}} = \frac{I_b}{\sqrt{2}} \qquad (14\text{-}24)$$

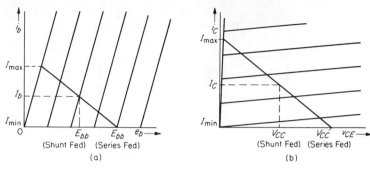

Fig. 14-4. a) Idealized triode plate characteristics. b) Idealized transistor collector characteristics.

since I_b falls halfway between I_{max} and I_{min} owing to the ideal linear plate characteristics. The power supplied to the plate circuit by E_{bb} is

$$P_{bb} = E_{bb} \times I_b \tag{14-25}$$

By definition, the plate-circuit efficiency is the ratio of the a-c output power to the d-c input power supplied by E_{bb}, or

$$\eta_p \triangleq \frac{\text{a-c power output to load}}{\text{d-c power into plate}} \times 100\% \tag{14-26}$$

$$\eta_p = \frac{I_p^2 R_i}{E_{bb}I_b} \times 100\% = \frac{I_p(I_p R_i)}{E_{bb}I_b} \times 100\%$$

where the term $I_p R_i$ is the a-c output voltage E_p, or

$$E_p = I_p R_i \tag{14-27}$$

If the a-c load resistance R_i is very large, the a-c load line is almost horizontal to the X axis. The maximum value of the a-c output voltage can approach E_{bb} when the a-c load resistance is large, or

$$E_p = \frac{E_{bb}}{\sqrt{2}} \bigg|_{R_i \to \infty} \tag{14-28}$$

where the $\sqrt{2}$ factor is necessary because E_p is an rms quantity and the value E_{bb} is the peak value that e_p can attain.

The maximum plate-circuit efficiency can now be calculated by substituting Eqs. 14-25 and 14-28 into Eq. 14-26,

$$\eta_{p\,max} = \frac{(I_b/\sqrt{2})(E_{bb}/\sqrt{2})}{E_{bb}I_b} \times 100\% = 50\% \tag{14-29}$$

Series-Fed Amplifier. If the plate circuit is series-fed, as shown in Fig. 14-5b, the maximum plate-circuit efficiency is only 25 percent. The reason for this is because the maximum value of the a-c output voltage

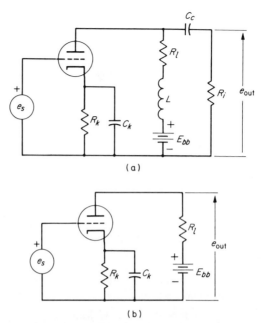

FIG. 14-5. a) A typical shunt-fed amplifier. b) A typical series-fed amplifier.

e_{out} can only approach $E_{bb}/2$ instead of E_{bb}. Proof of this is left as an exercise for the reader.

If transistors instead of vacuum tubes are utilized, the analysis and results for the collector-circuit efficiency are the same as for the vacuum-tube circuits just analyzed; namely, $\eta_{p_{max}}$ = 50 percent for the shunt-fed collector circuit and $\eta_{p_{max}}$ = 25 percent for the series-fed collector circuit.

The reader chould be careful not to confuse the plate- or collector-circuit efficiency with the overall efficiency of the electronic circuit. The plate- or collector-circuit efficiency is always greater than the overall efficiency, since the filament and screen (for the multigrid tubes) powers are omitted for the vacuum tube, and the base input power is omitted for the transistor circuit. These additional power terms must be included in the denominator of Eq. 14-26 if the overall efficiency is to be calculated.

Problems

14-1. Repeat the problem given in Example 14-1, but let e_s = 2 cos ωt volts.

14-2. Using the circuit and 2N238 common-emitter-collector characteristics shown in the accompanying illustrations, solve for the collector current i_C if $e_S = 1 \cos 10^5 t$ volts.

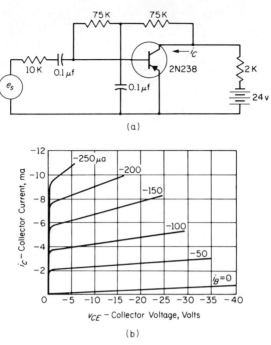

(a)

(b)

PROB. 14-2.

14-3. Repeat Prob. 14-2 if the two 75 KΩ biasing resistances are reduced to 50 KΩ each, and if e_s is increased to 1.25 cos $10^5 t$ volts.

14-4. A 6J5 triode is used in the shunt-fed amplifier shown in Fig. 14-5a. If E_{bb} is 200 volts, R_k = 1870 ohms, R_l = 12.5 KΩ, L = 10 henries, C_k = 10 μf, C_c = 0.1 μf, R_i = 25 KΩ, and E_s = 6 cos $10^6 t$ volts, a) determine i_b. b) Can the third and fourth harmonics be omitted as being negligibly small as compared to the fundamental and second harmonic terms in part a? (This approximation can often be made in circuits using triodes, but is is usually not acceptable for pentode or transistor circuits.)

14-5. Using the accompanying illustration, R_1 = 10 KΩ and R_2 = 20 KΩ. A nonlinear graphical analysis yields, for the plate current,

$$i_b = 12 + 2 + 8 \cos \omega t + 2.5 \cos 2\omega t \text{ ma}$$

and the third and fourth harmonics are considered to be negligibly small. a) Determine E_{out}. b) If E_{bb} = 250 volts, what is the power supplied to the plate circuit? c) If R_k = 1.2 KΩ, what is the value of the grid-bias voltage? d) Should the minimum wattage rating of R_1 be ½, 1, or 2 watts? e) Should the minimum wattage rating of R_2 be ½, 1, or 2 watts? f) What is the plate-circuit efficiency if both R_1 and R_2 are considered as belonging to the a-c load impedance?

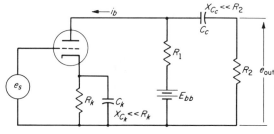

PROB. 14-5.

14-6. In a common-emitter circuit similar to Fig. 14-3a, a graphical analysis determines that

$$i_C = 3 + 0.6 + 3.2 \cos \omega t + 0.6 \cos 2\omega t + 0.2 \cos 3\omega t - 0.3 \cos 4\omega t \text{ ma}$$

a) What are the percentage second-, third-, and fourth-harmonic distortions? b) What is the total distortion? c) If $V_{CC} = 25$ volts, what is the collector input power? d) If the a-c load resistance is 2000 ohms, what is the collector-circuit efficiency?

14-7. A graphical analysis on a series-fed triode amplifier, as shown in Fig. 14-5b, yields the following Fourier series for the plate current:

$$i_b = 5 + 0.6 + 4 \cos \omega t + 0.35 \cos 2\omega t \text{ ma}$$

a) What is the percentage distortion? b) If $R_L = 20$ KΩ and $E_{bb} = 220$ volts, what is the plate-circuit efficiency?

14-8. Using a triode circuit similar to Fig. 14-5a and the circuit values $E_{bb} = 250$ volts, $R_l = 5$ KΩ, $L = 10$ henries, $R_i = 3$ KΩ, and $f = 10^4$ Hz, and if all capacitive reactances are negligibly small, determine the plate-circuit efficiency if a graphical analysis yields the following value for i_b:

$$i_b = 45 + 6 + 38 \cos \omega t + 7 \cos 2\omega t + 2 \cos 3\omega t + 0.8 \cos 4\omega t \text{ ma}$$

15/ *Power Supplies and*
Filters

In Chapter 5 the physical characteristics of the diode were discussed, but only rather brief mention was made as to how the diode might be used in a circuit (such as the shunt regulator or digital logic circuits). The diode is noted for its ability to rectify electrical signals. We recall that all circuits using tubes and transistors must have a d-c power supply (V_{CC} or E_{bb}). This power supply is a chemical battery for most portable equipment, but for nonportable electronic equipment the power supply is usually electronic in nature.

Since almost all electric power is generated as alternating current for reasons of economy (60 Hz is predominant in this country), it is necessary to convert the a-c voltage and current to a d-c voltage and current in order to operate electronic equipment (the radio and TV set are typical examples of electronic equipment that must be able to convert alternating to direct current). The electronic device that converts alternating to direct current is commonly called a *power supply.* The shunt regulator circuit presented in Chapter 5 is a simple but reasonably effective electronic power supply.

In a power supply a diode converts the alternating current to a pulsating unidirectional current, and the filter smooths the pulsating unidirectional current so that, ideally, a constant or ripple-free d-c voltage appears at the output of the power supply.

The qualitative explanation of the operation of a power supply and associated filters is quite easy, but an accurate quantitative description is sometimes very involved. Understanding and proper handling of the boundary and initial conditions of the circuit perhaps causes the greatest difficulty in the analysis.

In this chapter a study of the half-wave and full-wave rectifiers, along with the basic series-inductor and shunt-capacitor filters, will be undertaken.

15-1. Half-Wave Rectification

In Chapter 5 the use of a diode as a rectifier was stressed. Figure 5-10a, repeated here as Fig. 15-1, is the basic half-wave rectifier circuit.

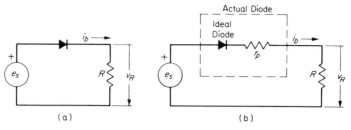

FIG. 15-1. a) Basic half-wave rectifier circuit. b) Part a repeated with the actual diode replaced by and ideal diode and a
series resistance.

Although the voltage e_s is considered to be sinusoidal in this chapter, in general, e_s could be any wave shape. The diode shown in Fig. 15-1a is not to be considered ideal, that is, a short circuit to current flow in one direction and an open circuit to current flow in the opposite direction. However, for analytical purposes, it is desirable to consider the diode as an ideal device. This can be done if the actual diode is replaced by an equivalent circuit consisting of an ideal diode in series with the dynamic plate resistance of the diode. (Exactly the same type of substitution was made in the case of the transistor or tube.) The forward dynamic anode (or plate) resistance r_p of most semiconductor diodes is under 50 ohms, and it is under 500 ohms for vacuum tube diodes. The reverse resistance for both types of diodes usually exceeds 10 megohms, which is so large that, for all practical purposes, the diode does not allow any current flow in the reverse direction. Thus the actual diode shown in Fig. 15-1a can be replaced by an ideal diode and a series resistance r_p (Fig. 15-1b).

If $e_s = E_m \sin \omega t$ in the half-wave circuit of Fig. 15-1b, the current i_b through R is

$$
\left.
\begin{aligned}
i_b &= \frac{E_m}{R + r_p} \sin \omega t \qquad && 0 \le \omega t \le \pi \\
&= 0 \qquad && \pi \le \omega t \le 2\pi
\end{aligned}
\right\}
\tag{15-1}
$$

and the instantaneous output voltage v_R is

$$
\left.
\begin{aligned}
v_R &= i_b R \\
&= \frac{R E_m}{R + r_p} \sin \omega t \qquad && 0 \le \omega t \le \pi \\
&= 0 \qquad && \pi \le \omega t \le 2\pi
\end{aligned}
\right\}
\tag{15-2}
$$

Thus the instantaneous output voltage v_R and current i_b are both half-wave rectified sine waves.

The average or d-c value of a half sinusoid is

$$
I_{\text{d-c}} = \frac{1}{2\pi} \int_0^{2\pi} i_b \, d\omega t
$$

$$= \frac{1}{2\pi} \int_0^\pi \frac{E_m}{R + r_p} \sin \omega t \, d\omega t + \frac{1}{2\pi} \int_\pi^{2\pi} 0 \, d\omega t \qquad (15\text{-}3)$$

Upon integrating and substituting in limits, we obtain

$$I_{\text{d-c}} = \frac{1}{\pi} \frac{E_m}{R + r_p} \qquad (15\text{-}4)$$

The d-c output voltage $V_{\text{d-c}}$ is

$$V_{\text{d-c}} = R I_{\text{d-c}}$$
$$= \frac{R}{\pi} \frac{E_m}{R + r_p} \qquad (15\text{-}5)$$

Although Eqs. 15-4 and 15-5 show that a d-c component of current or voltage is available at the output terminals, it must be remembered that the actual output current and voltage are pulsating in nature (a half sinusoid), so the rectifier circuit cannot yet be considered a d-c power supply. Somehow, the d-c components of the half-sinusoid voltage and current wave forms must be separated from other components if a usable d-c power supply is to be obtained. Before proceeding further, we must determine just what the other components are. The mathematical method for determining the mathematical description of periodic, well-behaved signals (such as the half-wave sinusoid) is termed a Fourier series analysis.[1] As shown in Chapter 14 a Fourier series is a trigonometric series involving a d-c term, a fundamental component, and an infinite number

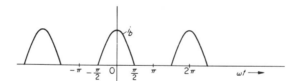

FIG. 15-2. The half-wave rectified sine wave.

of harmonic terms.[2] For the half sinusoid with the coordinate axes chosen as shown in Fig. 15-2, the Fourier series expansion for i_b is

$$i_b = \frac{E_m}{R + r_p} \left[\frac{1}{\pi} + \frac{1}{2} \cos \omega t + \frac{2}{\pi} \sum_{\substack{n = 2,4,6,\dots}}^{n \text{ even}} \frac{\sin (n - 1)\frac{\pi}{2} \cos n\omega t}{(n + 1)(n - 1)} \right] \qquad (15\text{-}6)$$

where the first term on the right-hand side of Eq. 15-6 is identical to the d-c term already determined by Eq. 15-4. This first term is the average or d-c term because it is not a function of ωt.

[1] A brief discussion of Fourier series and the derivation of the Fourier series for a half-wave and full-wave rectified sine wave is given in Appendix D.
[2] The nth harmonic is equal to n (an integer) times the fundamental frequency ω_1, or $\omega_n = n\omega_1$.

Equation 15-6 is very enlightening in that it shows the nature of the terms that must be eliminated in order to isolate the d-c term at the output terminals. Since all terms except the d-c terms are sinusoidal, one could make an educated guess as to the correct form of network that would block the alternating current and pass the direct current. Such a network appears in Fig. 15-3.[3] From conventional a-c circuit theory, recall that

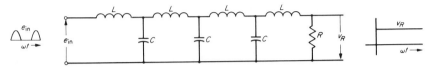

FIG. 15-3. An *L-C* ladder type of network used for filtering the alternating current from the direct current in a power supply.

$X_L = \omega L$ and $X_C = 1/\omega C$; thus the series inductances in the filter of Fig. 15-3 tend to block (offer a high impedance) the a-c terms and to pass (offer low impedance) the d-c terms. On the other hand, the shunt capacitors tend to bypass or short to ground the a-c terms but block the d-c component. In practice, a filter like that shown in Fig. 15-3 is very effective in passing the d-c terms while rejecting the a-c terms. However, as in all engineering practice, cost is an ever-present factor, and it is not economical to do any more filtering than is necessary. It is beyond the scope of this text to analyze mathematically the somewhat complex filter that is shown in Fig. 15-3.

Half-wave rectification is rarely used in power supplies because it is only half as efficient as full-wave rectification. For this reason, it is much more important to focus attention on full-wave rectifiers and the associated filter circuits.

15-2. Full-Wave Rectification

Figure 15-4a shows the basic full-wave rectifier circuit, which is the same as two half-wave rectifiers sharing a common load resistance R. A transformer is almost always used with power-supply circuits which are, in turn, used to supply direct current to vacuum-tube devices. There are several reasons for this: 1) to step up the line voltage in order to obtain the 200 to 300 volts of direct current necessary to operate vacuum tubes (a transformer turns ratio of 1:5 or 1:7 is common), 2) to provide the convenient center tap on the secondary so that balanced sine-wave voltages that are 180° out of phase (using the center tap as reference) will be applied to the two rectifier diodes, and 3) to isolate conductively the power-

---

[3]The input to the filter e_{in} need not be sinusoidal in nature for the filter to produce a d-c output. A sufficient condition is that the waveshape be pulsating and unidirectional.

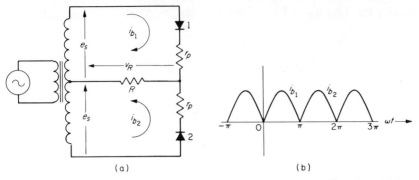

FIG. 15-4.   a) Basic full-wave rectifier circuit. b) Full-wave rectified sinusoidal current wave form i_b.

line supply from the rest of the electronic circuit. For the low-voltage power supplies used in transistor circuits, a transformer is not needed, except that it is still desirable to isolate conductively the power line from the rest of the network. The full wave bridge rectifier employing four diodes is used when a transformer is undesirable or unnecessary.   The reader will recall that the full wave bridge rectifier was used in the design example of Chapter 6 that was concerned with building a strain gauge indicator. It is a little easier to explain how the two-diode-with-transformer full wave rectifier operates and so only this type of rectifier will be considered in this chapter.

As shown in Fig. 15-4b, one of the diodes conducts for 180° and then the other diode conducts for the other 180° of the 360° input cycle, as is indicated by the current pulses i_{b_1} and i_{b_2}.

Assuming that the diodes in Fig. 15-4a are identical and that $e_s = E_m \sin \omega t$, the respective diode currents are

$$\left. \begin{aligned} i_{b_1} &= \frac{E_m}{R + r_p} \sin \omega t \\ i_{b_2} &= 0 \end{aligned} \right\} \quad 0 \le \omega t \le \pi \qquad (15\text{-}7)$$

$$\left. \begin{aligned} i_{b_1} &= 0 \\ i_{b_2} &= -\frac{E_m}{R + r_p} \sin \omega t \end{aligned} \right\} \quad \pi \le \omega t \le 2\pi \qquad (15\text{-}8)$$

for the first cycle. The limits on ωt change in steps of 2π for succeeding cycles.

The reader should carefully consider Fig. 15-4a so as to understand why the current pulses i_{b_1} and i_{b_2} flow through R in the same direction.

Since the full-wave circuit is essentially the superposition of two half-wave circuits, the average or d-c current flowing through R is twice that given by Eq. 15-4, or

$$I_{d\text{-}c} = \frac{2}{\pi} \frac{E_m}{R + r_p} \tag{15-9}$$

and the d-c output voltage $V_{d\text{-}c}$ is

$$V_{d\text{-}c} = RI_{d\text{-}c}$$
$$= \frac{2R}{\pi} \frac{E_m}{R + r_p} \tag{15-10}$$

A Fourier series analysis on the full-wave rectified sine wave (see Appendix D) yields

$$i_b = \frac{E_m}{R + r_p} \left[\frac{2}{\pi} - \frac{4}{\pi} \sum_{n=2,4,6,\ldots}^{n \text{ even}} \frac{\cos n\omega t}{(n+1)(n-1)} \right] \tag{15-11}$$

where i_b is the current flowing through R and is the sum of i_{b_1} and i_{b_2}. As should be expected, the first or d-c term on the right-hand-side of Eq. 15-11 is identical to Eq. 15-9. In fact, the d-c term in Eq. 15-11 is twice the value given for the d-c term in i_b in Eq. 15-6. (The reason that a $\cos \omega t$ term appears in Eq. 15-6 but not in Eq. 15-11 is because of the difference in symmetry between the half- and full-wave rectified sinusoids.)

It is easier to filter the d-c from the a-c current for the full-wave rectified sine wave than for the half-wave case. This can be explained qualitatively by observing that the half-wave case presents a more severely irregular wave than does the full-wave rectified current. The greater ease in filtering the full-wave over the half-wave rectified wave can be quantitatively explained by observing Eqs. 15-11 and 15-6. Equation 15-11 for the full-wave current waveform shows a fundamental frequency component ω that is twice that for the half-wave current given in Eq. 15-6. As will be seen in a later section, the higher frequencies are easier to filter from the direct current than are the lower frequencies. Because the full-wave rectifier makes more efficient use of the input sine wave, and because the rectifier output is easier to filter, almost all power supplies use the full-wave rectifier. Only full-wave rectifiers will be considered in analyzing two of the basic filter circuits—the series-inductor filter and the shunt-capacitor filter. This choice of rectifiers and filters is made because they are both basic and yet relatively easy to analyze. The half-wave rectifier and associated filters are not quite so easy to analyze as in the full-wave case.

15-3. The Series-Inductor Filter

Figure 15-5 shows a full-wave rectifier with an inductor connected in series with the load resistance. One of the fundamental properties of an inductor is that the inductance always tries to resist a change in current flow through the inductor. Because of this inertia-like property of an inductance, one of the two diodes is forced to conduct when a series inductor is used in the output of a full-wave rectifier. In other words, the filter input voltage e_{in} is a pure, full-wave rectified sine wave.

To simplify the argument somewhat, let us consider that the diodes in

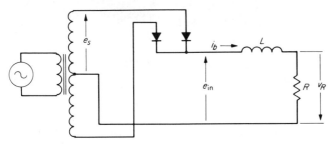

FIG. 15-5. Full-wave rectifier with a series-inductor filter.

Fig. 15-5 are ideal and that the resistance of the inductor is negligibly small in comparison to the series resistance R. The supply voltage e_s can be considered an infinite source, which means that, no matter what the output current is, e_s is always a perfect sine wave. If $e_s = E_m \sin \omega t$, the filter input voltage is a full-wave rectified sinusoidal voltage which can be expressed in a Fourier series as

$$e_{in} = E_m \left[\frac{2}{\pi} - \frac{4}{\pi} \sum_{n=2,4,6,\ldots}^{n\ even} \frac{\cos n\omega t}{(n+1)(n-1)} \right] \qquad (15\text{-}12)$$

Equation 15-12 is quite similar in form to Eq. 15-11. Upon expanding Eq. 15-12,

$$e_{in} = \frac{2E_m}{\pi} - \frac{4E_m}{3\pi} \cos 2\omega t - \frac{4E_m}{15\pi} \cos 4\omega t$$
$$- \frac{4E_m}{35\pi} \cos 6\omega t - \cdots - \frac{4E_m}{(n+1)(n-1)} \cos n\omega t \qquad (15\text{-}13)$$

Since R and L are linear elements, the superposition theorem can be employed and the series current i_b obtained. From conventional a-c circuit analysis, the impedance Z of a series R-L circuit is

$$Z = R + j\omega L \qquad (15\text{-}14)$$

Upon making a term-by-term calculation, the series current i_b becomes

$$i_b = \frac{2E_m}{\pi R} - \frac{4E_m}{3\pi(R + j2\omega L)} \cos 2\omega t - \frac{4E_m}{15\pi(R + j4\omega L)} \cos 4\omega t$$
$$- \frac{4E_m}{35\pi(R + j6\omega L)} \cos 6\omega t - \cdots \qquad (15\text{-}15)$$

Equation 15-15 is not in very good form, since instantaneous and phasor quantities should not be mixed or confused. A better form for Eq. 15-15 is

$$i_b = \frac{2E_m}{\pi R} - \frac{4E_m}{3\pi[R^2 + (2\omega L)^2]^{1/2}} \cos(2\omega t - \phi_2)$$
$$- \frac{4E_m}{15\pi[R^2 + (4\omega L)^2]^{1/2}} \cos(4\omega t - \phi_4) - \cdots \quad (15\text{-}16)$$

where

$$\left.\begin{array}{l} \phi_2 = \tan^{-1}\dfrac{2\omega L}{R} \\[2mm] \phi_4 = \tan^{-1}\dfrac{4\omega L}{R} \end{array}\right\} \quad (15\text{-}17)$$

Usually, the phase angles are of little interest as far as the filtering properties of the inductor are concerned.

The output voltage is defined as

$$v_R = i_b R \quad (15\text{-}18)$$

and, substituting Eq. 15-16 into Eq. 15-18,

$$v_R = \frac{2E_m}{\pi} - \frac{4E_m R}{3\pi[R^2 + (2\omega L)^2]^{1/2}} \cos(2\omega t - \phi_2)$$
$$- \frac{4E_m R}{15\pi[R^2 + (4\omega L)^2]^{1/2}} \cos(4\omega t - \phi_4) - \cdots \quad (15\text{-}19)$$

The rms output voltage $V_{R\,\text{rms}}$ is the square root of the sum of the square of the individual rms voltages given in Eq. 15-19 or

$$(V_{R\,\text{rms}})^2 = \left(\frac{2E_m}{\pi}\right)^2 + \left[\frac{4E_m R}{\sqrt{2}\,3\pi[R^2 + (2\omega L)^2]^{1/2}}\right]^2$$
$$+ \left[\frac{4E_m R}{\sqrt{2}\,15\pi[R^2 + (4\omega L)^2]^{1/2}}\right]^2 + \cdots \quad (15\text{-}20)$$

For a perfect d-c power supply it would be desirable to filter all of the a-c terms from the d-c component of voltage, but this is impossible in a practical filter. The ratio of the second harmonic to the d-c component of output voltage indicates the amount of second-harmonic ripple that is present in the output voltage.

$$\frac{V_2}{V_{\text{d-c}}} = \frac{\dfrac{4E_m R}{\sqrt{2}\,3\pi[R^2 + (2\omega L)^2]^{1/2}}}{\dfrac{2E_m}{\pi}}$$
$$= \frac{2R}{3\sqrt{2}\,[R^2 + (2\omega L)^2]^{1/2}} \quad (15\text{-}21)$$

Similar expressions could be obtained using the fourth, sixth, and so on, harmonics.

15-4. Ripple Factor

To reduce as much as possible the undesirable ripple effects of the harmonics in the case of the series inductor filter, we make $\omega L \gg R$. The a-c components in the output voltage are considered undesirable primarily because of the audible hum they introduce in audio or sound circuits. The overall ripple caused by all of the a-c voltages is defined as follows:

$$\text{Ripple factor} \triangleq \frac{\text{rms value of all a-c components}}{\text{d-c component}} = \frac{V_{\text{a-c}}}{V_{\text{d-c}}} \quad (15\text{-}22)$$

From Eq. 15-20, $V_{\text{d-c}} = 2 E_m / \pi$ and $V_{\text{a-c}}$ is the square root of the sum of the squares of all the a-c terms, or

$$(V_{\text{a-c}})^2 = \left[\frac{4 E_m R}{\sqrt{2}\, 3\pi [R^2 + (2\omega L)^2]^{1/2}} \right]^2$$

$$+ \left[\frac{4 E_m R}{\sqrt{2}\, 15\pi [R^2 + (4\omega L)^2]^{1/2}} \right]^2 + \cdots \quad (15\text{-}23)$$

Compare these a-c and d-c relations with those given by Eq. 6-45.

EXAMPLE 15-1. Calculate the ripple factor of the half-wave and full-wave rectifier output when no filter is used.

Solution: The rms value of the output voltage $V_{R\,\text{rms}}$ is obtained by a straightforward integration. (We assume that $e_s = E_m \sin \omega t$ and that the diodes are ideal.)

Half-wave rectifier:

$$V_{R\,\text{rms}} = \left[\frac{1}{2\pi} \int_0^\pi (E_m \sin \omega t)^2 \, d\omega t \right]^{1/2}$$

and, upon integrating and substituting in limits,

$$V_{R\,\text{rms}} = \frac{E_m}{2} \quad (15\text{-}24)$$

The d-c output voltage is

$$V_{\text{d-c}} = \frac{E_m}{\pi} \quad (15\text{-}25)$$

which is the same as Eq. 15-5 with $r_p = 0$.

Full-wave rectifier:

$$V_{R\,\text{rms}} = \left[\frac{2}{2\pi} \int_0^\pi (E_m \sin \omega t)^2 \, d\omega t \right]^{1/2}$$

$$= \frac{E_m}{\sqrt{2}} \quad (15\text{-}26)$$

The d-c output voltage is

$$V_{\text{d-c}} = \frac{2 E_m}{\pi} \quad (15\text{-}27)$$

which is the same as the first term in Eq. 15-19.

The general expression for the ripple factor, as defined by Eq. 15-22, can be written as

$$\text{Ripple factor} = \frac{V_{a\text{-}c}}{V_{d\text{-}c}} = \frac{\sqrt{V_{R\,\text{rms}}^2 - V_{d\text{-}c}^2}}{V_{d\text{-}c}}$$

$$= \sqrt{\left(\frac{V_{R\,\text{rms}}}{V_{d\text{-}c}}\right)^2 - 1} \qquad (15\text{-}28)$$

Using Eq. 15-28 the half-wave ripple factor is

$$\text{Ripple factor (half wave)} = \sqrt{\left(\frac{E_m/2}{E_m/\pi}\right)^2 - 1} = 1.21$$

The full-wave ripple factor is

$$\text{Ripple factor (full wave)} = \sqrt{\left(\frac{E_m/\sqrt{2}}{2E_m/\pi}\right)^2 - 1} = 0.47$$

As might be expected from observation of the two wave forms, the full-wave rectified voltage has less ripple than does the half-wave.

15-5. The Shunt-Capacitor Filter

The simple shunt-capacitor filter applied to a full-wave rectifier is shown in Fig. 15-6. The behavior and analysis of the shunt-capacitor filter

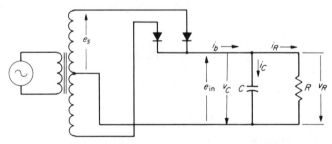

FIG. 15-6.   Full-wave rectifier and shunt-capacitor filter.

differ markedly from those of the series-inductor filter. One reason for this is because the capacitor resists a change in voltage while the inductor opposes a change in current. The voltage input e_{in} to the filter is no longer purely a rectified sine wave as it was for the series-inductor filter. The input voltage e_{in} (and also v_R, since $e_{in} = v_R$) for the shunt-capacitor filter is shown in Fig. 15-7, which will be an aid in presenting a qualitative explanation of the behavior of the shunt-capacitor filter. First, let us assume that the capacitor is initially uncharged, that the diodes are ideal, and that the input voltage e_s is a source of infinite power capacity. At $\omega t = 0$, we apply a sinusoidal input voltage of $e_s = E_m \sin \omega t$. One of the diodes begins to conduct, and the sinusoidal voltage is applied to the parallel *R-C* circuit, with the result that the capacitor charges following the relation $q = Cv$. A short time after the input voltage e_s passes its peak value, the input voltage e_s drops slightly below the capacitor voltage. This point is largely determined by the RC time constant and is denoted θ_a in Fig. 15-7a.

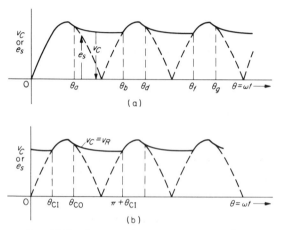

FIG. 15-7. Output voltage v_R versus ωt for a shunt-capacitor filter ($v_R = v_C$).

When this point is reached, the conducting diode is immediately cut off (since the plate voltage is lower than the cathode voltage), and the capacitor continues discharging through the load resistance *R*. The input voltage e_s continues as a pure sinusoidal voltage, as is indicated by the dotted lines in Fig. 15-7. When the voltage on the plate of the other diode just equals the voltage on the capacitor, the second diode begins to conduct, and the pure sinusoidal voltage of e_s appears across the *R-C* circuit (this point is marked θ_b). The capacitor again charges, and at θ_d a separation of the v_C and e_s voltage curves occurs just as it did at point θ_a. It might be rightly reasoned that if *R* and *C* were both large, the output voltage v_R would almost be a straight line (of a purely d-c nature) with a value of E_m.

A quantitative approach to the solution for the output voltage can be obtained by using Figs. 15-6 and 15-7b, ideal diodes, and an infinite source for e_s. When one of the diodes is conducting,

$$i_b = i_C + i_R = C \frac{dv_C}{dt} + \frac{v_R}{R} \tag{15-29}$$

which occurs between the intervals θ_b to θ_d, θ_f to θ_g, and so on. When the diodes are cut off, $i_b = 0$, as occurs between θ_a and θ_b, θ_d and θ_f, and so on. Also, when the diodes are conducting, the source voltage e_s is applied to the parallel *R-C* circuit so that

$$e_s = v_C = v_R = E_m \sin \omega t \qquad (15\text{-}30)$$

Upon substituting Eq. 15-30 into Eq. 15-29,

$$i_b = CE_m\omega \cos \omega t + \frac{E_m}{R} \sin \omega t \qquad \theta_{CI} \leq \omega t \leq \theta_{CO} \qquad (15\text{-}31)$$

where θ_{CI} is the cut-in angle ($\theta_{CI} = \theta_b, \theta_f$, and so on), and θ_{CO} is the cut-off angle ($\theta_{CO} = \theta_d, \theta_g$, and so on).

To find the cutoff angle θ_{CO} in terms of the circuit constants, we substitute $i_b = 0$, $\omega t = \theta_{CO}$ (the definition of cutoff is when $i_b = 0$) into Eq. 15-31, or

$$\omega CE_m \cos \theta_{CO} = -\frac{E_m}{R} \sin \theta_{CO}$$

and, solving for θ_{CO},

$$\theta_{CO} = \tan^{-1}(-\omega RC) \qquad (15\text{-}32)$$

One might ask if i_b is not also equal to zero at $\omega t = \theta_{CI}$. The answer is no—at least not under the assumptions made in this analysis. At the cut-in angle θ_{CI}, one of the diodes suddenly connects a sinusoidally changing source voltage e_s to the capacitor, and, since the current through the capacitor is the time derivative of the voltage, a discontinuous but finite jump in diode current i_b can occur. Between the cutoff angle θ_{CO} and the cut-in angle ($\theta_{CI} + \pi$) (note Fig. 15-7b, which more clearly shows the reason for using the angular relations employed in this analysis), $i_b = 0$, and so Eq. 15-29 becomes

$$0 = i_C + i_R = \frac{C\,dv_C}{dt} + \frac{v_C}{R} \qquad (15\text{-}33)$$

The solution to the first-order differential equation given in Eq. 15-33 is

$$v_C = Ae^{-\frac{1}{RC}t} \qquad (15\text{-}34)$$

The constant A in Eq. 15-34 can be determined by recalling that at $\omega t = \theta_{CO}$, $v_C = E_m \sin \theta_{CO}$, or

$$E_m \sin \theta_{CO} = Ae^{-\frac{1}{RC}\frac{\theta_{CO}}{\omega}}$$

thus

$$A = E_m \sin \theta_{CO}\, e^{\frac{1}{RC}\frac{\theta_{CO}}{\omega}} \qquad (15\text{-}35)$$

Equation 15-34 can now be rewritten, using the value of A given in Eq. 15-35,

$$v_C = E_m \sin \theta_{CO}\, e^{\frac{1}{RC}\frac{\theta_{CO}}{\omega} - \frac{1}{RC}t}$$

$$= E_m \sin \theta_{CO}\, e^{\frac{-(\omega t - \theta_{CO})}{\omega RC}} \qquad \theta_{CO} \leq \omega t \leq (\theta_{CI} + \pi) \qquad (15\text{-}36)$$

Equation 15-36 defines the behavior of the voltage across the capacitor (and hence the output voltage v_R) during the time that the diodes are not conducting. The relationship between θ_{CO} and θ_{CI} can be obtained by evaluating Eq. 15-36 at $\omega t = \theta_{CI} + \pi$; that is, the point at which one of the diodes starts conducting and applies the source voltage e_s to the capacitor-resistance circuit, or

$$E_m \sin \theta_{CI} = E_m \sin \theta_{CO}\, e^{\frac{-(\pi + \theta_{CI} - \theta_{CO})}{\omega RC}}$$

and

$$\sin \theta_{CI} = \sin \theta_{CO}\, e^{\frac{-(\pi + \theta_{CI} - \theta_{CO})}{\omega RC}} \tag{15-37}$$

Equation 15-37 is a transcendental relation that is best solved graphically. First, we use Eq. 15-32 in order to calculate θ_{CO} as a function of ωRC, and then we use Eq. 15-37 to calculate θ_{CI} as a function of ωRC. A plot of θ_{CI} and θ_{CO} versus ωRC for a full-wave rectifier with a shunt-capacitor filter is shown in Fig. 15-8. A similar calculation could be made

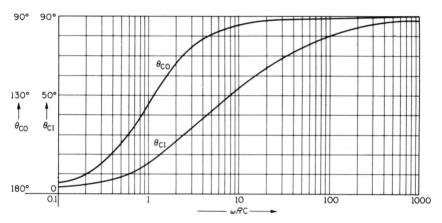

FIG. 15-8. Cut-in θ_{CI} and cutoff θ_{CO} angles versus ωRC for a full-wave rectifier with a shunt-capacitor filter.

for the half-wave rectifier and shunt-capacitor filter, but this will not be presented here.

One important precaution that should be noted can be seen by observing Eq. 15-31. If ωCE_m is made too large, there is a possibility of drawing an excessively large and damaging diode current; thus, for a given supply voltage and frequency, there is a practical limit as to the size of the shunt-filter capacitor.

The d-c output voltage $V_{d\text{-}c}$ can be obtained by integrating Eq. 15-30 and Eq. 15-36 over the proper intervals, or

$$V_{\text{d-c}} = \frac{1}{\pi}\left[\int_{\theta_{CI}}^{\theta_{CO}} E_m \sin \omega t\, d\omega t + \int_{\theta_{CO}}^{\pi+\theta_{CI}} E_m \sin \theta_{CO}\, e^{\frac{-(\omega t - \theta_{CO})}{\omega RC}}\, d\omega t\right]$$

$$= \frac{E_m}{\pi}\left[\cos \theta_{CI} - \cos \theta_{CO} - \omega RC \sin \theta_{CO}\, e^{\frac{\theta_{CO}}{\omega RC}}\left(e^{\frac{-(\pi+\theta_{CI})}{\omega RC}} - e^{\frac{-\theta_{CO}}{\omega RC}}\right)\right]$$

or

$$V_{\text{d-c}} = \frac{E_m}{\pi}\left[\cos \theta_{CI} - \cos \theta_{CO} + \omega RC \sin \theta_{CO}\left(1 - e^{\frac{\theta_{CO}-\theta_{CI}-\pi}{\omega RC}}\right)\right] \quad (15\text{-}38)$$

Likewise, the rms output voltage $V_{R\text{rms}}$ is calculated by

$$V_{R\text{rms}} = \left\{\frac{1}{\pi}\left[\int_{\theta_{CI}}^{\theta_{CO}} (E_m \sin \omega t)^2\, d\omega t \right.\right.$$

$$\left.\left. + \int_{\theta_{CO}}^{\theta_{CI}+\pi}\left(E_m \sin \theta_{CO}\, e^{-\frac{(\omega t - \theta_{CO})}{\omega RC}}\right)^2 d\omega t\right]\right\}^{1/2}$$

and, upon integrating and substituting in limits,

$$V_{R\text{rms}} = \frac{E_m}{\sqrt{2\pi}}\left[\theta_{CO} - \theta_{CI} - \frac{1}{2}(\sin 2\theta_{CO} - \sin 2\theta_{CI})\right.$$

$$\left. - \sin^2 \theta_{CO}\,\omega RC\left(e^{\frac{2(\theta_{CO}-\theta_{CI}-\pi)}{\omega RC}} - 1\right)\right]^{1/2} \quad (15\text{-}39)$$

Equations 15-38 and 15-39 could be simplified further by using various trigonometric identities, but this will not be done here.

EXAMPLE 15-2. Calculate the d-c output voltage and ripple factor for the shunt-capacitor filter shown in Fig. 15-6 if $R = 2000$ ohms, $C = 10$ μf, and $e_s = 350 \sin 2\pi 60t$ volts. Assume ideal diodes and a source of infinite power capacity for e_s.

Solution: First, we calculate ωRC,

$$\omega RC = 2\pi \times 60 \times 2000 \times 10 \times 10^{-6} = 7.54$$

Using Fig. 15-8, we read θ_{CI} and θ_{CO} at $\omega RC = 7.54$, or $\theta_{CI} = 48°$ and $\theta_{CO} = 97°$. Next, we calculate $V_{\text{d-c}}$, using Eq. 15-38,

$$V_{\text{d-c}} = \frac{350}{\pi}\left[\cos 48° - \cos 97° + 7.54 \sin 97°\left(1 - e^{\frac{\frac{97-48}{57.3}-\pi}{7.54}}\right)\right]$$

$$= \frac{350}{\pi}[0.67 + 0.122 + 7.54 \times 0.99(1 - 0.738)]$$

$$= 306 \text{ volts}$$

We calculate $V_{R\text{rms}}$, using Eq. 15-39,

$$V_{R\text{rms}} = \frac{350}{\sqrt{2\pi}}\left[\frac{97-48}{57.3} - \frac{1}{2}(\sin 194° - \sin 96°) - \sin^2 97°\,(7.54)(e^{-0.606} - 1)\right]^{1/2}$$

$$= \frac{350}{\sqrt{2\pi}} \left[0.855 - \frac{1}{2}(-0.242 - 0.99) - (0.99)^2 \, 7.54(0.545 - 1) \right]^{1/2}$$

$$= 308 \text{ volts}$$

The ripple factor can now be calculated, using Eq. 15-28,

$$\text{Ripple factor} = \sqrt{\left(\frac{308}{306}\right)^2 - 1}$$

$$= 0.106$$

which is a marked improvement over the unfiltered full-wave rectified sine wave.

Problems

15-1. A half-wave diode rectifier circuit similar to Fig. 15-1 uses a diode with a forward dynamic resistance r_p of 40 ohms. If $e_s = 280 \sin \omega t$ and the load resistance $R = 4000$ ohms, a) what is the peak value of the diode current? b) What is the average value of the diode current? c) What is the d-c voltage drop across R?

15-2. Calculate the diode power loss in Prob. 15-1. [*Hint:*

$$P = I_{rms}^2 \, r_p \quad \text{and} \quad I_{rms} = \left(\frac{1}{2\pi} \int_0^{2\pi} i_b^2 \, d\omega t \right)^{1/2}$$

where I_{rms} is the total rms value of the output current.]

15-3. A full-wave diode rectifier similar to Fig. 15-4a uses two vacuum tube diodes with forward dynamic resistances r_p of 500 ohms. If the load resistance R is 2500 ohms and $e_s = 350 \sin \omega t$, a) what is the peak current value passing through one of the diodes? b) What is the average value of one of the diode currents? c) What is the average value of the current through R? d) What is the d-c voltage drop across R?

15-4. Calculate the diode power loss in Prob. 15-3. Note the hint in Prob. 15-2.

15-5. A full-wave rectifier with a series-inductor filter, as shown in Fig. 15-5, has the following circuit parameters: $L = 5$ henries, $R = 4000$ ohms, and $e_s = 500 \sin 2\pi \, 60t$ volts. If the diodes are considered ideal, a) what is the d-c output voltage V_{d-c} across R? b) What is the rms output voltage $V_{R_{rms}}$ across R? c) What is the a-c output voltage V_{a-c} across R? d) What is the ripple factor? e) Calculate the power loss in R. Consider harmonics up to and including the fourth.

15-6. Repeat Prob. 15-5 if L is changed from 5 to 10 henries. Comment on the result of increasing L.

15-7. Repeat Prob. 15-5 if the diodes are not ideal and $r_p = 500$ ohms.

15-8. Plot V_{d-c}/E_m versus $\log \omega RC$ from Eq. 15-38.

15-9. Check the correctness of Fig. 15-8 at $\omega RC = 1, 4,$ and 20 by using Eqs. 15-32 and 15-37.

15-10. A full-wave rectifier with a shunt-capacitor filter has the following parameters: $C = 10\ \mu f$, $R = 5000$ ohms, $f = 60$ cps, and $E_s = 300$ volts. a) Find the cut-in and cut-out angles. b) Find the d-c output voltage. c) Find the ripple factor.

15-11. Determine the minimum wattage rating of R in Prob. 15-10.

15-12. A full-wave rectifier uses a shunt-capacitor filter as shown in Fig. 15-6. If $R = 5000$ ohms, $C = 0.438\ \mu f$, and $e_s = 350 \sin 2\pi\ 400t$ (the diodes can be considered ideal), a) find the d-c output voltage. b) Find the ripple factor. c) Find the power loss in R.

15-13. Repeat Prob. 15-12 if R is reduced to 1000 ohms.

15-14. A full-wave rectifier using a shunt-capacitor filter has a cut-in angle of 50° as measured on an oscilloscope. If $e_s = 380 \sin 2500t$ and $R = 4000$ ohms, a) What is the value of C? b) What is the d-c current flowing through R?

16 / *Photosensitive Devices*

Certain materials release free electrons when they are bombarded by photons of light.[1] This effect was first observed by Hertz in 1887 and is known as the photoelectric effect. The photoelectric effect now includes 1) photoemission, 2) the photovoltaic effect, and 3) the photoconductive effect. Photoemission occurs when a photon of light strikes a surface and releases free electrons, that is, the principle by which the common photoelectric cell works. The photovoltaic effect occurs when a material converts the energy of the impinging light photons into electrical energy by generating a small voltage. This is the principle of operation of the solar cells that are used as power supplies for many of the satellites and other exploratory space vehicles. When the resistance of a material decreases upon being irradiated by light photons, the material is said to be photoconductive.

All three of the photoelectric effects are important and will be considered in this chapter.

16-1. Terminology

At the lower radio and audio frequencies, we usually refer to the frequency of a given signal in cycles per second, or Hertz (Hz). At the higher infrared and optical frequencies it is somewhat more convenient to talk about the wavelength of a signal or photon. The relationship between wavelength and frequency is

$$\lambda = \frac{v}{f} \qquad (16\text{-}1)$$

where v is the velocity of the photon electromagnetic (EM) energy in meters per second, λ is the wavelength in meters, and f is the frequency in cycles per second. In free space, $v = c = 3 \times 10^8$ m/sec. The earth's atmosphere is almost "free space" as far as the velocity of EM wave propagation is concerned.

The visible light spectrum covers a frequency range from 4.3×10^{14} to 7.5×10^{14} Hz. Using Eq. 16-1, these two frequencies correspond to a

---

[1]The term *light* does not simply refer to photons of radiant electromagnetic energy that are visible as measured by the human eye. The term generally includes the infrared and ultraviolet frequencies. (See Chapter 4 for a discussion of the frequency spectrum.)

wavelength of 0.7 micron or 7000 angstroms to 0.04 micron or 400 angstroms, respectively. By definition, a micron and an angstrom unit are

$$\text{Micron} = \mu = 10^{-6} \text{ m}$$
$$\text{Angstrom} = \overset{\circ}{A} = 10^{-10} \text{ m}$$

Some of the terms and concepts employed in photometric analysis are rather specialized and need to be reviewed before proceeding to a more detailed discussion of photoelectric devices.

A *candle* is a special type of lamp (usually an incandescent bulb) operated under carefully regulated conditions of temperature, current, and environment. A standard candle emits 4π lumens of luminous or visible flux. By definition, luminous flux is the rate of flow of *visible* radiant energy (photons per second) from a source and has the dimensions of power.

A *lumen* is the amount of luminous flux emitted per unit solid angle from a uniform (usually considered spherical) source of one standard candle.

A *foot-candle* is the amount of illumination falling on 1 sq ft of a spherical surface located at a radius of 1 ft from the standard candle. Since the inner area of a sphere 1 ft in radius is 4π sq ft, and since there are 4π unit solid angles surrounding a point, a foot-candle equals a lumen per square foot, or

$$\text{Foot-candle} = \frac{\text{lumens}}{\text{area in square feet}} \quad \text{or} \quad F = \frac{\Lambda}{A} \quad (16\text{-}2)$$

Sources of visible radiant energy are quite often rated in terms of candlepower. If a source can be considered to radiate energy uniformly in all directions and can also be considered a point source, then the candlepower rating of the source is

$$C = r^2 \frac{\Lambda}{A} \quad (16\text{-}3)$$

where C is the candlepower, r is the radius of any sphere of interest, and Λ is the illumination in lumens falling on an area A of the inner surface of the chosen sphere.

All of the usual light sources such as incandescent lamps, fluorescent lamps, arc lights, gaslights, and so forth, emit photons of visible light over a range or spectrum of frequencies. Only the recently developed laser light sources (see Chapter 17) emit photons at a single frequency. To find the overall visible radiant energy in a light source, it is necessary to integrate the number of photons at each individual frequency over the visible frequency range. The human eye is often used as the "integrator" to obtain the candlepower of an uncalibrated source. This is accomplished by visually comparing the output of a standard candle with the output of the uncalibrated source.

16-2. Photoemissive Cells

As shown in Fig. 16-1, the photoemissive cell (sometimes called the photoelectric tube) consists of two electrodes—a small-diameter wire used

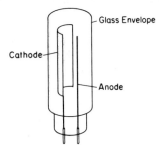

FIG. 16-1. The photoemissive-
tube construction.

for the anode or plate and a cylindrically shaped cathode constructed of a thin sheet of metal coated with a special photoemissive material. There are two general types of photocells—the vacuum tube and the gas tube. As will be seen later, the gas tube is more sensitive than the vacuum tube; that is, more electrons per lumen are released by the gas tube.

Vacuum Cells. The explanation of the operation of the vacuum photocell is quite simple. Figure 16-2a shows a photocell connected to a

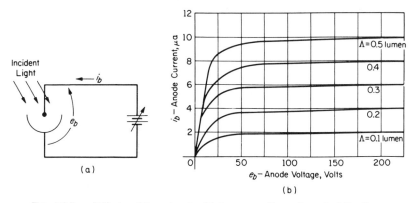

FIG. 16-2. a) Photocell in a circuit. b) Ampere-voltage characteristics for a
type 922 vacuum photocell.

variable-source voltage, and Fig. 16-2b shows the characteristic curves of the vacuum photocell, that is, i_b versus e_b with the light intensity in lumens as a third parameter. Although the photocell is a two-element tube, it is evident, from inspection of Fig. 16-2b, that the light source causes an i_b versus e_b behavior much like transistors or the grid-controlled vacuum-tube pentode.

The photoelectric effect was discussed in Chapter 4, but it is well to repeat some of the discussion here. Einstein's photoemission equation

$$hf = hf_w + \tfrac{1}{2}mv^2 \tag{16-4}$$

relates the energy of an incident photon hf both to the work function energy hf_w of the material on which the photon impinges and to the maximum kinetic energy $\tfrac{1}{2}mv^2$ of any released electrons. Only certain materials in the alkali-metal group of lithium, potassium, sodium, cesium, and rubidium have low enough work functions to allow photoemission by visible light photons. One of the most common materials used in the cathode construction of a photocell is a silver surface coated with a monatomic layer of cesium over a thin layer of cesium oxide. When visible light photons fall on the cathode of the photocell, electrons are emitted in direct proportion to the light intensity or the number of photons that are bombarding the photoemissive surface. If a positive potential is applied to the photocell anode, the released electrons will flow to the anode, thus producing a current flow in the external circuit. As the ordinate scale of Fig. 16-2b indicates, the current flow is only a few microamperes for the vacuum photocell. Because the number of released electrons is so small, there is a negligible space charge around the cathode, and the tube quickly saturates; that is, all of the released electrons are attracted to the plate. (This saturation effect is indicated by the straight, almost horizontal lines of i_b versus e_b for values of e_b above the knee of the curve.)

Circuit Application: Figure 16-3a shows a photocell used in a basic circuit configuration. The analysis of circuits using photocells is identical to the methods used in tube or transistor circuits; namely, by the graphical or equivalent-circuit techniques. To employ graphical methods to solve for the quiescent currents and voltages (for a given quiescent or biasing value for Λ), we simply construct a d-c load line on the photocell characteristics. Two such load lines for $R_l = 5$ and 20 megohms are shown on Fig. 16-3b.

EXAMPLE 16-1. A type 922 vacuum photocell is used in the circuit of Fig. 16-3a. If $\Lambda = 0.2 + 0.1 \sin \omega t$ lumen and $R_l = 20$ megohms, find e_{out}, the instantaneous a-c output voltage, and i_p the instantaneous a-c plate current.

Solution: First, we draw the d-c load line on the photocell characteristics, as indicated in Fig. 16-3b. Next, we find the Q point corresponding to a value of 0.2 lumen. As shown in Fig. 16-3b, the Q point is located at $e_b = 122$ volts, $i_b = 3.9 \ \mu a$. Next, we let Λ vary from a maximum value of 0.3 to a minimum value of 0.1 lumen. Remaining on the a-c (which is the same as the d-c, in this case) load line, we read $e_b = 86$, $i_b = 5.7 \ \mu2$ at $\Lambda = 0.3$ and $e_b = 160$, $i_b = 2 \ \mu a$ at $\Lambda = 0.1$

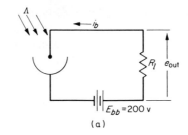

(a)

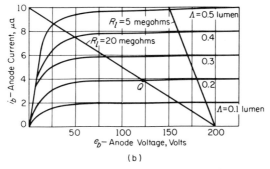

(b)

FIG. 16-3. a) Basic photocell circuit. b) Photocell characteristics for a type 922 vacuum
photocell.

lumen. The instantaneous output voltage is

$$e_{out} = \frac{-160 - 86}{2} \sin \omega t$$

$$= -37 \sin \omega t \text{ volts}$$

where the negative sign shows that the output voltage and lumen input are 180°
out of phase, just as they were for the grid and plate voltages in the vacuum tube.
It should be obvious that the plate current and lumen input are in phase, since the
number of released electrons depends directly on the lumen intensity of the incident photons, or

$$i_p = \frac{5.7 - 2}{2} \sin \omega t \ \mu a$$

$$= 1.85 \sin \omega t \ \mu a$$

Actually, there is a small lag of about 10^{-9} sec between the time when the
photon strikes the cathode and the time when an electron is released, but,
for frequencies under 100 KHz, this time lag causes negligible phase shift
between Λ and i_b.

If the photocell is operated within the linear region of its characteristics, an equivalent circuit for the photocell can be developed. Because the
i_b versus e_b curves are almost straight, horizontal lines in the linear region

of operation, a very simple a-c equivalent circuit in the form of a current generator can be used to replace the physical photocell. By choosing any Q point in the linear region of operation, we find that the a-c plate impedance r_p is so large (usually a few gigaohms) that it can be neglected if a current generator is used in the equivalent circuit. The choice of a current generator means that the a-c plate impedance will be connected in parallel with the current generator, and, since the load resistance R_l is usually much smaller than the a-c plate impedance r_p, the parallel equivalent of the two impedances is almost the same as R_l. Figure 16-4 shows the sim-

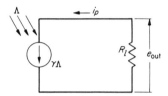

FIG. 16-4. A-c equivalent for the photocell circuit of Fig. 16-3a.

plified a-c equivalent circuit for the photocell circuit of Fig. 16-3a, where the photo amplification factor γ is measured in microamperes per lumen and is found in the same way g_m was found for the vacuum, tube, or

$$\gamma = \frac{\partial i_b}{\partial \Lambda} = \frac{\Delta i_b}{\Delta \Lambda} \bigg|_{e_b = \text{a constant}} \tag{16-5}$$

EXAMPLE 16-2.   Solve the problem given in Example 16-1, using equivalent-circuit methods.

Solution: Using the Q point as found in Example 16-1, we can determine γ from Eq. 16-5 and Fig. 16:3b. First, we draw a vertical line indicating a constant e_b that passes through the Q point. For convenience, we move to the $\Lambda = 0.1$ and $\Lambda = 0.3$ lumen, and read the corresponding values of i_b, and substitute into Eq. 16-5, or

$$\gamma = \frac{5.8 - 2}{0.3 - 0.1} = 19\ \mu a/\text{lumen}$$

Next, we use Fig. 16-4 and calculate i_p from

$$i_p = \gamma \Lambda_{\text{a-c}}$$

where $\Lambda_{\text{a-c}}$ is the a-c portion of Λ. Thus

$$i_p = 19 \times 0.1 \sin \omega t\ \mu a$$
$$= 1.9 \sin \omega t\ \mu a$$

Also using Fig. 16-4, the output voltage is defined as

$$e_{out} = -i_p R_l$$
$$= -(1.9 \times 10^{-6} \sin \omega t)(20 \times 10^6)$$
$$= -38 \sin \omega t \text{ volts}$$

The values for i_p and e_{out} in Example 16-2 check closely with those obtained by graphical means in Example 16-1. The primary reason that the two sets of answers differ slightly lies in the omission of the a-c plate impedance in Fig. 16-4.

Gas Cells. The photo amplification factor γ can be materially increased if a small amount of inert gas, such as argon, is enclosed within the tube envelope. Figure 16-5 shows the i_b versus e_b characteristics for a

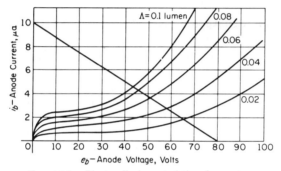

Fig. 16-5. Photocell characteristics for a type 930 gas-filled tube.

type 930 gas-filled photocell. The portion of the curve from zero to 25 volts is similar to the vacuum photocell; that is, from 0 to 5 volts a very small space charge exists, and then from 5 to 25 volts the photocell operates in its saturation region. After the anode voltage passes 25 volts or so, the released cathode electrons acquire enough energy on their way to the anode to ionize the inert gas atoms should a collision occur between the atoms and the electrons. The ionization produces additional free electrons, thus causing a net increase in the current flow. Unfortunately, the gas-photocell characteristics are quire nonlinear in nature and so cannot always be used in applications requiring a linear relation between lumen input and current output.

The safe anode voltage for most gas photocells should not exceed 100 volts and should be reduced for high-lumen intensities. The reason for this precaution is because high-speed positive gas ions are heavy enough to damage the thin monatomic film of cesium, and the velocity or energy of the ions is proportional to the applied cathode-anode voltage.

Because of the nonlinear or nonuniform nature of the gas-photocell characteristics, graphical but not equivalent-circuit techniques should be employed on circuits using gas photocells.

EXAMPLE 16-3.   A type 930 gas photocell is used in the circuit of Fig. 16-3a. a) Find the quiescent values of anode current and voltage if E_{bb} = 80 volts, R_l = 8 megohms, and Λ = 0.08 lumen. b) Find the power in the 8-megohm resistance.

Solution: We draw a d-c load line with an 8-megohm slope, as shown in Fig. 16-5. The Q-point voltage and current are

$$I_b = 4.15 \,\mu\text{a}$$
$$E_b = 47 \text{ volts}$$

$$\text{Power in 8 megohm resistance} = I_b^2 R = (4.15 \times 10^{-6})^2 \times 8 \times 10^6$$
$$= 0.1376 \text{ mwatts}$$

Photoemissive-Cell Applications.   Photoemissive cells can be used in a variety of ways, such as in burglar alarms, automatic control of doors or windows, control of street lighting, control of room illumination, and so forth. However, in almost every case the low energy level of the output of the photoemissive cell must be amplified before it can be put to use. This amplification can be obtained by employing the photocell in the grid circuit of a vacuum-tube amplifier, as shown in Fig. 16-6a.

Let us assume that the purpose of this circuit is to aid in controlling the level of illumination in a room that is subject to the fluctuating outside illumination of the sunshine. Before proceeding with a specific example, let us first consider the general behavior of the circuit of Fig. 16-6a. If the room is completely dark, then the relay should be closed in order for the room lights to come on. Since the relay is normally closed, E_{bb} and R_k should be adjusted so that the triode does not pass enough current i_{b_1} to cause the relay to be energized and thus open. As soon as the outside source of light becomes sufficient to illuminate the room without the artificial sources, the relay should open and turn off the lamps. The photocell causes the relay to open by decreasing the negative bias on the vacuum triode. Since the biasing voltage e_c is

$$e_c = i_{b_2} R_{l_2} - i_{b_1} R_k \tag{16-6}$$

as the photocell current increases with increasing room illumination, the grid bias becomes less negative and the triode conducts more current. This causes the relay to open, thereby turning off the room lights. Several such circuits could be used to give very uniform control over the level of room illumination by arranging the bias voltages so that only certain portions of the room lights would be turned on or off as the outside source of illumination varied (owing to causes such as change in cloud cover or during the early morning or evening hours).

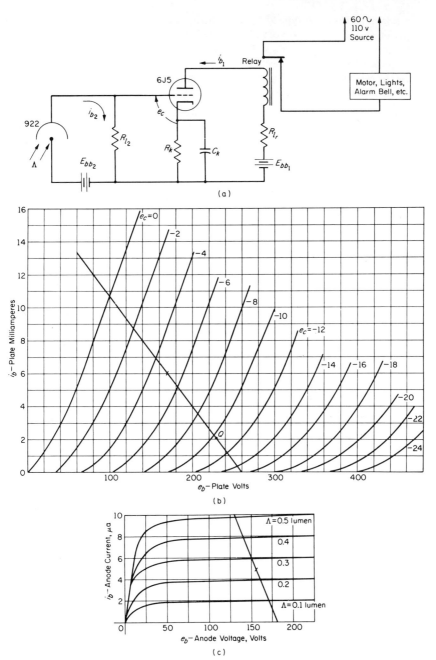

FIG. 16-6. a) A photoemissive tube used in conjunction with a grounded-cathode triode amplifier to control a relay circuit. b) Plate characteristics for the 6J5 triode. c) Plate characteristics for the 922 photocell.

EXAMPLE 16-4.   A type 922 photocell and 6J5 triode are used in the circuit of Fig. 16-6a. If E_{bb_1} = 260 volts, E_{bb_2} = 180 volts, the relay coil resistance R_{l_r} = 10,000 ohms, R_{l_2} = 5 megohms, R_k = 5000 Ω, and the relay drop-out and pickup currents are 3 and 6 ma, respectively, find the level of room illumination that will cause the relay to a) pick up or open and b) drop out or close. Let us assume that the area of the cathode of the photocell is 0.5 sq in.

Solution: Figures 16-6b and c show the respective plate characteristics for the 6J5 triode and the 922 photocell, repeated here for convenience. First, we draw the load lines on the two sets of characteristics, as indicated in parts b and c, noting that the d-c load line for the triode is $R_k + R_{l/r}$ = 5 + 10 = 15 K. Next, we assume that the room is dark, and we check to see if there is sufficient bias voltage to cause the relay to drop out; that is, the tube current must be less than 3 ma. By trial and error or by drawing a grid-bias voltage curve, we can determine that the quiescent triode plate current is about 2.1 ma at a quiescent grid voltage E_c = −10.6 volts. As a check,

$$-(R_k I_b) = E_c$$

or

$$-(5000 \times 2.1 \times 10^{-3}) = -10.6$$
$$-10.5 = -10.6$$

The relay current must increase to 6 ma in order to cause the relay to pick up. According to Fig. 16-6b, the grid voltage corresponding to this plate current is −5 volts. This means that, to close the relay, the illumination level in the room must increase so that Eq. 16-6 is satisfied at e_c = −5 volts, or

$$-5 = i_{b_2} \times 5 \times 10^6 - 6 \times 10^{-3} \times 5000$$
$$= i_{b_2} \times 5 \times 10^6 - 30$$

or

$$i_{b_2} = \frac{25}{5 \times 10^6}$$
$$= 5\,\mu a$$

Figure 16-6c indicates that when 0.25 lumen falls on the photocell, a photocell current of 5 μa is produced.

To determine the outside lumen illumination level that will produce a drop-out relay current of 3 ma, we read the value of e_c = −9 at i_{b_1} = 3 ma, and then substitute into Eq. 16-6 again to find the necessary i_{b_2} to satisfy Eq. 16-6, or

$$-9 = i_{b_2} \times 5 \times 10^6 - 3 \times 10^{-3} \times 5000$$
$$= i_{b_2} \times 5 \times 10^6 - 15$$
$$i_{b_2} = 6/(5 \times 10^6) = 1.2\,\mu a$$

Figure 16-6 shows that 0.06 lumen is necessary to produce a photocell current of 1.2 μa. Thus the pickup and drop-out illumination levels are

$$F = \frac{\Lambda}{A}\,(\text{pickup}) = \frac{0.25}{0.5} \times 144 = 72 \text{ foot-candles}$$

$$F = \frac{\Lambda}{A}\,(\text{drop out}) = \frac{0.06}{0.5} \times 144 = 17.3 \text{ foot-candles}$$

where foot-candles define the level of room illumination better than do lumens.

16-3. Photovoltaic Cells

The photovoltaic cell directly converts light energy into electrical energy. These cells are used in photographic exposure meters and other applications such as the conversion of the sun's inexhaustible light energy into electrical energy for use in space vehicles.

The physical construction of the photovoltaic cell consists of either a specially prepared semiconductor PN junction or a layer of semiconductor bonded to a metal base. The principle of operation is the same in either case. As pointed out in Chapter 5, an electric field is created across a PN junction because of the diffusion of the holes into the N and the electrons into the P regions of the diode. The same is true of the semiconductor and the metal. When light photons fall on either the PN junction (the diode must be constructed so that the junction can be illuminated) or the semiconductor-metal junction, hole-electron pairs are created. The released electrons are then moved across the junction by the action of the previously mentioned electric field. If an external resistance is placed across the two electrodes that are attached to the junction materials, a complete circuit is formed and current flows.

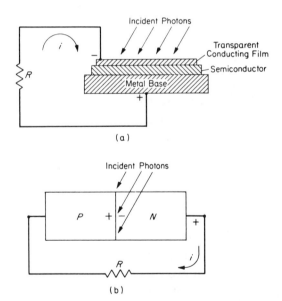

Fig. 16-7. Photovoltaic-cell construction a) using a metal base and a semiconductor junction, and b) using a germanium or silicon PN junction.

Figure 16-7a shows a photovoltaic cell constructed of either selenium on iron or copper oxide on copper. A thin, transparent, conductive film covers the semiconductor materials and has two functions: 1) to allow the light photons to pass through and strike the semiconductor and 2) to form an electrode so that current can flow into the junction. Figure 16-7b shows a simple germanium or silicon PN junction used as a photovoltaic cell.

Figure 16-8 shows a typical output characteristic for a photovoltaic cell, with the series resistance R as a parameter.

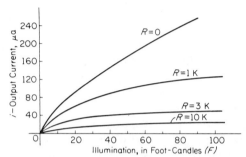

FIG. 16-8. Output current versus foot-candle illumination for a photovoltaic cell with various load resistances.

EXAMPLE 16-5. If the illumination level in a room is known to be 70 foot-candles, find the value of R needed in Fig. 16-7a or b to produce a full-scale deflection on a microammeter movement rated at 100 μa maximum. The area of the photovoltaic cell is known to be 1.2 sq in.

Solution: Assuming that Fig. 16-8 defines the output characteristic of the photovoltaic cell, we find the intersection of $F = 70$ and $I = 100 \, \mu$a, or

$$R = 1200 \, \text{ohms}$$

The microammeter scale can now be calibrated in terms of foot-candles (or lumens, since the area of the photovoltaic cell is known), and the device can be used as an illumination meter.

16-4. Photoconductive Cells

The earlier photoconductive cells usually employed a thin film of selenium sandwiched between two conductive electrodes. When light photons strike the selenium semiconductor, some of the filled-band electrons are given sufficient energy to become conduction-band electrons. (See Chapter 4 for a discussion of the energy levels in semiconductor materials.) This effectively reduces the resistance of the semiconductor.

By placing the photoconductive cell in a simple series circuit consisting of the cell, a battery, and a microammeter, a light detector can be constructed. Since the difference in energy between the filled and the conduction bands (that is, the forbidden-band energy) is only 0.5 ev or so, the photoconductive cell can be used to detect infrared photons.

A phototransistor or photodiode is another type of photoconductive cell. This type of diode is identical to the diode used for the photovoltaic cell but the application is different. Figure 16-9a shows a PN-junction diode, especially constructed so that the junction can be exposed to light photons, placed in a series circuit consisting of the diode, a resistance, and a back-biasing battery. The incident light photons create hole-electron pairs, causing a current flow that is almost directly proportional to the light intensity. Figure 16-9b shows the typical characteristics of a photodiode.

If a transistor is used in place of the diode, the photoconductive cell is called a phototransistor. Because of the inherent gain expected from the transistor circuit, the phototransistor is more sensitive than the photodiode.

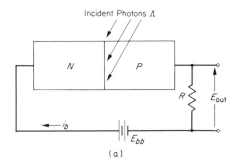

(a)

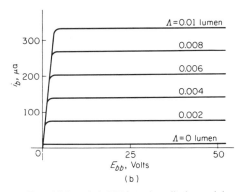

(b)

FIG. 16-9. a) A PN-junction diode used in a photodiode. b) Typical characteristics for a photodiode.

Problems

16-1. A certain photocell surface has a work function or threshold wavelength of 6400 Å. If the surface is bombarded with green light at a wavelength of 5250 Å, what will be the maximum velocity of the emitted electrons?

16-2. Plot i_b versus lumen illumination for a type 922 vacuum photocell for load resistances R_l of 5 and 20 megohms (see Fig. 16-3b). Designate the linear region of operation for the photocell.

16-3. Determine the dynamic plate resistance r_p of the 922 photocell at a Q point of E_b = 100 volts, Λ = 0.2 lumen. Is the value of r_p large enough to be neglected in the circuit of Fig. 16-4?

16-4. A 60-candlepower point source is situated 8 ft from a photocell whose photo amplification factor γ is 20. If the circular area of the photocell is 0.8 sq in., calculate the focal length of a lens 2 in. in diameter which must be placed between the point source and the photocell in order to produce a photocell output of 3 μa.

16-5. A type 930 gas-filled phototube is connected in series with a 90-volt battery and a 15-megohm resistance. If the phototube is 1.5 ft away from a point light source whose candlepower is C = 26 + 6.5 sin ωt, find the peak-to-peak output voltage across the 15-megohm resistance. The cathode area of the photocell is 0.4 sq in.

16-6. A 50-candlepower point source is used to illuminate a type 930 gas-filled photocell. If the photocell is connected in series with a 90-volt d-c source and a 10-megohm resistance, and if the current through the resistance is to be 3 μa, how far away is the photocell from the light source? The area of the photocell cathode is 0.4 sq in.

16-7. Using the circuit of Fig. 16-6a and other necessary curves, and if E_{bb_1} = 200 volts, E_{bb_2} = 150 volts, R_{l_1} = 19 KΩ, and R_k = 1 KΩ, find the foot-candle illumination that will cause a relay pickup current of 5 ma if a) R_{l_2} = 2 megohms and b) R_{l_2} = 10 megohms.

16-8. The resistance in the photovoltaic-cell circuit shown in Fig. 16-7a is 1000 ohms. a) What is the lumen intensity of the light falling on the cell surface if 0.05 volt is produced across R and the area of the active portion of the cell is 2.2 sq in.? b) What is the voltage output of the photovoltaic cell if 90 foot-candles of illumination fall on the cathode surface of the cell?

16-9. If 40 foot-candles of luminous flux fall on the photovoltaic cell shown in Fig. 16-7a, determine the value of resistance R that will produce the maximum cell power output. Also, what is the maximum power output? (*Hint:* Plot $i_b^2 R$ versus R for F = a constant.)

16-10. Repeat Prob. 16-9 for F = 70 foot-candles.

16-11. A photoconductive PN junction, whose characteristics are the same as those given in Fig. 16-9b, is used in a circuit similar to Fig. 16-9a. If E_{bb} = 50 volts and R = 250 KΩ, a) what is the voltage across the resistance if 3 millilumens fall on the PN junction? b) What is the rms value of the a-c component of the output voltage across the resistance if Λ = 3 + 2 sin ωt millilumens?

16-12. Using the photoconductive PN-junction characteristics of Fig. 16-9b, a) determine the dynamic collector resistance r_c (similar to r_p for the phototube). b) Determine the photo amplification factor γ. Use a Q point of E_h = 30 volts, I_b = 200 μa to make these calculations.

17 / *Radio Propagation and Modulation*

The phenomena of EM (electromagnetic) wave propagation and associated concepts, such as antenna theory, comprise one of the most highly developed scientific areas studied today. Many great men, such as Faraday, Gauss, Maxwell, Helmholtz, Hertz, Lorentz, Einstein, and so on, have contributed much to the development of EM field theory, but Maxwell's four equations and the Lorentz equation for the force on a moving charge are thought to describe fully the basic laws of electromagnetism.

Rapid communication between countries on this planet (and eventually between planets themselves) is a very necessary part of today's living. According to present-day theory and practice, the velocity of light is considered to be the maximum speed with which information can be conveyed from one point to another. All EM frequencies propagate through free space at the speed of light (about 3×10^8 m/sec) and so could be used as carriers of information. The familiar AM broadcast frequencies cover a spectrum from 500 to 1500 KHz, TV frequencies extend from 54 to 88 MHz and 174 to 216 MHz for the first 12 channels (these are the very-high-frequency or vhf channels) and from 470 to 890 MHz for channels 14 through 83 (the newer ultrahigh-frequency or uhf channels). Similar bands of frequencies are allocated by the Federal Communications Commission (FCC) for other uses, such as bands for the short waves, amateur radio (ham) operators, police, FM radio, the military, microwaves, and so on.

Although it may not yet be obvious why so many frequencies are necessary for the various bands, suffice it to say that different communication frequencies are needed to avoid objectionable interference with other channels of communication. (The reader might stop to consider that there are literally hundreds of different EM signals passing through his body at any time, and, if he had the proper receiver, he could listen to the information contained in these signals.)

Until the recent discovery of the maser-type device, there was real concern that there would not be a sufficient range of frequencies to meet future communication requirements. As will be discussed in a later section, the maser-type device extends the usable radio frequency (r-f) range even beyond the visible light frequencies. (The higher frequency masers are called lasers.)

17-1. Characteristic of EM Waves

Electromagnetic radiations of all types, such as X-rays, light waves, microwaves, radio waves, and so forth, travel at a velocity of 3×10^8 m/sec in free space and can be refracted, reflected, and diffracted. Figure 17-1 shows that an EM wave can be mathematically described in terms

Fig. 17-1. Representation of an EM wave, consisting of an electric field ε and magnetic field **H** vector.

of sinusoidally time-varying electric ε and magnetic **H** field-intensity vectors which move in the direction of $\varepsilon \times$ **H**. A plane containing the vectors ε and **H** is called the *wave front*.

The velocity of propagation of an EM wave is governed by the relation

$$\mathbf{v} = \frac{1}{\sqrt{\mu \epsilon}} \tag{17-1}$$

where μ is the permeability, in henries per meter, and ϵ is the permittivity or dielectric constant in farads per meter, of the material through which the wave is traveling. In free space, in the rmks system of units, $\mu = 4\pi \times 10^{-7}$ henries/m and $\epsilon = (1/36\pi) \times 10^{-9}$ farads/m. Substituting these values into Eq. 17-1, we obtain $\mathbf{v} = 3 \times 10^8$ m/sec.

Well-known and verified laws govern the propagation of EM waves through, and reflection or refraction from, boundaries of different types of material through which a wave is traveling. In general, if μ and ϵ for two materials differ greatly, the EM wave will be greatly disturbed in passing from one material into the other. As an extreme example, an EM wave coming from free space will be almost totally reflected upon striking a good conductor such as a sheet of metal (this occurs because the metal offers a short circuit to the electric field ε).

17-2. Terrestrial Propagation

The theory of radio propagation through the unbounded medium of free space is well established and relatively easy to understand. The application of EM wave theory to the phenomenon of the propagation of radio waves through the earth's atmosphere is a much more difficult problem.

For the purpose of discussing terrestrial radio propagation, the earth's atmosphere will be divided into two layers: 1) the troposphere, which extends from the earth's surface to a height of 50 miles or so, and 2) the ionosphere, which extends from 50 to about 300 miles above the earth. The troposphere consists of the usual oxygen, nitrogen, water vapor, and so forth, whereas the ionosphere is composed of free electrons, positive ions, and neutral gas atoms.

There are several reasons for the earth's atmosphere having a disturbing effect on radio propagation; for example: 1) The dielectric constant[1] of the troposphere is not uniform, because of air turbulence, temperature gradients, and the differing composition of the oxygen, nitrogen, water vapor, carbon dioxide, and so forth, of the various layers of the atmosphere. 2) The free electrons in the ionosphere produce a particularly disturbing effect on EM waves. Figure 17-2 shows the frequency bands that

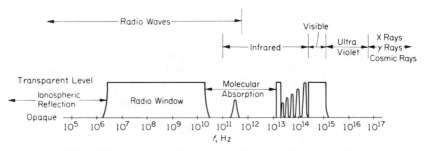

FIG. 17-2. EM wave attenuation of the earth's atmosphere versus frequency.

are passed and absorbed by the earth's atmosphere. For purposes of illustration, let us assume that a person on the moon is trying to communicate with a person on the earth. Figure 17-2 shows that frequencies below about 3 MHz are reflected by the earth's ionosphere. A person on the moon could neither receive signals transmitted from the earth nor transmit information to the earth using frequencies below 3 MHz. Radio frequencies from about 3 to 50,000 MHz can be transmitted back and forth through the earth's atmosphere with little loss in signal strength. This band of frequencies is often termed the *radio window*. From 50,000 MHz to about 300 tHz (t = tera = 10^{12}), the gas molecules in the earth's atmosphere absorb much of the EM radiation. Frequencies from 300 tHz to about 1000 tHz can pass through the atmosphere relatively unimpeded. This band of frequencies includes the visible light spectrum. Thus, for

---

[1]The dielectric constant ϵ of a material determines its ability to reflect, refract, transmit, and absorb EM energy. Also, the velocity of propagation of an EM wave is inversely proportional to the square root of the dielectric constant. The permeability μ of most non-ferrite materials is a constant μ_0, where $\mu_0 = 4\pi \times 10^{-7}$ henries/m.

purposes of transmitting information into and out of the earth's atmosphere, the choice of a frequency must be made carefully.

Since EM waves tend to travel in a straight-line path, r-f reception from point to point along the earth's surface is very reliable if the receiving and transmitting antennas are situated along a line of sight; that is, when a straight line can be drawn from one antenna to the other without the line passing through a dense obstacle such as a mountain, steel-structured building, or some such object.

A receiving antenna can pick up a signal from a transmitting antenna, located beyond the horizon by two general means: 1) a tropospheric wave or a ground wave, or 2) a sky wave reflected from the ionosphere. The tropospheric wave travels through the atmosphere and is refracted and reflected around the earth's curvature by the nonuniform dielectric properties of the atmosphere. This dielectric nonuniformity is caused primarily by masses of air that have different temperatures and moisture contents. The ground wave is a radio wave that tends to "hug" the earth's surface, because the earth acts like a conductor and thus somewhat guides the wave around the earth's curvature.

The Ionosphere. The earth's ionosphere deserves special attention in a discussion of wave propagation through the earth's atmosphere. The existence of the ionosphere is believed to be caused primarily by the sun's ultraviolet radiation, although meteorite showers contribute somewhat to the free electrons in the ionosphere. (Nuclear explosions in the upper atmosphere are also a contributing source of these free electrons.)

Figure 17-3 shows the principal layers of ionization that surround the

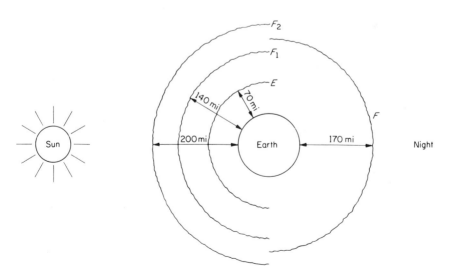

FIG. 17-3. The ionized layers surround the earth both at night and during the day.

earth. At night, only the F layer, which is situated about 170 miles above the earth, is of any real importance. During the day, several distinct layers are created by the sun's rays: 1) The F layer splits into the F_1 and F_2 layers, which are about 140 and 200 miles, respectively, above the surface of the earth. 2) The E layer is created about 70 miles above the earth and can exist only during the daytime, because the electrons and ions rather rapidly recombine in the somewhat lower, and hence denser, atmosphere. 3) Still another layer, called the D layer, is created below the E layer during the midday hours only.

Ionospheric Propagation. ⮕ The dielectric constant ϵ of a uniform volume of electrons is

$$\frac{\epsilon}{\epsilon_0} = 1 - \frac{Ne^2}{\epsilon_0 m \omega^2} \tag{17-2}$$

where N is the electron density in electrons per cubic meter, e is the magnitude of the electron charge, m is the electron mass, ω is the frequency of the incident EM radiation, in radians per second, and ϵ_0 is the permittivity or dielectric constant of free space. Since the derivation of Eq. 17-2 requires an understanding of appreciation of Maxwell's equations, the meaning of Eq. 17-2 will be stated without proof; that is, when $\epsilon/\epsilon_0 = 1$, the incident radiation experiences only free space and passes unimpeded through the ionized volume; as ϵ/ϵ_0 approaches zero, a greater portion of the incident radiation is reflected; and when ϵ/ϵ_0 equals zero or becomes negative, all the incident radiation is reflected and none is transmitted through the volume containing the electrons (see Fig. 17-4).

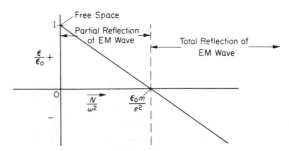

FIG. 17-4. The relative dielectric constant ϵ/ϵ_0 of an electron cloud as a function of N/ω^2.

Since the electron density of the ionosphere is rarely greater than $N = 10^{12}$ electrons/cu m, it is possible to calculate a maximum upper cut-off frequency ω_{co} by letting Eq. 17-2 equal zero, or

$$\frac{\epsilon}{\epsilon_0} = 1 - \frac{10^{12}(1.6 \times 10^{-19})^2}{\dfrac{1}{36\pi} \times 10^{-9} \times 9.1 \times 10^{-31} \omega_{co}^2} = 0$$

and

$$\omega_{co} = 5.67 \times 10^7 \text{ radians/sec}$$

All the standard AM broadcast-station frequencies, as well as many short-wave bands, fall below this upper cutoff frequency and are reflected back toward the earth by the ionospheric layers. Were it not for the reflecting characteristics of the ionosphere, worldwide radio communications from continent to continent would be impossible. In order to use the higher frequencies for global-communication purposes, it is necessary to use artificial earth satellites as reflecting or transmitting mediums (such as the Echo satellite balloons or the many receiver-transmitter satellites now orbiting the earth).

The earth's magnetic field produces a further interesting effect in ionospheric behavior. The magnetic field causes the electrons to spin (often referred to as *magnetic spin* or *cyclotron resonance*), and this spinning effect causes electrons to absorb more and more energy from the incident EM wave. When the spinning electrons collide with gas atoms or ions, the electrons' energy is dissipated, and the incident EM wave suffers much more attenuation than if the magnetic field were not present. This magnetic-spin resonance effect occurs when

$$\omega_s = \frac{e}{m} \mathbf{B} \qquad (17\text{-}3)$$

where ω_s is the magnetic-spin frequency, $\mathbf{B}$ is the component of the earth's magnetic field that is normal to the direction of propagation of the incident EM wave, and e and m are the electron charge magnitude and mass, respectively. Because of the magnetic-spin effect, certain frequencies in the AM broadcast band cannot be used for long-distance communications, since the EM waves are absorbed rather than reflected by the ionosphere.

17-3. Radio Astronomy

The layman usually considers astronomy as limited to viewing the universe through an optical telescope. Such a restriction is unjustified if we stop to consider that the visible light spectrum is a very small portion of the overall frequency spectrum. A radio telescope is used to detect frequencies other than those in the visible light spectrum.

The term *free space* is somewhat of a misnomer when applied to the reaches of outer space. Vast clouds of electrons, ions, and gaseous and particle matter are dispersed throughout the universe. It is estimated that over 90 percent of our own Milky Way galaxy is obscured from visual observation—even using the powerful 200-in. Mount Palomar telescope. Although the visual EM radiation is largely absorbed by the space debris, the lower-frequency radio waves are able to penetrate the cloud masses and reach the earth.

Radio astronomy is still in its infancy, although the first discovery of radio waves as emanating from a source in space was made by Jansky in 1931. It was not until after World War II that serious attention was focused on this area of science. The delay was caused partly by lack of good radio equipment—a lack that was quickly overcome by the rapid technological advances that were made during and after World War II.

The radio telescope uses both the radar and radio principles. Radar (an acronym for *ra*dio *d*etection *a*nd *r*anging) makes use of a transmitter and a receiver. The radar transmitter sends out pulses of energy which are aimed at various space objects, such as meteors, planets, and so forth, and the reflected pulses are returned to the radar receiver. The radar-type telescope can thus detect space objects that are unseen by visual means, and it can be used to determine the size, relative velocity, and distance of the objects.

The usual radio telescope employs an antenna and a receiver and simply detects the EM waves emanating from outer space. The antenna of a radio telescope performs the same function as the collecting lens of a visual telescope.

Both the radar and radio telescopes that are situated on the earth's surface are subject to rather severe limitations: 1) the atmospheric absorption of radio waves, and 2) other radiations, such as from radio, TV, microwave stations, automobile ignition systems, neon signs, mercury and sodium vapor street lights, and so forth, interfering with radiations received from galactic sources. An earth orbiting or moon-based radio (or optical) telescope should enable man to increase greatly his knowledge of the physical universe.

A radio telescope has two further limitations that are not dictated by its location: 1) inherent receiver noise (which has been much reduced by the maser device, to be discussed later), and 2) the angular resolving power of a telescope, which is inversely proportional to the frequency of the radiation being detected. In other words, lower-frequency radio waves are harder to focus than are the higher-frequency waves. For a microwave radio telescope to compare to an optical telescope in resolution, the radio telescope would need an antenna array several hundreds of miles in extent.

Only a scant two hundred or so point sources of radio activity have now been charted by the radio astronomers, and this number is insignificant when one considers the vast number of visible stellar bodies which give rise to starlight. The universe, as seen by the radio telescope, is largely populated by turbulent clouds of gas. Whether these turbulent gas clouds are stars in the process of being born and not yet hot enough to emit visible radiation, whether they are far-off visible stars that emit more radio energy than visible energy, or whether the sources are new types of stellar objects remains to be discovered. Perhaps the most interesting and perplexing phenomena recently discovered (about 1960) in the universe are the exceedingly distant and powerful EM sources called quasars. About one hundred quasars have been identified to date. One quasar, identified as 3C 273, emits more EM energy than our entire Milky Way galaxy, yet quasar 3C 273 seems to be about the same size as the earth-sun solar system. To emit such prodigious EM energies for such a small size (by astronomical standards) requires a concentration of mass and energy that is far larger than any previously known.

17-4. Maser-Type Devices

The word "maser" is coined from the words "*m*icrowave *a*mplification by *s*timulated *e*mission of *r*adiation." C. H. Townes and a group of Columbia University researchers proposed and developed the first maser in 1951–52. The maser is a device which operates as an oscillator or an amplifier of microwave frequencies. Similarly, the word "laser" means *l*ight *a*mplification by *s*timulated *e*mission of *r*adiation." Since the theory of operation of the maser and laser are so similar, the laser is often called an optical maser.[2]

The first maser that Townes and his group developed was a gaseous ammonia beam maser that was used primarily as a frequency standard. Although the power output of the maser is measured in nanowatts, this is sufficient for a frequency standard. A precise, quantitative discussion as to how the maser works is beyond the scope of this text. However, a qualitative description is sufficient to give a reasonably fair grasp of the principle of operation.

A knowledge of quantum theory is the key to understanding the operation of the maser-type device. Quantum theory is a rather high-level, involved subject and as the name implies has to do with quantized or discrete changes of some sort. These quantized, discrete changes are usually considered to be the changes in the energy levels in atomic structure. In the early 1900's Max Planck postulated that energy states in atoms existed in discrete steps and that the energy of an atom could not vary continuously over a range of energy levels. At the time, Planck was investigating the then perplexing discrepancy in interpreting the results of blackbody radiation as determined by classical theory and experimental results.

It is now known that a given atom or molecule can absorb or radiate

---

[2]Other names are employed for other frequencies of interest. At ultrahigh microwave frequencies ($f > 3 \times 10^{10}$ Hz) the term *ultramaser* is used, and at infrared frequencies ($f < 10^{13}$ Hz or so) the term *iraser* is used. More recently it has been suggested that the word *molecular* be substituted for the word *microwave* in the acronym *maser*.

energy only as a discontinuous process. Thus, according to Planck's theory, absorption of energy occurs only when an atom jumps from a lower to a higher energy level, and radiation of energy occurs when an atom jumps from a higher to a lower energy level. A given atomic or molecular structure can exhibit a multitude of these discrete energy levels or steps, but usually the atomic structure has "favored" energy levels that are more easily "jumped into" or "out of" than other energy levels. Some of the energy levels can be considered very stable, whereas others are so unstable that the atomic structure cannot or will not long remain at these levels. The maser-type devices take advantage of the discrete energy levels that exist in atomic structure.

The basic *modus operandi* of maser-type devices consists of exciting a well-chosen atomic structure with a high-energy source (called the *pump frequency*) and then inducing the atomic structure to radiate at the desired frequency. An excited atom usually returns rather rapidly to its original unexcited energy level. In atomic structures useful for masers, however, there is at least one so-called *metastable* state from which a spontaneous return to an unexcited state is unlikely. There are two ways by which an atom emits a photon of EM energy and returns to a lower energy level—by spontaneous and by induced emissions. Spontaneous emission is somewhat random in nature and is undesirable as far as maser operation is concerned. The spontaneous emission can be considered as a noise or unwanted emission of radiation. Fortunately, induced emission is always in phase with the emission causing radiation. This means that the induced emission acts in such a way as to cause amplification of the inducing radiation, which is the desired result.

It is usually very difficult for radio or microwave receivers to detect very weak signals, because of the rather high noise level present in conventional amplifying devices. To a large degree, the solid-state maser amplifier overcomes this limitation by being immersed in liquid helium. Both theoretical and experimental investigations show that the thermal noise of the maser, when operated at very low temperatures, is negligible. Also, since the spontaneous emission is very small at low temperatures, the maser can be considered almost a noiseless device. The use of the maser amplifier in weak-signal amplification is extremely important in such areas as radio astronomy or communication over long distances such as will be encountered in outer-space travel and planet-to-planet communications.

It is very difficult to generate frequencies above 10^{11} Hz (called *millimeter waves*) with conventional electronic equipment such as tubes or transistors, that is, the macroscopic type of generator. The molecular or microscopic approach to vhf generation, as typified by the maser-type generator, opens a new vista for investigating and using the submillimeter, infrared and visible light frequencies for communication purposes. The prospect of obtaining and using coherent[3] sources at these very high frequencies is extremely promising and fascinating.

Lasers. Laser radiation can be contained in a pencil-like beam that can be focused so as to illuminate an area only 2 miles in diameter by the time it reaches the moon. Such focusing and concentration of energy are impossible at the lower microwave frequencies. Use of the laser for communication purposes has a twofold meaning: 1) It will be possible to receive and transmit successfully radio, TV, and other signals over vast reaches of space, with a minimum expenditure of energy, and 2) large numbers of separate channels of information will be able to be carried on the laser beam.

The solid-state laser is basically a very simple device. A typical laser consists of a ruby rod several inches long and less than an inch in diameter. The ends of the rod are polished until very flat and parallel, and then the ends are coated with a thin, partially transparent layer of silver. The sides of the rod are left open to admit the exciting pump-frequency radiation. An intense flash of ordinary white light is suitable as a pump frequency. If one of the silvered ends of the ruby rod is more opaque than the other, an intense beam of red light is emitted from the more transparent end.

The purpose of the silvering is to form a cavity or echo-chamber effect within the ruby rod. This creates an amplification effect by reflecting some of the induced radiations, thereby causing the waves to bounce back and forth through the ruby rod. The partial trapping of the induced radiations further enhances the induced emissions of the excited atoms, creating an intense burst of coherent emission.

The gas laser is capable of larger power outputs than the solid state laser and so is more useful in some applications. The carbon dioxide (CO_2), helium-neon and argon-krypton lasers are currently the most popular gaseous lasers. The CO_2 laser is capable of delivering a high energy output capable of burning holes in bricks or sheet steel. The argon-krypton laser lases at up to 11 colors simultaneously and has a frequency stability that allows it to be modulated at rates up to 500 megabits per second (modulation is discussed in the next section). The degree of frequency stability determines how much useful information can be placed on the laser beam or color.

Although gas lasers are presently at a higher state of development

---

[3]Coherent signals have phase relationships such that interference patterns can be obtained. In incoherent signals, there is a random phase relation between signals. In order to be able to detect, amplify, and otherwise use a signal source for communication purposes, the source must produce a coherent signal.

than the solid state devices, the double-YAG solid state laser promises a much higher efficiency and frequency stability than its gas cousins. The double-YAG laser is a yttrium-aluminum garnet crystal that emits light at a wavelength of 10,600 Å that is frequency doubled to 5300 Å using a barium-sodium-niobate crystal. Several hundred milliwatts or even a few watts of power output at efficiencies of 10 percent or greater are expected. The output frequency is doubled (which means the wavelength is halved) because laser detectors or receivers are much more sensitive to light in the blue-green region of the visible light spectrum. The photomultiplier tube, which is one of the basic laser color detectors, is discussed in the next section.

Since a rather large amount of energy can be concentrated in a laser beam, the laser device may well be a future ray-beam weapon, or it may be useful for transmitting energy over long distances for use in communications, in charging batteries of spacecraft, supplying space stations with heat, and so forth. Additional aspects of using lasers for communication purposes will be discussed at the end of the next section.

17-5. Modulation and Demodulation

There are several ways of using EM radiation to convey information from one point to another, but the use of a carrier radiation frequency is by far the most effective method of transmitting intelligence. The desired intelligence is placed on the carrier by a process called modulation. There are many ways by which a carrier can be modulated, but two of the more common are amplitude modulation (AM) and frequency modulation (FM). Amplitude modulation and demodulation will be discussed in some detail in this section, but only a brief presentation of frequency modulation will be made.

Amplitude Modulation. A description of a typical AM broadcast transmitter-receiver is perhaps the best way to explain what is meant by this form of modulation. Figure 17-5 shows the barest essentials of a broadcast station and a radio receiver. The carrier frequency of a typical AM broadcast station can have any fixed value between 500 and 1500 KHz. This frequency is assigned by the FCC and is carefully chosen to avoid undue interference with other stations operating in the same area.

A microphone converts the acoustical audio frequency into an electrical signal frequency. These audio frequencies represent the information that is to be conveyed to the listener of the receiver set. The signal frequency can be almost any frequency but, for all practical purposes, is limited to 20 to 17,000 Hz—the audio-frequency range. (The average human ear can rarely detect audio signals beyond 17,000 Hz.)

The signal and carrier frequencies are then mixed together in a nonlinear device such as a vacuum tube or transistor. This nonlinear mixing causes the carrier frequency to be amplitude-modulated by the signal fre-

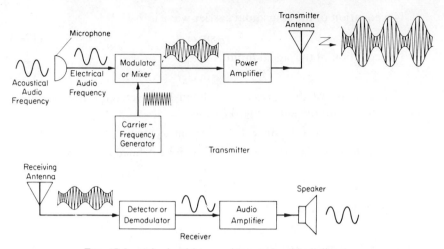

FIG. 17-5. A basic AM transmitter-receiver block diagram.

quency. In Fig. 17-6a, b, and c are shown the unmodulated carrier, the signal frequency, and the modulated carrier after mixing. Amplitude modulation is characterized by a constant carrier frequency, but the carrier amplitude varies in proportion to the amplitude of the audio signal.

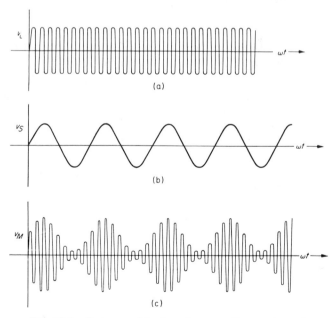

FIG. 17-6. Parts a and b show the respective carrier and signal frequencies and part c shows the result of nonlinear mixing to produce an AM carrier.

The equation of the sinusoidal carrier wave is

$$v_C = V_C \sin \omega_C t \qquad (17\text{-}4)$$

The equation of the signal input is

$$v_S = V_S \sin \omega_S t \qquad (17\text{-}5)$$

The equation of the carrier modulated by or mixed with the signal can be written with the aid of Fig. 17-6c as

$$v_M = V_C \sin \omega_C t + M V_C \sin \omega_S t \, \sin \omega_C t \qquad (17\text{-}6)$$

where M is the modulation factor, which can be defined as follows:

$$M \triangleq \frac{V_S}{V_C} \qquad (17\text{-}7)$$

By using the trigonometric identity

$$\sin x \sin y = \tfrac{1}{2} \cos (x - y) - \tfrac{1}{2} \cos (x + y)$$

Equation 17-6 can be rewritten as

$$v_M = V_C \sin \omega_C t + \frac{M V_C}{2} \cos (\omega_C - \omega_S) t - \frac{M V_C}{2} \cos (\omega_C + \omega_S) t \qquad (17\text{-}8)$$

It should be pointed out that Eqs. 17-6 and 17-8 represent the idealized AM wave rather than the actual output of the usual tube or transistor modulator. Frequency terms involving $n\omega_C$, $n\omega_S$, $\omega_C \pm n\omega_S$, where n is an integer number, would normally also appear in the equation for the modulated wave.

Even though Eq. 17-8 is a somewhat idealized equation, it does indicate the effect of the nonlinear mixing of two signals; that is, not only do the carrier and signal frequencies appear, but the sum and difference frequencies (called *side bands*) also appear in the output of the modulator. It is important to recognize that the signal information is contained in the side-band frequencies. These side-band frequencies are not simply the result of mathematical manipulation; they actually appear in the modulator output and can be filtered out by properly designed electric filter networks.

The standard AM broadcast station radiates the carrier ω_C and the two adjacent side-band frequencies $\omega_C + \omega_S$ and $\omega_C - \omega_S$, but this is a waste of energy, as the carrier frequency ω_C contains no signal information whatsoever. In fact, it is only necessary to radiate either the upper side band $\omega_C + \omega_S$ or the lower side band $\omega_C - \omega_S$ in order to transmit the desired signal information. The transmission of a single side band with suppressed carrier is now commonly employed in many radio transmitter-receivers.

Figure 17-5 shows an r-f power amplifier following the modulator stage. The purpose of this amplifier is twofold: 1) to amplify the carrier and the two adjacent side-band frequencies so that a relatively large amount of power is radiated by the antenna (by FCC regulation, 50,000 watts is the largest allowable power output of a civilian AM broadcast station), and 2) to filter out all the other multiple and side-band frequencies, because other radiated frequencies not only waste power but create interference with signals from other radio stations.

The study of antennas is quite involved and will not be considered here, so the reader must accept the fact that time-varying electrical signals can somehow escape from an antenna and radiate into space.

Let us next assume that a receiving antenna is situated so as to pick up a portion of the transmitted signal. The audio signal frequency ω_S must be separated from the incoming transmitted signal and converted from an electrical to an acoustical frequency by means of a loudspeaker.

Detection. The heart of the conventional AM radio receiving set is a rather simple device called a *detector* or a *demodulator*. Figure 17-7a

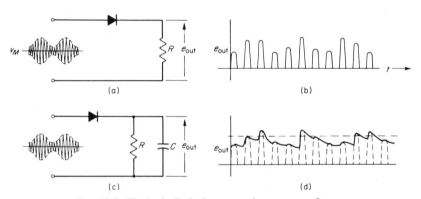

FIG. 17-7. The basic diode detector and output waveforms.

shows a diode connected in series with a resistance *R*. As explained in Chapter 5, a diode has a low impedance to current flow when the anode is positive with respect to the cathode, but the impedance is very high when the anode is negative with respect to the cathode. Since the incoming modulated signal voltage has both positive and negative variations about an average baseline value, the output voltage e_{out} across the resistance has only the positive variations (the negative variations appear across the diode). Figure 17-7b shows a plot of e_{out} versus *t*.

The output voltage e_{out}, shown in Fig. 17-7b, must be further modified before the desired audio signal can be recovered. If we were to amplify the output voltage e_{out} of Fig. 17-7b and then try to feed the signal to a loudspeaker, nothing would be heard because the frequency variations of ω_C are far above the audible range.

Figure 17-7c shows a shunt capacitor added to the circuit of Fig.

17-7a. This circuit is identical to the shunt-capacitor filter circuit discussed in Chapter 15. It will be recalled that the effect of the shunt capacitor is to smooth out the alternating variations of e_{out}. Figure 17-7d shows a rather distorted audio signal because ω_C is only 4 or 5 times as large as $1/RC$. In a standard AM broadcast station's modulated wave, the carrier ω_C is at least 100 times as large as $1/RC$, and so the filtering recovers an almost completely undistorted audio signal. Typical values for R and C in Fig. 17-7c are 200,000 ohms and 250 pf (picofarads), respectively.

Basic TRF Receiver. Figure 17-8 shows an elementary AM radio re-

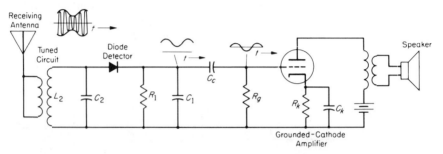

FIG. 17-8. Basic TRF (tuned radio frequency) AM radio receiver.

ceiver using the basic building blocks developed in this text. The antenna picks up the r-f energy of the incoming EM wave, and this energy is inductively coupled by transformer action into the tuned circuit of $L_2 C_2$. The tuned circuit enables the receiver to choose radio signals of differing frequencies, that is, permits frequency selectivity. From conventional a-c circuit theory, it is known that an L-C circuit resonates at a frequency $\omega = 1/\sqrt{LC}$. By letting L_2 be a fixed inductance and C_2 a variable capacitor, the receiver can be made so as to select frequencies from 500 to 1500 KHz.

The tuned-circuit output is then fed into the diode detector to be demodulated. After demodulation, the audio signal is passed through a blocking capacitor which blocks out the d-c portion of the demodulated signal. Although Fig. 17-8 shows a grounded-cathode triode amplifier, a common-emitter transistor amplifier could be used to drive the loudspeaker.

Frequency Modulation. Frequency modulation is usually considered a better method of modulation than amplitude modulation. The modulators and demodulators for frequency modulation are much more complex than those used in amplitude modulation and so will not be considered in detail in this text. Instead, attention will be focused on a description of the FM signal and on a very simple type of demodulator.

The carrier frequency of the standard FM radio station is about 90 MHz while the audio frequencies can again be considered the modulating signal. The appearance of the FM carrier differs markedly from the AM wave. Figure 17-9 shows that the modulated carrier is of constant ampli-

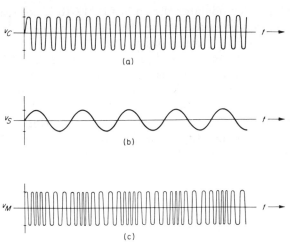

FIG. 17-9. An FM carrier. a) Carrier. b) Modulating
signal. c) Modulated carrier.

tude but that the frequency of the carrier is varied in proportion to the amplitude of the audio signal. The rate at which the carrier frequency changes is proportional to the rate of change (the frequency) of the audio signal.

A very simple demodulator, using a series inductance and resistance and a diode detector, is shown in Fig. 17-10. Since the inductive reactance ωL varies directly with frequency, the higher frequencies are blocked more

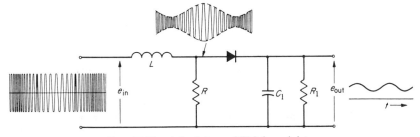

FIG. 17-10. A basic type of FM demodulator.

effectively than the lower frequencies. This results in the FM signal being converted into an amplitude modulated FM signal, as shown in Fig. 17-10. The envelope of the new FM signal is varying at the audio frequency; hence the envelope can be detected by the same type of diode used in detecting true AM signals. The fact that the carrier frequency inside the envelope is not a constant has very little effect on the output of the detector.

Although the demodulator shown in Fig. 17-10 is not really sensitive enough for use in a practical FM receiver, the principle of FM demodulation is amply illustrated by this basic circuit.

There are two reasons why FM reception is better than AM reception: 1) The carrier of the FM wave is transmitted at a constant amplitude. This means that externally introduced noise, such as atmospheric lightning, that may be superimposed on the EM wave can be removed from the incoming wave. The receiver does this by *clipping* the entire FM signal, thereby removing the noise. Clipping a signal is a straightforward operation which is illustrated in Fig. 17-11a. Noise almost

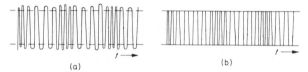

(a)   (b)

FIG. 17-11. a) Clipping an FM signal in order to remove the noise. b) Clipped signal.

always amplitude-modulates any EM radiation. On an FM receiver the FM wave is passed through a limiting circuit which clips the peaks off the signal. Figure 17-11b shows the effect of clipping an FM signal so that the noise voltages are removed thereby, leaving only a noise-free FM signal which can then be demodulated. Noise superimposed on an AM signal cannot be so removed because the clipping would distort the audio information contained in the AM envelope. 2) Frequency-modulation stations are allowed side-band frequency variations of 75 KHz, as compared to the 5 KHz for AM stations. This means that an FM station can transmit higher-fidelity audio information than can the AM station.

Laser Transmission. As pointed out in the introductory remarks, there is a tremendous demand for electromagnetic frequencies for communication purposes. To avoid interference and provide security, it is necessary that thousands of frequency bands or channels be available to transmit and receive information via EM waves. As of now, EM frequencies up to 10^{10} Hz are practically filled with allotted channels. Fortunately, the clever use of modulation-demodulation techniques and channel allocation from a geographical standpoint has allowed us to use simultaneously the same frequencies without interference or compromising security. Even with all of this cleverness, there is a pressing demand for more frequency bands to handle the growing communication needs.

The laser offers a way to meet this challenge in two ways: 1) the laser carrier frequency is about 10^{14} Hz which means that theoretically almost all of today's EM communication channels could be transmitted on one laser beam, and 2) the laser beam travels in a narrow, straight-line path which makes it rather easy both to avoid interfering reception from other laser transmitters and to provide security from unwanted reception by other laser receivers. Present laser technology indicates that a laser can be made to lase at several colors simultaneously. Each color can be separated by optics and modulated separately, as indicated in Fig. 17-12.

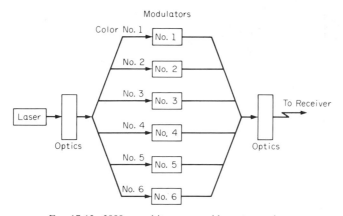

FIG. 17-12. 3000-megabit per second laser transmitter.

Six modulators, each capable of producing a 500 megabits per second digital signal can be paralleled to produce a very wide bandwidth of 3000 megabits per second in a single laser beam. Over one million telephone conversations could be transmitted simultaneously over such a laser beam. Hopefully, it will be possible in the future to modulate a single laser color at rates far above 5×10^8 bits per second since the laser frequency can accommodate the higher modulation rates if the stability of the mode frequency, the modulators, and detectors can be improved.

Laser Modulation. The three primary ways to modulate any EM wave are by amplitude, frequency, and pulse modulation. Both AM and FM methods were discussed in a previous section so only pulse modulation will be elaborated on here. Pulse code, pulse width, pulse position, and pulse frequency modulation are four of the most widely used methods of pulse modulating lasers. Although a discussion of PM methods is beyond the scope of this text, PM is essentially a transmission of short bursts of EM energy. The well known Shannon sampling theorem states that an EM signal must be pulse-sampled at least twice as often as the highest frequency component in the sampled EM wave to avoid loss of any information that may have been placed on the original wave. For instance the 4.2 MHz bandwidth of the conventional television EM wave must be sampled at a minimum rate of 8.4 megabits/second to make sure that all of the information in the transmitted wave is recovered by the receiver.

As an example of pulse modulation consider the three diagrams shown in Fig. 17-13. A laser color that is modulated by conventional

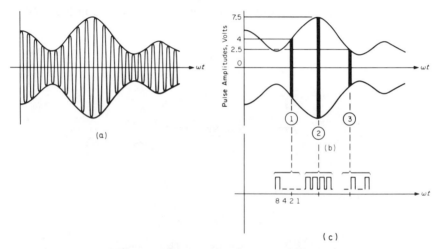

FIG. 17-13. a) Original continuous amplitude-modulated laser color. b) Sampled original signal in pulsed analog form. c) Sampled original signal in pulsed digital form.

AM is given in Fig. 17-13a, which is identical to Fig. 17-6c with the carrier frequency about 10^{14} Hz. The laser must be sampled at a frequency or rate at least twice as high as the signal frequency that is modulating the laser color. A sampling frequency that is about three times higher than the modulating frequency gives the series of pulses shown in Fig. 17-13b. Two points need to be made about the pulses shown in Fig. 17-13b: 1) the pulse amplitude can vary from 0 to 15 volts depending on when the modulated laser color is sampled, and 2) each pulse contains thousands of cycles of the laser color (for instance for a sampling rate of 10^7 Hz, a pulse width of 1/100 of the sampling rate and a laser color frequency of 10^{14} Hz each pulse would contain 10^5 cycles of the laser color frequency).

The pulses of Fig. 17-13b could be transmitted to a suitable receiver and the original intelligence of the laser modulating signal recovered, but unfortunately it is easy for noise or an uncalibrated receiver or transmitter to cause distortion of the original information. This distortion comes about because the amplitude of each pulse is an important part of the information contained in the overall signal, and noise or equipment calibration errors can cause the received pulse amplitude to differ from its actual value. To minimize these noise and calibration errors each analog pulse can be converted into a digital pulse train as shown in Fig. 17-13c. Assuming that an accuracy of ±0.5 volt in each analog pulse is sufficient to give good reproduction of the original modulating frequency and that the analog pulses can vary from 0 to 15 volts, a digital train of 4 uniform pulses can convey the desired information (recall from Chapter 11 that $2^0 + 2^1 + 2^2 + 2^3 = 15$). The first pulse in each train is always $2^0 = 1$, the second $2^1 = 2$, the third $2^2 = 4$, and the fourth $2^3 = 8$.

As a practical matter it is better to sample and digitally code the modulating signal before actually modulating the laser color. The discussion centering around Fig. 17-13 gives the same final results of a pulse-coded digital laser output for either method. It is beyond the scope and intent of the text to pursue further the modulation methods used in laser transmitter-receiver systems. The reader is encouraged to look in the many texts devoted to this subject for further information.

Laser Reception. To recover the information from incoming laser beams it is necessary to convert the EM light photons into electrical signals. This can be accomplished by either solid state or vacuum tube photo detectors. Although the vacuum or gas phototubes as well as the semiconductor photodiodes of Chapter 16 could be used for detection purposes, one of the most sensitive photodetectors is the photomultiplier tube. Whether the laser is modulated by AM, FM, or PM methods, the photomultiplier makes an excellent detector. It is interesting to note that a rather old vacuum tube device is still superior in some respects to its solid state cousin.

The photomultiplier is conceptually a rather simple device but is a rather expensive one as compared to solid state photodiodes. The photomultiplier shown in Fig. 17-14 illustrates that the photomultiplier is somewhat like a vacuum phototube except that additional plates called dynodes are placed between the cathode (C) and anode (A). The incoming photons strike the cathode releasing a few electrons. These electrons are attracted to dynode 1 which is 100 or so volts higher in potential than the cathode. Upon striking dynode 1 each electron from the cathode releases several additional electrons by the secondary emission process explained in Chapter 4. This process is repeated several times (commercial photomultipliers have up to 15 dynodes) so that current gains of 10^6 or so between the

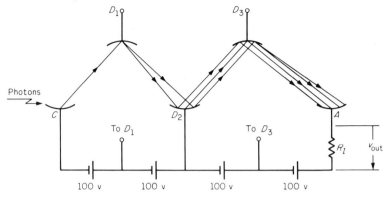

FIG. 17-14. Basic photomultiplier.

cathode and anode are realized. High quality photomultipliers can detect currents of less than 0.1 pico-ampere which is a small current indeed. The output of the photomultiplier rarely exceeds 100 μampere, which means that conventional solid state or vacuum tube amplifiers must usually be used to increase the photomultiplier output to a power level that is suitable to drive meters, speakers, recorders, etc.

The photomultiplier is particularly useful in detecting pulses of light, such as laser colors, and has a useful frequency range of several hundred megahertz.

18 / Gas Tubes and Controlled Rectifiers

Many industrial processes and applications require that large amounts of electrical energy be converted from alternating to direct current. In addition merely to converting alternating to direct current, various industrial uses in the field of automation often must have close control of the energy-conversion process. Both the gas tube and the solid-state rectifier are capable of controlled rectification, and the characteristics of each device will be presented in this chapter. The gas tube is capable of handling large amounts of energy more efficiently than can the vacuum tube. To a degree, the current flow through a gas tube can be controlled, but, as will be seen, the gas tube behaves in a manner much different from that of the vacuum tube.

The simplest gas tube is the diode, which consists of an anode and a cathode surrounded by a mercury vapor or some inert noble gas. Neon tubes are essentially gas diodes. The controlled-rectifier type of gas tube has a third element which controls the current flow in the tube. The thyratron and ignitron are two common commercial names for controlled-gas-tube rectifiers.

Semiconductor devices can also be used for controlled rectification, although the power-handling capacity of these rectifiers does not yet approach that of the ignitron type of mercury-vapor, mercury-cathode controlled rectifier. However, the controlled-junction rectifier does respond much faster, is more efficient, and is smaller in size than the medium- to low-power gas-tube rectifiers such as the thyratron. The silicon controlled rectifier (SCR) is a good example of a controlled-junction rectifier.

Actually the SCR is only one of a family of silicon devices (called thyristors) that are used to control electrical energy. Since the SCR is the best known and at present the most useful of the thyristors it is given the most attention in this chapter.

18-1. Electrical Discharges

A gaseous electrical discharge is a rather complex phenomenon. Although a gas tube usually operates as an arc-discharge device, there are

517

two other types of characteristic electrical discharges that must be considered if one is to appreciate electrical-discharge phenomena. The three regions of an electrical discharge are termed the *Townsend discharge*, the *glow discharge*, and the *arc discharge*. Figure 18-1a shows a cold-cathode-

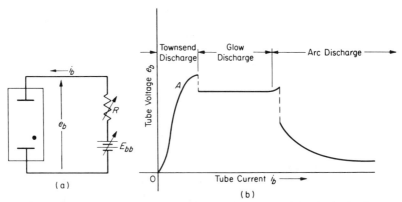

FIG. 18-1. a) A cold-cathode-discharge tube connected in the basic discharge circuit. b) Cold-cathode gaseous-discharge characteristics.

discharge tube connected in series with a battery and a resistance, both of which are variable. The dot inside the tube in Fig. 18-1a indicates that there is gas in the tube. A cold cathode rather than a thermionic cathode must be considered in order to explain satisfactorily the three-region discharge characteristics. Figure 18-1b shows the volt-ampere characteristics of the three discharge regions. The region from O to A is known as the "dark" or Townsend discharge. Theoretically, there should be no current flowing in this region, because the cold cathode does not emit any electrons. From a practical and experimental standpoint, however, there is a small current flow which is caused by external X-ray and cosmic radiations penetrating the tube envelope and partially ionizing the gas atoms by photo-ionization. The electrons thus freed by this photo-ionization proceed to the anode, while the positive ions drift toward the cathode. The rather steep slope shown in the Townsend-discharge region is due to an additional ionization of the gas by the photoelectrons which have acquired sufficient energy for ionization from the applied anode potential.

As the applied voltage E_{bb} is further increased, the Townsend current tends to show signs of saturation. Suddenly, the tube voltage e_b drops, and the current i_b increases. Because of the current-limiting resistance R, which must be rather critically adjusted, the discharge next stabilizes in a region termed the glow discharge. The glow discharge is typified by an almost constant tube voltage and a varying current. The tube current can be varied by varying R or E_{bb} over rather narrow limits. The glow discharge is maintained by the action of the positive ions striking the cathode and releasing secondary electrons, which then gain kinetic energy because of e_b and ionize additional atoms. The discharge is stabilized by the presence of R.

As the supply voltage E_{bb} is further increased or R is decreased, a second sharp discontinuity is reached. The voltage e_b again rapidly drops, and the current i_b rapidly increases. This region is called the arc discharge. The arc discharge is characteristic of the hot-cathode (thermionic) discharge typified by the usual gas diode, thyratron, ignitron, and so forth, for, in the arc-discharge region, there is a large enough positive-ion bombardment to cause sufficient heating of the cathode to produce thermionic emission. The arc discharge thus differs from the glow and Townsend discharges by virtue of a large number of free electrons being present at the cathode.

This chapter is primarily concerned with the arc-discharge tube, although the glow-discharge effect is important in relation to such applications as voltage-regulator tubes. The Townsend-discharge phenomenon is useful in measuring the strength of external X-ray, gamma-ray, and other radiations, but it will not be further pursued in this chapter.

The Glow Diode. The glow diode utilizes the typical cold-cathode type of electrical discharge. If the glow diode is operated in the glow-discharge region of the electrical-discharge characteristic, the voltage across the diode is almost constant over a fairly sizable current range (note Fig. 18-1b). This constant-voltage characteristic of the cold-cathode glow discharge makes the glow diode very useful as a voltage reference or a voltage regulator. The glow diode is thus the tube counterpart of the Zener solid-state diode.

Glow tubes are generally manufactured with constant-voltage ratings of 75, 90, 105, and 150 volts, although other voltages could be obtained if desired. As an example of how a glow tube can be utilized, let us consider the circuit shown in Fig. 18-2. The OC3/105 glow diode maintains a constant terminal voltage of 105 volts over a current range of 5 to 40 ma. If the current drops below 5 ma, the terminal voltage increases; if the current goes beyond 40 ma, there is danger that the glow discharge

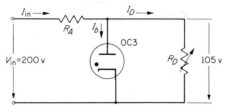

FIG. 18-2. An OC3/105 glow diode used as a voltage regulator.

will suddenly become an arc discharge, with a resultant drop in terminal voltage and possible tube damage.

Let us suppose that we want to maintain a constant voltage of 105 volts across the variable load resistance R_D. If the input voltage V_{in} is assumed to be 200 volts, we must know the range of the variable resistance R_D in order to determine the proper value of R_A. We assume that R_D varies from infinity to some limiting finite value.

From Fig. 18-2 it is obvious that the following current relation can be written

$$I_{in} = I_b + I_D \tag{18-1}$$

and the voltage relation

$$200 = I_{in} R_A + 105 \tag{18-2}$$

must be satisfied if the glow diode is to be useful as a voltage regulator. The maximum tube current I_b occurs when the load current $I_D = 0$ (or $R_D = \infty$), and the maximum allowable glow-diode current is 40 ma. Thus, using Eq. 18-2,

$$R_A = \frac{200 - 105}{0.04}$$
$$= 2375 \text{ ohms}$$

The minimum allowable tube current is 5 ma, which means that R_D cannot become too small in value, thereby shunting too much current through R_D. The minimum allowable value for R_D is

$$R_D = \frac{105}{0.040 - 0.005}$$
$$= 3000 \text{ ohms}$$

If R_D drops below 3000 ohms, the glow discharge will be extinguished, and the network will act as a simple voltage divider with no voltage regulation.

The Gas Diode. The thermionic gas diode is very similar in construction to the thermionic vacuum diode, except that an inert gas is placed within the tube envelope. A typical vacuum tube is operated in a high vacuum of only a few nanometers of mercury, whereas a gas tube has a higher pressure, ranging from a few micrometers (often called microns) to a few millimeters of mercury (atmospheric pressure is about 0.76 m of mercury). As in the case of the vacuum tube, electrons are emitted from a heated cathode and are attracted to a positive anode; however, a new phenomenon occurs in the gas tube that does not occur in the vacuum tube. As the electrons leave the cathode and travel toward the anode (assuming that the anode is positive with respect to the cathode), collisions with the rather abundant inert-gas atoms are going to occur. If the electrons have sufficient energy (about 10 to 20 ev for most gas tubes), the electron-atom collisions usually result in the ionization of the atom. The ionizing electron and the released electron proceed toward the anode, and, when each has acquired sufficient energy, two more atoms may become ionized. The net result is cumulative ionization until the gas breaks down, creating an arc discharge.

The small constant-voltage-drop characteristic of the arc discharge makes the thermionic gas diode an almost ideal device for rectifying alternating to direct current. Not only does the arc gas diode make an efficient rectifier because of its low voltage drop, but it also is capable of handling hundreds or thousands of amperes.

For currents of only a few amperes, the oxide-coated cathode type of thermionic gas diode is widely used for rectification purposes. Figure 18-3

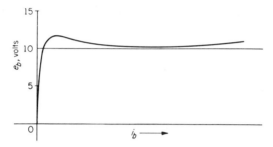

FIG. 18-3. The e_b versus i_b characteristics for a thermionic mercury-vapor diode.

shows the characteristic volt-ampere curve for a thermionic mercury-vapor diode. Although other inert gases could be and are used, mercury vapor has the advantage of a low ionization potential (10.4 volts), and the liquid mercury provides additional mercury atoms when the vapor is absorbed by the electrodes. One can tell a glass mercury rectifier tube from other glass tubes by observing whether any droplets of mercury are present inside the envelope.

Figure 18-3 illustrates that the ionization potential of the gas inside the tube must be exceeded before the tube actually "fires." After the tube fires and the arc discharge is established, the tube voltage drops to the ionization potential and remains at this voltage over a wide current range. Actually, the ionization potential is not a constant but depends on the temperature and pressure of the gas within the tube. Under normal operating conditions the arc voltage drop for a mercury-vapor tube can be considered constant at 10.4 volts.

Figure 18-4 shows the thermionic gas diode used in a basic rectifier circuit. When the anode is positive with respect to the cathode, the tube

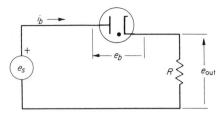

FIG. 18-4. A thermionic gas diode
used in a basic rectifier circuit.

can fire, and the following KVL can be written:

$$e_s = e_b + i_b R \tag{18-3}$$

Equation 18-3 shows that an arc-discharge tube must be connected in series with a resistance in order to limit the tube current i_b to a safe value. If R were zero, the tube current would increase until the diode was destroyed.

If the input voltage $e_s = E_m \sin \omega t$, and if the tube drop e_b is assumed to be a constant value ($e_b = E_t$), then Eq. 18-3 can be rewritten as

$$i_b = \frac{E_m \sin \omega t - E_t}{R} \tag{18-4}$$

The assumption of a constant tube voltage drop can be justified as long as E_m is much greater than E_t. Figure 18-5 shows the actual and assumed

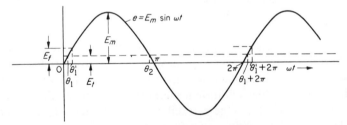

FIG. 18-5. Gas-diode voltage relations showing actual θ'_1 and assumed θ_1 firing angles.

firing angles θ'_1 and θ_1, respectively. Although the diode actually requires a firing potential of E_f volts, little error is introduced in Eq. 18-4 if it is assumed that the tube fires when $e_b = E_t$ volts. The assumed firing angle θ_1 is

$$\theta_1 = \arcsin \frac{E_t}{E_m} \tag{18-5}$$

When the tube voltage drops below E_t, the diode ceases current conduction, and so the cutoff angle θ_2 can be calculated as

$$\theta_2 = \pi - \arcsin \frac{E_t}{E_m} \tag{18-6}$$

The diode current i_b over a full cycle can now be mathematically expressed as

$$i_b = \frac{E_m \sin \omega t - E_t}{R} \qquad \theta_1 < \omega t < \theta_2$$

$$= 0 \qquad \theta_2 < \omega t < \theta_1 + 2\pi \tag{18-7}$$

The d-c or average current can be obtained by integrating i_b over a full cycle, or

$$I_{\text{d-c}} = \frac{1}{2\pi} \int_0^{2\pi} i_b \, d\omega t$$

$$= \frac{1}{2\pi} \int_{\theta_1}^{\theta_2} \frac{E_m \sin \omega t - E_t}{R} \, d\omega t$$

$$= \frac{1}{2\pi R} [E_m (\cos \theta_1 - \cos \theta_2) - E_t(\theta_2 - \theta_1)] \tag{18-8a}$$

and, making the substitution,

$$\theta_2 = \pi - \theta_1$$

from Eq. 18-6,

$$I_{\text{d-c}} = \frac{E_m}{2\pi R} \left[2 \cos \theta_1 - \frac{E_t}{E_m} (\pi - 2\theta_1) \right] \tag{18-8b}$$

As E_m becomes much larger than the tube drop E_t, both E_t/E_m and θ_1 approach zero. Equation 18-8b thus reduces to

$$I_{\text{d-c}} = \frac{E_m}{\pi R} \tag{18-9}$$

The d-c output voltage across the load resistance R is

$$V_{\text{d-c}} = I_{\text{d-c}} R = \frac{E_m}{\pi} \tag{18-10}$$

which is the same as Eq. 15-5 with $r_p = 0$.

18-2. Controlled Rectification

Although the gas diode is an important and much-used rectifying device, the three-element controlled gas tube makes a more interesting and versatile type of rectifier. The thyratron and ignitron typify the two basic kinds of controlled gas tubes. The thyratron uses a heated cathode which continuously emits electrons, but the addition of a grid prevents the tube from firing until desired, whereas in the ignitron the arc is initiated when desired. The ignitron uses a pool of liquid mercury for a cathode, and, although the mercury pool is not directly heated, a localized hot spot develops on top of the mercury, creating thermionic emission, which is necessary in an arc-discharge tube.

The Thyratron. Figure 18-6 shows a cutaway sketch of a typical

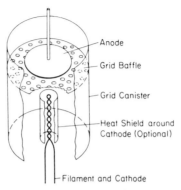

FIG. 18-6. Sketch of the structure of a typical thyratron.

thyratron. In order for the grid to control the firing of the tube, the thyratron grid structure must form an almost complete shield around the cathode.

As long as the grid-to-cathode potential is sufficiently negative, electrons are neither allowed to pass through the grid structure nor accelerated to an energy equal to or exceeding the gas-ionization potential; if, however, the negative grid voltage is reduced, some electrons may escape through the grid structure, become attracted by a positive anode voltage, and thereby reach the gas-ionization potential. The resulting positive gas ions are attracted by and tend to neutralize the negative grid field that exists near the openings in the grid structure. This neutralization process allows more cathode electrons to pass through the grid, thereby creating a condition that is favorable for establishing an arc discharge.

After the cumulative ionization effect causes the arc discharge, the positive gas ions surround the grid and there completely isolate or neutralize any effect that a negative grid voltage may have on the arc discharge. *Herein lies a fundamental phenomenon that is characteristic of all gas triodes such as the thyratron; namely the grid voltage can control the initiation of the arc, but it has no control over the arc discharge once the arc has been established.*

Thyratrons can be manufactured with either negative or positive grid-voltage characteristics. If small holes are drilled in the grid structure, the grid may so well shield the cathode from the anode that a positive grid voltage is required to initiate the arc. Larger holes cause less shielding, with the result that a negative grid voltage is required to suppress the firing of the tube. In the latter case, the required grid voltage to prevent the arc from starting is usually only a few volts for most thyratrons. For analytical purposes, it is convenient to assume that the thyratron will not fire as long as the grid voltage is negative, even though the anode is at a high positive potential. In other words, if the anode is positive, the arc can be initiated only if the grid voltage is zero or some positive value. Of course, if the anode is negative, it can be assumed that no amount of grid voltage will cause an arc that will create a cathode-to-anode current flow.

Although it takes time (a few microseconds), for the gas in the thyratron to ionize or deionize, this is of little importance in the low-frequency power rectification that is considered in this chapter.

To illustrate how a thyratron can be used in controlled rectification, let us consider the circuit shown in Fig. 18-7a. The object of the analysis to follow is to calculate the average or d-c current I_{d-c} flowing through the

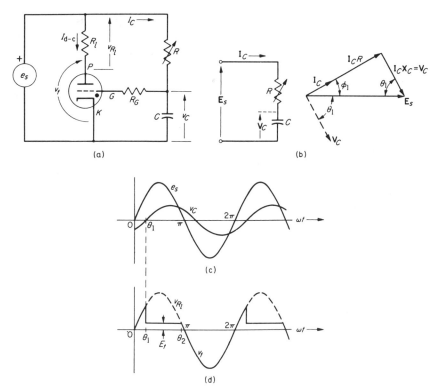

FIG. 18-7. Phase-shift control of a thyratron.

load resistance R_l. The source voltage e_s is assumed to sinusoidal; that is, $e_s = E_m \sin \omega t$.

The series R-C circuit that is connected in shunt or parallel with R_l and the thyratron is known as the *phase-control circuit*. It will be shown that this circuit controls the firing angle θ_1 of the thyratron. The resistance R_G is placed in series with the grid of the thyratron in order to limit the thyratron grid current after the arc discharge has been established.

Before continuing with the calculation for $I_{d\text{-}c}$, several assumptions and statements need to be listed. Finding the firing angle is the primary step that must be taken before $I_{d\text{-}c}$ can be calculated. To find the firing angle, we assume that: 1) There is no grid or plate thyratron current before firing. 2) If the plate voltage is positive, then, as soon as the grid voltage becomes zero (assuming that the grid voltage has been negative) the thyratron fires; if the grid is already positive, then as soon as the plate to cathode voltage becomes positive, the tube will fire. This means that the grid-cathode voltage must lag the plate-cathode if the grid is to control the firing angle.

Because of the assumptions in the preceding paragraph, the only current that flows in the circuit of Fig. 18-7a before firing is I_C, which flows through the series R-C control circuit. To understand the behavior of this circuit, we refer to the phasor diagram of Fig. 18-7b. Using E_s as the reference phasor, we know that I_C must lie in the first quadrant for the R-C circuit. Conventional a-c circuit theory shows that the current I_C leads the voltage E_s by an angle which is

$$\phi = \tan^{-1} \frac{1}{\omega RC} \qquad (18\text{-}11)$$

Figure 18-7c is a plot of the two voltages that must be considered when calculating the firing angle of the thyratron; namely, the plate-to-cathode voltage v_{pk} and the grid-to-cathode voltage v_{gk}. From Fig. 18-7a it is evident that before the tube fires

$$v_{pk} = e_s$$
$$v_{gk} = v_C$$

Figure 18-7c further illustrates that between $0 = \omega t < \theta_1$, v_{pk} is positive but v_{gk} is negative, and thus the thyratron is cut off. But at $\omega t = \theta_1$, v_{gk} becomes zero and v_{pk} is still positive, so the tube fires. The angle θ_1 is called the *firing angle* of the thyratron. In the control circuit of Fig. 18-7a it is defined as

$$\theta_1 = \frac{\pi}{2} - \phi$$

$$= \tan^{-1} \omega RC \qquad (18\text{-}12)$$

The thyratron cuts off when the plate-cathode voltage drops below the grid-cathode voltage. During conduction the thyratron behaves just like the thermionic gas diode, so the thyratron cutoff angle θ_2 is defined by Eq. 18-6, repeated here as Eq. 18-13:

$$\theta_2 = \pi - \arcsin \frac{E_t}{E_m} \tag{18-13}$$

Figure 18-7d shows a plot of the tube voltage v_t and load-resistance voltage v_{R_l} for a firing angle θ_1 of about 35°.

The average or d-c value of the current through the thyratron can be found by integrating the tube current over a full cycle, or

$$I_{\text{d-c}} = \frac{1}{2\pi} \int_{\theta_1}^{\theta_2} \frac{E_m \sin \omega t - E_t}{R_l} \, d\omega t \tag{18-14}$$

which is identical to Eq. 18-8a except that, ideally, θ_1 can have any angular value between 0 and 180°. Upon integrating Eq. 18-14,

$$I_{\text{d-c}} = \frac{E_m}{2\pi R_l} \left[(\cos \theta_1 - \cos \theta_2) - \frac{E_t}{E_m} (\theta_2 - \theta_1) \right] \tag{18-15}$$

and if it is assumed that $E_m \gg E_t$, then Eq. 18-15 reduces to

$$I_{\text{d-c}} = \frac{E_m}{2\pi R_l} (\cos \theta_1 + 1) \tag{18-16}$$

since $\theta_2 \approx 180°$ and $E_t / E_m \approx 0$ under this assumption.

The control circuit shown in Fig. 18-7a can control the firing angle only between 0 and 90°. Other circuits allowing full control over θ_1 from 0 to 180° will be discussed later.

The Controlled-Junction Rectifier. The controlled-junction rectifier is the solid-state equivalent of the thyratron. The SCR (*silicon controlled rectifier*) offers several advantages over the thyratron: It is smaller, reacts faster, has a much smaller voltage drop when conducting (and thus a higher efficiency), and has a larger current-carrying capacity. However, the SCR also has some disadvantages, such as a peak inverse voltage of only a few hundred volts (as compared to a few thousand for the thyratron), and the fact that the SCR is much more susceptible to damage owing to high temperatures (125°C or so is the allowable upper temperature limit).

The thyratron is a voltage-controlled device, whereas the SCR is controlled by a flow of current. Figure 18-8a shows the basic diagram for the four-layer PNPN controlled-junction rectifier, and Fig. 18-8c shows the conventional schematic symbol for the rectifier. Figure 18-8b is the same as part a, except that voltage sources are shown applied to the anode-cathode and gate-cathode terminals.

A qualitative explanation for the operation of the SCR can be made by noting that if the center NP diode is bypassed or shorted, the remaining PN-junction diode will be biased in the forward direction, and a siz-

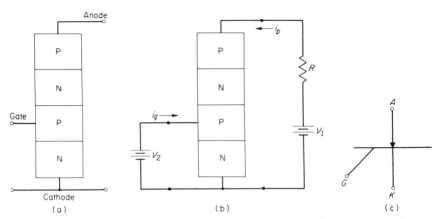

FIG. 18-8. A PNPN solid-state controlled rectifier. a) Pictorial diagram. b) Circuit Application. c) Schematic symbol for the solid-state controlled rectifier.

able current flow i_b will result. If the anode-cathode supply voltage V_1 is reversed, i_b will be a very small current. If the center NP diode is now reinserted into the circuit, no current will flow through the SCR, regardless of the polarity of V_1 (this assumes that V_2 is removed from the circuit and $i_g = 0$), because either the outer PN or the inner NP diodes will always be reverse-biased.

If V_2 is applied to the gate-cathode terminals so that the lower PN junction is forward-biased as shown, the resulting current i_g places free electrons in the lower PN junction. This causes electrons to diffuse into the upper N region, and these diffused electrons are immediately attracted to the positive anode, thereby causing an increase in i_b. Once an anode-cathode current flow i_b is established, the isolation effect of the inner NP diode is destroyed, and the four-layer SCR behaves exactly like a simple semiconductor diode. A more accurate description of how a SCR is "turned on" by using a two transistor equivalent circuit is given in a following subsection on thyristors.

As the grid has no control over a conducting thyratron, so the gate has no control over a conducting SCR but can only control the firing angle of the SCR.

The voltage drop across a conducting SCR is only 1 volt or so (as compared to the 10 to 20 volts for gas tubes), and, as with the thyratron, the SCR must always be protected with a current-limiting series resistance. Figure 18-9 shows an SCR used as a half-wave rectifier with an R-C phase-shifting control circuit. This circuit is the same as the one shown in Fig. 18-7a, except that a diode D is connected in series with the gate terminal. This diode D prevents damaging reverse gate current through the SCR when a negative voltage appears across the capacitance. When

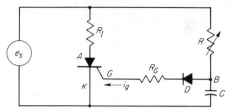

FIG. 18-9. A solid-state controlled
rectifier with phase-shift control.

the capacitive voltage v_{BK} is positive, the diode D conducts, and, when i_g becomes large enough, the SCR fires. The voltage-versus-time character-istic of the SCR circuit is so similar to the curves shown in Fig. 18-7b and d for the thyratron circuit that no corresponding curves for the SCR circuit will be drawn. Since the voltage drop across the conducting SCR is so small, the assumption that the cutoff angle θ_2 equals 180° is even more valid for the SCR than it is for the thyratron.

It can also be assumed, with little error, that the firing angle θ_1 can be calculated at the point where the gate-cathode voltage v_{GK} becomes zero (after being negative) when the anode-cathode voltage v_{AK} is positive. Typical SCR characteristics require that about 5 volts v_{GK} and 100 to 200 ma i_g are necessary to fire the SCR. The value of the series gate resistance R_G is determined by the maximum value of v_{BK} since i_g must be safely limited to a peak value of 2 amp for the typical SCR, or

$$R_G = \frac{(v_{BK})_{\max} - 5}{2} \quad \text{ohms} \tag{18-17}$$

The firing angle and d-c current for the SCR circuit shown in Fig. 18-9 are the same as Eqs. 18-12 and 18-16, respectively, for the thyratron.

Phase-Shift Control of Rectifiers. The phase-shift network shown in Fig. 18-10a gives a full 180° range for the firing angle θ_1. The phasor dia-gram shown in Fig. 18-10b presents the phasor-diagram development that will aid in calculating the firing angle θ_1. The analysis is the same whether a thyratron or an SCR is used in the circuit of Fig. 18-10a. As always *be-fore firing*, no current is flowing in the controlled-rectifier circuit. This means that $\mathbf{V}_{12} = \mathbf{V}_{AK}$ and $\mathbf{V}_{42} = \mathbf{V}_{GK}$ before firing. The source voltage $\mathbf{V}_{13}$ is supplied by a center-tapped transformer, which means that $\mathbf{V}_{12} = \mathbf{V}_{23}$. If $\mathbf{V}_{13}$ is used as a reference phasor, then the current $\mathbf{I}_{13}$ which flows through the series R-C circuit leads the reference voltage by some angle ϕ such that

$$\phi = \tan^{-1} \frac{1}{\omega RC} \tag{18-18}$$

which is the same as Eq. 18-11.

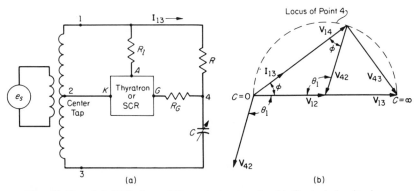

FIG. 18-10. a) A 180° phase-shift control network. b) Control-circuit phasor diagram.

To find θ_1, we note that the resistive voltage drop $\mathbf{V}_{14}$, which is in phase with $\mathbf{I}_{13}$, and the capacitive voltage drop $\mathbf{V}_{43}$ are 90° out of phase, and that $\mathbf{V}_{14} + \mathbf{V}_{43} = \mathbf{V}_{13}$. We can recall, from plane geometry, that two chords of a circle which are drawn from opposite ends of a diameter always intersect at right angles. Thus, as C is varied from zero to ∞ farads, the current $\mathbf{I}_3$ goes from

$$\frac{|\mathbf{V}_{13}| \underline{/0^\circ}}{R - j\infty} = 0\underline{/90^\circ}$$

to

$$\frac{|\mathbf{V}_{13}| \underline{/0^\circ}}{R - j0} = \frac{|\mathbf{V}_{13}|}{R} \underline{/0^\circ}$$

respectively. Since the locus of node 4 lies on a semicircle, $|\mathbf{V}_{12}| = |\mathbf{V}_{23}| = |\mathbf{V}_{42}|$, because each voltage is a radius of a circle.

Finally, $\mathbf{V}_{42}$ is seen to lag $\mathbf{V}_{12}$ by an angle θ_1, which is seen to be

$$\theta_1 = \pi - 2\phi$$

$$= \pi - 2\tan^{-1}\frac{1}{\omega RC} \qquad (18\text{-}19)$$

If the grid or gate to the cathode voltage leads instead of lags the plate-to-cathode voltage, the firing angle θ_1 will always be zero.

EXAMPLE 18-1. If a thyratron is used in the circuit of Fig. 18-10a and $v_{13} = 400 \sin 1000t$, $R = 10$ K, $C = 0.2$ μf, $R_l = 200$ Ω, and if R_G is adjusted for proper current limitation, find $I_{\text{d-c}}$.

Solution: The approximate solution for an idealized thyratron is given by Eq. 18-16, or

$$I_{\text{d-c}} = \frac{E_m}{2\pi R_l}(\cos \theta_1 + 1)$$

where

$$E_m = \frac{400}{2} = 200 \text{ volts}$$

$$\theta_1 = \pi - 2\tan^{-1}\frac{1}{\omega RC} = \pi - 2\tan^{-1}\frac{1}{1000 \times 10,000 \times 0.2 \times 10^{-6}}$$

$$= \pi - 2\tan^{-1}\tfrac{1}{2}$$

$$= \pi - 2(0.464)$$

$$= 2.21 \text{ radian} = 126.8°$$

$$I_{d\text{-}c} = \frac{200}{2\pi \times 200}(\cos 126.8° + 1)$$

$$= \frac{200}{400\pi}(-0.6 + 1)$$

$$= \frac{0.5 \times 0.4}{\pi} \text{ amp}$$

$$= 64 \text{ ma}$$

The phase-shift control circuit shown in Fig. 18-10a is not too practical for the SCR, since so much gate current is required for firing. In other words, instead of R and C being 10,000 Ω and 0.2 μf, respective values of 100 Ω and 20 μf would be more practical. This causes considerable wasted energy to be dissipated in the control circuit.

The Ignitron. The ignitron performs the same function as a thyratron or SCR, but it is capable of controlling and converting large amounts of a-c to d-c energy. The ignitron is a mercury-vapor gas tube with a mercury-pool cathode. The essential difference between the thyratron and the ignitron is that the thyratron grid *prevents* the arc from starting until desired, whereas in the ignitron the arc is *initiated* when desired. Figure 18-11 shows a basic ignitron. If a positive voltage is applied to the anode-

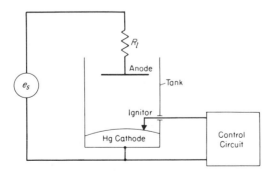

FIG. 18-11. The mercury-pool-cathode igni-tron used as a controlled rectifier.

cathode circuit, nothing will happen unless an arc or hot spot is present on the mercury-cathode surface. This is because the mercury pool is normally cool, and very little or no electron emission from the mercury can take place. If a localized hot spot is created the mercury can emit copious amounts of electrons and the tube can fire. Once the tube begins conduction, the cathode hot spot is self-sustaining. The cathode hot spot appears to be a bright spot of light that moves erratically over the mercury surface.

Actually, the theory of the cathode hot-spot formation and the resulting electron emission is not completely known or understood. The temperature at the hot spot is only 300 or 400 C which is too low to account for the thermionic emission. Apparently, high field emission (see Chapter 1) is the most logical and acceptable explanation for the emission, but even this theory does not adequately account for the very high current densities that are released from the cathode hot spot.

If the anode is positive, the ignitron can be fired by passing a current surge through it. The control circuit shown in Fig. 18-11 produces the current surge at the desired firing angle. Although the ignitron control circuit is more complex than the phase-shifting networks described earlier, Eq. 18-15 or Eq. 18-16 can be used to calculate I_{d-c} for any particular firing angle. The ignitron-tube voltage drop E_t is typically 15 to 25 volts.

Thyristors. As mentioned in the introduction the SCR is only one of a class of semiconductor switches called thyristor whose bistable behavior depends on PNPN regenerative feedback. The SCR is a three terminal unidirectional device but bidirectional devices are also available. In addition thyristors can have two, three, or even four terminals.

Along with the SCR the thyristor family includes the SCS (*s*ilicon *c*ontrolled *s*witch), the LASCR (*l*ight *a*ctivated *s*ilicon *c*ontrolled *r*ectifier), the SUS (*s*ilicon *u*nilateral *s*witch), the SBS (*s*ilicon *b*ilateral *s*witch), and the Triac (*tri*ode *a-c* switch). Since all thyristors have some common behavioral characteristics, a discussion of the SCS and Triac (in addition to the previously discussed SCR) will be considered sufficient to give the reader a reasonably good introduction to controlled rectification or switching.

The symbols for the SCS and Triac are given in Fig. 18-12 and it is obvious that the former is a 4 terminal device while the latter has 3 terminals.

The SCS. The operation of the SCS is similar to that of the SCR in that it conducts only when the anode to cathode voltage V_{AK} is positive. If the anode gate G_A is left unconnected then the SCS behaves just like a conventional SCR. If proper voltages are applied to the anode gate G_A and the cathode gate G_K, then the SCS becomes a very sensitive SCR in that the gate currents necessary to turn the SCS on are much less than the necessary switching current for the SCR. As shown in the pictorial sym-

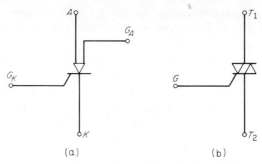

FIG. 18-12. a) Symbol for the 4-terminal
SCS. b) Symbol for the Triac.

bol of Fig. 18-13a the SCS is the same as the SCR except for the additional anode gate lead. To explain the operation of the SCS consider the often-used equivalent circuit of Fig. 18-13b. Although the SCS is not really made this way, the connection of a PNP and NPN BJT does explain adequately how the SCS (or the SCR if G_A is not connected to any external circuitry) behaves. To argue how this equivalent circuit works assume that a positive anode to cathode voltage V_{AK} is applied and that no connection is made to G_A. If now a current is sent into the cathode gate G_K (by application of a positive voltage between the G_K and the K nodes), the collector current of T_2 increases. Since the collector current of T_2 is the base current of T_1, an increase in the T_2 current turns on T_1 (causes an increase in the collector current T_1). Note that the collector current of T_1 feeds the base of T_2. This results in a whirlwind action (called regenerative feedback) that quickly turns the SCS fully on by allowing a low resistance current path between the anode and cathode.

By providing a voltage at the G_A node that is negative with respect to the A node the SCS can be made to switch even faster than before. One

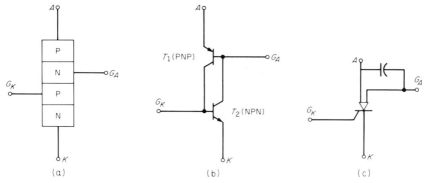

FIG. 18-13. a) Pictorial symbol for the SCS. b) Equivalent circuit for the SCS.
c) One way to connect the anode gate to get regenerative feedback.

way to provide a proper voltage to the G_A node to give fast, sensitive switching is to couple the anode A and anode gate G_A nodes through a coupling capacitor.

Triac. As is perhaps indicated by its symbol in Fig. 18-12b the Triac is capable of conducting current in both directions. To understand how this is accomplished recall that the SCR is first turned on if the anode to cathode voltage is positive and then the gate to cathode voltage is made positive such that a critical gate current is allowed. The SCR is turned off as soon as the anode to cathode voltage goes negative regardless of the value of the gate to cathode voltage. No reference was made earlier as to how fast the SCR could be turned on and off because normally this is no problem at the usual power frequencies of 25 Hz to 1000 Hz. Even with the gate current zero, there is a limit as to how fast a positive voltage can be reapplied to the anode after cutoff. The stored charges inside the SCR must be allowed to recombine (usually this takes only a few microseconds) before reapplication of a positive anode-cathode voltage in order to assume that the SCR will remain off until gate current is applied.

With the Triac the applied voltages are usually 60 Hz since this is the

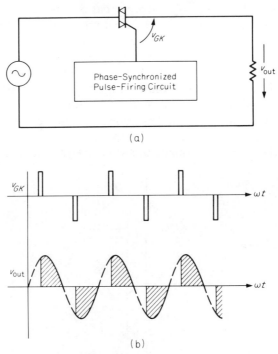

(a)

(b)

FIG. 18-14. a) Triac used in a full-wave control circuit. b) Pulse-firing v_{GK} and the resulting v_{out}.

predominate power frequency used in this country. Fortunately, if the gate current is zero or very small, the Triac turns off as the voltage $V_{T_1 T_2}$ passes through zero at reasonably low frequencies. This means that the Triac can make use of both halves of the 60 Hz sinusoidal source. As shown in Fig. 18-14a the Triac is connected into the circuit much like the SCR, but Fig. 18-14b shows that the Triac allows current to flow in both directions. Note that the Triac does not convert the input sinusoid into a controlled full-wave rectified signal at the output. Instead, the output voltage or current is simply a full-wave controlled sinusoid.

The best control circuit for the Triac or for almost all controlled switches or rectifiers is a pulse input to the gate circuits that is phase synchronized. There are two reasons for this choice: 1) pulses give precise control over when the firing is to take place, and 2) pulses provide current only for a short period of time. This second point is quite important because the average power in a pulse is small which means little power is expended in the gate-cathode region. As mentioned in an earlier discussion on the SCR, the SCR must be protected against excessive gate currents to avoid permanent damage. Pulse firing gives this protection.

It is a bit beyond the scope of this text to develop the phase-synchronized pulse firing circuits for controlled rectifiers and switches and the reader is advised to refer to such handbooks as the General Electric SCR Manual for details on this subject.

Problems

18-1. If an inductance replaces the capacitance in the control circuits of Figs. 18-7, 18-9, and 18-10, prove that the firing angle θ_1 is always 0°, regardless of the values of R or L.

18-2. The circuit shown in Fig. 18-2 has an input voltage of 250 volts and an R_D of 2000 ohms. Find the value of R_A if the OC3/105 tube current is equal to 25 ma.

18-3. The circuit of the accompanying figures uses two OD3/150 voltage-regulator tubes in series, thus providing two regulated voltages of 150 and 300

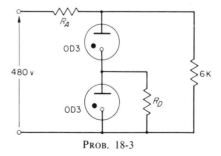

PROB. 18-3

volts. If the OD3 glow tube has the same rated 5- to 40-ma current range as the OC3 tube, find the minimum allowable value of R_D and the corresponding value of R_A that will not cause either tube to operate outside its rated current range.

18-4. In the half-wave diode circuit of Fig. 18-4, find the d-c current through the resistance if $e_s = 120 \sin \omega t$, $R = 1200$ ohms, and $E_t = 20$ volts. Assume that the firing potential E_f equals E_t.

18-5. Derive an expression for the total rms current flowing in the circuit of Fig. 18-4.

18-6. Using the I_{rms} expression derived in Prob. 18-5 and the parameter values of Prob. 18-4, find the power loss in the load resistance R.

18-7. Derive an equation for the average power loss in the diode of Fig. 18-4.

$$\left(Hint: \quad P = \frac{1}{2\pi} \int_0^{2\pi} e_b i_b \, d\omega t. \right)$$

18-8. In the circuit of the accompanying figure, calculate the firing angle. Sketch a phasor diagram, using V_{13} as a reference to aid in proving the resulting answer. Assume that the thyratron is ideal.

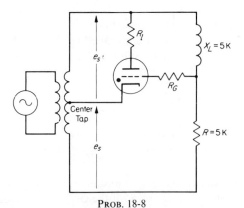

PROB. 18-8

18-9. Find the firing angle of the SCR circuit shown in the accompanying figure if $|V_{12}| = 3$ $|V_{23}|$ and $R = X_C$. Sketch a phasor diagram, with V_{13} as a reference, to obtain the answer.

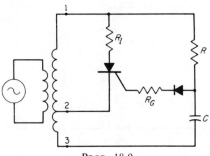

PROB. 18-9

18-10. If $e_s = 300 \sin \omega t$ volts and $R_l = 800$ ohms in Prob. 18-8, find $I_{\text{d-c}}$ flowing through the thyratron.

18-11. In the phase-shifting network shown in Fig. 18-10a, let $C = 0.5\ \mu\text{f}$ and R be a variable resistance. If the input frequency is 60 Hz, over what range must R be varied if it is desired to vary the firing angle θ_1 from $30°$ to $150°$?

19 / Laplace Transforms

Although the primary purpose of introducing Laplace transforms in the latter portion of this text is to support the material presented in the two chapters following on analog computers and servomechanisms, the Laplace transform method is a very general technique that is used to solve integro-differential equations. The Laplace transform method can be utilized very effectively in the study of electric circuits and as one studies the material in this chapter one should observe how Laplace transforms could have been used in analyzing the material in Chapter 13 on the frequency response of electronic circuits.

Unfortunately, one does not always have the time or, perhaps, feel it necessary to learn why a certain mathematical method works, or—even more important—when and under what conditions it can be applied; that is, the necessary and sufficient mathematical conditions for successful application of the method.

To develop an appreciation for the elegance of integral-transform methods in solving ordinary or partial integrodifferential equations, one should 1) understand how to solve ordinary differential equations by classical means, 2) learn complex variable theory, and 3) if possible, take formal instruction in Laplace-transform theory and application. Learning is a rather complicated process of progression, retrogression, and digression. The material in this text teaches *how* to use Laplace transforms in solving circuit equations, but it points out only *superficially why* transform methods are valuable and under what *conditions* they can be used; thus, Secs. 19-1 and 19-3 could be omitted by those already familiar with transform techniques.

The *how* to use Laplace transforms is developed in sufficient detail to allow the reader to solve many linear-circuit problems which can be described by ordinary linear integrodifferential equations. The *why* to use Laplace transforms can be given as a simple statement of fact: Laplace transforms allow one to solve high-order, linear, ordinary or partial differential equations by relatively simple algebraic methods and also allow the easy introduction of initial or boundary conditions. The necessary and sufficient *conditions* that must be met before transform techniques may be used in a problem solution are quite important and should not be ignored; however, since mathematicians have already established that almost all ordinary circuit problems can be solved by using Laplace transforms, one can simply accept this fact and proceed immediately to the task of solving problems. Additional comments on the necessary and sufficient conditions will be given in subsequent developments presented in this chapter.

19-1. Integral Transforms

An integral transform can be defined as follows:

$$F(\mathbf{x}) \triangleq \int_a^b f(y)K(\mathbf{x},y)\,dy \qquad (19\text{-}1)$$

where $K(\mathbf{x},y)$ is a known function of $\mathbf{x}$ and y and is called the kernel of the transform. When the limits a and b are finite, then $F(\mathbf{x})$ is the finite transform of $f(y)$; however, we shall let $a = 0$ or $-\infty$, and $b = \infty$, in all the discussions that follow.

Three kernels that are of interest to engineers are e^{-xy} sin xy or cos xy, and e^{-xy}, which are used in defining the Laplace transform, the Fourier sine, the Fourier cosine, and the complex Fourier transforms, respectively:

Laplace transform (single-sided):

$$F(\mathbf{x}) = \int_0^\infty f(y)e^{-xy}\,dy \qquad (19\text{-}2)$$

Fourier sine and Fourier cosine transforms:

$$F(\mathbf{x}) = \int_{-\infty}^\infty f(y)\sin(xy)\,dy \quad \text{and} \quad \int_{-\infty}^\infty f(y)\cos(xy)\,dy \quad (19\text{-}3)$$

Complex Fourier transform or double-sided Laplace transform:

$$F(\mathbf{x}) = \int_{-\infty}^\infty f(y)e^{-xy}\,dy \qquad (19\text{-}4)$$

In the usual engineering problem, y is a real quantity, but $\mathbf{x}$ can be either real, complex, or imaginary.

19-2. Laplace Transforms

The only integral transform that will be pursued in this chapter is the Laplace transform. All problems considered here will have time as the independent variable, or $y = t$. It is accepted practice to use the letter $\mathbf{s}$ rather than the letter $\mathbf{x}$ to distinguish the Laplace transform, so Eq. 19-1 becomes

$$F(\mathbf{s}) \triangleq \mathcal{L}[f(t)] = \int_0^\infty f(t)e^{-st}\,dt \qquad (19\text{-}5)$$

where, in general, **s** is a complex number $a + jb^1$ where a and b are real numbers unrelated to the limits in Eq. 19-1. In order for $f(t)$, in Eq. 19-5, to be Laplace transformable, $f(t)$ must satisfy the Dirichlet conditions, and the integral

$$\mathbf{F(s)} = \int_0^\infty f(t)e^{-st} \, dt$$

must exist. Functions which satisfy the Dirichlet conditions are those having only a finite number of maxima and minima within a finite interval and ordinary discontinuities. Equation 19-5 represents the single-sided Laplace transform and *by this concept it is not necessary to know or define $f(t)$ for* negative values of time. Alternatively, one can restrict analysis to so called *causal functions* which are functions that are zero for negative time. In this manner, if agreement is made to consider zero time as that instant at which the system is excited, all system variables will be causal functions since the system cannot respond before it is excited. In this manner almost all engineering system variables are causal functions and if the system is linear, the functions are generally Laplace transformable.

If the concept of an integral transformation seems difficult or mysterious, consider the many types of transformations used in mathematics; for example, the multiplication of a particular function by a constant or by a specified function, or the differentiating and integrating transformations of calculus. To give further motivation for the reason for using transformations, suppose that one knows how to add but does not know how to multiply. If one also knows how to take the log and antilog of a number, one never needs to know the process called multiplication. To multiply 10×1000, simply add log 10 + log $1000 = 1 + 3 = 4$, and then find the antilog of 4, or 10,000. The technique used to multiply is to first convert from the decimal-number domain to the log domain, then carry out the necessary addition in the log domain, and, finally, take the inverse transform to obtain the desired result in the decimal-number domain.

As brought out in the preceding paragraph, one must usually be able to perform an inverse transformation as well as the forward transformation in order to have a meaningful and useful analytical tool. Integral transformations do have well-defined inverse transformations, and, for the Laplace transformation, the inverse transformation is defined as

$$f(t) \triangleq \mathcal{L}^{-1}[\mathbf{F(s)}]$$

$$= \frac{1}{2\pi j} \int_{c-j\infty}^{c+j\infty} \mathbf{F(s)}e^{st} \, d\mathbf{s} \tag{19-6}$$

---

[1]Mathematicians usually use i for $\sqrt{-1}$, but in electric circuits, i is the symbol for current. To avoid confusion, j will be used to indicate the imaginary quantity $\sqrt{-1}$ as it was used in Chapter 12.

The evaluation of this integral is most easily accomplished by a contour integration in the complex plane. There are two reasons why Eq. 19-6 will not be utilized in this text: 1) there is no need to employ Eq. 19-6 in *those* Laplace-transform manipulations encountered in this text, and 2) since contour integration in the complex plane may be unfamiliar to the reader, an adequate development of the evaluation of Eq. 19-6 would be unnecessarily long and would probably not add much insight to the use of transform techniques. Perhaps the overriding reason for even mentioning the existence of the inverse transformation is to point out that there is a uniqueness or 1-to-1 correspondence between the forward $\mathbf{F(s)}$ and its inverse transform $f(t)$ for the functions encountered in most engineering systems.

Figure 19-1 illustrates the general procedure involved when solving a time-domain differential equation:

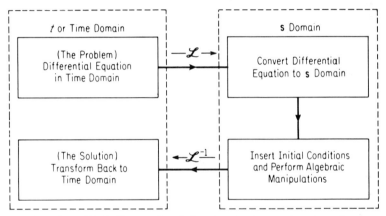

FIG. 19-1. Block-diagram description showing how a time-dependent differential equation is solved by using Laplace transforms.

1) Transform the equation from the time t domain to the $\mathbf{s}$ domain.

2) Insert the initial conditions and carry out the necessary algebraic manipulations.

3) Transform the new equation from the $\mathbf{s}$ domain to the t domain. (This process is called *taking the inverse transform.*)

Derivation of Some Basic Transforms. When confronted with the evaluation of integrals, the practicing engineer uses tables of integrals, whenever possible, rather than a formal mathematical solution. Likewise, when working with integral transforms, one tries to take advantage of the available tables of transform pairs $f(t)$ and the corresponding $\mathbf{F(s)}$.

The derivation of the transform $\mathbf{F(s)}$ of a given $f(t)$ is accomplished by performing the indicated integration given in Eq. 19-5. Three examples using explicit functions follow.

1. Let $f(t) = A$, a step function

$$F(s) = \int_0^\infty A e^{-st}\, dt$$

$$= -\frac{A}{s} e^{-st} \Big|_0^\infty$$

$$= \frac{A}{s}$$

if the real part of **s** is greater than zero, that is, $R(s) > 0$.

2. Let $f(t) = Be^{-at}$, an exponential function

$$F(s) = \int_0^\infty Be^{-at} e^{-st}\, dt$$

$$= \int_0^\infty Be^{-(a+s)t}\, dt$$

$$= -\frac{B}{s+a} e^{-(a+s)t} \Big|_0^\infty$$

$$= \frac{B}{s+a}$$

if $a + R(s) > 0$.

3. Let $f(t) = \cos \omega t$, a trigonometric function

$$F(s) = \int_0^\infty \cos(\omega t) e^{-st}\, dt$$

Using the trigonometric identity

$$\cos(\omega t) = \frac{e^{j\omega t} + e^{-j\omega t}}{2}$$

$$F(s) = \frac{1}{2} \int_0^\infty [e^{(j\omega - s)t} + e^{-(j\omega + s)t}]\, dt$$

$$= \frac{1}{2}\left[\frac{1}{j\omega - s} e^{(j\omega - s)t} - \frac{1}{j\omega + s} e^{-(j\omega + s)t} \right]\Bigg|_0^\infty$$

and allowing **s** to have a small positive real component so that the negative exponents damp out for large values of time,

$$F(s) = -\frac{1}{2}\left(\frac{1}{j\omega - s} - \frac{1}{j\omega + s} \right)$$

$$= -\frac{1}{2}\left[\frac{(j\omega + s) - (j\omega - s)}{(j\omega - s)(j\omega + s)} \right]$$

$$= -\frac{1}{2}\left[\frac{2s}{-(\omega^2 + s^2)} \right] = \frac{s}{s^2 + \omega^2}$$

if $R(s) > 0$.

By using Eq. 19-6, $f(t)$ could be evaluated from the above-calculated functions for $F(s)$, but this will not be shown here. The list of Laplace transforms for explicit functions of time could be continued almost indefinitely. Lengthy tables of Laplace-transform pairs appear in many texts, but only a short table, as given in Table 19-1, is sufficient for the usual problems encountered in linear-circuit theory. As will be shown later, complex polynomial expressions can be reduced to a summation of simple terms (such as those in Table 19-1) by partial-fraction-expansion techniques.

TABLE 19-1
TABLE OF LAPLACE-TRANSFORM PAIRS

$f(t)$	$F(s)$	$f(t)$	$F(s)$
1. 1	$1/s$	8. $f(t - a)u(t - a)$	$e^{-as}F(s)$
2. $u(t - a) = 0 \; t < a$ $= 1 \; t \geq a$	e^{-as}	9. $\sinh(at)$	$a/(s^2 - a^2)$
		10. $\cosh(at)$	$s/(s^2 - a^2)$
3. e^{-at}	$1/(s + a)$	11. $1 - e^{-at}$	$a/s(s + a)$
4. $\sin(\omega t)$	$\omega/(s^2 + \omega^2)$	12. $\dfrac{1}{b - a}(e^{-at} - e^{-bt})$	$\dfrac{1}{(s + a)(s + b)}$
5. $\cos(\omega t)$	$s/(s^2 + \omega^2)$	13. $\dfrac{1}{a - b}(ae^{-at} - be^{-bt})$	$\dfrac{s}{(s + a)(s + b)}$
6. $\dfrac{t^{n-1}}{(n - 1)!}$	$\dfrac{1}{s^n}\;(n = 1,2,\ldots)$	14. $\dfrac{1}{b}e^{-at}\sin bt$	$\dfrac{1}{(s + a)^2 + b^2}$
7. $\dfrac{t^{n-1}e^{-at}}{(n - 1)!}$	$1/(s + a)^n$	15. $e^{-at}\cos bt$	$\dfrac{s + a}{(s + a)^2 + b^2}$

Table 19-1 is not quite complete for our needs, because the Laplace transforms of the integral and differential operators are also necessary if integrodifferential equations are to be solved by utilizing the Laplace transform.

4a. Let $g(t)$ be the indefinite integral $\int f(t)\,dt$. The derivation of the Laplace transform of the indefinite integral can best be developed by redefining this integral expression as

$$g(t) \triangleq \int f(t)\,dt \triangleq \int_{\tau = -\infty}^{\tau = t} f(\tau)\,d\tau = \int_{-\infty}^{0} f(\tau)\,d\tau + \int_{0}^{t} f(\tau)\,d\tau$$

$$= g(0) + \int_{0}^{t} f(\tau)\,d\tau \qquad (19\text{-}7)$$

where τ is simply a convenient dummy variable.

The $g(0)$ expression on the right-hand side of Eq. 19-7 gives the past history of $f(t)$ over all negative time. As will be seen later, this is one of the important "initial-condition" terms which are a part of the solution to any time-dependent integrodifferential equation. Since an initial condition is a specified value or simply a known constant, the Laplace transform of Eq. 19-7 becomes

$$\mathcal{L}\left[g(t)\right] = \mathcal{L}\left[\int f(t)\, dt\right] = \mathcal{L}\left[g(0)\right] + \mathcal{L}\left[\int_0^t f(\tau)\, d\tau\right]$$

$$= \frac{g(0)}{s} + \int_0^\infty \left[\int_0^t f(\tau)\, d\tau\right] e^{-st}\, dt \quad (19\text{-}8)$$

The second expression in Eq. 19-8 can be integrated by parts by letting

$$u = \int_0^t f(\tau)\, d\tau \qquad\qquad dv = e^{-st}\, dt$$

$$du = f(t)\, dt \qquad\qquad v = -\frac{1}{s} e^{-st}$$

or

$$\int_0^\infty \left[\int_0^t f(\tau)\, d\tau\right] e^{-st}\, dt = \left. -\frac{1}{s} e^{-st} \int_0^t f(\tau)\, d\tau \right|_0^\infty + \frac{1}{s} \int_0^\infty f(t) e^{-st}\, dt$$

$$= 0 + \frac{F(s)}{s} \quad (19\text{-}9)$$

Now it should be obvious that the second term on the right-hand side of Eq. 19-9 is $F(s)/s$ simply by definition of the Laplace transform, but equating the first term to zero requires some additional explanation. The first term can be written

$$\left. -\frac{1}{s} e^{-st} \int_0^t f(\tau)\, d\tau \right|_0^\infty = -\frac{1}{s} e^{-s\infty} \int_0^\infty f(\tau)\, d\tau$$

$$+ \frac{1}{s} e^0 \int_0^0 f(\tau)\, d\tau \quad (19\text{-}10)$$

The term $\int_0^0 f(\tau)d\tau$, in Eq. 19-10, is identically zero for a Reimann integral, because no area can be contained in a zero-width increment. The first term $-(1/s)e^{-s\infty} \int_0^\infty f(\tau)d\tau$ is equal to zero only if $\int_0^\infty f(\tau)d\tau$ is a properly behaved function. In other words, if $f(\tau)$ is of exponential order such that

$$f(\tau) < M e^{\gamma t}$$

then, if the real part of s (or α is greater than γ, the term $-(1/s) \cdot e^{-s\infty} \int_0^\infty f(\tau)d\tau$ indeed vanishes. Thus, for a function to be Laplace transformable, the function $f(t)$ cannot increase in magnitude at a rate greater than a function of exponential order. Many functions, particularly those encountered in physical systems, are of exponential order or are bounded by an exponential-order function. Some common functions which are Laplace transformable are t^n, e^{at}, cos at, sin at, sinh at, and cosh at. However, functions such as tan at and cot at are not Laplace transformable, because the function approaches infinity at a faster rate than does a function of exponential order.

Combining Eqs. 19-8 and 19-9 results in

$$\mathcal{L}\left[\int f(t)\,dt\right] = \frac{F(s)}{s} + \frac{g(0)}{s} \qquad (19\text{-}11)$$

where $g(0) = \int_{-\infty}^{0} f(t)dt$, which accounts for any initial condition present at time $t = 0$.

4b. It can be shown that the Laplace transform of the double integration is

$$\mathcal{L}[\int \int f(t)\,dt^2] = \frac{F(s)}{s^2} + \frac{\displaystyle\int_{-\infty}^{0} f(\tau)\,d\tau}{s^2} + \frac{\displaystyle\int_{-\infty}^{0} \int_{-\infty}^{0} f(\tau)\,d\tau^2}{s} \qquad (19\text{-}12)$$

5a. Let $f(t)$ be the time derivative of $f(t)$ or $\dfrac{df(t)}{dt}$

$$F(s) = \mathcal{L}\left[\frac{df(t)}{dt}\right] = \int_{0}^{\infty} \frac{df(t)}{dt} e^{-st}\,dt$$

Again using integration by parts, let

$$u = e^{-st} \qquad dv = \frac{df(t)}{dt}\,dt$$

$$du = -se^{-st}\,dt \qquad v = f(t)$$

Thus,

$$\mathcal{L}\left[\frac{df(t)}{dt}\right] = e^{-st} f(t)\Big|_{0}^{\infty} + s\int_{0}^{\infty} f(t)\,e^{-st}\,dt$$

$$= \frac{1}{e^{s\infty}} f(\infty) - f(0^+) + sF(s)$$

$$= sF(s) - f(0^+) \qquad (19\text{-}13)$$

where it is assumed that $1/e^{s\infty} \rightarrow 0$ faster than $f(\infty) \rightarrow \infty$; that is; $\lim_{t \to \infty} [e^{-st} f(t)] = 0$ [again $f(t)$ is assumed to be of exponential order and the real part of s is greater than the real part of the exponent of $f(t)$]. Note $f(0^+)$ is the limit of $f(t)$ at $t = 0$ approaching the origin $(t \rightarrow 0)$ from the right. This defines the value to be used if $f(t)$ has a discontinuity at the origin $(t = 0)$.

5b. The Laplace transform for the second derivative *wrt* time is

$$\mathcal{L}\left[\frac{d^2f(t)}{dt^2}\right] = s^2 F(s) - sf(0^+) - f'(0^+) \qquad (19\text{-}14)$$

5c. The Laplace transform of the *n*th derivative *wrt* time is

$$\mathcal{L}\left[\frac{d^nf(t)}{dt^n}\right] = s^n F(s) - s^{n-1} f(0^+) - s^{n-2} f'(0^+) - \cdots - f^{n-1}(0^+) \quad (19\text{-}15)$$

where $f(0^+)$ is the initial value of the function $f(t)$ by approaching a zero value of time from positive time. Figure 19-2 shows that all time func-

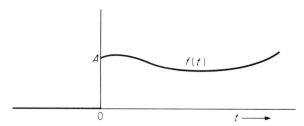

FIG. 19-2. A typical time function that is Laplace transformable. Note that the function is identically equal to zero for all time < 0.

tions, when transformed by using the single-sided Laplace transformation as given by Eq. 19-2 can be interpreted as though $f(t) = 0$ for all $t < 0$, i.e., are causal functions, since the lower limit on the integral is $t = 0$ and the integral is evaluated only over positive values of time t. As shown in Fig. 19-2, the proper initial value for $f(t)$ at $t = 0$ is A and not 0. Thus, $f'(0^+), f''(0^+)$, and so forth, are the initial values of the first, second, and so on, derivatives of $f(t)$ evaluated again as $t \to 0$ from positive values of time.

EXAMPLE 19-1. In the *R-L* circuit shown in Fig. 19-3a, the switch *S* has been in position *A* for a long time. (A long time means that the current has had sufficient time to reach its final or steady-state value.) At $t = 0$, the switch *S* is instantaneously thrown from position *A* to position *B*. Find the expression for the current in the *R-L* circuit after the switch has been thrown from *A* to *B*.

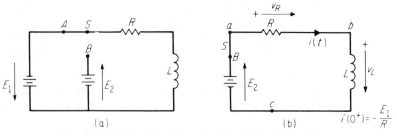

FIG. 19-3. a) Circuit for Example 19-1. b) Part a, simplified.

Solution: Figure 19-3b shows the circuit from which it is desired to solve for $i(t)$. Since one is asked to solve for the current flowing through the circuit, how does one proceed? Before writing the necessary equations, note that the voltage source and the assumed current flow are written as functions of time, whereas, in the examples in Chapter 2, the variables e, v, and i were often written as E, V, and I. *In resistive networks involving only d-c sources (such as batteries), the instantaneous quantities, which are written in lowercase letters, are the same as the average[2] quantities, which are written in uppercase letters.* A purely resistive network does not contain any energy-storage elements such as inductance or capacitance, and hence the current in the circuit *immediately* reaches its steady-state value upon application of an electric-energy source. If the circuit does contain an energy-storage element, then it usually takes a finite amount of time for the circuit voltages or currents to reach their steady-state values. *The general study of network behavior is usually divided into a transient and a steady-state analysis and for purely resistive circuits the transient solution is identically equal to zero.* The distinction between the transient and steady-state analyses will be emphasized in Sec. 19-4.

Using the differential-equation form of Kirchhoff's voltage law $v_{ac} = v_{ab} + v_{bc}$ and using the relationships given by Eqs. 12-27 and 12-43,

$$E_2 = \underbrace{Ri}_{v_R} + \underbrace{L\frac{di}{dt}}_{v_L} \qquad (19\text{-}16)$$

The first step, in using Laplace transforms, is to transform each term in the differential equation. This is accomplished by using the tables of Laplace-transform pairs (such as Table 19-1) for the various time functions that appear in the equation. For instance, the time function E_2, which is a constant, is transformed by using the first transform pair in Table 19-1 while $i(t)$ transforms directly into $I(s)$ by definition. The first derivative term $L(di/dt)$ transforms according to derivation 5a (Eq. 19-13). The transform of Eq. 19-16 is thus

$$\frac{E_2}{s} = R\,I(s) + L\,[s\,I(s) - i(0^+)] \qquad (19\text{-}17)$$

The second step is to substitute the initial conditions into the transformed equation. Since current cannot change instantaneously through an inductance (because $L\,di/dt$ is constrained to be finite in a circuit), the initial current flowing through the inductance is the same both the instant before and the instant after the switch S is thrown from position A to position B. (Of course, one must assume perfect or ideal switching to achieve this result; otherwise, arcing across the switch would occur.) Before switch S is thrown the steady state d-c current $i(0^+)$ is limited only by the resistance R or

$$i(0^+) = -\frac{E_1}{R} \qquad (19\text{-}18)$$

---

[2] As used in this text, the average, steady-state, and maximum values of a function are written as uppercase letters.

where the minus sign is necessary because $i(0^+)$ flows in the opposite direction to the assumed direction of $i(t)$, as shown in Fig. 19-3b. The complete Laplace-transformed equation, with initial conditions properly substituted into the equation, is

$$\frac{E_2}{s} = R\,\mathbf{I}(s) + L\left[s\,\mathbf{I}(s) - \left(-\frac{E_1}{R}\right)\right] \tag{19-19}$$

The third step is to solve for the unknown variable, which, in this case, is $\mathbf{I}(s)$.

$$(R + Ls)\,\mathbf{I}(s) = \frac{E_2}{s} - \frac{LE_1}{R}$$

or

$$\mathbf{I}(s) = \frac{E_2}{s(Ls + R)} - \frac{LE_1}{R(Ls + R)} \tag{19-20}$$

The fourth and final step is to take the inverse transform, or

$$i(t) = \mathcal{L}^{-1}[\mathbf{I}(s)] \tag{19-21}$$

To find the inverse transform, the transformed equation must be rearranged into the proper "form," so that the table of transform pairs can be matched with the terms in the transformed equation. Reorganizing Eq. 19-20, one obtains

$$\mathbf{I}(s) = \frac{E_2}{R}\left[\frac{R/L}{s(s + R/L)}\right] - \frac{E_1}{R}\left(\frac{1}{s + R/L}\right) \tag{19-22}$$

where the first term on the right-hand side of Eq. 19-22 matches transform pair 11, and the second term matches transform pair 3 (in Table 19-1). The constants outside the parentheses in each term go through the transformation unchanged, as any constant does in the process of an integration. The inverse transform of Eq. 19-22 becomes

$$i(t) = \mathcal{L}^{-1}\left[\mathbf{I}(s)\right] = \frac{E_2}{R}\left(1 - e^{-\frac{R}{L}\cdot t}\right) - \frac{E_1}{R}e^{-\frac{R}{L}\cdot t} \tag{19-23}$$

The terms which damp out after a period of time (the terms involving $e^{-R/L\,t}$) are the transient portion of the solution, and the remaining term (E_2/R) is the steady-state portion. One should be able to reason that the final value of $i(t)$ after a long time $[\lim_{t\to\infty} i(t)]$ is (E_2/R) by inspection of Fig. 19-3b. Also notice that Eq. 19-23 gives Eq. 19-18 when $t = 0$ as it should.

Partial-Fraction Expansion. Example 19-1 illustrates the easiest way to find the inverse transform of a given function; that is, look it up in a table of transform pairs. Even lengthy tables, however, are not necessary for the solution of a great many engineering problems. Perhaps the most useful single transform pair in Table 19-1 is pair 3, as will be demonstrated in this section.

A common mathematical manipulation used in performing integrations on polynomial-type integrands is to reduce the integrand to a series of simpler terms by the use of partial fractions, perform the integration term by term, and then add the resultant terms to obtain the solution. It is assumed that the reader is already familiar with the partial-fraction-expansion technique, so only a review of the method will be given here. Consider a ratio of 2 polynomials with the order of the denominator polynomial being equal to, or greater than, the order of the numerator polynomial (this is a necessary mathematical condition almost always met by equations describing practical engineering systems), for such is the case in many circuit problems. The following example illustrates the partial-fraction method of finding the inverse transform of a function expressed as the ratio of 2 polynomials.

EXAMPLES 19-2. Find the inverse transform of $a^3/s(s + a)^3$.

Solution: First, one tries to find this particular transform in the table of transform pairs. Since it is not listed in Table 3-1, partial fractions can be used to reduce the given transform to a series of simpler terms which can be found in Table 19-1. One way to do this is as follows:

$$\frac{a^3}{s(s + a)^3} = \frac{A}{s} + \frac{B}{(s + a)^3} + \frac{C}{(s + a)^2} + \frac{D}{s + a} \qquad (19\text{-}24)$$

where A, B, C, and D are constants that must be evaluated. Four arbitrary constants are necessary, because the denominator polynomial is of fourth order (the order is the highest power of s present in the denominator polynomial). These arbitrary constants could have been introduced in other ways, such as

$$\frac{a^3}{s(s + a)^3} = \frac{A}{s} + \frac{Es^2 + Fs + G}{(s + a)^3}$$

or

$$\frac{a}{s(s + a)^3} = \frac{A}{s} + \frac{H}{(s + a)^3} + \frac{Ms + N}{(s + a)^2}$$

but Eq. 19-24 is the form that is usually the most desirable. (The primary reason for this choice is that terms of the form $Es^2/(s + a)^3$ or $Ms/(s + a)^2$ are not listed in Table 19-1.)

If the constants A, B, C, and D in Eq. 19-24 can be evaluated, transform pairs 1 and 7 can be used to determine the inverse transform of each term individually and the resultant inverse transforms summed to give the time-domain expression, as the inverse transformation is often termed. The evaluation of the constants A, B, C, and D can be done in either of two ways. Both methods will be presented, as it is difficult to say which is the better. (Usually, a combination of the two methods will yield the quickest results.)

Method 1. Repeating Eq. 19-24 and finding a common denominator for all terms on the right-hand side of Eq. 19-24,

$$\frac{a^3}{s(s + a)^3} = \frac{A(s + a)^3 + Bs + Cs(s + a) + Ds(s + a)^2}{s(s + a)^3} \qquad (19\text{-}25)$$

Upon multiplying both sides of Eq. 19-25 by $s(s + a)^3$ and expanding,

$$a^3 = A(s^3 + 3s^2a + 3sa^2 + a^3)$$
$$+ Bs + Cs(s + a) + Ds(s^2 + 2as + a^2) \qquad (19\text{-}26)$$

and collecting the coefficients of like powers of s on either side of Eq. 19-26,

$$a^3 = (A + D)s^3 + (3aA + C + 2aD)s^2$$
$$+ (3a^2A + B + Ca + Da^2)s + a^3A \qquad (19\text{-}27)$$

To satisfy Eq. 19-27, the coefficients of like powers of s on either side of Eq. 19-27 must be equal, or

$$A + D = 0 \qquad (19\text{-}28a)$$
$$3aA + C + 2aD = 0 \qquad (19\text{-}28b)$$
$$3a^2A + B + Ca + Da^2 = 0 \qquad (19\text{-}28c)$$
$$a^3A = a^3 \qquad (19\text{-}28d)$$

Equation set 19-28 is a set of four simultaneous linear algebraic equations with four unknowns. The set is quite easy to solve by using the substitution method. Using Eq. 19-28d and substituting the result into Eq. 19-28a, one obtains

$$A = 1 \quad \text{and} \quad D = -1$$

Next, using Eq. 19-28b,

$$3a(1) + C + 2a(-1) = 0$$
$$C = -a$$

Finally, using Eq. 19-28c,

$$3a^2(1) + B + (-a)a + (-1)a^2 = 0$$
$$B = -a^2$$

Now that the constants have been evaluated, one can write Eq. 19-24 as

$$\frac{a^3}{s(s + a)^3} = \frac{1}{s} - \frac{a^2}{(s + a)^3} - \frac{a}{(s + a)^2} - \frac{1}{s + a} \qquad (19\text{-}29)$$

and the inverse transform of Eq. 19-29 can be accomplished by a term-by-term comparison with a table of transform pairs. An inspection of Eq. 19-29 reveals that only transform pairs 1 and 7 (from Table 19-1) are needed. Thus,

$$\mathcal{L}^{-1}\left[\frac{a^3}{s(s + a)^3}\right] = 1 - \frac{a^2 t^{3-1}e^{-at}}{(3 - 1)!} - \frac{a\,t^{2-1}e^{-at}}{(2 - 1)!} - \frac{t^{1-1}e^{-at}}{(1 - 1)!}$$

Gathering terms and simplifying gives

$$\mathcal{L}^{-1}\left[\frac{a^3}{s(s + a)^3}\right] = 1 - \left(\frac{a^2 t^2}{2} + at + 1\right)e^{-at} \qquad (19\text{-}30)$$

Method 2. This is probably the most popular (because it is often the quicker) of the two methods except, perhaps, when repeated roots are encountered, such as the $(s + a)^3$ term in this problem. Again repeating Eq. 19-24,

$$\frac{a^3}{s(s + a)^3} = \frac{A}{s} + \frac{B}{(s + a)^3} + \frac{C}{(s + a)^2} + \frac{D}{s + a} \qquad (19\text{-}31)$$

It should be pointed out that the`s is a variable and can have any value—real or complex. To evaluate the constant A, we multiply both sides of Eq. 19-31 by s, or

$$\frac{a^3}{(s+a)^3} = A + \frac{Bs}{(s+a)^3} + \frac{Cs}{(s+a)^2} + \frac{Ds}{s+a} \qquad (19\text{-}32)$$

Since Eq. 19-32 is valid for any value of s, let $s = 0$ for convenience, because this eliminates all but one term on the right-hand side of Eq. 19-32, or

$$\frac{a^3}{a^3} = A$$

and

$$A = 1$$

Next, multiply both sides of Eq. 19-31 by $(s+a)^3$ or

$$\frac{a^3}{s} = \frac{A(s+a)^3}{s} + B + C(s+a) + D(s+a)^2 \qquad (19\text{-}33)$$

If one lets $s = -a$, all terms on the right-hand side of Eq. 19-33 involving the $(s+a)$ term equal 0, and thus,

$$\frac{a^3}{-a} = B$$

or

$$B = -a^2 \qquad (19\text{-}34)$$

The evaluation of the constants C and D could be made by going back to method 1 or by taking successive derivatives of Eq. 19-33. Taking the derivative of both sides of Eq. 19-33 with respect to s gives

$$\frac{-a^3}{s^2} = \frac{A[3s(s+a)^2 - (s+a)^3]}{s^2} + 0 + C + 2D(s+a) \qquad (19\text{-}35)$$

Again letting $s = -a$,

$$\frac{-a^3}{(-a)^2} = C$$

or

$$C = -a \qquad (19\text{-}36)$$

To evaluate D, we take the derivative of Eq. 19-35 with respect to s, or

$$\frac{2a^3}{s^3} = A\left[\frac{6s(s+a)}{s^2} - \frac{6s^2(s+a)^2 - 2s(s+a)^3}{s^4}\right] + 2D \qquad (19\text{-}37)$$

Again letting $s = -a$,

$$\frac{2a^3}{(-a)^3} = A[0] + 2D$$

or

$$D = -1 \qquad (19\text{-}38)$$

The values of the constants A, B, C, and D are seen to be the same, regardless of whether method 1 or method 2 is used. It is usually advantageous, with respect to the amount of computational time involved, to use a combination of these methods in the determination of the constants in a partial-fraction expansion.

Several more examples will be given to illustrate the use of Laplace-transform techniques.

EXAMPLE 19-3. Find the inverse Laplace transform of

$$\frac{3s^2 + 9s + 5}{(s + 3)(s^2 + 3s + 2)}$$

Solution: Using the partial-fraction-expansion method 2 and rewriting the denominator polynomial into its individual roots,

$$\frac{3s^2 + 9s + 5}{(s + 3)(s + 2)(s + 1)} = \frac{A}{s + 3} + \frac{B}{s + 2} + \frac{C}{s + 1}$$

and solving for the constants A, B, and C (which are called residues).

$$A = \left.\frac{3s^2 + 9s + 5}{(s + 2)(s + 1)}\right|_{s = -3} = \frac{3(-3)^2 + 9(-3) + 5}{(-3 + 2)(-3 + 1)} = \frac{5}{2}$$

$$B = \left.\frac{3s^2 + 9s + 5}{(s + 3)(s + 1)}\right|_{s = -2} = \frac{3(-2)^2 + 9(-2) + 5}{(-2 + 3)(-2 + 1)} = 1$$

$$C = \left.\frac{3s^2 + 9s + 5}{(s + 3)(s + 2)}\right|_{s = -1} = \frac{3(-1)^2 + 9(-1) + 5}{(-1 + 3)(-1 + 2)} = -\frac{1}{2}$$

Thus,

$$\mathcal{L}^{-1}\left[\frac{3s^2 + 9s + 5}{(s + 3)(s^2 + 3s + 2)}\right] = \mathcal{L}^{-1}\left(\frac{5/2}{s + 3}\right) + \mathcal{L}^{-1}\left(\frac{1}{s + 2}\right) + \mathcal{L}^{-1}\left(\frac{-1/2}{s + 1}\right)$$

$$= \frac{5}{2}e^{-3t} + e^{-2t} - \frac{1}{2}e^{-t}$$

where it is seen that transform pair 3 is used to find the inverse.

EXAMPLE 19-4. Derive the transform for the pulse function $e(t)$ shown in Fig. 19-4.

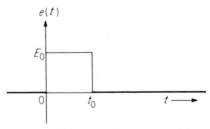

FIG. 19-4.   Relationship of $e(t)$ versus t for Example 19-4.

Solution:

$$\mathcal{L}\ [e(t)] = \int_0^{t_0} E_0 e^{-st}\,dt + \int_{t_0}^{\infty} 0e^{-st}\,dt$$

$$= -\frac{1}{s}E_0 e^{-st}\bigg|_0^{t_0} + 0$$

$$= -\frac{E_0}{s}(e^{-st_0} - 1)$$

or

$$\mathcal{L}\ [e(t)] = \frac{E_0}{s}(1 - e^{-st_0}) \tag{19-39}$$

EXAMPLE 19-5. If $E_0 = 1/a$ and $t_0 = a$, find the Laplace transform of Eq. 19-39 as $a \to 0$.

Solution: Figure 19-5 shows the plot of this newly defined pulse function. Note that as $a \to 0$, the amplitude of the pulse approaches ∞ and the width

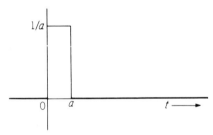

FIG. 19-5. Plot of a unit pulse function (reduces to a unit impulse function as $a \to 0$).

approaches zero, but the area remains unity ($a \times 1/a = 1$). This particular function, as just defined, is termed a unit impulse function. The Laplace transform for the unit impulse function can be obtained by rewriting Eq. 19-39 as

$$\mathcal{L}\ [\text{unit impulse function}] \triangleq \lim_{a \to 0} \left[\frac{1}{as}(1 - e^{-as})\right] \tag{19-40}$$

and, using the power-series expansion of e^{-as},

$$e^{-as} = 1 - as + \frac{(as)^2}{2!} - \frac{(as)^3}{3!} + \cdots \tag{19-41}$$

As $a \to 0$, the magnitudes of all terms of second order and above become negligibly small compared with the magnitude of the linear term. Upon substituting Eq. 19-41 into Eq. 19-40 and taking the limit,

$$\mathcal{L}\ [\text{unit impulse function}] = \lim_{a \to 0} \left\{\frac{1}{as}[1 - (1 - as)]\right\} = 1 \tag{19-42}$$

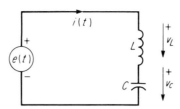

FIG. 19-6. Circuit diagram for
Example 19-6.

EXAMPLE 19-6. If $e(t)$, in Fig. 19-6, is a unit impulse function, find $i(t)$. All initial voltages and currents are zero.

Solution: Kirchhoff's voltage law in integral-differential-equation form, for the circuit in Fig. 19-6, is

$$e(t) = \underbrace{L\frac{di}{dt}}_{v_L} + \underbrace{\frac{1}{C}\int_{-\infty}^{t} i\,dt}_{v_C} \qquad (19\text{-}43)$$

Taking the Laplace transform of Eq. 19-43,

$$1 = L[sI(s) - i(0^+)] + \frac{1}{C}\left[\frac{I(s)}{s} + \frac{\int_{-\infty}^{0} i(t)\,dt}{s}\right] \qquad (19\text{-}44)$$

and, since all initial conditions are zero, $i(0^+)$ and $q(0^+) = \int_{-\infty}^{0} i(t)\,dt = 0$.
Next, solving for $I(s)$,

$$1 = \left(Ls + \frac{1}{Cs}\right)I(s)$$

or

$$I(s) = \frac{s/L}{s^2 + \dfrac{1}{LC}} \qquad (19\text{-}45)$$

The inverse transform is found by using transform pair 5 in Table 19-1, or

$$i(t) = \frac{1}{L}\cos\left(\sqrt{\frac{1}{LC}}\,t\right) \qquad (19\text{-}46)$$

Several important aspects of the preceding discussions and examples need to be emphasized at this point, as these observations will be of paramount interest in many of the following analyses.

1. Recall that the denominators of the ratios of s-domain polynomials completely determine the essential characteristics of the time-domain expressions, whereas the numerator polynomials merely affect the value of the coefficients (or residues). The denominator of a ratio of polynomials

representing the s-domain expression, when set equal to zero, is often aptly called the *characteristic* equation, because the interesting or characteristic behavior of the inverse or time-domain expression is contained in this denominator polynomial.

2. Because of its simplicity, the unit impulse is often used as a forcing or excitation function in analyzing the behavior of circuits or systems. A transfer function $T(s)$ is the ratio of some transformed output variable over a transformed input variable, or

$$T(s) = \frac{\text{output quantity (s)}}{\text{input quantity (s)}}$$
$$= \frac{a_m s^m + a_{m-1} s^{m-1} + \cdots + a_0}{b_n s^n + b_{n-1} s^{n-1} + \cdots + b_0} \qquad (19\text{-}47)$$

where $n \geq m$ for all problems treated in this text. This is also true for the majority of problems of engineering interest. Thus, if the input is a unit impulse, the output is

$$\begin{aligned} \text{Output (s)} &= T(s)\,[\text{Input (s)}] \\ &= T(s)[1] \\ &= T(s) \end{aligned}$$

or, in the time domain,

$$\text{Output}(t) = \mathcal{L}^{-1}\,[T(s)]$$

The above equation shows that the time-domain output is simply the inverse transform of the transfer function itself. The unit impulse response of a circuit or system indicates only the characteristic behavior of the circuit or system under consideration; that is, the unit impulse response can be considered the homogeneous or complementary solution to the differential equation which describes the system.

Initial- and Final-Value Theorems. Quite often one is interested only in the initial or the final value of the time-domain response. One can find these values in a straightforward way by letting $t = 0$ or $t = \infty$ in the time-domain expression, but two theorems concerning Laplace transforms allow one to determine these values while the system-response function is still in the s domain. Obviously, these theorems can save a great deal of time and effort if the system response is such that the inverse transformation is tedious or difficult to obtain.

Since a rigorous proof of these two theorems is a bit involved, only the statement and application of each theorem will be offered here. The initial-value theorem states that

$$\lim_{s \to \infty} [sF(s)] = \lim_{t \to 0} [f(t)]$$

with the stipulation that $f(t)$ and its derivative $f'(t)$ are Laplace transformable, that $f(t)$ has the transform $F(s)$, and that $\lim_{s \to \infty} [sF(s)]$ exists.

The final-value theorem states that

$$\lim_{s \to 0} [\mathbf{sF(s)}] = \lim_{t \to \infty} [f(t)]$$

with the stipulation that $f(t)$ and its derivative $f'(t)$ are Laplace transformable, that $f(t)$ has the transform $\mathbf{F(s)}$, and that the poles (roots of the denominator polynomial in systems treated in this text) of the function $\mathbf{F(s)}$ lie *inside* the left half of the complex s plane. This last restriction is necessary to ensure that $f(t)$ decreases exponentially with time.

EXAMPLE 19-7. If the transfer function $\mathbf{V_{out}(s)}/\mathbf{V_{in}(s)}$ of the network shown in Fig. 19-7 is as indicated, find the initial and final values of the output voltage v_{out} if v_{in} is a unit-step function.

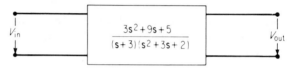

FIG. 19-7. Figure to accompany Example 19-7.

Solution: With v_{in} a unit step of 1-volt amplitude, $\mathbf{V_{in}(s)} = 1/s$. From Fig. 19-7,

$$\frac{\mathbf{V_{out}(s)}}{\mathbf{V_{in}(s)}} = \frac{3s^2 + 9s + 5}{(s + 3)(s^2 + 3s + 2)}$$

or

$$\mathbf{V_{out}(s)} = \frac{1}{s} \frac{3s^2 + 9s + 5}{(s + 3)(s^2 + 3s + 2)}$$

The initial value is

$$\lim_{s \to \infty} [\mathbf{sF(s)}] = \lim_{s \to \infty} [\mathbf{sV_{out}(s)}] = \lim_{s \to \infty} \left[\frac{3s^2 + 9s + 5}{(s + 3)(s^2 + 3s + 2)} \right] = 0$$

Thus,

$$\lim_{t \to 0} [v_{out}(t)] = 0$$

The final value is

$$\lim_{s \to 0} [\mathbf{sF(s)}] = \lim_{s \to 0} [\mathbf{sV_{out}(s)}] = \lim_{s \to 0} \left[\frac{3s^2 + 9s + 5}{(s + 3)(s^2 + 3s + 2)} \right] = \frac{5}{6}$$

Thus,

$$\lim_{t \to \infty} [v_{out}(t)] = \tfrac{5}{6} \text{ volt}$$

19-3. Operational Impedances

The text material and examples given in Chapter 2 on loop-current and node-voltage methods employed only resistors as the elements to impede the flow of current, and in the preceding discussion of Laplace-transform techniques the differential equations for the R, L, and C elements were used primarily to illustrate the solution of differential equations by Laplace transforms. In general, the three circuit qualities— resistance, inductance, and capacitance—must be incorporated into the study of circuits and circuit analysis. The circuit of Fig. 19-8 shows a

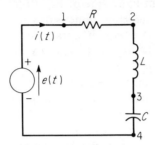

FIG. 19-8. A series *R-L-C* circuit.

resistor, an inductor, and a capacitor connected in series across a voltage source.

The current flowing in this circuit can be solved for by using Kirchhoff's voltage law and the relations given in Eqs. 12-27, 12-43, and 12-47. In general terms,

$$v_{14} = e(t) = v_{12} + v_{23} + v_{34} \qquad (19\text{-}48)$$

where

$$v_{12} = v_{12}(t) = i(t)\,R$$

$$v_{23} = v_{23}(t) = L\frac{d[i(t)]}{dt}$$

$$v_{34} = v_{34}(t) = \frac{1}{C}\int_{-\infty}^{t} i(t)\,dt$$

Equation 19-48 becomes

$$e(t) = Ri(t) + L\frac{d[i(t)]}{dt} + \frac{1}{C}\int_{-\infty}^{t} i(t)\,dt \qquad (19\text{-}49)$$

which is a linear integral-differential equation with constant coefficients. Until $e(t)$ is specified, a solution cannot be obtained, and even if $e(t)$ is assumed to be a step function (or d-c source), the current $i(t)$ can take on many values, depending on the relative values of the R, L, and C parameters. The solution to Eq. 19-49 can be obtained by classical means or by using Laplace transforms. Using Laplace transforms,

$$\mathbf{E(s)} = R\,\mathbf{I(s)} + L[s\mathbf{I(s)} - i(0^{+})] + \frac{1}{C}\left[\frac{\mathbf{I(s)}}{s} + \frac{\int_{-\infty}^{0} i(\tau)\,d\tau}{s}\right] \qquad (19\text{-}50a)$$

$$E(s) = \left(R + Ls + \frac{1}{Cs}\right)I(s) - Li(0^+) + \frac{\int_{-\infty}^{0} i(\tau)d\tau}{Cs} \qquad (19\text{-}50b)$$

Equation 19-50b suggests an interesting mathematical concept. Note that the first term on the right-hand side of Eq. 19-50b contains three entries, all of which must have the same dimensions in order to be compatible. Since the resistance R is a circuit parameter which *impedes* the flow of current, it seems only logical to call Ls and $1/Cs$ by some name which connotates an *impedance* to current flow. When using Laplace transforms, these impedances are commonly called operational impedances. Table 19-2 summarizes the operational-impedance relationships. The remaining two terms in Eq. 19-50b are due to the initial conditions in the circuit. *By definition or by concept, it should be understood that initial conditions do not affect the impedance of a linear circuit.*

TABLE 19-2

Element	Operational Impedance
Resistance R	R
Inductance L	Ls
Capacitance C	$1/Cs$

19-4. Circuit Analysis

The solution to a linear, time-dependent differential equation with constant coefficients (all circuit equations encountered in this text fall in this category) consists, in general, of two parts—the transient and the steady-state portions. From the mathematician's viewpoint, the transient and steady-state solutions are closely akin to the complementary and particular solutions, respectively. Since some confusion might exist over the meaning of the terms transient and steady state, the following definitions will be adhered to in this text:

1. For either a d-c or a periodic input excitation (such as a sinusoidal function), the transient portion of a response function is that portion which, for all practical purposes, decays to zero after a finite period of time has elapsed.

2. For either a d-c or a periodic input excitation, that portion of a response function which remains over all positive values of time is called the steady-state portion.

The current $i(t)$ in a series R-L circuit with a sinusoidal-voltage excitation is known to be

$$i(t) = A\left(-\cos \omega t + \frac{R}{\omega L}\sin \omega t + e^{-(R/L)t}\right)$$

(Compare Fig. 19-14 and Eq. 19-72). The transient portion $i_T(t)$ of the response function $i(t)$ is

$$i_T(t) = Ae^{-(R/L)t}$$

which decays almost to zero after a reasonable period of time. [A reasonable period of time might be for $t \geq 5 \cdot L/R$ because $e^{-R/L \cdot 5L/R} = e^{-5} = 1/150$ which means that $i_T(t = 5L/R)$ equals only $A/150$ of its initial value at $t = 0$.]

The steady-state portion $i_{ss}(t)$ of the response function $i(t)$ is

$$i_{ss}(t) = A\left(-\cos \omega t + \frac{R}{\omega L}\sin \omega t\right)$$

which remains for all values of time.

It is quite easy to determine and write the necessary independent equations for a circuit solution, but the actual solving of the equations can be a very difficult or tedious task. Although the proper introduction of initial conditions into the circuit solutions is important, the vast majority of circuit problems usually encountered assume that all initial conditions are zero. This is especially true when the steady-state solution (all transient terms become negligible) is desired for a particular excitation function. For this reason, with the exception of this chapter, almost all the examples and problems assume zero for the initial conditions.

The examples given in the section on Laplace transforms illustrated solutions to relatively simple circuits. The circuit shown in Fig. 19-8 is a somewhat more complicated problem, even though only a single mesh is encountered. Let us assume that $e(t)$ is a step function of magnitude E_1 and that all initial conditions are zero. The concept of the step function in a circuit is the same as showing a d-c source and a switch. In other words, the voltage source $e(t)$, in Fig. 19-8, could be replaced by a d-c source and a switch (which is closed at $t = 0$) if $e(t)$ is specified as a step function (refer to Fig. 19-9). For a unit step function, E_1 would equal 1

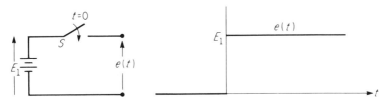

FIG. 19-9. Comparison of a d-c source and switch and a step function.

volt. Using these stipulations, and under the condition of zero initial current and zero initial capacitor voltage, Eq. 19-50b becomes

$$\frac{E_1}{s} = \left(R + Ls + \frac{1}{Cs}\right)I(s) \tag{19-51}$$

and, solving for $\mathbf{I(s)}$,

$$\mathbf{I(s)} = \frac{E_1}{\mathbf{s}\left(\dfrac{LC\mathbf{s}^2 + RC\mathbf{s} + 1}{C\mathbf{s}}\right)}$$

$$= \frac{E_1/L}{\mathbf{s}^2 + \dfrac{R}{L}\mathbf{s} + \dfrac{1}{LC}} \tag{19-52}$$

The current $i(t)$ can be obtained by taking the inverse transformation. This can be accomplished by factoring the denominator polynomial of Eq. 19-52 and using transform pair 12 (given in Table 19-1). The factors of the denominator polynomial are

$$\mathbf{s}^2 + \frac{R}{L}\mathbf{s} + \frac{1}{LC} = \left[\mathbf{s} + \frac{R}{2L} + \sqrt{\left(\frac{R}{2L}\right)^2 - \frac{1}{LC}}\right]$$

$$\left[\mathbf{s} + \frac{R}{2L} - \sqrt{\left(\frac{R}{2L}\right)^2 - \frac{1}{LC}}\right] \tag{19-53}$$

Transform pair 12 indicates that

$$\mathcal{L}^{-1}\left[\frac{1}{(\mathbf{s} + a)(\mathbf{s} + b)}\right] = \frac{1}{b - a}(e^{-at} - e^{-bt}) \tag{19-54}$$

Comparison of Eqs. 19-53 and 19-54 shows that

$$a = +\frac{R}{2L} + \sqrt{\left(\frac{R}{2L}\right)^2 - \frac{1}{LC}}$$

$$b = +\frac{R}{2L} - \sqrt{\left(\frac{R}{2L}\right)^2 - \frac{1}{LC}}$$

Thus, the inverse transform of Eq. 19-52 is

$$i(t) = \left(\frac{E_1}{L}\right)\left[\frac{1}{-2\sqrt{\left(\frac{R}{2L}\right)^2 - \frac{1}{LC}}}\right]$$

$$\left\{e^{\left[-\frac{R}{2L} - \sqrt{\left(\frac{R}{2L}\right)^2 - \frac{1}{LC}}\right]t} - e^{\left[-\frac{R}{2L} + \sqrt{\left(\frac{R}{2L}\right)^2 - \frac{1}{LC}}\right]t}\right\}$$

$$= \left[\frac{E_1}{L\sqrt{\left(\frac{R}{2L}\right)^2 - \frac{1}{LC}}}\right]\left(e^{-\frac{R}{2L}t}\right)\left\{\frac{e^{\left[\sqrt{\left(\frac{R}{2L}\right)^2 - \frac{1}{LC}}\right]t} - e^{-\left[\sqrt{\left(\frac{R}{2L}\right)^2 - \frac{1}{LC}}\right]t}}{2}\right\}$$

$$= \left[\frac{E_1}{L\sqrt{\left(\frac{R}{2L}\right)^2 - \frac{1}{LC}}}\right]\left(e^{-\frac{R}{2L}t}\right)\sinh\left[\sqrt{\left(\frac{R}{2L}\right)^2 - \frac{1}{LC}}\right]t \tag{19-55}$$

Equation 19-55 represents the overdamped solution to $i(t)$ for $(R/2L)^2 > 1/LC$.

If $(R/2L)^2 = 1/LC$, Eq. 19-53 reduces to

$$s^2 + \frac{R}{L}s + \frac{1}{LC} = \left(s + \frac{R}{2L}\right)^2$$

and so Eq. 19-52 becomes

$$\mathbf{I(s)} = \frac{E_1/L}{\left(s + \frac{R}{2L}\right)^2}$$

Thus, $\mathcal{L}^{-1}[\mathbf{I(s)}]$ is

$$i(t) = \mathcal{L}^{-1}\left[\frac{E_1/L}{\left(s + \frac{R}{2L}\right)^2}\right]$$

Using transform pair 7 of Table 19-1,

$$i(t) = \frac{E_1}{L}te^{-\frac{R}{2L}t} \tag{19-56}$$

Equation 19-56 is called the critically damped solution.

Finally, if $(R/2L)^2 < 1/LC$, the hyperbolic function becomes a trigonometric relation (recall the identity $\sinh jat = j \sin at$), so that Eq. 19-55 becomes

$$i(t) = \frac{E_1}{L\sqrt{\frac{1}{LC} - \left(\frac{R}{2L}\right)^2}}e^{-\frac{R}{2L}t}\sin\left[\sqrt{\frac{1}{LC} - \left(\frac{R}{2L}\right)^2}\right]t \tag{19-57}$$

Equation 19-57 is the underdamped solution to the problem and is probably the solution of most interest in many circuit-analysis problems. Figure 19-10 shows a sketch of the solutions represented by Eqs. 19-55, 19-56, and 19-57.

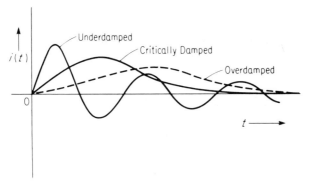

FIG. 19-10.   The current $i(t)$ for a series R-L-C circuit with a step input voltage.

Actually, Eq. 19-57 can be obtained more quickly and easily by using transform pair 14 of Table 19-1. First complete the square in the denominator of Eq. 19-52 and obtain

$$I(s) = \frac{E_1/L}{\left(s + \dfrac{R}{2L}\right)^2 + \dfrac{1}{LC} - \left(\dfrac{R}{2L}\right)^2}$$

and note that $a = R/2L$ and $b = \sqrt{1/LC - (R/2L)^2}$. Thus,

$$i(t) = \frac{E_1}{L} \frac{1}{\sqrt{\dfrac{1}{LC} - \left(\dfrac{R}{2L}\right)^2}} e^{-\frac{Rt}{2L}} \sin \left[\frac{1}{LC} - \left(\frac{R}{2L}\right)^2\right]^{1/2} t$$

which is identical to Eq. 19-57.

It is always wise to see if a qualitative interpretation of the solution can be made. In this example the current begins at zero and ends at zero. Is this to be expected? Yes, because the initial current $i(t = 0)$ is known to be zero, and, because of the series inductance, the current must also be zero at the instant that the voltage source is applied to the circuit. Because of the series capacitance, the current must eventually return to zero, as a capacitor blocks the flow of d-c current. The reader should be able to present arguments as to why the voltage across the capacitor starts at 0 at $t = 0$ and approaches E_1 volts at $t \to \infty$.

Several additional examples will serve to illustrate the solution to circuits by using Kirchhoff's laws, Ohm's law, and Laplace transforms.

EXAMPLE 19-8. Determine the current $i(t)$, in the circuit of Fig. 19-11, if the capacitor is initially charged with a voltage V_{co}. The current begins to flow when switch S is closed.

Solution: This problem is almost identical to the one just described. As will be seen, the initial voltage on the capacitor produces almost the same solution for $i(t)$ as the one obtained for $i(t)$ in using Fig. 19-8 with $e(t)$ a step function and all initial conditions equal to zero.

Obviously, the current $i(t)$ is equal to zero until switch S is closed. At the instant of closure of switch S, a loop-current equation results in

$$v_{ab} + v_{bd} + v_{da} = 0$$

and, writing the differential-equation quantities for v_{ab}, v_{bd}, and v_{da},

$$R i + L\frac{di}{dt} + \frac{1}{C} \int_{-\infty}^{t} i \, dt = 0 \tag{19-58}$$

Taking the Laplace transform of Eq. 19-58,

$$R\,\mathbf{I}(s) + L\,[s\,\mathbf{I}(s) - i(0^+)] + \frac{1}{C}\left[\frac{\mathbf{I}(s)}{s} + \frac{\int_{-\infty}^{0} i\,dt}{s}\right] = 0 \qquad (19\text{-}59)$$

Before solving for $I(s)$, it is necessary to substitute the proper initial conditions into Eq. 19-59. Of course, $i(0^+)$ is equal to zero, but what is the evaluation of $\int_{-\infty}^{0} i\,dt$? Since $i = dq/dt$, the integral must have the dimensions of charge—but what charge? The limits on the integral $\int_{-\infty}^{0} i(t)\,dt$ indicate that the integral should be evaluated for all values of time *before* switch S is closed.[3] In other words the past history showing the net charge placed on the capacitor should be accounted for. The net charge on the capacitor at $t = 0^+$ (the very instant the switch is closed) is noted as Q_0. Solving Eq. 19-59 for $I(s)$, one obtains

$$\left(R + Ls + \frac{1}{Cs}\right)\mathbf{I}(s) = -\frac{Q_0}{Cs} \qquad (19\text{-}60)$$

The term Q_0 is substituted into Eq. 19-59 as a positive quantity because of the *assumed* polarity chosen for V_{C0} and the *assumed* direction for $i(t)$. Note that $i(t)$ is chosen in such a manner that, for a positive $i(t)$, the charge on the capacitor would increase in the positive sense. Had V_{C0} or $i(t)$ been chosen in the opposite

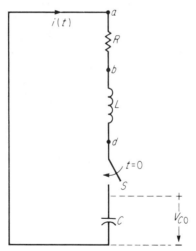

FIG. 19-11. Figure to accompany
Example 19-8.

[3]The question of a finite discontinuity of charge at $t = 0$, such as that illustrated in the accompanying figure, might arise in the mind of the reader. This *cannot* occur, because it violates the law of conservation of momentum in that the charge on a capacitor cannot be changed instantaneously. Hence the charge curve, q versus t, must be continuous over all time, including $t = 0$.

sense, Q_0 in Eq. 19-59 would have been entered as a negative quantity. Using the relation $Q = Cv$, V_{C0} can be substituted for Q_0/C. $\mathbf{I(s)}$ therefore becomes

$$\mathbf{I(s)} = \frac{-V_{C0}}{s\left(R + Ls + \dfrac{1}{Cs}\right)}$$

$$= \frac{-V_{C0}C}{LCs^2 + RCs + 1}$$

$$= -\frac{V_{C0}/L}{s^2 + \dfrac{R}{L}s + \dfrac{1}{LC}} \qquad (19\text{-}61)$$

Equation 19-61 is the same as Eq. 19-52 except for the minus sign. The solution to Eq. 19-61 is therefore identical to Eqs. 19-55, 19-56, and 19-57 if V_{C0} is substituted for E_1 and a minus sign is affixed to each equation.

Up to this point, all the examples have used single-loop circuits to illustrate the techniques of problem solution. It should be obvious that since just a single loop can create a fair amount of mathematical manipulations needed to effect a solution, a multiloop circuit can present a very formidable array of equations to solve. Some simple multiloop circuits will now be used to illustrate the solution techniques which can be used on larger problems. Although, as mentioned earlier, it is important for one to know how to handle properly the initial conditions in circuit problems, the majority of desired problem solutions do not include the effect of initial conditions. The reason that initial conditions are not so terribly important in linear networks is that the homogeneous equation of the network is not affected by the initial conditions. The initial conditions affect only the particular solution and can be considered a part of the initial forcing functions. Of course, making the initial conditions equal to zero does tend to detract somewhat from the reasons for using Laplace-transform solutions, but Laplace transformations still present a much easier and more orderly way to solve many of the differential equations encountered in circuit theory, as compared to classical means.

EXAMPLE 19-9. The 2-mesh circuit in Fig. 19-12a has both capacitors initially uncharged. Determine the currents flowing through R_1 and R_2.

Solution: By choosing the mesh currents i_1 and i_2 as indicated in Fig. 19-12a, one can observe that $i_{R_1} = i_1$ and $i_{R_2} = i_2$. Using Kirchhoff's laws, Ohm's law, and the operational-impedance concept, the loop-current equations can be written, with the aid of Fig. 19-12b, as

$$\left.\begin{array}{l} \left(R_1 + \dfrac{1}{C_1 s}\right)\mathbf{I}_1(s) - \dfrac{1}{C_1 s}\mathbf{I}_2(s) = \mathbf{E(s)} \\[3mm] -\dfrac{1}{C_1 s}\mathbf{I}_1(s) + \left(R_2 + \dfrac{1}{C_1 s} + \dfrac{1}{C_2 s}\right)\mathbf{I}_2(s) = 0 \end{array}\right\} \qquad (19\text{-}62)$$

The reader should derive Eq. set 19-62 by writing the loop-current equations to

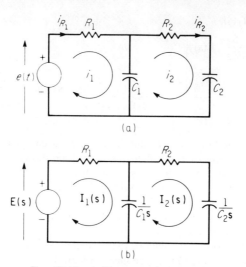

Fig. 19-12. a) Figure used to accompany Example 19-9. b) Network of part a, set up for Laplace-transform solution.

Fig. 19-12a in differential-equation form, taking the Laplace transform of each equation, setting all initial conditions equal to zero, and collecting terms. The transformation from Fig. 19-12a to Fig. 19-12b and the resultant loop-current equations (Eq. 19-62) should be thoroughly studied and understood, as all networks will be solved in this fashion henceforth.

Solving for $\mathbf{I}_1(\mathbf{s})$,

$$
\mathbf{I}_1(\mathbf{s}) = \frac{\begin{vmatrix} \mathbf{E(s)} & -\dfrac{1}{C_1 \mathbf{s}} \\[3mm] 0 & R_2 + \dfrac{1}{C_1 \mathbf{s}} + \dfrac{1}{C_2 \mathbf{s}} \end{vmatrix}}{\begin{vmatrix} R_1 + \dfrac{1}{C_1 \mathbf{s}} & -\dfrac{1}{C_1 \mathbf{s}} \\[3mm] -\dfrac{1}{C_1 \mathbf{s}} & R_2 + \dfrac{1}{C_1 \mathbf{s}} + \dfrac{1}{C_2 \mathbf{s}} \end{vmatrix}}
$$

$$
= \frac{\left(R_2 + \dfrac{1}{C_1 \mathbf{s}} + \dfrac{1}{C_2 \mathbf{s}}\right)\mathbf{E(s)}}{\left(R_1 + \dfrac{1}{C_1 \mathbf{s}}\right)\left(R_2 + \dfrac{1}{C_1 \mathbf{s}} + \dfrac{1}{C_2 \mathbf{s}}\right) - \left(\dfrac{1}{C_1 \mathbf{s}}\right)^2}
$$

$$
= \frac{(R_2 C_1 C_2 \mathbf{s}^2 + C_2 \mathbf{s} + C_1 \mathbf{s})\,\mathbf{E(s)}}{C_1 C_2 \mathbf{s}^2 \left[\dfrac{R_1(R_2 C_1 C_2 \mathbf{s}^2 + C_2 \mathbf{s} + C_1 \mathbf{s})}{C_1 C_2 \mathbf{s}^2} + \left(\dfrac{1}{C_1 \mathbf{s}}\right)\dfrac{(R_2 C_2 \mathbf{s} + 1)}{C_2 \mathbf{s}}\right]}
$$

$$
= \frac{[R_2 C_1 C_2 \mathbf{s}^2 + (C_1 + C_2)\mathbf{s}]\mathbf{E(s)}}{R_1 R_2 C_1 C_2 \mathbf{s}^2 + (R_1 C_2 + R_1 C_1 + R_2 C_2)\mathbf{s} + 1} \qquad (19\text{-}63)
$$

Solving for $I_2(s)$,

$$I_2(s) = \frac{\begin{vmatrix} R_1 + \dfrac{1}{C_1 s} & E(s) \\[2ex] -\dfrac{1}{C_1 s} & 0 \end{vmatrix}}{\Delta} = \frac{s C_2 E(s)}{R_1 R_2 C_1 C_2 s^2 + (R_1 C_2 + R_1 C_1 + R_2 C_2)s + 1}$$

$$(19\text{-}64)$$

A particular solution for $i_1(t)$ and $i_2(t)$ can be accomplished by choosing a desired functional form for $E(s)$. Before this is done, however, several observations are in order concerning Eqs. 19-63 and 19-64. First, note the identical denominators of the two equations. In taking the inverse of Eqs. 19-63 and 19-64 [after a functional form is chosen for $e(t)$ or $E(s)$], it should be recalled that the *characteristic* behavior of $i_1(t)$ and $i_2(t)$ is determined by the denominator polynomial. The numerator polynomials contribute only to the coefficients or *residues* of the denominator roots. In general, these coefficients or residues are complex numbers. A lot of information can be gleaned from the denominator expressions concerning the time-domain expressions $i_1(t)$ and $i_2(t)$ without actually performing the inverse transformation. For instance, can the expressions for $i_1(t)$ and $i_2(t)$ be underdamped or oscillatory for any chosen values of R_1, R_2, C_1, and C_2? The answer to this can be learned by recalling that the roots of the quadratic expression

$$as^2 + bs + c = 0 \tag{19-65}$$

are

$$s = -\frac{b}{2a} \pm \frac{\sqrt{b^2 - 4ac}}{2a}$$

If $4ac > b^2$, then the roots are complex, and the circuit variable (such as i or v) undergoes damped oscillations. Comparing the coefficients of Eq. 19-65 with the like coefficients of the denominator polynomial of Eq. 19-63, one observes that

$$\begin{aligned}
b^2 - 4ac &= (R_1 C_2 + R_1 C_1 + R_2 C_2)^2 - 4R_1 R_2 C_1 C_2 \\
&= [R_1 (C_1 + C_2)]^2 + 2R_2 C_2 (R_1 C_1 + R_1 C_2) \\
&\qquad\qquad + (R_2 C_2)^2 - 4R_1 R_2 C_1 C_2 \\
&= R_1^2 (C_1 + C_2)^2 + 2R_2 R_1 C_2^2 - 2R_1 R_2 C_1 C_2 + (R_2 C_2)^2 \\
&= R_1^2 C_1^2 + 2R_1^2 C_1 C_2 + R_1^2 C_2^2 + 2R_2 R_1 C_2^2 \\
&\qquad\qquad + R_2^2 C_2^2 - 2R_1 R_2 C_1 C_2 \\
&= (R_1 C_1 - R_2 C_2)^2 + 2R_1^2 C_1 C_2 + R_1^2 C_2^2 + 2R_1 R_2 C_2^2 \quad (19\text{-}66)
\end{aligned}$$

Equation 19-66 is greater than zero for any combination of values for the resistances and capacitances. This means that $i_1(t)$ and $i_2(t)$ cannot convert the energy in $E(s)$ or $e(t)$ into a damped oscillatory function unless, of course, $E(s)$ or $e(t)$ contains an oscillatory forcing function.

Second, consider the ratio $E(s)/I_1(s)$, from Eq. 19-63:

$$\frac{E(s)}{I_1(s)} = \frac{R_1 R_2 C_1 C_2 s^2 + (R_1 C_2 + R_1 C_1 + R_2 C_2)s + 1}{s(R_2 C_1 C_2 s + C_1 + C_2)} \tag{19-67}$$

Since Eq. 19-67 is the ratio of a voltage to a current, the ratio can be properly termed an operational impedance. Because $I_1(s)$ can be considered the total current delivered by $E(s)$ (refer to Fig. 19-12b or 19-12a), the impedance function of Eq. 19-67 is usually called a *driving-point operational impedance*. If the ratio $E(s)/I_2(s)$ from Eq. 19-64 were formed, the resultant impedance would be called a *transfer operational impedance*. Obviously, the ratios $I_1(s)/E(s)$ or $I_2(s)/E(s)$ could also be obtained and the respective ratios would be termed the *driving-point or transfer admittance*. It should be apparent that the various driving-point or transfer ratios are, in general, ratios of polynomials.

Third, consider the initial and steady-state values of the currents for a particular $E(s)$ or $e(t)$. The fundamental law of conservation of momentum demands that the charge—and therefore the voltage ($v = q/c$)—across a capacitor cannot change instantaneously. If the capacitors are initially uncharged at the instant $e(t)$ takes a value, the voltage across the capacitors remains at zero at $t = 0^+$. Thus the current $i_2(t)|_{t=0^+}$ must start at zero, since the voltage across capacitor C_1 is still zero at $t = 0^+$. The current $i_1(t)|_{t=0^+}$ has the value

$$i_1(t)|_{t=0^+} = \frac{e(t)|_{t=0^+}}{R_1}$$

and

$$i_2(t)|_{t=0^+} = 0$$

The steady-state value (as $t \to \infty$) also depends on the functional form of $e(t)$. If $e(t)$ is a step function, then both $i_1(t)$ and $i_2(t)$ approach zero as $t \to \infty$, because both capacitors finally charge to the value of the step-function voltage, thus blocking any further flow of current.

We shall now present a numerical example for the preceding symbolic example.

EXAMPLE 19-10.   In the circuit of Fig. 19-12a, let $R_1 = 100$ KΩ, $R_2 = 1$ MΩ, $C_1 = 1$ μf, and $C_2 = 0.1$ μf, and solve for the currents $i_1(t)$ and $i_2(t)$ if $e(t)$ is a step function with an amplitude of 10 volts.

Solution: Using Eq. 19-63,

$$I_1(s) = \frac{[10^6 \cdot 10^{-6} \cdot 10^{-7}s^2 + 1.1 \cdot 10^{-6}s] \cdot 10/s}{10^5 \cdot 10^6 \cdot 10^{-6} \cdot 10^{-7}s^2 + (10^5 \cdot 10^{-7} + 10^5 \cdot 10^{-6} + 10^6 \cdot 10^{-7})s + 1}$$

$$= \frac{(s + 11)10^{-6}}{10^{-2}s^2 + (10^{-2} + 10^{-1} + 10^{-1})s + 1}$$

$$= \frac{10^{-6}(s + 11)}{0.01s^2 + 0.21s + 1}$$

$$= \frac{10^{-4}(s + 11)}{s^2 + 21s + 100}$$

$$= \frac{10^{-4}(s + 11)}{(s + 13.705)(s + 7.295)}$$

$$= 10^{-4}\left[\frac{s}{(s + 13.705)(s + 7.295)} + \frac{11}{(s + 13.705)(s + 7.295)}\right]$$

Using transform pairs 12 and 13 of Table 19-1, where $a = 13.705$ and $b = 7.295$.

$$i_1(t) = 10^{-4} \left(\frac{1}{+ 13.705 - 7.295} \right)$$

$$\times (+13.705e^{-13.705t} - 7.295e^{-7.295t} - 11e^{-13.705t} + 11e^{-7.295t})$$

$$= \frac{10^{-4}}{6.41} (2.705e^{-13.705t} + 3.705e^{-7.295t})$$

$$= 42.25e^{-13.705t} + 57.75e^{-7.295t} \; \mu\text{amps} \tag{19-68}$$

Using Eq. 19-64,

$$\mathbf{I}_2(\mathbf{s}) = \frac{10^{-7}\mathbf{s}}{0.01\mathbf{s}^2 + 0.21\mathbf{s} + 1} \cdot \frac{10}{\mathbf{s}}$$

$$= \frac{10^{-4}}{(\mathbf{s} + 13.705)(\mathbf{s} + 7.295)}$$

Employing transform pair 12,

$$i_2(t) = \frac{10^{-4}}{7.295 - 13.705} (e^{-13.705t} - e^{-7.295t})$$

$$= 15.6(e^{-7.295t} - e^{-13.705t}) \; \mu\text{amps} \tag{19-69}$$

As expected, $i_1(0^+) = 10 - 10^6/R_1 = 10 - 10^6/10^5 = 42.25 + 57.75 = 100$ μamps, and $i_2(0^+) = 0$. The entire time-domain curves for $i_1(t)$ and $i_2(t)$ are plotted in Fig. 19-13.

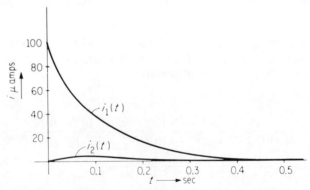

FIG. 19-13.   A plot of $i_1(t)$ and $i_2(t)$ from Eqs. 19-68 and 19-69.

The primary reasons for choosing the impulse or step-excitation functions are simplicity and ease of calculation. Usually, it is desirable to know the *characteristic* behavior of a network, and *any* forcing function which energizes the circuit (including initial conditions) suffices to determine this characteristic behavior; hence the reason for choosing the simple step or impulse function.

Another very important class of forcing functions consists of the trigonometric sinusoidal functions. These forcing functions are important because of the sinusoidal oscillations which are so frequently observed in natural phenomena. Consider the *R-L* network shown in Fig. 19-14.

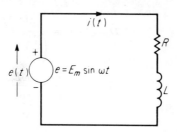

FIG. 19-14.   Series *R-C* circuit
with sinusoidal excitation.

Using either the differential-equation approach or the operational-impedance approach $I(s)$ (assuming zero initial current) can be determined as $I(s) = E(s)/(R + Ls)$, and substituting the Laplace transform for the sinusoidal function $e(t)$ (see Table 19-1, transform pair 4), one obtains

$$I(s) = \frac{\omega E_m}{(s^2 + \omega^2)(Ls + R)}$$

$$= \frac{\omega E_m / L}{(s^2 + \omega^2)(s + R/L)} \qquad (19\text{-}70)$$

The inverse transform of Eq. 19-70 can be determined by the use of partial-fraction-expansion techniques (method 2):

$$\frac{\omega E_m / L}{(s^2 + \omega^2)(s + R/L)} = \frac{As + B}{s^2 + \omega^2} + \frac{D}{s + R/L}$$

$$\frac{\omega E_m}{L} = (As + B)(s + R/L) + D(s^2 + \omega^2)$$

$$= As^2 + \left(B + \frac{AR}{L}\right)s + \frac{BR}{L} + Ds^2 + D\omega^2$$

Equating coefficients of like powers of s,

$$A + D = 0$$

$$B + \frac{AR}{L} = 0$$

$$\frac{BR}{L} + D\omega^2 = \frac{\omega E_m}{L}$$

Solving for the constants D and B from the first two equations and substituting into the third and solving,

$$-\frac{AR}{L}\frac{R}{L} - A\omega^2 = \frac{\omega E_m}{L}$$

$$A = -\frac{\omega E_m/L}{\left(\dfrac{R}{L}\right)^2 + \omega^2}$$

$$D = -A = \frac{\omega E_m/L}{\left(\dfrac{R}{L}\right)^2 + \omega^2}$$

$$B = -A\frac{R}{L} = \frac{\omega E_m R/L^2}{\left(\dfrac{R}{L}\right)^2 + \omega^2}$$

Equation 19-70 becomes

$$\mathbf{I(s)} = \frac{A\mathbf{s}}{\mathbf{s}^2 + \omega^2} + \frac{B}{\mathbf{s}^2 + \omega^2} + \frac{D}{\mathbf{s} + R/L}$$

$$= \frac{\omega E_m/L}{\left(\dfrac{R}{L}\right)^2 + \omega^2}\left(\frac{-\mathbf{s}}{\mathbf{s}^2 + \omega^2} + \frac{R/L}{\mathbf{s}^2 + \omega^2} + \frac{1}{\mathbf{s} + R/L}\right) \qquad (19\text{-}71)$$

The inverse transform of Eq. 19-71 is

$$i(t) = \frac{\omega E_m/L}{\left(\dfrac{R}{L}\right)^2 + \omega^2}\left(-\cos \omega t + \frac{R}{\omega L}\sin \omega t + e^{-\frac{R}{L}t}\right) \qquad (19\text{-}72)$$

Equation 19-72 is a good example to show the distinction between the transient and the steady-state solutions. After a reasonable time (after $t \geq 4L/R$, the exponential term can be considered to be negligible since $e^{-4} = 0.0183$, which is less than 2 percent of the initial value), Eq. 19-72 reduces to a periodic value (current repeats itself in a regular manner) involving the sine and cosine terms. Equation 19-72 is the general solution to the problem, the term involving the damped exponential $e^{-(R/L)t}$ is the transient portion of the solution, and the remainder is the steady-state periodic solution.

The study of the steady-state periodic response to sinusoidal excitations is a very important and major part of the study of network analysis. As might be anticipated, special analytic tools have been developed to handle the particular case involving sinusoidal excitation. Chapter 12 was devoted entirely to the steady-state (in a periodic sense) analysis of the sinusoidally excited circuit. The phasor concept and complex algebraic mathematics were introduced as the computational tools which cleverly develop and exploit the subject of alternating current or a-c circuits. (Alternating current is a synonymous term for sinusoidal excitation.)

Although Laplace-transform notation can be, and is, used extensively in analyzing circuits which are sinusoidally excited, it is almost mandatory that the reader be familiar with the terminology and manipulations of the phasor concepts before using Laplace-transform methods on sinusoidally excited circuits.

In Chapter 2 it was shown how one could set up the necessary mesh-current or node-voltage equations required for a circuit solution. For simplicity, all the circuit elements were chosen as resistances or conductances. As shown in Chapter 12 the addition of capacitance and inductance imparts little to the complexity of writing the necessary independent simultaneous equations, but the solution of the equations may be quite a formidable task. In fact, without the aid of machine computation, the solution of a large circuit problem is impractical—if not impossible—in terms of the effort required.

The network shown in Fig. 19-15 is an arbitrary arrangement of R-L-C elements with a single independent forcing function $e(t)$. Suppose

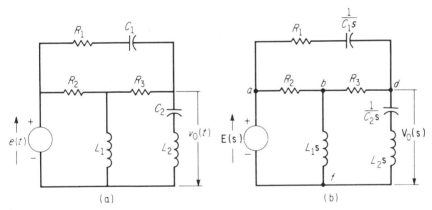

FIG. 19-15. a) A 3-mesh network. b) The network of part a, properly labeled for node-voltage equations using operational impedances.

one is asked to solve for $v_0(t)$ for a given $e(t)$. How should one proceed? Should one use mesh currents or node voltages? Should the equations be written in differential-equation form, or should operational impedances be employed? In this example it is best to use the node-voltage method and operational impedances. The reasons for these choices are twofold: 1) Only 2 node-voltage equations[4] are required to solve for $v_0(t)$, whereas 3 mesh-current equations plus an Ohm's-law equation are necessary if

---

[4]Only 2 node-voltage equations are required if the node between R_1 and C_1 and the node between L_2 and C_2 are eliminated by combining the series branch impedances: namely, $R_1 + 1/C_1 s$ and $L_2 s + 1/C_2 s$. Justification for this step was given in Chapter 2.

mesh currents are used. 2) If the solution is going to be obtained by Laplace transforms, it is simpler to use operational impedances rather than to write the differential equation and then take the Laplace transform. Using Fig. 19-5b and choosing node f as the reference or datum node, the node-voltage equations are as follows:

Node a: No equation need be written, since $V_{af}(s) = E(s)$, a known voltage.

Node b:

$$\frac{V_{bf}(s) - V_{af}(s)}{R_2} + \frac{V_{bf}(s) - V_{df}(s)}{R_3} + \frac{V_{bf}(s)}{L_1 s} = 0$$

Node d:

$$\frac{V_{df}(s) - V_{bf}(s)}{R_3} + \frac{V_{df}(s) - V_{af}(s)}{R_1 + \dfrac{1}{C_1 s}} + \frac{V_{df}(s)}{L_2 s + \dfrac{1}{C_2 s}} = 0$$

(19-73a)

Making the substitution $V_{af}(s) = E(s)$ and $V_{df} = V_0(s)$ and rearranging, one obtains

$$\left(\frac{1}{R_2} + \frac{1}{R_3} + \frac{1}{L_1 s}\right)V_{bf}(s) - \frac{1}{R_3}V_0(s) = \frac{E(s)}{R_2} \qquad (19\text{-}73b)$$

$$-\frac{1}{R_3}V_{bf}(s) + \left(\frac{1}{R_3} + \frac{1}{\dfrac{1}{C_2 s} + L_2 s} + \frac{1}{R_1 + \dfrac{1}{C_1 s}}\right)V_0(s)$$

$$= \frac{E(s)}{R_1 + \dfrac{1}{C_1 s}} \qquad (19\text{-}73c)$$

Note that the major diagonal terms contain all the admittance terms connected to the chosen nodes, and the off-diagonal terms contain the common admittances shared between the nodes. All the independent forcing functions (and initial conditions, if any) are placed on the right-hand side of the equations. The indicated solution for $V_0(s)$, in determinant form, is

$$V_0(s) = \frac{\begin{vmatrix} \dfrac{1}{R_2} + \dfrac{1}{R_3} + \dfrac{1}{L_1 s} & \dfrac{E(s)}{R_2} \\[3mm] -\dfrac{1}{R_3} & \dfrac{E(s)}{R_1 + \dfrac{1}{C_1 s}} \end{vmatrix}}{\begin{vmatrix} \dfrac{1}{R_2} + \dfrac{1}{R_3} + \dfrac{1}{L_1 s} & -\dfrac{1}{R_3} \\[3mm] -\dfrac{1}{R_3} & \dfrac{1}{R_3} + \dfrac{1}{\dfrac{1}{C_2 s} + L_2 s} + \dfrac{1}{R_1 + \dfrac{1}{C_1 s}} \end{vmatrix}} \qquad (19\text{-}74)$$

Since Eq. 19-74 is a 2 × 2 determinant, it can be expanded by inspection, as follows:

$$V_0(s) = \frac{\left(\dfrac{1}{R_2} + \dfrac{1}{R_3} + \dfrac{1}{L_1 s}\right)\left(\dfrac{1}{R_1 + \dfrac{1}{C_1 s}}\right) + \dfrac{1}{R_2 R_3}}{\left(\dfrac{1}{R_2} + \dfrac{1}{R_3} + \dfrac{1}{L_1 s}\right)\left(\dfrac{1}{R_3} + \dfrac{1}{\dfrac{1}{C_2 s} + L_2 s} + \dfrac{1}{R_1 + \dfrac{1}{C_1 s}}\right) - \dfrac{1}{R_3^2}} \cdot E(s)$$

$$= \frac{\left[\left(\dfrac{R_3 L_1 s + R_2 L_1 s + R_2 R_3}{R_2 R_3 L_1 s}\right)\left(\dfrac{C_1 s}{R_1 C_1 s + 1}\right) + \dfrac{1}{R_2 R_3}\right] \cdot E(s)}{\left(\dfrac{R_3 L_1 s + R_2 L_1 s + R_2 R_3}{R_2 R_3 L_1 s}\right)\left(\dfrac{1}{R_3} + \dfrac{C_2 s}{1 + L_2 C_2 s^2} + \dfrac{C_1 s}{R_1 C_1 s + 1}\right) - \dfrac{1}{R_3^2}}$$

$$V_0(s) = [(R_3 L_1 s + R_2 L_1 s + R_2 R_3)(C_1 s)(R_2 R_3)$$
$$+ R_2 R_3 L_1 s(R_1 C_1 s + 1)]E(s)$$

$$\div R_2 \left\{ \frac{\begin{array}{l}(R_3 L_1 s + R_2 L_1 s + R_2 R_3)[(1 + L_2 C_2 s^2)(R_1 C_1 s + 1) \\ \quad + R_3 C_2 s(R_1 C_1 s + 1) + R_3 C_1 s(1 + L_2 C_2 s^2)]R_3^2 \\ \quad - R_3(1 + L_2 C_2 s^2)(R_1 C_1 s + 1)R_2 R_3 L_1 s\end{array}}{R_3(1 + L_2 C_2 s^2)} \right\}$$

$$(19\text{-}75)$$

As can be seen, Eq. 19-75 is becoming a bit unwieldy because of the symbolic expressions. Before simplifying further, numerical substitutions will be introduced for the parameters in Eq. 19-75. Let all the resistances be 1 ohm, all the inductances be 1 henry, and all the capacitances be 1 farad. Making these substitutions, Eq. 19-75 becomes

$$V_0(s) = \frac{[(2s + 1)s + s(s + 1)](1 + s^2)}{\left\{\begin{array}{l}[(2s + 1)[(1 + s^2)(s + 1) + (s + 1)s + (1 + s^2)(s)] \\ \quad - (1 + s^2)(s + 1)(s)\end{array}\right\}} \cdot E(s)$$

$$= \frac{s(3s + 2)(1 + s^2)}{(2s + 1)(2s^3 + 2s^2 + 3s + 1) - (s^3 + s^2 + s + 1)s} \cdot E(s)$$

$$= \frac{s(3s + 2)(s^2 + 1)}{(4s^4 + 6s^3 + 8s^2 + 5s + 1) - (s^4 + s^3 + s^2 + s)} \cdot E(s)$$

$$= \frac{s(3s + 2)(s^2 + 1)}{3s^4 + 5s^3 + 7s^2 + 4s + 1} \cdot E(s) \qquad (19\text{-}76)$$

Making the rather impractical substitutions in Eq. 19-75 (which led to Eq. 19-76) serves only to illustrate an important point in the analysis; namely, that $V_0(s)$ *can be reduced to a ratio of 2 polynomials in* s. *If* $e(t)$ *is chosen as a sinusoidal forcing function* $e(t) = E_m \sin \omega t$, *then* $E(s) =$

$\omega E_m/s^2 + \omega^2$, and Eq. 19-76 becomes

$$V_0(s) = \frac{s(3s + 2)(s^2 + 1)\,\omega E_m}{(3s^4 + 5s^3 + 7s^2 + 4s + 1)(s^2 + \omega^2)} \qquad (19\text{-}77a)$$

$$V_0(s) = \frac{s(3s + 2)(s^2 + 1)\,\omega E_m}{3s^6 + 5s^5 + (7 + 3\omega^2)s^4 + (4 + 5\omega^2)s^3 + (1 + 7\omega^2)s^2 + 4\omega^2 s + \omega^2}$$
$$(19\text{-}77b)$$

In order to solve for $v_0(t)$, one must choose a value of ω, and perhaps of E_m, and then take the inverse transform of Eq. 19-77. Even though the circuit shown in Fig. 19-15 appears to be rather simple, the solution to Eq. 19-77 is not simple. To solve Eq. 19-75, one must factor the roots of the 6th-order denominator polynomial. Of course, the $(s^2 + \omega^2)$ term can be factored from Eq. 19-77b, which returns us to Eq. 19-77a. Even so, the quartic polynomial of Eq. 19-77a can be somewhat difficult and tedious to factor by hand computation. It is quite unpleasant to attempt to factor the higher-order polynomials by hand calculation. Using machine computation, such factoring is usually routine, quick, and accurate. In all fairness to the reader, it must be pointed out that, for a complicated network, it is a tedious chore to reduce the desired circuit unknown (such as a node voltage or a mesh current) to a ratio of 2 polynomials in s, as illustrated in Eq. 19-77.

An appropriate question at this point might be: How can a complicated circuit be solved with only a minimum of hand computation? One answer lies in using the host of network reduction techniques that have been evolved over the past six decades or so and which are a tremendous aid to the person doing hand calculations. Some of these reduction techniques were discussed in Chapter 2. Another answer lies in making full utilization of the capabilities of computers. The remainder of this chapter will be devoted to a discussion of machine solution to circuits such as that illustrated by Fig. 19-15.

Machine Solution. Before "jumping in" and beginning a problem solution, one needs to reflect on the purpose of or the use that is to be made of a given solution. Sometimes it is desirable to solve a problem completely; that is, determine the complete solution to all or part of the circuit variables. But sometimes a problem is considered solved if only the *characteristic behavior* is known. Also, one needs to stipulate the accuracies required to give a satisfactory solution. Are accuracies of 0.1 percent, 1 percent, 10 percent, or what, required to give satisfactory insight to the problem variables? For most circuit problems, 1 to 10 percent tolerance on the accuracy of the solutions is quite acceptable. In this chapter, only the complete solution or a portion of the complete solution to a problem will be discussed. In subsequent chapters, additional attention will be focused on the information which can be gleaned from the *characteristic equation* of the circuit variables.

If machine computation is to be utilized, should a digital or an analog computer be employed? Aside from the important factors of availability and cost of the machine and of operator experience in its use, there is no ready rule that states which machine is best for a given problem—unless accuracy is a major consideration. If solution accuracies of less than 1 percent are required, then the digital computer is superior to the analog, but, for many problems, it is up to the engineer to decide which machine to use.

To solve the problem given above (Fig. 19-15), digital computers can be used in two general ways: 1) The computer can be programmed to solve for the roots of polynomials such as those given in Eq. 19-75. Of course, considerable hand calculation or reduction is necessary to reduce the unknown variable to a ratio of 2 polynomials. As the problem complexity increases, the necessary hand calculations may prove too burdensome to proceed along this course. 2) The computer can be programmed to solve the original set of simultaneous equations by repeated substitution or by using problem oriented languages such as ECAP, PCAP or NET1. This method is the more promising for large sets of equations.

Analog computers can be programmed to solve the original simultaneous differential equations and present the unknown variables as a voltage output in continuous data form (see Chapter 20). To write the original equations in differential-equation form, again consider Fig. 19-15 and Eq. 19-73. Recall that s can be considered a derivative operator and $\frac{1}{s}$ an integral operator, or

$$\left. \begin{array}{l} s = \dfrac{d(\ )}{dt} \\[2mm] \dfrac{1}{s} = \displaystyle\int (\ )\, dt \end{array} \right\} \qquad (19\text{-}78)$$

Using these relations, Eq. 19-73a can be written as

$$\left(\frac{1}{R_2} + \frac{1}{R_3}\right) v_{bf}(t) + \frac{1}{L_1}\int v_{bf}(t)\, dt - \frac{1}{R_3} v_0(t) = \frac{e(t)}{R_2} \qquad (19\text{-}79)$$

Equation 19-73b is a bit more complex than this and cannot immediately be written in differential-equation form by using the relations given in Eq. 19-78, because of the terms

$$\frac{1}{\dfrac{1}{C_2 s} + L_2 s} \qquad \text{and} \qquad \frac{1}{R_1 + \dfrac{1}{C_1 s}}$$

One way of handling these terms is by a bit of manipulation. Consider the term

$$X_1(s) = \frac{1}{\dfrac{1}{C_2 s} + L_2 s} \cdot Y_1(s) = \frac{C_2 s}{1 + L_2 C_2 s^2} \cdot Y_1(s)$$

or

$$(1 + L_2C_2s^2)X_1(s) = C_2s\,Y_1(s)$$

which, in differential equation form, becomes

$$x_1(t) + L_2C_2\frac{d^2[x_1(t)]}{dt^2} = C_2\frac{dy_1(t)}{dt} \qquad (19\text{-}80)$$

Similarly, the term

$$X_2(s) = \frac{1}{R_1 + \dfrac{1}{C_1s}} \cdot Y_2(s)$$

$$= \frac{C_1s}{R_1C_1s + 1} \cdot Y_2(s)$$

or

$$(R_1C_1s + 1)X_2(s) = C_1s\,Y_2(s)$$

can be written in differential-equation form as

$$R_1C_1\frac{dx_2(t)}{dt} + x_2(t) = C_1\frac{dy_2(t)}{dt} \qquad (19\text{-}81)$$

Equation 19-73b can then be written as

$$-\frac{1}{R_3}v_{bf}(t) + \frac{1}{R_3}v_0(t) + x_1(t) + x_2(t) = x_3(t) \qquad (19\text{-}82)$$

where

$x_1(t)$ is solved from Eq. 19-80 with $y_1(t) \triangleq v_0(t)$
$x_2(t)$ is solved from Eq. 19-81 with $y_2(t) \triangleq v_0(t)$
$x_3(t)$ is solved from an equation similar to Eq. 19-81 with $y_2(t) \triangleq e(t)$ and with $X_3(s) = E(s)/(R_1 + 1/C_1s)$

This example will be developed in detail in Chapter 20.

Problems

Laplace Transforms

19-1. Find the Laplace transform of the following functions:
a) $\sin(\omega t)$ b) $\cosh(at)$
c) t, t^2, t^n d) $1 - e^{-at}$
e) $\sin(\omega t + \theta)$ f) $\cos(\omega t + \theta)$
g) $e^{-at}\sin(\omega t)$ h) $e^{-at}\cos(\omega t)$

19-2. Expand the following functions in a partial-fraction expansion.

a) $\dfrac{2s}{s^2 - 1}$ b) $\dfrac{7s + 1}{s^3 + 3s^2 + 2s}$

c) $\dfrac{s^2}{s - 1}$ d) $\dfrac{3(s + 2)}{s^2 + 1}$

e) $\dfrac{s^3 - 6s^2 + 8s + 3}{s^2(s^2 + 9)}$ f) $\dfrac{1}{s^3 + 2s^2 + 2s + 1}$

Is there any need to expand parts a and b? Is there any problem associated with part c?

19-3. Show that the following relationships are true.

a) $\mathcal{L}^{-1}\left[\dfrac{3s}{(s^2+1)(s^2+4)}\right] = \cos t - \cos 2t$

b) $\mathcal{L}^{-1}\left(\dfrac{s+1}{s^2+2s}\right) = 0.5 + 0.5e^{-2t}$

c) $\mathcal{L}^{-1}\left[\dfrac{1}{(s+1)(s+2)^2}\right] = e^{-t} - e^{-2t}(1+t)$

d) $\mathcal{L}^{-1}\left[\dfrac{s^2}{(s^2+1)^2}\right] = 0.5t\cos t + \dfrac{1}{2}\sin t$

19-4. Solve the following differential equations for the indicated time-variable quantity, assuming that all initial conditions are zero.

a) $\dfrac{d^2x(t)}{dt^2} + \dfrac{dx(t)}{dt} = t^2 + 2t$　　b) $L\dfrac{di(t)}{dt} + Ri(t) = E$

c) $L\dfrac{d^2q(t)}{dt^2} + R\dfrac{dq(t)}{dt} = E$　　d) $Ri(t) + \dfrac{1}{C}\int_{-\infty}^{t} i(t)dt = E$

c) $\dfrac{R\,dq(t)}{dt} + \dfrac{q(t)}{c} = E$　　f) $\dfrac{1}{L}\int_{-\infty}^{t} v(t)\,dt + \dfrac{v(t)}{R} = I$

g) $C\dfrac{dv(t)}{dt} + \dfrac{v(t)}{R} = I$

19-5. Derive the Laplace transform of the following functions:

a) $f(t) = 1, \quad 0 < t < a$
$\quad = -1, \quad a < t < 2a$
$\quad = 0, \quad 2a < t$

b) $f(t) = \sin \omega t, \quad 0 < t < \dfrac{2\pi}{\omega}$
$\quad = 0, \quad \dfrac{2\pi}{\omega} < t$

c) $f(t) = |\sin \omega t|$

19-6. A periodic function is defined by $f(t) = f(t+T)$, where T is the period of the function. Show that, for any periodic function,

$$\mathcal{L}[f(t)] = F(s) = \dfrac{1}{1-e^{-Ts}}\int_0^T f(t)e^{-st}\,dt$$

Circuit Analysis

19-7. In the circuit shown in the accompanying figure, switch S_1 is closed at $t = 0$. After 5 msec (milliseconds), switch S_2 is opened. Find the current flowing

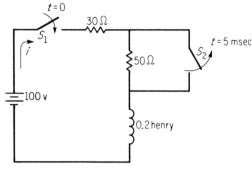

PROB. 19-7

through the 50-ohm resistance and sketch $i(t)$ versus t for the intervals $0 < t <$ t_1 and $t_1 < t$, where $t_1 = 5$ msec.

19-8. A constant voltage is applied to a series R-L circuit by closing a switch. The voltage across L is 80 volts at $t = 0$ and drops to 10 volts at $t = 0.4$ sec. If $R = 1000$ Ω, what must be the value of L?

19-9. In the circuit shown in the accompanying figure, the switch is closed on position 1 for a long period of time (compared to the circuit time constants) and then moved to position 2. Find $i(t)$ and the time at which the current is zero and reversing its direction.

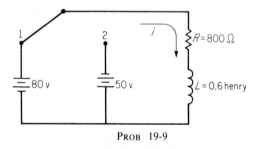

PROB 19-9

19-10. For the circuit of Prob. 19-9 find the energy dissipated in the 800-ohm resistance during the transient portion of the current solution. (Assume that the transient decays to zero after 5 time constants, where 1 time constant $= L/R$ sec.)

19-11. The R-C circuit shown in the accompanying figure has an initial charge on the capacitor of $q_0 = 800 \times 10^{-6}$ coulomb with the polarity shown in the

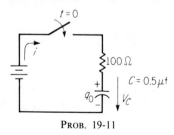

PROB. 19-11

diagram. Find the current $i(t)$ and capacitor voltage $v_C(t)$ transients which result when the switch is closed.

19-12. Repeat Prob. 19-11 after reversing the polarity of the initial charge q_0 on the capacitor.

19-13. A capacitor of 1 μf with an initial charge q_0 = 100 microcoulombs is connected across the terminals of a 1000-ohm resistor at t = 0. Calculate the time it takes the transient voltage across the resistor to drop from 40 to 10 volts.

19-14. In the circuit shown in the accompanying figure, the switch is closed on position 1 at t = 0 and then moved to position 2 after 1 time constant. Find the transient current expression in the R-C circuit for all time (1 time constant = $R \cdot C$).

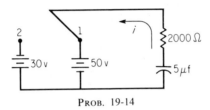

PROB. 19-14

19-15. In the circuit shown in the accompanying figure, capacitor C_1 has an initial charge q_0 = 300 microcoulombs. If the switch is closed at t = 0, find the current transient and final voltage across capacitor C_1. Sketch i.

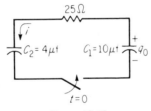

PROB. 19-15

19-16. A series R-L-C circuit with R = 200 ohms, L = 1 henry, and C = 10 μf has a step-function voltage V = 100 volts applied at t = 0. Find the current, assuming that all initial conditions are zero.

19-17. An R-L series circuit with R = 100 ohms and L = 2 henries has a sinusoidal excitation voltage of v = 100 sin $(100t + \theta)$ applied across the terminals. If the switch is closed at t = 0 for θ = 0°, obtain the resulting current transient.

19-18. Repeat Prob. 19-17 if θ = $\tan^{-1} 200/100$. Comments on the results.

19-19. In the 2-mesh network shown in the accompanying figure, find the mesh currents i_1 and i_2 after the switch is closed at t = 0. Sketch the current through the capacitor.

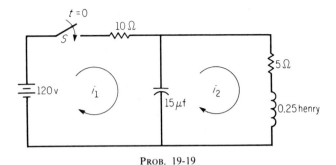

PROB. 19-19

19-20. In the network shown in the accompanying figure, all initial conditions are zero. Set up the indicated loop-current equations and solve for the voltage $V_{R_3}(s)$, where $e(t)$ is an unknown voltage. (*Note:* the capacitors are rather large, for easier evaluation of the determinant expression.)

$$Ans.: \quad V_{R_3}(s) = I_3(s)R_3 = \frac{15(16s^4 + 240s^3 + 12s^2 + 20s + 1)}{320s^4 + 5012s^3 + 6600s^2 + 801s + 25} E(s)$$

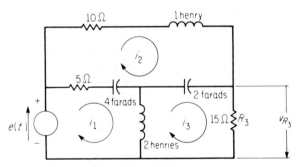

PROB. 19-20

19-21. Using the answer to Prob. 19-20, find the steady-state value of i_3 if $e(t)$ is a step function of 20 volts. (*Hint:* Use the final-value theorem.) Does this answer agree with physical reasoning?

19-22. Using the answer to Prob. 19-20, find the initial value of i_3 if $e(t)$ is a step function of 20 volts. (*Hint:* Use the initial value theorem.) Does this answer agree with physical reasoning?

20 / *Analog Computers*

As described in Chapter 11, a digital computer solves differential equations by indirect methods. Various numerical analytic techniques convert the original differential equation into a form which is amenable to digital solutions; that is, the differential equation is changed so that the high-speed adding capability of the digital computer can be used in effecting the problem solution. The problem solution can take several forms, depending upon the needs and desires of the engineer or analyst. One form might be the solution to the roots of the characteristic equation of the network determinant; another might be the solution of the problem variables under sinusoidal steady-state excitation; a third might be the complete solution for the variables in the time domain. In any case, the output of the digital computer is a series of numbers. Again, depending upon the type of solution desired, these numbers could represent the roots of the characteristic equation, or the complex phasor quantities for the problem variable, or the value of the problem variables at discrete instants of time. Oftentimes, a presentation of the problem-dependent variables as a function of a continuous independent variable, such as time, is desired. If this is so, then a digital-to-analog (D-to-A) converter is needed to give this presentation. (A human being could perform this conversion function by laboriously plotting each point, but usually some sort of electronic D-to-A converter is utilized.)

In contrast to the discrete-data output of the digital computer, the analog computer gives a continuous-data solution to the dependent variables in a differential equation. Thus, for problem solutions involving an actual plot of a dependent variable versus an independent variable, such as time or displacement, an analog computer may be far superior to a digital computer. A medium-sized analog computer can solve certain classes of very difficult (from an analytical viewpoint) linear or nonlinear differential equations more easily than can a large digital computer. In addition, the analog computer can often give the engineer a better "feel" for certain problem solutions. The roots of the characteristic equation of the network determinant can sometimes be estimated from the plots given by the analog computer, but this is not always easy to accomplish.

From the preceding discussion, it should be apparent that it is not easy to draw a sharp line of demarcation between the determinations that dictate whether an analog or a digital computer is to be used in a given problem solution. The engineer should be familiar with the attributes and limitations of both types of computers, for two reasons: 1) Certain problem solutions can be obtained more easily on an analog than on a digital computer, and vice versa; and 2) some problems require the simultaneous use of analog and digital computers (this is called *hybrid computation.*)

Analog computers fall into two general categories—the direct and the operational. Both types are important and should be studied; however, it is probably easier to learn how to use the operational computer. Use of the direct analog computer requires more finesse and a much greater understanding and appreciation of circuit theory than does the operational computer. Also, the direct analog computer introduces considerable error into problem solution, because it is difficult to build near perfect electric components—particularly resistanceless inductors.

The direct analog computer compares the integrodifferential equation of an electric circuit to a similar equation in some other system, such as a mechanical or hydraulic system. Perhaps one of the main reasons that analog computers are popular is that it is difficult, tedious, costly, and time-consuming to build a model of either an actual mechanical or a hydraulic system and then change the parameters until the best or optimum system behavior is achieved. On the other hand, the electric circuit is easy to construct. Decade resistors, inductors, and capacitors are relatively inexpensive and present the designer with a wide range of parameter values. It is also easy to obtain a variety of forcing-function generators such as sine-wave, square-wave, triangular-wave, pulse, and so on, generators. The experimenter can work with low energy levels and can use a variety of instruments for recording the behavior of the variables. Oscilloscopes, chart and X-Y recorders, voltmeters, galvanometers, ammeters, and so on, are convenient and easy to use, cover a wide frequency spectrum (up to 1,000 MHz or so for the oscilloscope), and usually require a very small input energy.

The electronic operational analog computer approaches the solution to a differential equation in a more straightforward manner than does the direct analog computer. The process of integration is the basic operation of the operational analog computer. The main component of the electronic operational computer is a high-gain d-c amplifier. These high-gain d-c amplifiers, along with resistors and capacitors, are used to perform the summing, sign-changing, and integrating functions necessary to solve differential equations. Information is taken from the computer by sensing the voltage output, at various points in the computer, by means of strip chart or X-Y recorders, voltmeters, or oscilloscopes.

20-1. The Direct Analog Computer

An example is, perhaps, the best way to explain the concept and use of the direct analog computer. Consider the damped spring-mass mechanical system shown in Fig. 20-1. The differential equation describing

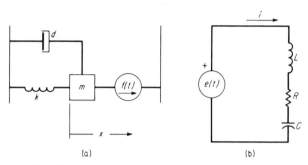

FIG. 20-1. a) A damped spring-mass mechanical system. b) An electrical analog to the system shown in part a.

the summation of forces in terms of the displacement x is

$$m\ddot{x} + d\dot{x} + kx = f(t) \tag{20-1}$$

Figure 20-1b shows a series R-L-C circuit. Using Kirchhoff's voltage law, the differential-equation summing voltage drops around the loop are

$$L\frac{di}{dt} + Ri + \frac{1}{C}\int i\,dt = e(t) \tag{20-2}$$

and, upon substituting $i = dq/dt$ into Eq. 20-2,

$$L\frac{d^2q}{dt^2} + R\frac{dq}{dt} + \frac{1}{C}q = e(t)$$

or

$$L\ddot{q} + R\dot{q} + \frac{1}{C}q = e(t) \tag{20-3}$$

It is obvious that Eqs. 20-1 and 20-3 are identical, except for the dimensions of the constants and the variables q and x. The solutions for Eqs. 20-1 and 20-3 are likewise identical, except for the dimensions. Thus, we can observe the behavior of q and directly relate this to the behavior of x. This observation is usually made by means of a voltmeter, an oscilloscope, or a chart recorder. Since $q = \int i\,dt$, the voltage across the capacitor is, in reality, q/C, and so, to record q, the input leads of the recording device should be attached to the capacitor terminals.

The methods used to find a suitable electrical analog for various systems require diligent study and practice before one can become adept in finding the analogous circuit. For this reason, the direct analog computer will not be studied any further.

20-2. The Electronic Operational Computer

The electronic analog computer that uses a high-gain d-c amplifier (called the operational amplifier) is termed either an *operational computer*, an *electronic analog computer*, or sometimes, a *differential analyzer*. This type of computer is best adapted to solving simultaneous linear or nonlinear differential equations. The computer performs the necessary functions for problem solutions such as summing, integrating, and differentiating. Other associated equipment allows the computer to perform such functions as multiplying, dividing, and resolving [a resolver transforms a polar quantity to a rectangular one; that is, we recall from a-c phasor-circuit analysis that $A \underline{/\theta} = A e^{j\theta} = A(\cos \theta + j \sin \theta)$ by Euler's rule], and the analog computer can be used to synthesize motors, generators, amplidynes, and any other piece of equipment for which a transfer function can be written.

Although the analog computer has inherent accuracy limitations, it is an almost invaluable tool for the engineer. The analog computer, like many other computing aids, saves the engineer a tremendous amount of time. For instance, consider the difficulty of solving 10 or 15 simultaneous linear or nonlinear differential equations with constant or nonconstant coefficients. Such a task would be tedious, unpleasant, and, perhaps, impossible, if it were not for computers.

The Basic Computing Circuit. Figure 20-2 shows the block-diagram

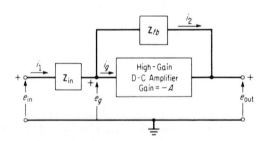

FIG. 20-2. Block diagram of the basic analog computing circuit.

representation of the basic analog computing circuit. It is beyond the scope of this text to show the circuit diagram or explain in any detail how the high-gain d-c amplifier is designed. Suffice it to say that a high-quality, commercial, direct-coupled (d-c) operational amplifier has the following external characteristics:

1. The input impedance is quite high, being in the order of 100

megohms. This means that the input current i_g to the amplifier is very small.

2. The forward open-circuit voltage gain of the amplifier is very large, being in the range of 10^6 to 10^8.

3. The output-voltage range for linear operation lies between plus and minus ($\pm$) 100 volts for most computer amplifiers, although some manufacturers use a ± 10-volt nominal output voltage. The maximum output current usually lies between 20 and 30 ma.

4. The voltage reference level of the direct-coupled amplifier has a tendency to drift over a period of time, due to temperature variations, aging of components, and so on. This drift can be minimized by using what is known as a "chopper-stabilized" amplifier. A balancing circuit is usually built into these amplifiers to allow one to rebalance the amplifier.

5. The operational amplifier usually has an inherent odd number of stages, so that a negative-going output results from a positive-going input. This $180°$ phase shift between the output and input is primarily necessary in order to achieve electrical stability in the amplifier, but is also advantageous to use as a sign-changing amplifier, as will be seen later.

Both transistors and vacuum tubes are used in present-day analog computers, but the advantages of one device over the other will not be dwelt on here.

The transfer function e_{out}/e_{in} for the amplifier configuration of Fig. 20-2 can be derived quite easily if certain factors are kept in mind; these include the facts that the gain magnitude of the amplifier is 10^6 or greater, the input current i_g is negligibly small, and the maximum output or input voltages are ± 100 or ± 10 volts. With these qualifications in mind, the transfer function is obtained as follows:

$$i_1 = i_2 \qquad (20\text{-}4)$$

since i_g is negligible.

$$i_1 = \frac{e_{in} - e_g}{Z_{in}} \qquad (20\text{-}5)$$

$$i_2 = \frac{e_g - e_{out}}{Z_{fb}} \qquad (20\text{-}6)$$

However, since $e_{out}/e_g = -A$ and $A > 10^6$, e_g is much less than either e_{out} or e_{in} and thus can be neglected in Eqs. 20-5 and 20-6. Upon substituting Eqs. 20-5 and 20-6 into Eq. 20-4.

$$\frac{e_{in}}{Z_{in}} = -\frac{e_{out}}{Z_{fb}}$$

$$\frac{e_{out}}{e_{in}} = - \frac{Z_{fb}}{Z_{in}} \qquad (20\text{-}7)$$

If the operational methods of Laplace transforms are used, Eq. 20-7 becomes

$$\frac{E_{out}(s)}{E_{in}(s)} = - \frac{Z_{fb}(s)}{Z_{in}(s)}$$

Equation 20-8 is the important form of the transfer function for the operational amplifier. The new circuit diagram replacing the circuit of Fig. 20-2 is shown in Fig. 20-3. The common ground or reference termi-

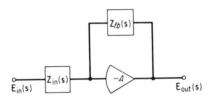

FIG. 20-3. The computing amplifier with associated operational input and feedback impedances.

nal is always understood to be present in a computer circuit, even though it is not explicitly shown in the diagram of Fig. 20-3.

Three of the most useful computing functions—the sign changer, the summer, and the integrator—will now be presented.

The Sign Changer. If the input and feedback impedances are replaced by resistances, as shown in Fig. 20-4, the transfer function, as

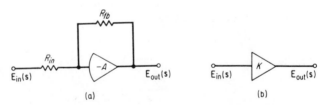

(a) (b)

FIG. 20-4. a) The sign changer. b) Shorthand symbol for the sign changer of part a. $K = R_{fb}/R_{in}$.

defined by Eq. 20-8 becomes

$$\frac{E_{out}(s)}{E_{in}(s)} = - \frac{R_{fb}}{R_{in}} \qquad (20\text{-}9)$$

The circuit of Fig. 20-4a is known as a sign changer, but, in addition to sign changing, this computing circuit multiplies the input voltage $E_{in}(s)$

by the constant factor R_{fb}/R_{in} in order to obtain the output voltage $E_{out}(s)$. Figure 20-4b shows the shorthand symbol for a sign changer with an R_{fb}/R_{in} ratio of K. It is common practice not to show the negative sign explicitly as it is always present in relating the output to the input.

The Summer. Figure 20-5a shows the schematic diagram for the

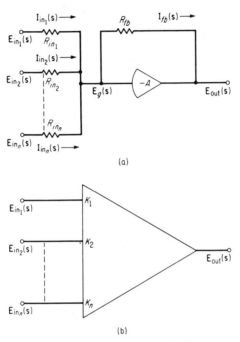

(a)

(b)

Fig. 20-5. a) The summer. b) Short-hand symbol for the summing circuit of part a.

summing computer function. Again making the usual assumption that A is so large that $E_g \approx 0$ and summing currents at the grid node,

$$\mathbf{I}_{in_1}(s) + \mathbf{I}_{in_2}(s) + \cdots + \mathbf{I}_{in_n}(s) = \mathbf{I}_{fb}(s) \qquad (20\text{-}10)$$

or

$$\frac{\mathbf{E}_{in_1}(s)}{R_{in_1}} + \frac{\mathbf{E}_{in_2}(s)}{R_{in_2}} + \cdots + \frac{\mathbf{E}_{in_n}(s)}{R_{in_n}} = -\frac{\mathbf{E}_{out}(s)}{R_{fb}} \qquad (20\text{-}11)$$

and solving for $\mathbf{E}_{out}(s)$

$$\mathbf{E}_{out}(s) = -\frac{R_{fb}}{R_{in_1}}\mathbf{E}_{in_1}(s) - \frac{R_{fb}}{R_{in_2}}\mathbf{E}_{in_2}(s) - \cdots - \frac{R_{fb}}{R_{in_n}}\mathbf{E}_{in_n}(s) \qquad (20\text{-}12)$$

Equation 20-12 shows that the output voltage is the negative sum of the input voltages times their respective scale factors R_{fb}/R_{in}. Figure 20-5 is the shorthand symbol for the summing operational amplifier circuit.

The Integrator. Figure 20-6a shows the schematic diagram for the

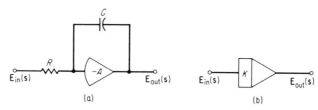

(a)   (b)

FIG. 20-6.   a) The integrator.   b) Shorthand symbol for the integrator circuit of part a.

integrator circuit. To illustrate that this is a pure integrator, we substitute the proper values for the operational impedances $\mathbf{Z}_{fb}(\mathbf{s})$ and $\mathbf{Z}_{in}(\mathbf{s})$, or

$$\mathbf{Z}_{fb}(\mathbf{s}) = \frac{1}{C\mathbf{s}} \tag{20-13}$$

$$\mathbf{Z}_{in}(\mathbf{s}) = R \tag{20-14}$$

and, substituting Eqs. 20-13 and 20-14 into Eq. 20-8,

$$\frac{\mathbf{E}_{out}(\mathbf{s})}{\mathbf{E}_{in}(\mathbf{s})} = -\frac{1/C\mathbf{s}}{R} = -\frac{1}{RC}\frac{1}{\mathbf{s}} \tag{20-15}$$

The term $1/\mathbf{s}$ is recognized, from material in Chapter 19, as the integrating operator, and the factor $1/RC$ is a constant multiplier. Figure 20-6b shows the shorthand symbol for the integrator with a gain $K = 1/RC$.

Other Transfer Functions. Although only summing and integrating circuits will be employed in this chapter, other $\mathbf{E}_{out}(\mathbf{s})/\mathbf{E}_{in}(\mathbf{s})$ transfer functions, sometimes extremely useful, can be found by substituting operational impedances into the input and feedback paths. This technique makes it rather easy to synthesize a given transfer function with an electrical network.

Let us consider the combination of feedback and input elements shown in Fig. 20-7a. The operational feedback impedance $\mathbf{Z}_{fb}(\mathbf{s})$ is

$$\mathbf{Z}_{fb}(\mathbf{s}) = R_2 + \frac{1}{C_2\mathbf{s}} \tag{20-16}$$

and the operational input impedance $\mathbf{Z}_{in}(\mathbf{s})$ is

$$\mathbf{Z}_{in}(\mathbf{s}) = R_1 + \frac{1}{C_1\mathbf{s}} \tag{20-17}$$

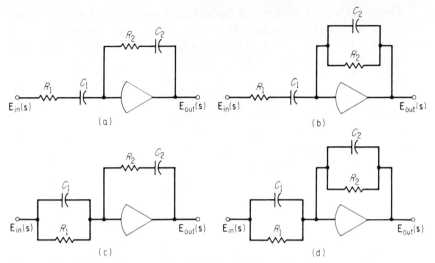

FIG. 20-7.   Other operational circuits useful in analog-computer studies.

Upon substituting Eqs. 20-16 and 20-17 into Eq. 20-18,

$$\frac{\mathbf{E}_{out}(\mathbf{s})}{\mathbf{E}_{in}(\mathbf{s})} = - \frac{R_2 + \dfrac{1}{C_2 \mathbf{s}}}{R_1 + \dfrac{1}{C_1 \mathbf{s}}}$$

$$= - \frac{C_1}{C_2} \frac{R_2 C_2 \mathbf{s} + 1}{R_1 C_1 \mathbf{s} + 1} \tag{20-18}$$

By a similar analysis, the respective transfer functions for parts b, c, and d of Fig. 20-7 are

$$\frac{\mathbf{E}_{out}}{\mathbf{E}_{in}}(\mathbf{s}) = - \frac{R_2 C_1 \mathbf{s}}{(R_1 C_1 \mathbf{s} + 1)(R_2 C_2 \mathbf{s} + 1)} \tag{20-19}$$

$$\frac{\mathbf{E}_{out}}{\mathbf{E}_{in}}(\mathbf{s}) = - \frac{(R_1 C_1 \mathbf{s} + 1)(R_2 C_2 \mathbf{s} + 1)}{R_1 C_2 \mathbf{s}} \tag{20-20}$$

$$\frac{\mathbf{E}_{out}}{\mathbf{E}_{in}}(\mathbf{s}) = - \frac{R_2(R_1 C_1 \mathbf{s} + 1)}{R_1(R_2 C_2 \mathbf{s} + 1)} \tag{20-21}$$

The reader should verify the correctness of Eqs. 20-19, 20-20, and 20-21.

Differential-Equation Solutions.   The operational analog computer is particularly valuable when one must solve simultaneous nonlinear differential equations; however, only linear differential equations will be considered in this text.   Consider Eq. 20-1 for the damped spring-mass system, repeated here as Eq. 20-22

$$m\ddot{x} + d\dot{x} + kx = f(t) \tag{20-22}$$

The usual way to prepare an equation for an analog-computer solution is to rearrange the differential equation so that the highest-order derivative is on one side and all other terms are on the other side of the equation. Rearranging Eq. 20-22

$$\ddot{x} = -\frac{d}{m}\dot{x} - \frac{k}{m}x + \frac{1}{m}f(t) \qquad (20\text{-}23)$$

The next step is to set up an analog-computer block diagram that will show the solution for the highest-order derivative. In Eq. 20-23, it is evident that the second derivative $\ddot{x}$ is the sum of three terms: $(-d/m)\dot{x}$, $(-k/m)x$, and $(1/m)f(t)$. Figure 20-8a shows the three terms (which are

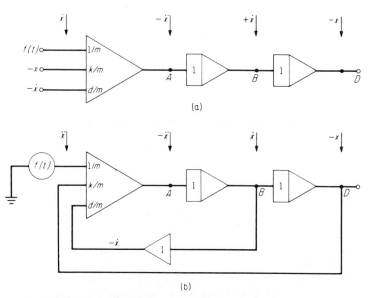

(a)

(b)

Fig. 20-8. a) Analog-computer diagram, showing the solution to Eq. 20-23. b) Same as part a, except for more detail. x and $\dot{x}$ are shown fed back into the input of the summer.

assumed to be available from some source) $(-d/m)\dot{x}$, $(-k/m)x$, and $(1/m)f(t)$ being summed and inverted to yield $-\ddot{x}$; then $-\ddot{x}$ is integrated and inverted to give $+\dot{x}$; finally, $\dot{x}$ is integrated and inverted to yield $-x$. Figure 20-8b shows how both x and $\dot{x}$ (with proper sign changes) are fed back into the summer along with $f(t)$ to produce $\ddot{x}$. Although the coefficient factors $1/m$, k/m, and d/m are shown on the summing amplifier, this is not necessarily the general or best way to set up the problem on the computer. Other ways will be shown later in this chapter.

The computer hookup in Fig. 20-8b is continuously solving for $\ddot{x}$, $\dot{x}$, and x for any value of $f(t)$ being introduced into the summer. It is not yet obvious, but initial conditions can also be introduced into the computer solution. To "see" the solutions for $\ddot{x}$, $\dot{x}$, or x, we simply connect a voltmeter, an *X-Y* or strip-chart recorder, or an oscilloscope to terminals *A*, *B*, or *D*, respectively. A voltmeter or a mechanical recorder can be used only if the frequencies present in the solution, if any, are within the frequency-response limitations of the output recording device. For the usual voltmeter or mechanical recorder, this means only a few hertz. Although oscilloscopes will easily follow frequencies of 500 KHz or more, the frequency bandwidth of most of the d-c amplifiers used in analog computers is limited to a few thousand hertz. The voltage levels at terminals *A*, *B*, or *D* can vary from $+100$ to -100 volts, or $+10$ to -10 volts, so that a sizeable voltage is usually available for monitoring purposes.

A more detailed schematic diagram, corresponding to the block diagram of Fig. 20-8, is shown in Fig. 20-9. The element values for the resistors are in megohms and for the capacitors in microfarads.

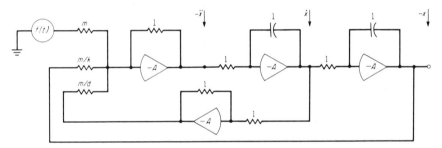

FIG. 20-9.   A more detailed schematic diagram for the block diagram of Fig. 20-8.

In Figs. 20-8 and 20-9, $\ddot{x}$ is explicitly available for readout purposes. If $\ddot{x}$ is not specifically desired in the solution, it is possible to combine the summing and integrating functions on one amplifier and obtain a new

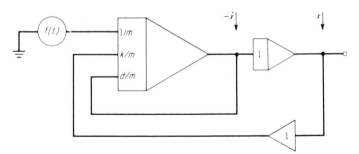

FIG. 20-10. Solution to Eq. 20-23 with $\dot{x}$ being explicitly solved for.

computer diagram, as shown in Fig. 20-10. Since the circuit of Fig. 20-10 contains one less amplifier than that of Fig. 20-9, it is a more desirable circuit, because indiscriminate use of amplifiers can lead to a shortage of same when setting up a large problem. There are other advantages in using fewer amplifiers, such as less likelihood of hookup errors and greater ease in time- and amplitude-scaling of the problem. The methods of and reasons for time- and amplitude-scale changes will be taken up later in this chapter.

Thus far, the solution of only one differential equation has been presented. Now suppose that one has two simultaneous differential equations to solve, as are obtained when one solves for the instantaneous loop

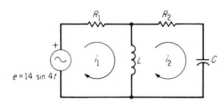

FIG. 20-11. Circuit used in writing the loop-current equations shown in Eq. 20-24.

currents i_1 and i_2 from Fig. 20-11. The two KVL equations are

$$\left.\begin{array}{l} 14 \sin 4t = R_1 i_1 + L \dfrac{di_1}{dt} - L \dfrac{di_2}{dt} \\[3mm] 0 = -L \dfrac{di_1}{dt} + L \dfrac{di_2}{dt} + R_2 i_2 + \dfrac{1}{C} \int i_2 dt \end{array}\right\} \quad (20\text{-}24)$$

and, substituting $i = dq/dt$ so that the integral expression in Eq. set 20-24 is eliminated,

$$\left.\begin{array}{l} 14 \sin 4t = R_1 \dfrac{dq_1}{dt} + L \dfrac{d^2 q_1}{dt^2} - L \dfrac{d^2 q_2}{dt^2} \\[3mm] 0 = -L \dfrac{d^2 q_1}{dt^2} + L \dfrac{d^2 q_2}{dt^2} + R_2 \dfrac{dq_2}{dt} + \dfrac{q_2}{C} \end{array}\right\} \quad (20\text{-}25)$$

Without regard for any scale changes, we rearrange the equations by solving for $\ddot{q}_1$ and $\ddot{q}_2$, or

$$\left.\begin{array}{l} \ddot{q}_1 = \dfrac{14}{L} \sin 4t - \dfrac{R_1}{L} \dot{q}_1 + \ddot{q}_2 \\[3mm] \ddot{q}_2 = \ddot{q}_1 - \dfrac{R_2}{L} \dot{q}_2 - \dfrac{1}{LC} q_2 \end{array}\right\} \quad (20\text{-}26)$$

Next, we set up the necessary summers and integrators to solve for each equation shown in Eq. set 20-26. Both equations involve the second derivative of the other term, and so $\ddot{q}_1$ and $\ddot{q}_2$ must be explicitly solved for, as is shown in Fig. 20-12a. Figure 20-12b shows the interconnections that are necessary in order to satisfy Eq. set 20-26. Note that if q_1 is not to be monitored, one of the integrators can be omitted, since q_1 is not fed back into the computer solution.

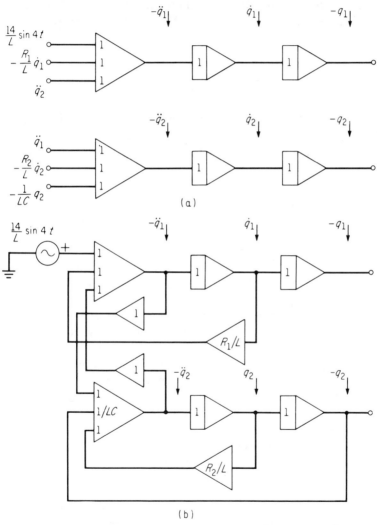

FIG. 20-12. a) A computer block diagram showing the independent circuits necessary to solve for q_1 and q_2 from Eq. set 20-26. b) Complete computer diagram necessary to solve Eq. set 20-26.

Initial Conditions. Almost every physical system is subject to initial or boundary conditions. These must be accounted for, whether differential equations are solved by analytical means or by a computer. Fortunately, initial conditions are rather easily introduced on the analog computer. Consider the equation

$$\ddot{y} = -8y \tag{20-27}$$

which is recognized as the equation describing simple harmonic motion. If the initial conditions $y(0^+) = 5$ units and $\dot{y}(0^+) = -7$ units are imposed, Fig. 20-13 shows how the initial conditions are placed on the computer.

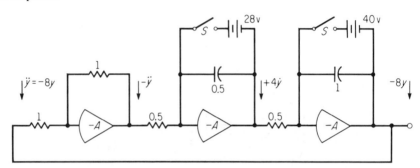

Fig. 20-13.   Computer solution to Eq. 20-27 with initial conditions included.

The choice of the R and C values is rather arbitrary (but not completely, as will be seen later) as long as the differential equation, as defined by Eq. 20-27, is satisfied. However, once the choices have been made, the initial conditions must be carefully inserted into the computer solution by placing an initial charge or voltage on the proper integrator capacitors. For instance, the output of the second integrator in Fig. 20-13 is seen to be $-8y$. Since the initial value for y is 5 units, it is necessary to start with an initial voltage of -8×5 or -40 volts at the output of the second integrator (assuming that 1 volt on the computer equals 1 problem unit). The 40 volts are applied across the capacitor by closing switch S. We recall that the input voltage to the amplifier is a very low level ($E_g < 10^{-3}$ volt), and thus the voltage across the capacitor is essentially equal to the output-to-ground voltage of the amplifier. The reader should verify that the 28-volt battery is of the proper magnitude and polarity for $\dot{y}(0^+) = -7$ units. In practice, the initial-condition switch S is closed until it is desired to start the problem solution, but, at the moment when the compute switch is thrown and the problem solution is begun, switch S (which is either a mechanical or an electronic relay) is opened.

Time- and Amplitude-Scale Changes.   It is indeed rare if a given differential equation can be immediately set up on the computer without any changes being made in the original equation. Usually, a scale model (the scale factor can be either large or small) of the original equation must be developed. Scaling has two important functions: 1) The frequency limitations of the operational amplifiers, and particularly the mechanical recording devices, must not be exceeded. This usually means that about 2 to 10 Hz is the highest frequency that can be tolerated in the solution if mechanical pen or hot-wire recorders are used as the output recording device. 2) To avoid undue errors in recording and to minimize the inherent noise voltages that are introduced in electronics components,[1] it is desirable to keep the output voltages of all the amplifiers at or near their maximum allowable values of 10 or 100 volts.

Time Scaling. The calculations involved in making a time-scale change are quite simple. All that is necessary is to make the substitution

$$t = \frac{\tau}{\alpha} \tag{20-28}$$

where t is the independent variable in the problem (t is always considered to be time in this chapter), τ is the new independent variable with the same dimensions as t, and α is a positive real constant. If $\alpha > 1$, the computer solution is slowed; if $\alpha < 1$, the solution is speeded up. To illustrate the performance of the time-scale change, consider the second-order differential equation

$$2\frac{d^2y}{dt^2} + 10\frac{dy}{dt} + 200y = 20\sin 40t \tag{20-29}$$

with the initial conditions

$$y(0^+) = 4 \quad \text{and} \quad \dot{y}(0^+) = 5 \tag{20-30}$$

The undamped natural frequency ω_n of the system, as described by Eq. 20-29, is the square root of the ratio of the coefficient of the zero-order term to the coefficient of the second-order term, or

$$\omega_n = \sqrt{\frac{200}{2}} = 10 \text{ radians/sec}$$

To slow the problem solution by a factor of 2, the variable substitution $t = \tau/2$ is necessary. Introducing this substitution into Eq. 20-29 yields

$$2\frac{d^2y}{\left(d\frac{\tau}{2}\right)^2} + 10\frac{dy}{d\left(\frac{\tau}{2}\right)} + 200y = 20\sin 40\frac{\tau}{2}$$

---

[1]Tubes, transistors, resistors, and so on, generate small but significant random and unwanted voltages called noise. Actually, noise voltages are any voltages that are not part of the desired signal voltages.

where $d(\tau/2) = \frac{1}{2}d\tau$. (A constant term can be brought outside of a differential.) Upon simplifying,

$$8 \frac{d^2y}{d\tau^2} + 20 \frac{dy}{d\tau} + 200y = 20 \sin 20\tau \qquad (20\text{-}31)$$

Now $t = 0^+$ implies $\tau = 0^+$ since $t = \tau/2$ so that

$$y(t = 0^+) = y(\tau = 0^+) = 4$$

and

$$\frac{dy}{dt}(t = 0^+) = \frac{dy}{d(\tau/2)}(\tau = 0^+) = 5$$

from which

$$y(\tau = 0^+) = 4$$

and

$$\frac{dy}{d\tau}(\tau = 0^+) = \frac{5}{2}$$

The undamped natural frequency of the new equation (Eq. 20-31) is

$$\omega_n = \sqrt{\frac{200}{8}} = 5 \text{ radians/sec}$$

Thus, both the forced (the sine term on the right-hand side of Eqs. 20-29 and 20-31) and the natural frequencies of Eq. 20-29 have been reduced by a factor of 2, as was desired.

In reading out the derivatives in a solution to a problem where a time-scale change has been performed, it is necessary to reinsert the time-scale factor in reverse order before labeling the ordinates of the graphical presentation. An example of this will be shown after the amplitude-scale-factor change has been explained.

Amplitude-Scale Factors. There are several ways to perform the amplitude-scale-factor change. *One method* is simply to multiply the entire differential equation by a real constant. Unfortunately, it is not always obvious what the constant should be, just as it is not always obvious what the time-scale factor α should be. The main reason for performing the amplitude-scale change is to try to make sure that the output of each amplifier is as high as possible and yet not overloaded; that is peak voltages of ± 50 to ± 100 or ± 5 to ± 10 volts should be maintained if possible.

Thus far, nothing has been said about how the voltage output of an amplifier can be related to the problem units. The easiest scheme is to let one problem unit equal 1 volt on the computer. In other words, 1 volt might equal 1 in., 1 in./sec, 1 ft/sec², 1 radian, and so on. An example will help to clarify the necessity for, and one method of performing, both the time- and the amplitude-scale changes.

EXAMPLE 20-1. It is desired to solve, on an analog computer, the second-order differential equation

$$26 \frac{d^2y}{dt^2} + 1134 \frac{dy}{dt} + 14{,}976y = 0 \tag{20-32}$$

with the initial conditions

$$y(0^+) = 2 \text{ units} \qquad \text{and} \qquad \dot{y}(0^+) = 0$$

The dimensions of y and the constant coefficients are not particularly important and will be referred to simply as units.

Solution: First, we check to see if a time-scale change is necessary or desirable. To be able to make an intelligent choice as to the magnitude of the time-scale change, it is necessary to know the form of the problem solution, that is, the exponential time constants and any frequencies that may be present in the solution. (For one who is familiar with the solution to differential equations, such information is often obtainable with little effort.)

It is assumed that the reader is familiar with the solution to the linear second-order differential equation; therefore, only a review of the more salient points in the solution to Eq. 20-32 will be presented here.

1. The highest value of the natural frequency in Eq. 20-32 is the undamped natural frequency ω_n, where

$$\omega_n = \sqrt{\frac{14{,}976}{26}} = 24 \text{ radians/sec} \tag{20-33}$$

The viscous-damping term $1134 \, (dy/dt)$ produces an actual damped frequency ω_d of

$$\omega_d = \sqrt{\frac{14{,}976}{26} - \frac{(1134)^2}{4 \times 26^2}} = \sqrt{576 - 476} = 10 \text{ radians/sec} \tag{20-34}$$

which is obviously less than the undamped natural frequency.

2. One term in the general solution of Eq. 20-32 is

$$y = Ae^{-at} \cos \omega t \tag{20-35}$$

where A depends on the initial conditions, a is the damping factor, and ω is the damped natural frequency ω_d.

From the initial condition of $y(0^+) = 2$ units, it is evident that $A = 2$, and a can be calculated from inspection of Eq. 20-32:

$$a = \frac{1134}{2 \times 26} = 21.8$$

Thus, Eq. 20-35 can be written as

$$y = 2e^{-21.8t} \cos 10t \tag{20-36}$$

However, it is a bit bothersome to evaluate the a's and ω's, since we really are concerned only with an approximate solution for y. Equations 20-35, and 20-36 indicate that, as time progresses, the magnitude of the peak values of y becomes less and less, owing to the exponential damping term e^{-at}.

At this point, it is desirable to recall that both the frequency and the peak amplitudes of the integrator and summer output voltages are of the most importance when scaling a problem. If one can estimate the worst possible conditions that will be encountered, an intelligent time- and amplitude-scale change can be made. Looking at Eq. 20-35 or Eq. 20-36, it is evident that the largest magnitude of y occurs at $t = 0$, and the largest possible value for ω is the undamped natural frequency ω_n. Using these conditions and approximations, Eq. 20-35 becomes

$$y = 2 \cos 24t \qquad (20\text{-}37)$$

Keep in mind that Eq. 20-37 is only a simple approximation to Eq. 20-35.

Equation 20-37 can be used to illustrate the frequency and magnitude problems often encountered in analog-computer work: 1) The frequency $\omega = 24$ radians/sec is too high for many recording devices, particularly for large-deviation plotting devices such as the X-Y recorder. A frequency of 1 to 4 radians/sec can be considered a good choice for many problems. 2) If we take successive derivatives of y, or

$$\dot{y} = -2 \times 24 \sin 24t$$
$$= -48 \sin 24t$$

and

$$\ddot{y} = -48 \times 24 \cos 24t$$
$$= -1152 \cos 24t$$

the magnitude problem becomes apparent. If 1 volt on the computer equals one problem unit, then, for a peak variation for y of only 2 volts, a peak variation of 1152 volts occurs for $\ddot{y}$. The 2-volt amplifier output for y_{max} is really too small (50 to 100 or 5 to 10 volts is preferable), and the 1152 volts for $\ddot{y}_{max}$ is intolerable, since extreme nonlinear distortion will be introduced into the problem solution (the amplifiers will saturate and clip all voltage outputs that go much beyond 100 or 10 volts).

There is an obvious necessity for making some scale changes. Attention is first focused on the time-scale change. If it is assumed that the recording device is such that about a 3-radians/sec maximum frequency input is most desirable, then a slowing of the computer solution by a factor of $\omega_n/3 = 24/3 = 8$ is called for. Substituting $t = \tau/8$ into Eq. 20-32 yields

$$26 \frac{d^2 y}{\dfrac{d\tau^2}{64}} + 1134 \frac{dy}{\dfrac{d\tau}{8}} + 14{,}976y = 0 \qquad (20\text{-}38)$$

with initial conditions

$$y(\tau = 0^+) = 2 \text{ units} \qquad \text{and} \qquad \dot{y}(\tau = 0^+) = 0$$

Upon simplifying Eq. 20-38,

$$26 \times 64 \frac{d^2 y}{d\tau^2} + 1134 \times 8 \frac{dy}{d\tau} + 14{,}976y = 0 \qquad (20\text{-}39)$$

As a check, the new value for ω_n is

$$\omega_n = \sqrt{\frac{14{,}976}{26 \times 64}} = 3 \text{ radians/sec}$$

If one problem unit is to equal one machine unit, Eq. 20-39 must undergo a magnitude-scale change, for if $y = 2$ volts at $t = 0$, then $14{,}976 \times 2$ is much too large a voltage output for an amplifier. The same is true for the other two terms in Eq. 20-39. After a few trials, we choose $1/576$ as a multiplying factor for Eq. 20-39, or

$$\frac{1}{576}\left(26 \times 64 \ \frac{d^2 y}{d\tau^2} + 1134 \times 8 \ \frac{dy}{d\tau} + 14{,}976y\right) = 0 \qquad (20\text{-}40)$$

Carrying out the multiplication, we obtain

$$\frac{26}{9} \frac{d^2 y}{d\tau^2} + \frac{63}{4} \frac{dy}{d\tau} + 26y = 0 \qquad (20\text{-}41)$$

Now we check to see what the maximum amplifier outputs will be if Eq. 20-41 is set up on a computer with ± 100 volt amplifiers. It is again anticipated that the maximum outputs will occur at $t = 0$. The third term on the right-hand side of Eq. 20-41 has a maximum value of $26 \times 2 = 52$; the second term has a maximum value of $(63/4) \times 3 \times 2 = 94.5$ volts; the first term has a maximum value of $(26/9) \times 3 \times 3 \times 2 = 52$ volts. Since these maximum expected voltages lie between 50 and 100 volts, the chosen time- and amplitude-scale changes will be satisfactory.

To place Eq. 20-41 on the computer, we solve for the term with the highest-order derivative, or

$$\frac{26}{9} \ddot{y} = -\frac{63}{4} \dot{y} - 26y \qquad (20\text{-}42)$$

where y is a function of τ. If it is desired to obtain explicitly the second derivative of y with respect to τ, Fig. 20-14 shows the computer diagram that will solve Eq. 20-42. Both axes of the recorded plots for $y(t)$, $\dot{y}(t)$, or $\ddot{y}(t)$ must be labeled properly to account for the time- and amplitude-scale changes.

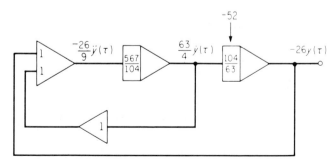

FIG. 20-14. Analog-computer solution to Eq. 20-42.

It should be emphasized that the method presented in Example 20-1 is intended only to illustrate one way to apply the time- and amplitude-scale changes but does not necessarily represent the most orderly method for applying these changes; for instance, the integrator gains of 567/104 and 104/63 may not be convenient choices for a computer that allows only gain factors of 1, 2, 5, and 10 to be introduced by the integrators. However, this presents no particular difficulty, except that it does take a little more time and care to become acquainted with, and master, the methods and techniques of time and amplitude scaling.

Another example will be used to illustrate the simulation of the equations of an electrical network by means of an analog computer.

EXAMPLE 20-2. In Chapter 19, an example was introduced which indicated that an analog-computer solution was feasible. Figure 20-15 is identical to Fig. 19-15, and the associated node-voltage equations are the same as Eqs. 19-73a and

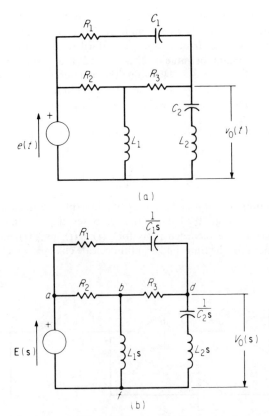

FIG. 20-15. a) Figure 19-15a repeated.
b) Figure 19-15b repeated.

19-73b. Using an analog computer, solve this set of simultaneous equations for the unknown output voltage.

Solution: The node-voltage equations of Eq. set 19-73, somewhat rearranged, are

$$\left(\frac{1}{R_2} + \frac{1}{R_3} + \frac{1}{L_1 s}\right) V_{bf}(s) - \frac{1}{R_3} V_0(s) = \frac{E(s)}{R_2} \qquad (20\text{-}43a)$$

$$-\frac{1}{R_3} V_{bf}(s) + \left(\frac{1}{R_3} + \frac{C_2 s}{1 + L_2 C_2 s^2} + \frac{C_1 s}{R_1 C_1 s + 1}\right) V_0(s) = \frac{C_1 s}{R_1 C_1 s + 1} E(s)$$

$$(20\text{-}43b)$$

Equation 20-43a can be simulated quite easily without added manipulation. The differential equation for $v_0(t)$ is, by inspection

$$v_0(t) = \left(\frac{R_3}{R_2} + 1\right) v_{bf}(t) + \frac{R_3}{L_1} \int v_{bf}(t)\, dt - \frac{R_3}{R_2} e(t) \qquad (20\text{-}44)$$

Equation 20-44 can be simulated (without regard to scale changes) as indicated in Fig. 20-16 by assuming that $v_{bf}(t)$ and $e(t)$ are known or obtainable quantities.

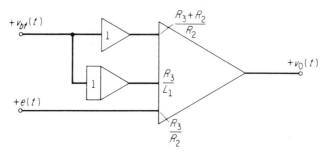

FIG. 20-16. A simulation of Eq. 20-44.

The realization of the rather large R_3/L_1 factor at the summer would probably not be practical, and, if necessary, the integrator could absorb much of the gain. However, from a purely theoretical standpoint, Fig. 20-16 is correct.

Equation 20-43b requires a little more finesse to simulate and can be handled in one of two ways. Since $v_0(t)$ was solved for in Eq. 20-43a, it is necessary to solve for $v_{bf}(t)$, in Eq. 20-43b, regardless of which method is used.

$$V_{bf}(s) = \left(1 + \frac{R_3 C_2 s}{1 + L_2 C_2 s^2} + \frac{R_3 C_1 s}{R_1 C_1 s + 1}\right) V_0(s) - \frac{R_3 C_1 s}{R_1 C_1 s + 1} E(s) \qquad (20\text{-}45)$$

Equation 20-45 can be solved for, as indicated in block-diagram form, in Fig. 20-17, where

$$\frac{N_1}{N_2}(s) = \frac{R_3 C_2 s}{1 + L_2 C_2 s^2}$$

$$\frac{N_3}{N_4}(s) = \frac{R_3 C_1 s}{R_1 C_1 s + 1}$$

$$\frac{N_5}{N_6}(s) = \frac{R_3 C_1 s}{R_1 C_1 s + 1}$$

To obtain the complete simulation in Eqs. 20-43a, and 20-43b, combine Figs.

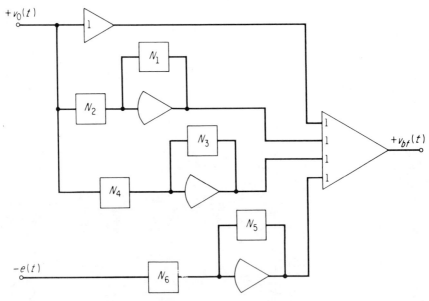

FIG. 20-17. A solution to Eq. 20-45.

20-16 and 20-17, as shown in Fig. 20-18. For a given $e(t)$, the voltage $v_0(t)$ or $v_{bf}(t)$ can easily be computed and recorded.

Aside from the probable scale changes which are necessary in most problem simulations, it is necessary to determine the networks N_1, N_2, N_3, N_4, N_5, and N_6 if the simulation is to be successful. The transfer function

$$\frac{N_3}{N_4}(s) = \frac{N_5}{N_6}(s) = \frac{R_3 C_1 s}{R_1 C_1 s + 1}$$

can be rather easily simulated by use of the circuit shown in Fig. 20-19. The transfer function $E_0(s)/E_{in}(s)$ is

$$\frac{E_0}{E_{in}}(s) = -\frac{Z_{fb}(s)}{Z_{in}(s)}$$

$$= -\frac{R_3}{R_1 + \dfrac{1}{C_1 s}}$$

$$= -\frac{R_3 C_1 s}{R_1 C_1 s + 1}$$

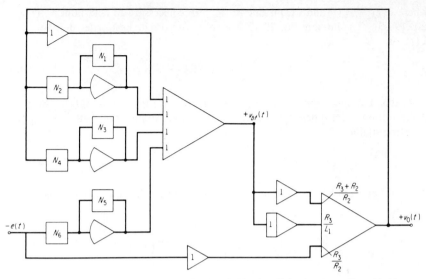

FIG. 20-18. Combination of Figs. 20-16 and 20-17, which simulates Eqs. 20-43a and 20-43b and obtains the solution for $v_0(t)$ for any $e(t)$.

Unfortunately, the transfer function

$$\frac{N_1}{N_2}(s) = \frac{R_3 C_2 s}{1 + L_2 C_2 s^2} \qquad (20\text{-}46)$$

cannot be simulated in this manner unless an inductor is used, and, since ideal inductors are difficult to fabricate (unless operating at cryogenic temperatures), it is desirable to try another means of simulating Eq. 20-46.

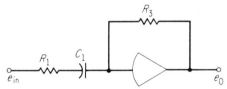

FIG. 20-19. Simulation of the transfer function $\dfrac{E_0}{E_{in}}(s) = -\dfrac{R_3 C_1 s}{R_1 C_1 s + 1}$.

As was pointed out in the latter part of Chapter 19 an s-domain transfer function of the form

$$\frac{X(s)}{Y(s)} = \frac{R_3 C_2 s}{1 + L_2 C_2 s^2}$$

can be written in differential-equation form as

$$x(t) + L_2 C_2 \frac{d^2 x(t)}{dt^2} = R_3 C_2 \frac{dy(t)}{dt} \qquad (20\text{-}47)$$

where $y(t) \overset{\Delta}{=} v_0(t)$ and $x(t)$ is one of the terms necessary to determine $v_{bf}(t)$ (cf Eq. 19-80). To simulate Eq. 20-47, we solve for the highest-order derivative of $x(t)$, or

$$\ddot{x}(t) = \frac{R_3 C_2}{L_2 C_2} \dot{v}_0(t) - \frac{1}{L_2 C_2} x(t) \qquad (20\text{-}48)$$

Note that Eq. 20-48 solves for $x(t)$ in terms of $\dot{v}_0(t)$ instead of $v_0(t)$. Because differentiation of a function is usually undesirable, we integrate both sides of the equation so that

$$\dot{x}(t) = \frac{R_3}{L_2} v_0(t) - \frac{1}{L_2 C_2} \int x(t) \qquad (20\text{-}49)$$

The simulation of Eq. 20-49 is shown in Fig. 20-20b. The complete simulation of

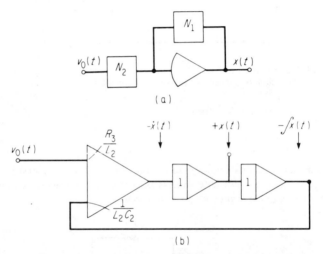

FIG. 20-20. a) Block-diagram simulation of Eq. 20-45, using the impedance method. b) Simulation of Eq. 20-45 (and Eq. 20-49), using the direct solution of the differential-equation method.

Eqs. 20-43a and 20-43b, using only R-C networks with the operational amplifiers, is shown in Fig. 20-21. No attempt has been made to minimize the number of amplifiers necessary in the simulation. The reader is encouraged to check and verify each of the connections shown in Fig. 20-21 and if possible to reduce the number of amplifiers required.

The unknown node voltages $v_0(t)$ and $v_{bf}(t)$ can be seen and recorded by connecting an oscilloscope or pen recorder to amplifiers 1 or 2. The forcing function $e(t)$ can be chosen as desired. Of course, amplitude- and time-scale changes should be expected. These changes are a function of the R-L-C network parameters and the forcing function $e(t)$.

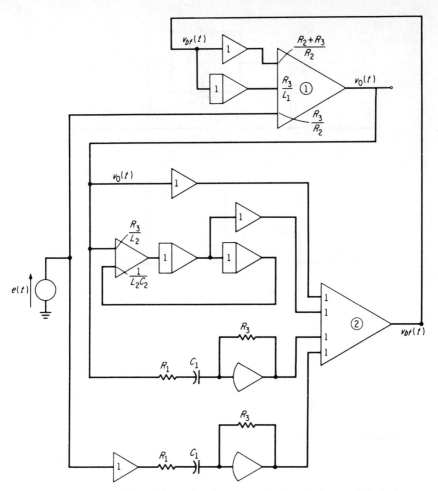

FIG. 20-21. Simulation of the node-voltage equations (Eqs. 20-43a and 20-43b).

Problems

Other Transfer Functions

20-1. Verify Eqs. 20-19, 20-20 and 20-21.

20-2. Find the transfer function of the circuits shown in the accompanying illustration (page 606).

20-3. a) If it were possible to make (and practical to use) ideal inductors in computer circuits, what would be the transfer function of the circuits shown in parts a, b, c, and d, of the accompanying illustration? b) What function would the circuits in parts a and b perform?

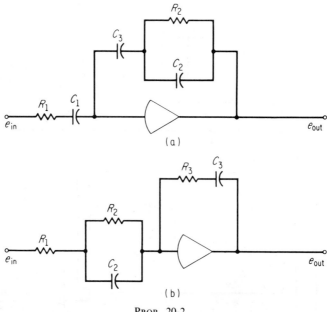

PROB. 20-2

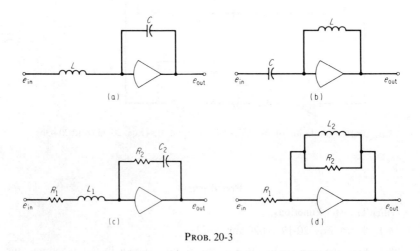

PROB. 20-3

20-4. Use node-voltage analysis to show that the transfer function for the circuit shown in the accompanying figure is

$$\frac{E_{out}}{E_{in}}(s) = -\frac{R_3/R_1}{R_2 R_3 C_1 C_2 s^2 + (R_2 R_3/R_1 + R_2 + R_3)C_2 s + 1}$$

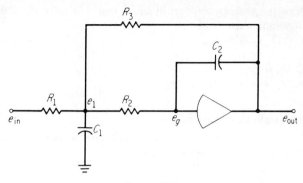

PROB. 20-4

20-5. Finish setting up the solution to the differential equation, shown in the accompanying figure, on the analog computer. Give values to all the components you use except those given here.

$$\frac{5d^3x}{dt^3} + \frac{10\,d^2x}{dt^2} + \frac{2\,dx}{dt} + 4x = 0$$

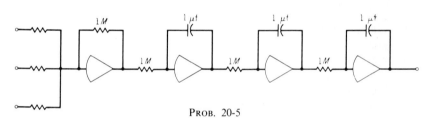

PROB. 20-5

20-6. Solve for $5(y)$ from the analog-computer hookup shown in the accompanying figure. (All R's are in megohms and C's in microfarads.)

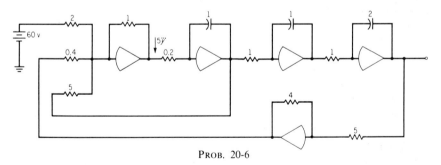

PROB. 20-6

20-7. Show the analog-computer hookup that will solve each of the following equations, with all initial conditions being equal to zero:

a) $\dddot{y} - 2\ddot{y} - 4y = 8$

b) $\ddot{x} - 3\dot{x} + 5x = 4\sin 3t$

c) $\ddot{x} + 6\dot{x} + 15x = 7$

20-8. Draw the analog-computer diagram that will solve the following sets of simultaneous equations:

a) $\ddot{y}_1 + y_1 + y_2 = 0$

$\ddot{y}_2 + \dot{y}_1 - \dot{y}_2 = 1$

b) $\ddot{x}_1 + 3\dot{x}_1 + 5x_1 = 6\ddot{x}_2 + 15$

$\ddot{x}_2 + 2\ddot{x}_2 + 8x_2 = 4\dot{x}_1 + 7x_1$

Time- and Amplitude-Scale Changes

20-9. Rewrite the equations and initial conditions, after slowing the solution by a factor of 10, in the following equations:

a) $0.05\ddot{y} + 0.2\dot{y} + 3y + 10\sin 80t = 0$

$\dot{y}(0^+) = 1.4$

$y(0^+) = 0$

b) $\ddot{x} + 4\ddot{x} + 20\dot{x} + 500x = 48e^{-15t}$

$\ddot{x}(0^+) = 8$

$\dot{x}(0^+) = x(0^+) = 0$

20-10. Draw the analog-computer diagram, showing the correct polarities of all initial conditions, for the following equations:

a) $\dddot{x} + 5\ddot{x} + 3\dot{x} + 15x = 10$

$\ddot{x}(0^+) = 1.5$

$\dot{x}(0^+) = 4$

$x0^+) = 7$

b) $\dddot{x} + 2\ddot{x} + 3\ddot{x} + 5\dot{x} + 20x = 0$

$\dddot{x}(0^+) = -3$

$\ddot{x}(0^+) = 0$

$\dot{x}(0^+) = 5$

$x(0^+) = -6$

20-11. Perform a speeding-up time-scale change of 0.5 and an amplitude-scale change that will normalize the coefficient of the highest-order derivative to unity in the following equations:

a) $3\ddot{x} + 5\dot{x} + 2x + 5\sin 4t = 0$

$\dot{x}(0^+) = 1$

$x(0^+) = 2.5$

b) $\dddot{x} + 2.5\ddot{x} + 2.5\ddot{x} + 7.6\dot{x} + 15x = 8e^{-0.6t}\cos(3\omega t + 40°)$

$\ddot{x}(0^+) = 2$

$\dot{x}(0^+) = 0$

$x(0^+) = 5$

General

20-12. Show an analog-computer simulation for the circuit given in Prob. 19-14.

20-13. Show an analog-computer simulation for Prob. 19-19.

20-14. Show an analog-computer simulation for Prob. 19-20.

20-15. a) Using Fig. 20-2 show that

$$\frac{e_{out}}{e_{in}} = \frac{-Z_{fb}/Z_{in}}{1 + (Z_{in} + Z_{fb})/AZ_{in}}$$

if e_g is not considered negligibly small. b) Under what condition does the expression in part a reduce to Eq. 20-7? c) State why it is necessary that the operational amplifier have a gain of $-A$ rather than $+A$ if the operational amplifier is to function properly. [*Hint:* Consider that resistors are used in place of the Z's and note that the amplifier can saturate (which means that the amplifier gain A can drop from a very large value to a small value).]

21/ Feedback and Servomechanisms

Feedback is a general term used in describing a property of a system in which a portion of a given output is combined with or compared to a given input quantity. Information fed back from an output to an input can be useful in improving the behavior of electrical, mechanical, pneumatic, or hydraulic systems, and hence is important in many engineering areas.

Historically, the electrical engineer first employed feedback to improve the performance of the early vacuum-tube amplifiers. It was discovered that by feeding back a portion of the amplifier output to the input, the overall gain of the amplifier could be increased. This caused greater distortion and instability, but it did increase the volume of an amplifier, such as might be found in a radio receiver. When the feedback signal tends to increase the overall gain, the feedback is termed *positive* or *regenerative*.

Later, it was discovered that negative feedback could be used to increase amplifier stability and reduce distortion and noise, and could, in general, improve other amplifier characteristics such as frequency response and input and output impedances. *Negative* feedback occurs when the feedback signal combines with the input signal in such a way that the overall gain is reduced.

The study of servomechanisms (abbreviated servos) is limited to feedback systems that are concerned with the positioning of an object. Although this restriction greatly narrows the scope of the application of feedback control, servo systems play a very important role in industrial and military areas. Aircraft autopilots, gunfire control, missile control, rudder control on ships, remote control of valves, power steering on automobiles, control of rolling mills in the steel and paper industry, control of machine tools, and automation in general all depend on the ability of a piece of equipment to be positioned accurately and quickly. Ideally, a servo system should operate so that the position output exactly and instantaneously follows the reference input signal. Obviously, this is impossible, since any mechanical system has mass or inertia and energy-storage properties. These combine to produce time delays, oscillatory conditions, and so forth, in any system. Ultimately, this means that the servo designer must design the servo system so that it gives the best or optimum response as determined by performance specifications.

The first portion of this chapter considers feedback in relation to electronic amplifiers; the second portion considers servomechanism theory and a few of the devices of particular interest in servo applications.

FEEDBACK

21-1. Feedback Theory

The block diagram of Fig. 21-1 can be used to derive the basic feedback relations that are common to all types of feedback theory. Here,

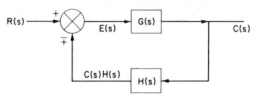

FIG. 21-1. Basic feedback block diagram.

$G(s)$ and $H(s)$ describe the transfer functions of the elements represented by the blocks. A transfer function is the ratio of the output response over the input excitation for a given system and, as explained in Chapters 13, 19, and 20, is generally expressed as a ratio of polynomials. The transfer-function concept will be further developed in the section on servomechanisms. The definitions of the symbols shown in Fig. 21-1 are:

$$
\begin{aligned}
\otimes &= \text{a summing junction} \\
R(s) &= \text{the reference input} \\
E(s) &= \text{the error signal} \\
C(s) &= \text{the controlled output} \\
G(s) &= \text{the forward transfer function} \\
H(s) &= \text{the feedback transfer function} \\
G(s)H(s) &= \text{the open-loop transfer function}
\end{aligned}
$$

The ratio $C(s)/R(s)$ is called the closed-loop transfer function and is derived as follows:

$$
G(s) = \frac{C(s)}{E(s)} \tag{21-1}
$$

$$
R(s) \mp C(s)H(s) = E(s) \tag{21-2}
$$

and, upon substituting Eq. 21-1 into Eq. 21-2 and solving for $C(s)/R(s)$,

$$R(s) = \pm C(s)H(s) + \frac{C(s)}{G(s)}$$

$$= C(s) \left[\pm H(s) + \frac{1}{G(s)} \right]$$

or

$$\frac{C(s)}{R(s)} = \frac{G(s)}{1 \pm G(s)H(s)} \tag{21-3}$$

It is evident that the $\pm$ sign in the denominator of Eq. 21-3 is dependent upon the arbitrary choice of the $\mp$ sign of the feedback signal at the summing junction of Fig. 21-1. In servomechanism analysis it is conventional to use the minus sign in Fig. 21-1 and thus the plus sign in Eq. 21-3. Not only is the opposite choice made in feedback amplifiers, but, in addition, the symbols A and β are used for G and H, respectively. Usually, only the steady-state sinusoidal analysis is considered in conventional amplifier analysis, which leads to the substitution of $s = j\omega$ in Eq. 21-3. In feedback amplifier analysis, Eq. 21-3 becomes

$$\frac{C(jw)}{R(jw)} = A'(j\omega) = \frac{A(j\omega)}{1 - A(j\omega)\beta(j\omega)} \tag{21-4}$$

A', A, and β are complex numbers and are defined as follows:

A' = the gain of the amplifier with feedback
A = the gain of the amplifier without feedback
β = the feedback factor defined as

$$\beta = \frac{\text{phasor feedback voltage in terms of the output voltage}}{\text{phasor output voltage}}$$

Since almost all amplifiers have a common connection between the output and input terminals, the block diagram of a feedback amplifier is as shown in Fig. 21-2. Usually, $R(j\omega)$ and $C(j\omega)$ are phasor voltages,

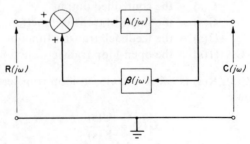

FIG. 21-2. Block diagram for a feedback
amplifier.

but they could be currents as well. In this section, for simplicity and brevity, $\mathbf{R}\,(j\omega)$ and $\mathbf{C}(j\omega)$ will be voltages.

Equation 21-4 can be used in defining negative and positive feedback, which were mentioned in the introduction.

$|1 - \mathbf{A}\beta| > 1$ defines the condition of negative feedback
$|1 - \mathbf{A}\beta| < 1$ defines the condition of positive feedback

21-2. Feedback Amplifier Circuits

As an example of feedback, let us consider the *R-C*-coupled amplifier of Fig. 21-3a. The fraction of the output voltage that is returned to the

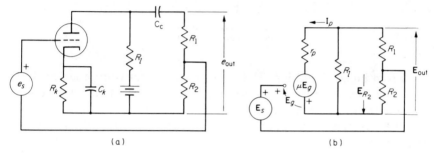

(a)   (b)

FIG. 21-3. a) An *R-C*-coupled amplifier with negative feedback. b) A-c equivalent circuit of part a.

grid-cathode circuit is

$$\beta = \frac{e_{\text{out}} R_2/(R_1 + R_2)}{e_{\text{out}}}$$

$$= \frac{R_2}{R_1 + R_2} \tag{21-5}$$

If $R_1 + R_2 \gg R_l$ and $X_{C_c} \ll R_1 + R_2$, then $\mathbf{A}$, the gain of the amplifier without feedback, can be derived from the equivalent circuit of Fig. 21-3b.

$$\mathbf{A} = \frac{\mathbf{E}_{\text{out}}}{\mathbf{E}_g} \simeq \frac{-\mu R_l}{r_p + R_l} \tag{21-6}$$

and $\mathbf{A}'$, the gain with feedback, is obtained from Eq. 21-4:

$$\mathbf{A}' = \frac{\mathbf{E}_{\text{out}}}{\mathbf{E}_s} = \frac{-\mu R_l/(r_p + R_l)}{1 - [-\mu R_l/(r_p + R_l)]\beta}$$

$$= \frac{-\mu R_l}{r_p + R_l + \mu\beta R_l}$$

$$\mathbf{A}' = \frac{-\mu R_l/(1 + \mu\beta)}{r_p/(1 + \mu\beta) + R_l} \tag{21-7}$$

From the form of Eq. 21-7, it is evident that the equivalent circuit of Fig. 21-3b can be replaced by an equivalent circuit of a conventional grounded-cathode amplifier, as shown in Fig. 21-4.

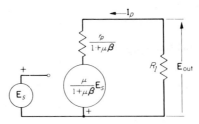

FIG. 21-4.   Grounded-cathode equivalent circuit for 21-3b.

From Eq. 21-7 and Fig. 21-4, it is apparent that voltage feedback reduces the internal resistance of the tube by the factor $1 + \mu\beta$. The equivalent μ of the tube is also reduced by the same factor.

If $R_1 + R_2$ is not much greater than R_l, it can be shown that the gain expression is

$$
\begin{aligned}
\mathbf{A'} &= \frac{\mathbf{E}_{out}}{\mathbf{E}_s} \\
&= \frac{-\left(\mu R_l/[R_l(1 + \mu\beta) + r_p]\right)(R_1 + R_2)}{\left(R_l r_p/[R_l(1 + \mu\beta) + r_p]\right) + R_1 + R_2} \\
&= \frac{-\mu R_l(R_1 + R_2)}{R_l r_p + (R_1 + R_2)r_p + R_l(R_1 + R_2)(1 + \mu\beta)}
\end{aligned}
\tag{21-8}
$$

It will now be shown that Eq. 21-8 is identical to that obtained when using conventional circuit analysis. From Fig. 21-3b, the necessary voltage-current relations are

$$
\mathbf{E}_s = \mathbf{E}_g + \mathbf{E}_{R_2}
\tag{21-9}
$$

$$
\mathbf{I}_p = \frac{\mu \mathbf{E}_g}{r_p + R_l R_g/(R_l + R_g)}
\tag{21-10}
$$

$$
\mathbf{E}_{R_2} = \left(\mathbf{I}_p \frac{R_l}{R_l + R_g}\right) R_2
\tag{21-11}
$$

$$
\mathbf{E}_{out} = -\mathbf{I}_p \frac{R_g R_l}{R_l + R_g}
\tag{21-12}
$$

where $R_g = R_1 + R_2$.

From these equations we solve for $\mathbf{E}_{out}/\mathbf{E}_s$ by first substituting Eq. 21-11 into Eq. 21-9.

$$\mathbf{E}_s = \mathbf{E}_g + \mathbf{I}_p \frac{R_l}{R_l + R_g} R_2 \qquad (21\text{-}13)$$

Now substituting $\mathbf{E}_g$ from Eq. 21-10 into Eq. 21-13,

$$\mathbf{E}_s = \left[\frac{r_p R_l + r_p R_g + R_l R_g + \mu R_l R_2}{\mu(R_l + R_g)} \, \mathbf{I}_p \right] \qquad (21\text{-}14)$$

Substituting $\mathbf{I}_p$ from Eq. 21-12 into Eq. 21-14,

$$\mathbf{E}_s = \left[\frac{r_p R_l + r_p R_g + R_l R_g + \mu R_l R_2}{\mu(R_l + R_g)} \right] \left[\frac{\mathbf{E}_{\text{out}}(R_l + R_g)}{R_l R_g} \right] \qquad (21\text{-}15)$$

Finally, we solve for $\mathbf{E}_{\text{out}} / \mathbf{E}_s$.

$$\mathbf{A}' = \frac{\mathbf{E}_{\text{out}}}{\mathbf{E}_s} = \frac{-\mu R_g R_l}{R_l(r_p + R_g + \mu R_2) + r_p R_g}$$

Upon substituting,

$$\beta = \frac{R_2}{R_g} = \frac{R_2}{R_1 + R_2} \qquad \text{and} \qquad R_g = R_1 + R_2$$

$$\mathbf{A}' = \frac{-\mu R_l(R_1 + R_2)}{r_p R_l + (R_1 + R_2)r_p + R_l(R_1 + R_2)(1 + \mu\beta)} \qquad (21\text{-}16)$$

which is identical to Eq. 21-8. If $R_g \gg R_l$, the reader should verify that Eq. 21-16 reduces to Eq. 21-7.

Frequency Response Improvement. As was brought out in Chapters 7 and 13, the frequency response of an amplifier is usually defined in terms of the -3-db bandwidth or the $(f_2 - f_1)$ frequency difference. An example will illustrate how negative feedback increases f_2 and decreases f_1, thus increasing the -3-db bandwidth.

EXAMPLE 21-1. An *R-C*-coupled amplifier has an f_1 of 100 Hz and an f_2 of 14,000 Hz. If the mid-frequency gain is -85, find the new f_1 and f_2 if a $\beta = 0.1$ is included.

Solution:

$$\mathbf{A}' = \frac{\mathbf{A}}{1 - \mathbf{A}\beta}$$

$$\mathbf{A}'_{\text{mid}} = \frac{\mathbf{A}_{\text{mid}}}{1 - \mathbf{A}_{\text{mid}}\beta} = \frac{-85}{1 + 8.5}$$

$$= -8.96$$

At 100 Hz,

$$\mathbf{A}_{100} = -85 \times 0.707 \underline{/45^\circ} \qquad (\text{refer to Eq. 13-43})$$

$$\mathbf{A}'_{100} = \frac{\mathbf{A}_{100}}{1 - \mathbf{A}_{100}\beta} = \frac{-60 \underline{/45^\circ}}{1 + 6 \underline{/45^\circ}} = -8.9 \underline{/6.2^\circ}$$

Similarly,

$$\mathbf{A}_{14,000} = -60 \underline{/-45^\circ} \qquad \text{(refer to Eq. 13-45)}$$

$$\mathbf{A'}_{14,000} = \frac{-60 \underline{/-45^\circ}}{1 + 6 \underline{/-45^\circ}} = -8.9 \underline{/-6.2^\circ}$$

By definition, the -3-db frequencies occur when the amplifier gain is reduced by a factor of 0.707 from the mid-frequency gain. For the new amplifier (the amplifier with feedback), it is necessary to find the new f_1 and f_2 frequencies corresponding to a gain of $-8.96 \times 0.707 \underline{/\perp45^\circ} = -6.32 \underline{/\pm45^\circ}$. Obviously, the gains at 100 and 14,000 Hz are well above this new value of -3-db gain.

By choosing values of $\mathbf{A}$, corresponding values of $\mathbf{A'}$ can be calculated.

$$\mathbf{A}_{50} = \frac{-85}{\sqrt{1 + 2^2}} \underline{/\tan^{-1}2} = -38 \underline{/63.4^\circ}$$

$$\mathbf{A'}_{50} = \frac{-38 \underline{/63.4^\circ}}{1 + 3.8 \underline{/63.4^\circ}} = -8.45 \underline{/14.2^\circ}$$

Table 21-1 is a tabulation of a few corresponding values of $\mathbf{A}$ and $\mathbf{A'}$.

TABLE 21-1

TABULATION OF $\mathbf{A}$ AND $\mathbf{A'}$ VERSUS f FOR EXAMPLE 21-1

Frequencies in Hertz	A	A' for $\beta = 0.1$
10	$-8.5 \underline{/84^\circ}$	$-6.15 \underline{/46^\circ}$
25	$-20.7 \underline{/76^\circ}$	$-8.22 \underline{/23^\circ}$
50	$-38 \quad \underline{/63.4^\circ}$	$-8.45 \underline{/14.2^\circ}$
100	$-60 \quad \underline{/45^\circ}$	$-8.9 \underline{/6.2^\circ}$
14,000	$-60 \quad \underline{/-45^\circ}$	$-8.9 \underline{/-6.2^\circ}$
28,000	$-38 \quad \underline{/-63.4^\circ}$	$-8.45 \underline{/-14.2^\circ}$
56,000	$-20.7 \underline{/-76^\circ}$	$-8.22 \underline{/-23^\circ}$
140,000	$-8.5 \underline{/-84^\circ}$	$-6.15 \underline{/-46^\circ}$

Figure 21-5 is a plot of $|\mathbf{A}|$ and $|\mathbf{A'}|$ versus f. From this curve, $f_2 = 130,000$ Hz and $f_1 = 20$ Hz. The bandwidth has increased from $(14,000 - 100)$ to $(130,000 - 20)$ Hz. Notice that for small values of $\mathbf{A}$ such that $|\mathbf{A}\beta| \ll 1$, the gain $\mathbf{A'}$ approaches $\mathbf{A}$. It is shown in Appendix F that $f'_1 = f_1/(1 - A_{mid}\beta)$ and $f'_2 = f_2(1 - A_{mid}\beta)$ where f'_1 and f'_2 are the -3 db frequencies with feedback and f_1 and f_2 are the -3db frequencies without feedback. Using these equations $f'_1 = 100/(1 + 8.5) = 10.5$ Hz and $f'_2 = 14,000 (1 + 8.5) = 133,000$ Hz which are reasonably close to the values obtained by graphical means. The graphical approach is always valid and easily remembered, even for complicated feedback functions, so it will be the method of attack emphasized here.

For large values of $\mathbf{A}\beta$, the gain with feedback becomes

$$\mathbf{A'} \approx \frac{\mathbf{A}}{-\mathbf{A}\beta} = -\frac{1}{\beta} \qquad (21\text{-}17)$$

which illustrates that the gain can be made dependent only upon β, and fluctuations in $\mathbf{A}$ are unnoticed. If the feedback network consists only of accurately known, constant-valued resistances, the amplifier gain will be extremely stable.

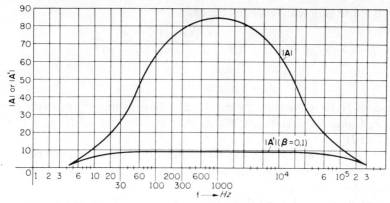

FIG. 21-5. Frequency-response curve of an *R-C*-coupled amplifier with and without feedback.

The Cathode Follower. In Chapter 10 the gain of the cathode follower was derived by using voltage-current equations, but the gain can also be derived by using the basic feedback equation. Figures 21-6a and b

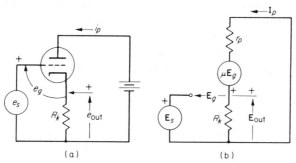

(a)   (b)

FIG. 21-6. a) Cathode-follower circuit.   b) A-c equivalent circuit of the cathode follower.

show the actual and the a-c equivalent circuits, respectively, of the basic cathode-follower circuit. The gain without feedback considered is

$$\mathbf{A} = \frac{e_{\text{out}}}{e_g} = \frac{\mathbf{E}_{\text{out}}}{\mathbf{E}_g} = \frac{+\mu R_k}{r_p + R_k} \qquad (21\text{-}18)$$

The feedback factor is

$$\beta = \frac{-e_{\text{out}}}{e_{\text{out}}} = -1$$

since all of e_{out} is fed back to the grid to cathode but is 180° out of phase with e_{out}, the cathode-to-ground voltage. (It is easy to see the validity of this relationship if we note that the grid and ground are tied together when

$e_s = 0$.) The gain with feedback becomes

$$A' = \frac{A}{1 - A\beta} = \frac{\mu R_k / (r_p + R_k)}{1 + \mu R_k / (r_p + R_k)}$$

$$= \frac{\mu R_k}{r_p + (\mu + 1) R_k} \tag{21-19}$$

Equation 21-19 is the same as Eq. 10-22. Two problems involving feedback in transistor circuits are left as exercises for the reader (see Probs. 21-6 and 21-7).

Negative feedback also exhibits other desirable characteristics such as the reduction of noise and distortion in amplifiers, as will now be explained.

Reduction of Noise and Distortion. Noise is defined as any undesired voltage that is superimposed upon the desired voltage signal. Noise can be either man-made, such as a 60-Hz power-frequency hum or pickup, ignition noise, and so forth, or it can be caused by natural disturbances, such as thermal noise in resistances and transistors, shot noise due to electron bombardment of the plate in a vacuum tube, lightning, cosmic radiation, and so on. Distortion is usually caused by the nonlinear characteristics of the device through which the signal voltage is passed. A signal is generally considered to be distorted if the output and input voltages and/or currents of an electronic device are not related by a fixed constant such as a constant gain factor.

Figures 21-7a and b illustrate how negative feedback reduces the effect

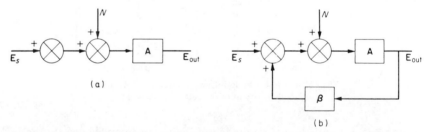

(a)

(b)

FIG. 21-7. a) An amplifier circuit showing both a signal and a noise voltage being introduced into the amplifier input. b) A feedback loop is added to the amplifier circuit of part a.

of a noise voltage N that is assumed to occur at the input of amplifier of gain A. Figure 21-7a shows an amplifier without feedback. The output voltage E_{out} is the sum of E_s and N times the amplifier gain A, or

$$E_{out} = (E_s + N) A \tag{21-20}$$

The signal-to-noise ratio is defined as the ratio of the desired output signal to the noise voltage appearing at the output, or

$$\frac{\mathbf{A}\mathbf{E}_s}{\mathbf{A}N} = \frac{\mathbf{E}_s}{N} \tag{21-21}$$

where $\mathbf{A}\mathbf{E}_s$ is the desired signal output voltage and $\mathbf{A}N$ is the undesired noise output voltage. Assuming N is a constant the signal-to-noise ratio can be improved simply by increasing $\mathbf{E}_s$. Because all amplifiers eventually reach a point where the output voltage can no longer increase for an increasing input voltage (the amplifier "saturates"), there is a practical upper bound on $\mathbf{E}_s$.

Figure 21-7b shows the same amplifier with a negative-feedback path added. To derive the equation for the output voltage, we use the relation

$$[(\mathbf{E}_s + \beta\mathbf{E}_{\text{out}}) + N]\mathbf{A} = \mathbf{E}_{\text{out}} \tag{21-22}$$

where $\mathbf{E}_s$, $\beta\mathbf{E}_{\text{out}}$, and N are the three voltages that comprise the input to amplifier $\mathbf{A}$. Solving for $\mathbf{E}_{\text{out}}$ from Eq. 21-22,

$$\mathbf{E}_{\text{out}} = \frac{\mathbf{A}}{1 - \mathbf{A}\beta}(\mathbf{E}_s + N) \tag{21-23}$$

At first glance, there seems to be no advantage to negative feedback, since, in Eq. 21-23, both $\mathbf{E}_s$ and N are multiplied by the same factor. The signal-to-noise ratio would seem to be the same as without feedback. However, let us assume that $\mathbf{E}_{\text{out}}$ is to be of the same magnitude both before and after negative feedback is applied. Since $|1 - \mathbf{A}\beta| > 1$ for negative feedback, it is necessary to increase the noise-free signal voltage $\mathbf{E}_s$ in order to increase $\mathbf{E}_{\text{out}}$ to its original value. Since the noise voltage N remains essentially constant, the signal-to-noise ratio is increased (which is the desired result).

EXAMPLE 21-2. An amplifier with a gain of 1200 has an internally introduced hum output of 0.5 volt when the signal output is 100 volts. Determine what value of β would be required to improve the signal-to-noise ratio to 1000:1 for the same signal output.

Solution:

$$\mathbf{A}\mathbf{E}_s = \mathbf{A}'\mathbf{E}_s' = 100 \text{ volts} \quad \text{and} \quad \mathbf{E}_s = \frac{100}{1200} = \frac{1}{12} \text{ volt}$$

Also, before feedback is added,

$$\frac{\mathbf{E}_s}{N} = \frac{100}{0.5} = 200$$

To improve the signal-to-noise ratio to 1000:1, the noise voltage must be effectively reduced to 0.1 volt, or

$$N' = \frac{N}{1 - \mathbf{A}\beta} \tag{21-24}$$

where N is the noise output voltage before feedback, and N' is the noise voltage output after feedback. Upon substituting in the proper values in Eq. 21-24,

$$0.1 = \frac{0.5}{1 - 1200\beta} \quad \text{or} \quad \beta = \frac{-4}{1200} = \frac{-1}{300}$$

and the new value of $\mathbf{E}_s$ (or $\mathbf{E}'_s$) required to return the amplifier output with feedback to 100 volts is

$$\mathbf{E}'_s = \frac{100}{\mathbf{A}'} = \frac{100}{240} = \frac{5}{12} \text{ volt}$$

Other applications of feedback will be studied in the following section on servomechanisms.

SERVOMECHANISMS

A somewhat specialized application of feedback theory is found in control-system analysis. The general servo problem and the straightforward mathematical solution will first be presented. In addition, the particular analytic techniques peculiar to servo theory, such as Nyquist, Bode, and pole-zero plots, will be explained. It should be emphasized that all these mathematical techniques are applicable also to the problems encountered in the study of feedback amplifiers.

The basic problem confronting the servo designer is to develop a system that has a satisfactory response within prescribed limits. As with many engineering problems, there is a great deal of "approximating" when considering either the analysis or the synthesis of a servo problem. In analysis, one begins with a given system and then determines the transfer function, or perhaps the output response for a particular input signal. In synthesis, one begins with a transfer function, or the desired output-response characteristics, and then determines the system that will satisfy these requirements. Only the tools of servo-system analysis will be presented here, since it is beyond the scope of this text to develop the techniques of synthesis.

In addition to limiting the discussion to analytic techniques, only continuous, linear, d-c systems will be considered. Other servo systems such as sampled-data, adaptive, nonlinear, and a-c control systems employ some or all of the mathematical methods used in the study of the linear, continuous, d-c servo systems. For this reason, it is imperative that the reader grasp the fundamentals of problem solution for this basic type of system.

21-3. The Second-Order System

In the study of electrical or dynamic systems, it is almost mandatory that one fully understand the mathematical significance of the second-order, linear, time-dependent, differential equation with constant coefficients. The reason for the emphasis on this particular type of equation is twofold: 1) The equation is easy to manipulate. 2) The second-order equation is the most basic equation that can describe the interchange of energy within a system, that is, the oscillation effect. This second point is by far the more important, for the mathematical description of a great many quite complicated physical systems is, to a large extent, adequately approximated by the second-order equation. The general form of the second-order differential equation that is found in most servo and some circuits texts is

$$\frac{d^2y}{dt^2} + 2\zeta\omega_n \frac{dy}{dt} + \omega_n^2 y = f(t) \tag{21-25}$$

where ζ is called the damping ratio, ω_n is the undamped natural frequency of the system, and $f(t)$ is an externally applied forcing function.

Two systems that have the typical second-order differential equation described by Eq. 21-25 are shown in Figs. 21-8a and b. Figure 21-8a

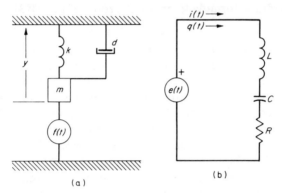

FIG. 21-8. a) A damped spring-mass system.
b) A series R-L-C circuit.

shows the familiar dynamic problem involving a damped spring-mass system with only one degree of freedom. The differential equation in terms of displacement y as a function of time t is

$$m\ddot{y} + d\dot{y} + ky = f(t) \tag{21-26}$$

Figure 21-8b shows an equally familiar series R-L-C circuit. The differential equation in terms of current is (sum of the voltage drops around a closed loop)

$$L\frac{di}{dt} + \frac{1}{C}\int i\,dt + iR = e(t) \tag{21-27}$$

and, rewriting Eq. 21-27 in terms of charge q by substituting $i = dq/dt$,

$$L\frac{d^2q}{dt^2} + R\frac{dq}{dt} + \frac{1}{C}q = e(t) \tag{21-28}$$

Both Eqs. 21-26 and 21-28 can be compared to Eq. 21-25, and, if desired, ζ and ω_n can be written in terms of m, d, and k of Eq. 21-26 or R, L, and C of Eq. 21-28.

Taking the Laplace transform of Eq. 21-25, with all initial conditions equal to zero,

$$s^2 Y(s) + 2\zeta\omega_n s Y(s) + \omega_n^2 Y(s) = F(s) \tag{21-29}$$

and, upon solving for $Y(s)$,

$$Y(s) = \frac{F(s)}{s^2 + 2\zeta\omega_n s + \omega_n^2} \tag{21-30}$$

The characteristic equation for the system described by either Eq. 21-29 or Eq. 21-30 is

$$s^2 + 2\zeta\omega_n s + \omega_n^2 = 0 \tag{21-31}$$

Using the quadratic formula to obtain the roots of Eq. 21-31,

$$s = -\zeta\omega_n \pm \omega_n \sqrt{\zeta^2 - 1} \tag{21-32}$$

where $\zeta\omega_n$ is the damping factor and for $\zeta < 1$, $\omega n \sqrt{\zeta^2 - 1}$ is called the damped natural frequency, ω_D, of the system described by Eq. 21-32.

These roots have two important interpretations: When $\zeta > 1$, both roots are real; when $\zeta < 1$, both roots are complex conjugate quantities. From a consideration of inverse Laplace transform theory and the partial-fraction expansion of ratios of polynomials, it is apparent that exponential time functions result from such denominator roots. Since complex roots *always* occur in conjugate pairs—that is, $(a + jb)$ and $(a - jb)$—it is possible to have trigonometric terms involving sines and cosines as well as wholly real exponential functions occurring in the time-domain expression.[1] Several examples of this were encountered in Chapter 19.

As will be seen later, a servo system—even a complex one—very often behaves like the damped spring-mass system. In almost every case, a properly designed servo system is underdamped and so it will respond in such a way that the output variable will oscillate around a given reference before settling down.

Let us consider briefly the spring-mass system. If the system is perturbed, how long is it before equilibrium is again established? If the damping factor is too small, the system responds quickly but oscillates for too long a period before settling down, whereas an overdamped system is too sluggish. From experience, we find that the best response is obtained if the system is a bit underdamped and is allowed to oscillate somewhat about the equilibrium point. Figure 21-9 shows the typical time domain

[1] Recall the identity

$$e^{-(a+jb)t} + e^{-(a-jb)t} = e^{-at}(e^{jbt} + e^{-jbt}) = 2e^{-at}\cos bt$$

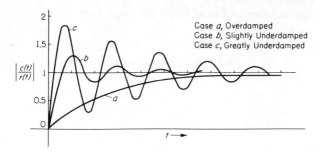

FIG. 21-9. The $c(t)$ response for a second-order, type 0
servo system with $r(t) = 1$.

response $c(t)$ for a unit step input $r(t) = 1$ for three values of the damping
factor. The "best" response curve is b, since it is generally accepted prac-
tice that $|c(t)/r(t)|_{max}$ should rarely exceed 1.5 or should be less than
1.1 for a suitable servo system.

21-4. Block-Diagram Representation

A typical servo system is shown in block-diagram form in Fig. 21-10.
This particular system is used to maintain automatically a constant output

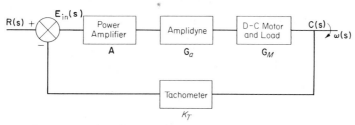

FIG. 21-10. Block diagram of a servo system involving the speed
control of a d-c motor.

speed, in spite of a varying load demand on the d-c motor. The reference
input $r(t)$ [or $R(s)$] is usually some fixed voltage that is required to start
the system operating. This reference voltage is applied to a vacuum tube
or transistor power amplifier which amplifies a few milliwatts to several
watts. The output of this power amplifier is then fed into an amplidyne,
which is a special type of d-c generator, where the few watts are amplified
to several thousand watts. This drives the d-c motor and connected load.
The tachometer, which is a fixed-field d-c generator, produces a voltage
directly proportional to the speed unit. This voltage is then compared to
the reference input to see if any speed correction is necessary.

To describe the system behavior, let us assume that the reference
$E_{in}(s)$ is set to a value that gives a particular motor speed ω. At this

speed the tachometer will feed back a voltage that is subtracted from the reference to produce an error voltage that determines the speed of the motor. If the motor slows down, the tachometer feeds back less voltage to be subtracted from the reference; hence more power is produced to drive the motor, and the speed increases again. Should the motor speed increase beyond the equilibrium speed, the reverse process occurs.

Although this descriptive analysis is indicative of how the system behaves, it does not tell much about the overall response of the system; in fact, the behavior of the system might be completely unsatisfactory. The analysis of the system could be ascertained by setting up an actual system, a scaled-down model, an analog or digital computer solution, or a completely mathematical analysis. The first two methods are costly and time-consuming and thus are considered impractical, but the latter two are useful means of analysis.

Before showing the computer and mathematical solutions, it is necessary to describe each unit of the block diagram in terms of a mathematical equation or description. The usual mathematics employed in servo work is the operational calculus of Laplace transforms. The next step is to develop further the transfer-function concept.

21-5. Transfer Functions of Circuits

The transfer-function concept, defined earlier, will be further developed in this section. In electrical networks a transfer function usually has the dimensions of ohms or mhos or is dimensionless, but, in general, can have any combination of the dimensions commonly accepted in engineering practice. Several standard and often-used R-C networks are shown in Fig. 21-11, along with their respective $E_{out}(s)/E_{in}(s)$ transfer functions.

Figure 21-11c will be used to illustrate the derivation of these transfer functions. First, we note that the desired transfer function is a ratio of the unknown output voltage to the known input voltage. To relate these two quantities it is necessary to solve for the current passing through R_2 in terms of the input voltage, $E_{in}(s)$. This current is

$$I(s) = \frac{E_{in}(s)}{Z_{in}(s)} \tag{21-33}$$

and

$$E_{out}(s) = I(s)R_2$$

$$= E_{in}(s)\frac{R_2}{Z_{in}(s)}$$

or

$$\frac{E_{out}}{E_{in}}(s) = \frac{R_2}{Z_{in}(s)} \tag{21-34}$$

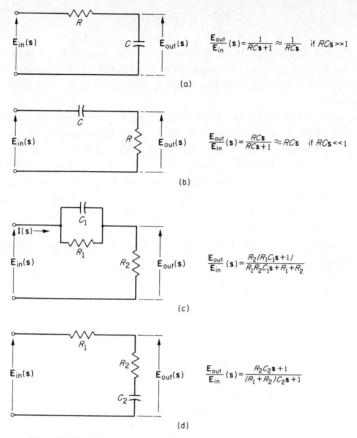

$$\frac{E_{out}}{E_{in}}(s) = \frac{1}{RCs+1} \approx \frac{1}{RCs} \quad \text{if } RCs \gg 1$$

(a)

$$\frac{E_{out}}{E_{in}}(s) = \frac{RCs}{RCs+1} \approx RCs \quad \text{if } RCs \ll 1$$

(b)

$$\frac{E_{out}}{E_{in}}(s) = \frac{R_2(R_1C_1s+1)}{R_1R_2C_1s+R_1+R_2}$$

(c)

$$\frac{E_{out}}{E_{in}}(s) = \frac{R_2C_2s+1}{(R_1+R_2)C_2s+1}$$

(d)

FIG. 21-11. a) Integrating or phase-lag network. b) Differentiating or phase-lead network. c) Phase-lead network with fixed d-c attenuation. d) Phase-lag network with fixed high-frequency attenuation.

where

$$\begin{aligned}
Z_{in}(s) &= \frac{R_1(1/C_1s)}{R_1 + (1/C_1s)} + R_2 \\
&= \frac{R_1 + R_2 + R_2R_1C_1s}{R_1C_1s + 1}
\end{aligned}$$

Therefore,

$$\frac{E_{out}}{E_{in}}(s) = \frac{R_2R_1C_1s + R_2}{R_2R_1C_1s + R_1 + R_2} \tag{21-35}$$

Equation 21-34 establishes a definite pattern for the calculation of the other transfer functions

$$\frac{E_{out}}{E_{in}}(s) = \frac{Z_{out}(s)}{Z_{in}(s)} \tag{21-36}$$

where $Z_{out}(s)$ is the impedance across which the output voltage is taken, and $Z_{in}(s)$ is the total input impedance of the circuit. This relation is valid only if the input current and the current through the output impedance are the same.

Figures 21-11a and b are interesting in that, under certain conditions, the output voltage is the integral or derivative, respectively, of the input voltage. To understand what the inequality $RCs \gg 1$ or $RCs \ll 1$ means, we simply substitute $j\omega$ for s. Figure 21-11a is a good integrator as long as the lowest frequency components of $E_{in}(j\omega)$ cause $RC\omega \gg 1$. Similarly, Fig. 21-11b is a good differentiator as long as the highest non-negligible frequency component of $E_{in}(j\omega)$ does not invalidate $RC\omega \ll 1$. Recalling the Fourier sine and cosine expansion for a periodic non-sinusoidal function is helpful in understanding the operation of these two networks.

The terms *phase lead* and *phase lag* imply that one is considering steady-state sinusoidal excitation and response functions. Figure 21-11a is a phase-lag network, because the phase angle associated with the phasor transfer function is negative when the substitution $s = j\omega$ is made,

$$\frac{E_{out}}{E_{in}}(j\omega) = \frac{1}{1 + j\omega RC} = \frac{1}{\sqrt{1 + \omega RC^2}} \underline{/-\tan^{-1}\omega RC}$$

or, in other words, the output voltage E_{out} lags the input voltage E_{in}. Similar arguments can be made for the positive phase angle of the phase-lead networks.

21-6. Transfer Function of Electromechanical Devices

When using electric motors, generators, and so on, in conjunction with the usual R-L-C parameters in circuits, it is necessary to develop the integrodifferential equations that mathematically describe these devices. The transfer function is, in reality, the ratio of the Laplace-transformed integrodifferential equation describing the output function to the Laplace-transformed integrodifferential equation describing the input function, with all initial conditions equal to zero.

Four electromechanical devices commonly used in servo systems are the tachometer, the generator, the motor, and the amplidyne. Although only the transfer function of d-c equipment will be considered here, it is possible to derive similar transfer functions for a-c devices.

Tachometer. A tachometer is a fixed-field d-c generator and is used as a speed sensor. The input is speed, which is usually expressed in radians per second, and the output is a d-c voltage which is directly proportional to the speed. The operation of the tachometer can be explained

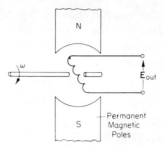

FIG. 21-12. Tachometer generator with a transfer function $E_{out}(s)/\omega(s) = K_T$. (Constant magnetic flux and no armature resistance or inductance are assumed.)

by recalling that the voltage generated in a coil which is rotating in a magnetic field is

$$e_G = K\phi\omega \qquad (21\text{-}37)$$

where K is a constant depending on the particular configuration of the generator, such as the number of turns in the coil, the length of the active conductor, and so on; ϕ is the magnetic flux in which the coil turns; and ω is the rotational speed of the coil, in radians per second.

As Fig. 21-12 shows, the flux is produced by permanent magnets, making ϕ a constant. Thus

$$\mathbf{E}_{out}(\mathbf{s}) = K_T\omega(\mathbf{s})$$

where K_T is the tachometer constant with dimensions of volts per radian per second. The transfer function for the tachometer is simply

$$\frac{\mathbf{E}_{out}}{\omega}(\mathbf{s}) = K_T \qquad (21\text{-}38)$$

Shunt-Field D-C Generator. The separately excited shunt-field d-c generator, like the tachometer, generates a voltage that is directly pro-

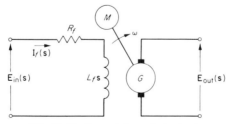

FIG. 21-13. Separately excited shunt-field d-c generator with a transfer function $E_{out}(s)/E_{in}(s) = (K_g/R_f)(1 + T_f s)$. (Constant speed, no armature resistance or inductance, and linear operation are assumed.)

portional to the air-gap flux and the speed of the rotor. In this particular case it is desired to hold the speed ω constant and to vary the flux ϕ. From Fig. 21-13, we see that the shunt-field current $\mathbf{I}_f(\mathbf{s})$ produces the air-gap flux. Equation 21-37 becomes

$$\mathbf{E}_{out}(\mathbf{s}) = K_g \mathbf{I}_f(\mathbf{s}) \tag{21-39}$$

where K_g is the generator constant with dimensions of volts per ampere. Keeping in mind that the desired transfer function is $\mathbf{E}_{out}(\mathbf{s})/\mathbf{E}_{in}(\mathbf{s})$, we write a KVL equation involving loop current $\mathbf{I}_f(\mathbf{s})$

$$\mathbf{E}_{in}(\mathbf{s}) = (R_f + L_f \mathbf{s})\mathbf{I}_f(\mathbf{s}) \tag{21-40}$$

where R_f and L_f are the resistance and inductance, respectively, of the shunt field. By dividing Eq. 21-40 into Eq. 21-39, the transfer function is

$$\frac{\mathbf{E}_{out}}{\mathbf{E}_{in}}(\mathbf{s}) = \frac{K_g}{R_f + L_f \mathbf{s}}$$

$$= \frac{K_g/R_f}{1 + \tau_f \mathbf{s}} \tag{21-41}$$

where $\tau_f = L_f/R_f$, the time constant of the generator.

Direct-Current Motor. In the case of the tachometer and the d-c generator, it seemed obvious as to what the desired transfer function should be, but this is not the case for the d-c motor. Figure 21-14 shows an armature-fed, constant-field, d-c motor where the desired transfer function is $\omega(\mathbf{s})/\mathbf{E}_{in}(\mathbf{s})$. Why not choose the transfer function $\theta(\mathbf{s})/\mathbf{E}_{in}(\mathbf{s})$ or why not feed a constant current to the armature and apply $\mathbf{E}_{in}(\mathbf{s})$ to the shunt-field terminals and thus obtain still another transfer function? The choice as to which particular transfer function is preferable depends entirely on the manner in which the equipment is to be utilized. For our purposes, the transfer function $\omega(\mathbf{s})/\mathbf{E}_{in}(\mathbf{s})$ from Fig. 21-14 is the preferred choice. The necessary equations to solve for the $\omega(\mathbf{s})/\mathbf{E}_{in}$ transfer function are developed as follows:

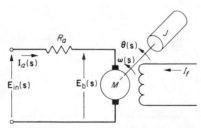

FIG. 21-14. A d-c motor with a transfer function $\omega(s)/E_{in}(s) = K_M(1 + \tau_M s)$. (Assume no damping, inertial load J is referred to the motor shaft, no armature inductance, linear operation, and constant-field excitation I_f.)

$$\mathbf{E}_b(\mathbf{s}) = K_b\omega(\mathbf{s}) \tag{21-42}$$

where K_b is the back emf, or generator constant of the motor in volts per radian per second. This equation again stems from Eq. 21-37 with ϕ a constant. Using loop current $\mathbf{I}_a(\mathbf{s})$,

$$\mathbf{E}_{in}(\mathbf{s}) = \mathbf{I}_a(\mathbf{s})R_a + \mathbf{E}_b(\mathbf{s}) \tag{21-43}$$

It is also necessary to relate mechanical and electrical torque terms. Electrical torque is proportional to the product of flux and current. Since the flux is constant, the electrical torque $\mathbf{T}_E$ developed by the motor is directly proportional to the armature current

$$\mathbf{T}_E(\mathbf{s}) = K_T\mathbf{I}_a(\mathbf{s}) \tag{21-44}$$

where K_T is the motor-torque constant, in pound-feet per ampere. From mechanics, mechanical torque is equal to

$$T_M = J\frac{d^2\theta}{dt^2} = J\frac{d\omega}{dt}$$

$$\mathbf{T}_M(\mathbf{s}) = Js\omega(\mathbf{s}) \tag{21-45}$$

where J is the inertia of the motor shaft and the load referred to the motor shaft, in slug-feet^2. In taking the Laplace transform of a differential equation, we recall that all initial conditions are set equal to zero when we develop the transfer-function concept.

Equations 21-42, 21-43, 21-44, and 21-45 can now be solved for $\omega(\mathbf{s})/\mathbf{E}_{in}$. Upon substituting Eq. 21-42 into Eq. 21-43,

$$\mathbf{E}_{in}(\mathbf{s}) = \mathbf{I}_a(\mathbf{s})R_a + K_b\omega(\mathbf{s}) \tag{21-46}$$

We equate Eqs. 21-44 and 21-45, since the electrical torque produces the mechanical output

$$K_T\mathbf{I}_a(\mathbf{s}) = Js\omega(\mathbf{s}) \tag{21-47}$$

Substituting Eq. 21-47 into Eq. 21-46,

$$\mathbf{E}_{in}(\mathbf{s}) = \frac{Js\omega(\mathbf{s})}{K_T}R_a + K_b\omega(\mathbf{s})$$

Therefore,

$$\frac{\omega(\mathbf{s})}{\mathbf{E}_{in}(\mathbf{s})} = \frac{1}{K_b + (JR_a/K_T)\mathbf{s}} = \frac{1/K_b}{1 + (JR_a/K_TK_b)\mathbf{s}}$$

$$\frac{\omega}{\mathbf{E}_{in}}(\mathbf{s}) = \frac{K_M}{1 + \tau_M\mathbf{s}} \tag{21-48}$$

where $K_M = 1/K_b$, the motor constant in radians per second per volt, and $\tau_M = JR_a/K_TK_b$, the motor time constant, in seconds.

Amplidyne. The amplidyne is a very rugged electromechanical power amplifier capable of the controlled transformation of a few watts into thousands of watts. The amplidyne is essentially a d-c generator with an extra set of brushes. In Fig. 21-15 the control field windings and the

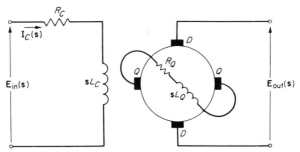

FIG. 21-15. Amplidyne power amplifier with a transfer function (E_{out}/E_{in}) $(s) = K_a/(1 + \tau_c s)$ $(1 + \tau_Q s)$. (Constant speed and linear operation are assumed.)

quadrature field (or QQ) brushes are used in much the same way as the similar elements on a d-c shunt generator. However, the two quadrature field brushes are literally shorted together, which creates a rather large current in this quadrature-axis winding. One would never knowingly short the terminals of a usual d-c shunt generator that is delivering rated voltage, because the generator would exceed its rated current capacity and would overheat. One way in which the amplidyne differs from the usual d-c generator is that the input power to the control field is only a few watts instead of 100 watts or so. This means that the short-circuited quadrature-axis windings generate a small voltage, which limits the quadrature-axis current to a safe value.

In spite of this small induced voltage, a maximum rated current of about 100 amp may flow through the quadrature-axis rotor winding. This current sets up a very strong quadrature-axis magnetic field. The phenomenon of an armature current setting up a magnetic field of its own is usually undesirable in a machine, but in the amplidyne this armature-reaction field is quite necessary and desirable. The vector direction of this armature reaction or quadrature-axis magnetic flux is normal to the control-field flux. The same rotor windings that generate the quadrature voltage cut through this quadrature-axis flux and generate another voltage. This new voltage E_{out} is then picked off the commutator by the direct-axis brushes DD, which are necessarily placed at right angles to the quadrature-axis brushes QQ. A compensating winding is placed in series with the direct-axis field circuit to cancel the undesirable armature-reaction effects of the direct-axis current. The net result is that a few watts delivered to the control windings result in thousands of watts being delivered to the output terminals. The energy for this power amplifier comes from the drive motor for the amplidyne. This motor is usually the rugged and reliable a-c induction motor.

Two of the three equations necessary to derive the amplidyne transfer function $E_{out}/E_{in}(s)$ are

$$E_{in}(s) = (R_C + L_C s)I_C(s) \tag{21-49}$$

$$E_Q(s) = K_Q I_C(s) = (R_Q + L_Q s)I_Q(s) \tag{21-50}$$

Equation 21-50 states that a voltage $E_Q(s)$ is generated in the rotor windings because of the flux set up by the control-field current $I_C(s)$. K_Q is the quadrature-field generator constant, in volts per ampere. Because the quadrature-axis brushes are shorted, this entire voltage is dropped across the internal resistance R_Q and the operational reactance $L_Q s$ of the quadrature-axis windings.

The output voltage in terms of the quadrature-axis current is the third necessary equation

$$E_{out}(s) = K_D I_Q(s) \tag{21-51}$$

where K_D is the direct-axis generator constant, in volts per ampere. The flux is set up by I_Q, which produces E_{out}.

To solve for $E_{out}(s)/E_{in}(s)$, we first substitute Eq. 21-51 into Eq. 21-50

$$K_Q I_C(s) = (R_Q + L_Q s) \frac{E_{out}(s)}{K_D} \tag{21-52}$$

and then substitute $I_C(s)$ from Eq. 21-52 into Eq. 21-49

$$E_{in}(s) = (R_C + L_C s)(R_Q + L_Q s) \frac{E_{out}(s)}{K_Q K_D}$$

$$\frac{E_{out}(s)}{E_{in}(s)} = \frac{K_Q K_D / R_C R_Q}{[1 + (L_C/R_C)s][1 + (L_Q/R_Q)s]} = \frac{K_a}{(1 + \tau_C s)(1 + \tau_Q s)} \tag{21-53}$$

where K_a is the amplidyne constant and is dimensionless, and τ_C and τ_Q are the control- and quadrature-axis-field time constants, respectively.

These transfer functions can now be used in developing the analog-computer and mathematical solutions for a given system. The analog-computer solution will be presented first.

21-7. The Analog-Computer Solution

In Chapter 20 the transfer function of the operational amplifier was shown to be $E_{out}/E_{in}(s) = -Z_{fb}(s)/Z_{in}(s)$. The transfer function of the

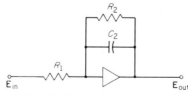

Fig. 21-16. An operational amplifier configuration synthesizing the transfer function $E_{out}(s)/E_{in}(s) = -K\alpha/(1 + \tau_\alpha s)$.

circuit in Fig. 21-16 is

$$\frac{\mathbf{E}_{out}}{\mathbf{E}_{in}}(\mathbf{s}) = -\frac{R_2(1/C_2\mathbf{s})}{R_1(R_2 + 1/C_2\mathbf{s})} = -\frac{R_2/R_1}{R_2C_2\mathbf{s} + 1} \qquad (21\text{-}54)$$

This transfer function has the form

$$\frac{-K_\alpha}{1 + \tau_\alpha\mathbf{s}} \qquad (21\text{-}55)$$

where $K_\alpha = R_2/R_1$ and $\tau_\alpha = R_2C_2$. From the preceding section we know that a motor has a transfer function of the form of Eq. 21-55 and that the amplidyne has two time constants in the denominator. These two time constants can be represented by the product of two such functions as given by Eq. 21-55. The block-diagram summation symbol can be replaced by the analog-computer summer, and the power amplifier and tachometer can be represented by the sign changer. One form of the computer setup that will synthesize the block-diagram components of Fig. 21-10 is shown in Fig. 21-17.

The motor has a transfer function

$$\frac{K_M}{1 + \mathbf{s}\tau_M}$$

where

$$K_M = \frac{R_6}{R_5} \qquad \text{and} \qquad \tau_M = R_6C_6$$

The amplidyne has a transfer function

$$\frac{K_a}{(1 + \tau_C\mathbf{s})(1 + \tau_Q\mathbf{s})}$$

where

$$K_a = \frac{R_2}{R_1} \times \frac{R_4}{R_3}$$

$$\tau_C = R_2C_2$$

$$\tau_Q = R_4C_4$$

The power amplifier gain A, and the tachometer gain K_T are self explana-

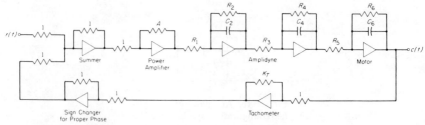

FIG. 21-17. Analog-computer synthesization of the block diagram of Fig. 21-10.

tory. The sign changer is necessary so that a negative feedback voltage is introduced for a positive input or reference voltage. In actual practice the gain of the power amplifier could be incorporated into the gain constant for the amplidyne, and then the phase sign changer could be eliminated.

To test the behavior of the system, various reference functions could be applied and the resulting output response recorded. Since it is rather difficult to change the various constants of the mechanical portion of the system, the most obvious controllable variable is the gain of the power amplifier. This can be varied in an actual system merely by turning the knob of a potentiometer. By changing this gain factor A in the computer simulation, a great many curves showing the output of the motor can be displayed. The servo designer then picks the one he wants and is ready to build the actual system.

This method, however, is not a panacea for the servo designers' problems. Although it is not yet evident, this system is totally unstable for certain values of A and may not yield a completely satisfactory system for any value of A. Even though the analog computer can show the solution for a particular system, it cannot tell the designer how to improve the system behavior. Even with the aid of the computer, a hit-or-miss approach to system design is obviously unsuitable. A computer is a tool much like the slide rule in that it provides the designer quicker solutions but does not tell how to improve the solution. For this reason, mathematical tools of analysis are vitally necessary. These will be presented in the following section.

21-8. Mathematical Analysis

From the basic feedback equation

$$G'(s) = \frac{C}{R}(s) = \frac{G(s)}{1 + G(s)H(s)}$$

it is important to consider the denominator, as the term $1 + G(s)H(s)$ determines the characteristic behavior of the system. In general, the form of the open-loop transfer function is

$$G(s)H(s) = \frac{K(1 + s\tau_1)(1 + s\tau_2) \cdots (1 + s\tau_m)}{s^N(1 + s\tau_a)(1 + s\tau_b) \cdots (1 + s\tau_n)}$$

while the time constants τ_1, τ_a, τ_m, and τ_n are usually real but can be complex numbers. The N defines the type of system under consideration. If $N = 0, 1, 2, 3$, and so on, a type 0, 1, 2, 3, and so on, system is defined, respectively. K is the open-loop gain factor and is usually considered as a positive real number.

Since $G(s)H(s)$ is a ratio of two polynomials, it is evident that $1 + G(s)H(s)$ will also be a ratio of two polynomials. The *numerator* polynomial of $1 + G(s)H(s)$ remains in the denominator of $G'(s)$, while the *denominator* of $1 + G(s)H(s)$ goes into the *numerator* of $G'(s)$ and completely cancels the denominator polynomial of $G(s)$.

As soon as the roots of the numerator polynomial or the roots of the characteristic equation $1 + G(s)H(s) = 0$ are known, the interesting portion of the output response $C(s)$ for a given $R(s)$ is also known. These roots determine whether or not the system is stable, the system damping factors, and the system oscillation frequencies. From the partial-fraction expansion methods on a ratio of two polynomials, we recall that the expansion is carried out in terms of the roots of the denominator polynomial. The numerator polynomial merely determines the coefficient multipliers or constants of the individual roots in the expansion.

If $c(t)$ is desired, we multiply $G'(s)$ by $R(s)$ and take the inverse transform, or

$$c(t) = \mathcal{L}^{-1}\left[\frac{G(s)}{1 + G(s)H(s)} R(s)\right]$$

In practice, it is not always necessary to find $c(t)$, since the roots of $1 + G(s)H(s) = 0$ usually give the desired information.

To illustrate this mathematical approach, the servo system of Fig. 21-10 will be used. The overall transfer function of this system is

$$G'(s) = \frac{C}{R}(s) = \frac{\omega(s)}{E_{in}(s)} = \frac{A\,G_a(s)G_M(s)}{1 + A\,G_a(s)G_M(s)K_T}$$

and, substituting in the values of G_M and G_a from Eqs. 21-48 and 21-53, respectively,

$$\frac{\omega(s)}{E_{in}(s)} = \frac{A\,[K_a/(1+\tau_C s)(1+\tau_Q s)][K_M/(1+\tau_M s)]}{1 + A\,[K_a/(1+\tau_C s)(1+\tau_Q s)][K_M/(1+\tau_M s)]K_T}$$

$$= \frac{AK_aK_M}{(1+\tau_C s)(1+\tau_Q s)(1+\tau_M s) + AK_aK_MK_T} \qquad (21\text{-}56)$$

Equation 21-56 is the closed-loop transfer function of the system, and it is here that the difficulty begins. Although the roots of the polynomials comprising $G(s)H(s)$ are usually evident, since the polynomials are in factored form, the roots of the numerator of $1 + G(s)H(s)$ must be calculated. This is done by obtaining a common denominator for $1 + G(s)H(s)$, expanding the two numerator polynomials of $1 + G(s)H(s)$, collecting the coefficients of like powers of s, and then finding the new roots of the numerator polynomial of $1 + G(s)H(s)$. Upon expanding the denominator of Eq. 21-56,

$$\frac{\omega}{E_{in}}(s)$$

$$= \frac{AK_aK_M}{\tau_C\tau_Q\tau_M s^3 + (\tau_C\tau_Q + \tau_C\tau_M + \tau_Q\tau_M)s^2 + (\tau_C + \tau_Q + \tau_M)s + 1 + AK_aK_MK_T}$$

$$= \frac{AK_aK_M/\tau_C\tau_Q\tau_M}{s^3 + \left(\dfrac{1}{\tau_M} + \dfrac{1}{\tau_Q} + \dfrac{1}{\tau_C}\right)s^2 + \left(\dfrac{1}{\tau_C\tau_Q} + \dfrac{1}{\tau_C\tau_M} + \dfrac{1}{\tau_Q\tau_M}\right)s + \dfrac{1 + AK_aK_MK_T}{\tau_C\tau_Q\tau_M}}$$

$$(21\text{-}57)$$

Finding the roots for even a cubic equation can be a tedious calculation. This task is greatly magnified if the gain factor A is varied over a wide range of values, thus necessitating a new evaluation of the roots for each value of A.

EXAMPLE 21-3. Let the forward transfer function of the amplifier, amplidyne, and motor combination be

$$G(s) = \frac{N}{(s + 2)(s + 3)(s + 5)}$$

where $N = AK_aK_M/\tau_C\tau_Q\tau_M$, as in Eq. 21-57, and the feedback transfer function be

$$H(s) = K_T$$

Determine stability, damping factors, and oscillation frequencies of the closed-loop system.

Solution: The closed-loop transfer function becomes

$$G'(s) = \frac{\omega}{E_{in}}(s) = \frac{N}{(s + 2)(s + 3)(s + 5) + D}$$

where $D = AK_aK_MK_T/\tau_C\tau_Q\tau_M$, as in Eq. 21-57. It is now necessary to expand the factored expression into expanded polynomial form

$$(s + 2)(s + 3)(s + 5) + D = s^3 + 10s^2 + 31s + 30 + D$$

To continue with the analysis, two values of D will be chosen. First, we let $D = 90$. Is the system stable? Are there any frequencies of oscillation? What are the damping factors? To answer these questions, we substitute $D = 90$ into the characteristic equation and find the new roots. The equation

$$s^3 + 10s^2 + 31s + 30 + 90 = s^3 + 10s^2 + 31s + 120 = 0 \qquad (21\text{-}58)$$

has three roots since it is a cubic, but there are only two combinations which the roots may have: 1) all roots can be real, or 2) one root can be real and the other two can be a complex conjugate pair. There are exact equations which can be used to solve for these roots, but a trial-and-error solution is more easily remembered. A trial-and-error division method is widely used. (Lin's[2] method is a well-known division process, in which the remainder is used in choosing the next trial divisor.) Since one of the roots is real, let us try $s = -8$ as a root of the polynomial:

---

[2]Lin, S. N., "Methods of Successive Approximations of Evaluating the Real and Complex Roots of Cubic and Higher Order Equations," *J. Math. Phys.*, Vol. 20, No. 3, 1941.

$$
\begin{array}{r}
s^2 + 2s + 15 \\
s + 8\overline{)s^3 + 10s^2 + 31s + 120} \\
s^3 + 8s^2 \\
\hline
2s^2 + 31s \\
2s^2 + 16s \\
\hline
15s + 120 \\
15s + 120
\end{array}
$$

Of course, this process usually has to be repeated a few times, since it is unlikely that one would always guess the correct root value the first time.

It has now been ascertained that $s = -8$ is one root. The other two roots are contained in the remaining factor $s^2 + 2s + 15$ and can easily be found by use of the quadratic formula:

$$
s = \frac{-2 \pm \sqrt{4 - 60}}{2} = -1 \pm j\sqrt{14}
$$

The final factored form of the denominator polynomial of Eq. 21-58 is

$$(s^3 + 10s^2 + 31s + 120) = (s + 8)(s + 1 - j\sqrt{14})(s + 1 + j\sqrt{14})$$

The closed-loop transfer function becomes

$$
\frac{\omega}{E_{in}}(s) = \frac{N}{(s + 8)(s + 1 - j\sqrt{14})(s + 1 + j\sqrt{14})}
$$

It is now possible to state positively whether or not the system is stable, and to find the damping factors and the frequency of oscillation. From Laplace transform analysis we recall that, from a partial-fraction expansion of the above and a subsequent inverse transform into the time domain, terms involving e^{-8t}, $e^{-(1-j\sqrt{14})t}$, and $e^{-(1+j\sqrt{14})t}$ appear along with appropriate constant multipliers. The factors 8 and 1 are real exponents of e and are the damping factors. The factor $j\sqrt{14}$ is an oscillatory term, as indicated by the j factor. The $\sqrt{14}$ represents a radian frequency of $\sqrt{14}$ radians/sec. As to whether or not the system is stable, we merely note the sign of the damping factors. Negative signs indicate that, as time progresses, the exponential term vanishes and the system reaches some steady-state or equilibrium condition. However, if the exponents are positive, then the exponential term increases for increasing time, equilibrium is not obtained, and the system is unstable. In this first case, all damping constants appear as negative exponents, and the system is stable.

Now, we choose $D = 630$. The denominator of Eq. 21-57 becomes

$$s^3 + 10s^2 + 31s + 30 + 630 = s^3 + 10s^2 + 31s + 660$$

The factored form of this polynomial is

$$s^3 + 10s^2 + 31s + 660 = (s + 12)(s - 1 + j\sqrt{55})(s - 1 - j\sqrt{55})$$

Here the exponential time-domain forms are e^{-12t}, $e^{-(-1+j\sqrt{55})t}$, and $e^{-(-1-j\sqrt{55})t}$. As indicated by the e^{+1t} term, the system is unstable. The natural frequency present is $\sqrt{55}$ radians/sec.

From Example 21-3 it is evident that there is considerable work involved in calculating the roots of the denominator polynomial of the closed-loop transfer function as the gain factor A is varied. The next question that should be asked is, "What value of A is best?" For this phase of the problem, the analog computer can be an invaluable tool. In optimizing the design of a given servo system, the variable A can be given a large number of values simply by changing a potentiometer setting on the computer. The servo designer can then choose the best response by watching the output response characteristic as it is presented on an oscilloscope or chart recorder.

It is entirely possible that no desirable solution is obtainable with the original system, regardless of the value given to the amplifier gain A. In this case, corrective networks may be required. These are almost always inserted into the electrical portion of the system (assuming that the system is electromechanical) and are usually combinations of resistances and capacitances only as inductors are bulky, do not exhibit purely inductive properties, cannot be obtained off-the-shelf in a wide variety of sizes, and cannot be easily fabricated using thin film or integrated circuit techniques. Electrical networks are chosen primarily for convenience, since it is often cumbersome to work with or change the mechanical portion of the servo loop.

The preceding analytic technique and Example 21-3 illustrate the straightforward or "brute-force" method of servo analysis. As might be expected, the engineer is always looking for easier and quicker methods of analysis. Several of the popular and standard servo-design techniques will be presented in the following section. First, the reduction of a complex feedback system into one overall transfer function will be discussed. Then the Bode, Nyquist, and root-locus plots will be presented. Unfortunately, it is not possible to develop in a short chapter the underlying theory behind the use of these plots, and the reader is encouraged to investigate a few of the many available texts on servos for further enlightenment in these areas.

21-9. Block Diagram Reduction

A complicated servo system can have many feedback paths, and it is often desirable and/or necessary to reduce the resulting block diagram to a single block representing the closed-loop transfer function. As an example, let us consider the block diagram of Fig. 21-18a. By starting with the inner loop and using the basic feedback formula $\mathbf{G'} = \mathbf{G}/(1 + \mathbf{GH})$, it is possible to reduce the diagram to Fig. 21-18b, where

$$\mathbf{G_4} = \frac{\mathbf{G_2 G_3}}{1 + \mathbf{G_2 G_3 H_1}}$$

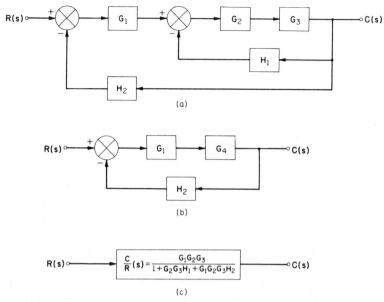

FIG. 21-18. Steps in the reduction of a block diagram to one overall transfer function.

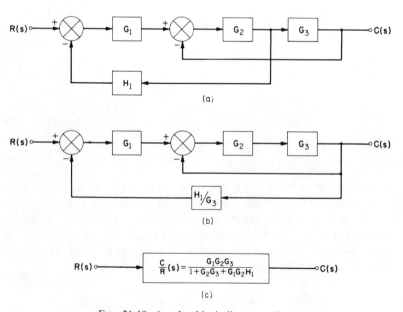

FIG. 21-19. Another block-diagram reduction.

Another reduction results in

$$G'(s) = \frac{C}{R}(s) = \frac{G_1 G_4}{1 + G_1 G_4 H_2} = \frac{G_1 G_2 G_3}{1 + G_2 G_3 H_1 + G_1 G_2 G_3 H_2}$$

where the G's and the H's are assumed to be functions of s.

Sometimes, adjustments must be made in the block diagram before the reduction process is possible. One of the most common problems is illustrated in Fig. 21-19. As the diagram is drawn in Fig. 21-19a, it is not possible to reduce either of the feedback loops, since this would eliminate a necessary node of the remaining feedback loop. One way to adjust the block diagram is to move the pick-off point for H_1 ahead of G_3. To compensate for this change, it is necessary to introduce a $1/G_3$ transfer function in the H_1 feedback loop so that the same information as before is fed back to the input summing junction. This step is shown in Fig. 21-19b. Figure 21-19c shows the block diagram for one overall transfer function.

21-10. Root-Locus Plots

The root-locus method, in which the poles and zeros of the open-loop transfer function are plotted on the complex s plane, is perhaps the most straightforward and logical method of analyzing a servo problem (but not necessarily the easiest or quickest). By definition, the zeros of $G(s)H(s)$ are simply the roots of the numerator polynomial of $G(s)H(s)$, whereas the poles are the roots of the denominator polynomial. As an example, let us consider the type 1 system

$$G(s)H(s) = \frac{K_1(s + 1)(s - 3)}{s(s + 3 - j5)(s + 3 + j5)(s + 4)(s - 5)}$$

where the zeros occur at $s = -1, 3$; and the poles occur at $s = 0, -4, 5$ and $-3 \pm j5$. When plotting these roots on the complex s plane, the symbol 0 is used for zeros and X is used for poles (see Fig. 21-20). Poles and zeros off the real axis always occur in complex conjugate pairs.

The root locus itself is a graphical plot showing all of the roots that satisfy the equation $1 + G(s)H(s) = 0$, with the gain factor K_1 being the running variable. Thus, by graphical means, we can find the individual roots or poles of the closed-loop transfer function for any value of K_1. This is certainly a very valuable piece of information, since all the frequencies of oscillation and the damping factors can be easily obtained, and corrections can be made if necessary. The actual plotting of the root locus is facilitated by use of a mechanical contrivance known as a *Spirule*[3] or by a digital or analog computer.

[3]The Spirule can be obtained from the Spirule Company, 9728 El Venado Avenue, Whittier, California.

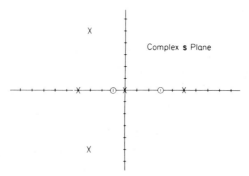

FIG. 21-20. A pole-zero plot of $G(s)H(s) = K_1(s + 1) (s - 3)/s (s + 3 - j5) (s + 3 + j5) (s + 4) (s - 5)$.

The root locus of the previously used open-loop transfer function

$$G(s)H(s) = \frac{K_1}{(s + 2)(s + 3)(s + 5)} \qquad (21\text{-}59)$$

is shown in Fig. 21-21. As pointed out in the preceding paragraph, the

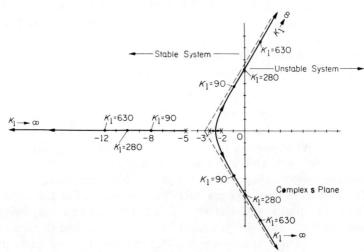

FIG. 21-21. Root locus plot of $s^3 + 10s^2 + 31s + 30 + K_1 = 0$ as K_1 is varied from 0 to ∞

root locus solves the characteristic equation $1 + G(s)H(s) = 0$ as K_1 is varied from 0 to ∞. Substituting Eq. 21-59 into the characteristic equation, we have

$$1 + \frac{K_1}{(s + 2)(s + 3)(s + 5)} = 0$$

or

$$s^3 + 10s^2 + 31s + 30 + K_1 = 0 \qquad (21\text{-}60)$$

Although thorough instruction in the overall plotting and use of the root locus is a bit too involved for this text, the following pertinent facts will be emphasized:

1. The root locus begins on poles. This means that at $K_1 = 0$ the roots of the characteristic equation are the original poles of $G(s)H(s)$. We let $K_1 = 0$, in Eq. 21-60, to check this fact.

2. The locus on the real axis exists only to the left of an odd number of explicit poles and zeros. Since Fig. 21-21 shows three explicit poles and no explicit zeros, the real-axis root locus exists between $s = -2$ and $s = -3$ and between $s = -5$ and $s = -\infty$.

3. The root locus terminates on the zeros of $G(s)H(s)$. In Fig. 21-21 there are no explicit zeros, but there are three implicit zeros at $s = \infty$, thus three branches of the locus proceed toward $s = \infty$.

4. Since the locus is calibrated in terms of the open-loop gain factor K_1, the locus proceeds from the poles at $K_1 = 0$ to terminate at the zeros at $K_1 = \infty$.

Several other rules are useful and necessary to complete the plot of the root locus, but these will not be presented here.

To illustrate the use of the root locus, we note the values of the roots of $1 + G(s)H(s)$ for Eq. 21-60 at $K_1 = 90$ and $K_1 = 630$. At $K_1 = 90$, the roots of Eq. 21-60 are $s = -1 \pm j\sqrt{14}$ and -8. At $K_1 = 630$, the roots are $s = 1 \pm j\sqrt{55}$ and -12. These same values were determined by a division process in an earlier calculation. However, the root locus has a distinct advantage over the rather cumbersome division method in that the root locus presents a picture, or graphical display, of all the roots of the characteristic equation as K_1 is varied over all possible values. In other words, the old saying that "a picture is better than a thousand words" is very applicable when referring to the root-locus plot.

As already pointed out, if any of the roots of the characteristic equation have positive real parts, there will be a positive exponent in the exponential time functions in $c(t)$. This means that the time response $c(t)$ increases without limit, and this condition defines system instability. Thus the root locus graphically shows that if the locus exists anywhere in the right half of the complex s plane, the system will be unstable if the value of K_1 is such that right-half plane roots are produced. As shown in Fig. 21-21, $K_1 = 630$ produces a complex conjugate pair of roots in the right half of the s plane, so the system defined by Eq. 21-60, is unstable when the loop is closed. For $K_1 = 90$, the system is stable, since only left-half plane roots exist.

Although whether or not a system is stable is perhaps the most important single aspect of servo analysis, there are many other factors besides absolute stability that are of concern to the servo designer. Such facts as the degree of stability, peak overshoot, settling time, rise time, sensitivity, system optimization, and so on, must be considered. The root-locus plot is a great aid in solving these problems, but such topics cannot be presented in a brief exposition on servo theory.

EXAMPLE 21-4. If $G(s) = K_1/(s + 2)(s + 3)(s + 5)$ and $H(s) = 1$, find the output response $c(t)$ with $r(t)$ a unit step input for $K_1 = 90$ and 280.

Solution: Using the root-locus plot in Fig. 21-21, we can determine the closed-loop poles for $K_1 = 90$ or 280. First, we let $K_1 = 90$,

$$\frac{C}{R}(s) = \frac{G(s)}{1 + G(s)H(s)} = \frac{90}{(s + 8)(s + 1 - j\sqrt{14})(s + 1 + j\sqrt{14})}$$

and, for $r(t)$ a unit step,

$$R(s) = \frac{1}{s}$$

Thus

$$C(s) = \frac{90}{s(s + 8)(s + 1 - j\sqrt{14})(s + 1 + j\sqrt{14})}$$

$$= \frac{A}{s} + \frac{B}{s + 8} + \frac{C}{s + 1 - j\sqrt{14}} + \frac{D}{s + 1 + j\sqrt{14}}$$

$$= \frac{3/4}{s} - \frac{9/504}{s + 8} - \frac{15/(28 + j7\sqrt{14})}{s + 1 - j\sqrt{14}} - \frac{15/(28 - j7\sqrt{14})}{s + 1 + j\sqrt{14}}$$

and, taking the inverse transform,

$$c(t) = \frac{3}{4} - \frac{9}{504} e^{-8t} - \frac{105}{1470} e^{-t}[4(e^{j\sqrt{14}t} + e^{-j\sqrt{14}t}) - j\sqrt{14}(e^{j\sqrt{14}t} - e^{j\sqrt{14}t})]$$

$$= \frac{3}{4} - \frac{1}{56} e^{-8t} - \frac{e^{-t}}{7}[4\cos\sqrt{14}t + \sqrt{14}\sin\sqrt{14}t] \qquad (21\text{-}61)$$

By inspection of Eq. 21-61, it is evident that as $t \to \infty$, $c(t) \to 3/4$, and the system is stable. The damping factors are -8 and -1, and the damped natural frequency of oscillation is $\sqrt{14}$.

The solution for $c(t)$ with $K_1 = 280$ is particularly interesting. From the calibrated root-locus plot,

$$\frac{C}{R}(s) = \frac{280}{(s + 10)(s + j\sqrt{31})(s - j\sqrt{31})} = \frac{280}{(s + 10)(s^2 + 31)}$$

$$C(s) = \frac{280}{s(s + 10)(s^2 + 31)} = \frac{A}{s} + \frac{B}{s + 10} + \frac{Cs}{s^2 + 31} + \frac{D}{s^2 + 31}$$

$$= \frac{28/31}{s} - \frac{28/131}{s + 10} - \frac{(2800/4061)s}{s^2 + 31} - \frac{280/131}{s^2 + 31}$$

and, taking the inverse transform,

$$c(t) = \frac{28}{31} - \frac{28}{131} e^{-10t} - \frac{2800}{4061} \cos \sqrt{31}\, t - \frac{280}{131 \sqrt{31}} \sin \sqrt{31}\, t \quad (21\text{-}62)$$

Equation 21-62 indicates that the $\cos \sqrt{31}\, t$ and $\sin \sqrt{31}\, t$ terms remain undamped as $t \to \infty$. The servo system thus behaves much like an oscillator with an undamped natural frequency of $\sqrt{31}$. (An oscillator is a generator of sinusoidal frequencies.)

Instead of further pursuing root-locus methods, the Bode-Nyquist criteria will be presented in the following sections. The Bode method is particularly easy to use, since it is not difficult to learn how to sketch the necessary Bode plots. However, the serious student of servo design should by all means learn the root-locus approach to servo analysis.

21-11. Bode-Nyquist Criteria

The Bode and Nyquist stability criteria tend to complement each other, although each method can be proved or used independently of the other. Both methods let $\mathbf{s} = j\omega$ in the $\mathbf{G(s)H(s)}$ transfer function, which means that only the steady-state sinusoidal response is considered. These two methods then take advantage of the rather tenuous relationship that exists between the time and real frequency domains, so that the results of the steady-state sinusoidal analysis may be used to reveal the overall transient performance of the system.

The Nyquist plot is a polar plot of the phasor quantity $\mathbf{G}(j\omega)\mathbf{H}(j\omega) = |\mathbf{GH}| \underline{/\theta}$, as ω varies from $-\infty$ to $+\infty$. Another way of presenting the quantity $|\mathbf{GH}| \underline{/\theta}$ consists of two rectangular coordinate plots: 1) a log-modulus versus log-frequency plot or 20 log $|\mathbf{GH}|$ versus log ω, and 2) a phase-angle versus log-frequency plot or θ versus log ω. Both plots are usually referred to as the Bode plot, and only positive values of frequency ω are considered.

To illustrate the mechanics of plotting these curves, let us consider the example of a previous section where

$$\mathbf{G(s)H(s)} = \frac{N}{(s + 2)(s + 3)(s + 5)}$$

and

$$\mathbf{G'(s)} = \frac{N}{(s + 2)(s + 3)(s + 5) + D}$$

and where $D = 90$ or 630.

The first step in plotting either the Nyquist or the Bode plot is to substitute $j\omega$ for $\mathbf{s}$. This automatically means that we are considering only steady-state sinusoidal excitation and response functions. For the first case of $D = 90$ and $K_T = 1$ (this is called direct feedback) so that $N = D$,

$$G(j\omega)H(j\omega) = \frac{90}{(j\omega + 2)(j\omega + 3)(j\omega + 5)}$$

$$= \frac{90}{\sqrt{\omega^2 + 4}\sqrt{\omega^2 + 9}\sqrt{\omega^2 + 25}}$$

$$\Bigg/ -\left(\tan^{-1}\frac{\omega}{2} + \tan^{-1}\frac{\omega}{3} + \tan^{-1}\frac{\omega}{5}\right)$$

and thus

$$|\,GH\,| = \frac{90}{\sqrt{\omega^2 + 4}\sqrt{\omega^2 + 9}\sqrt{\omega^2 + 25}}$$

$$\theta = -\left(\tan^{-1}\frac{\omega}{2} + \tan^{-1}\frac{\omega}{3} + \tan^{-1}\frac{\omega}{5}\right)$$

It is convenient to start the Nyquist plot by letting $\omega = 0$ and then $\omega = \pm\infty$. We then choose a few other values of ω, as necessary. At $\omega = 0$,

$$GH = \frac{90}{2 \times 3 \times 5}\,\underline{/0°} = 3\,\underline{/0°}$$

At $\omega = -\infty$,

$$GH = 0\,\underline{/-(-270°)} = 0\,\underline{/-90°}$$

At $\omega = +\infty$,

$$GH = 0\,\underline{/-270°} = 0\,\underline{/+90°}$$

The Nyquist plot of $G(j\omega)H(j\omega)$ for $D = 90$ is shown as curve a in Fig. 21-22. If D is chosen as 630, the Nyquist plot appears as curve b, which is

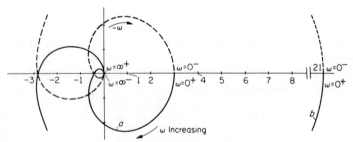

FIG. 21-22. Nyquist plot of $G(j\omega)H(j\omega) = N/(2 + j\omega)(3 + j\omega)(5 + j\omega)$ for a) $N = 90$ and b) $N = 630$.

the same as curve a magnified by a factor of 7. The dashed lines represent the locus of $GH(j\omega)$ for negative frequencies or $-\omega$. It is to be expected that $GH(j\omega)$ and $GH(-j\omega)$ are complex conjugates.

Before proceeding with the discussion of the Nyquist plot, let us recall that $G'(s) = G(s)/[1 + G(s)H(s)]$, and, if $1 + G(s)H(s) = 0$, it is obvious that theoretically the output signal becomes infinite, regardless of how small the input signal is. (Since noise and other extraneous signals are always present in a practical system, there is never just a zero input signal.) *It can be shown that if* $G(j\omega)H(j\omega)$ *passes through or encircles the* $-1 + j0$ *point, the closed-loop servo system will be unstable.* If the locus of $G(j\omega)H(j\omega)$ does not pass through or encircle the $-1 + j\ 0$ point, then the system is stable, but it is more underdamped the nearer the locus passes this $-1 + j0$ point. In other words, for a step input signal, the system output signal tends to have more oscillations about the equilibrium point the closer the locus of $GH\ (j\omega)$ approaches $-1 + j0$.

Although no proof will be given, the following rules are used in applying the Nyquist stability criterion:

1. Plot the $G(s)H(s)$ function as a polar plot of $s = j\omega$ for all values of ω from $-\infty$ to $+\infty$.

2. Trace the phasor $1 + G(j\omega)H(j\omega)$ by tracing along the locus of $G(j\omega)H(j\omega)$, using the $-1 + j0$ point as the origin of the phasor. Start the tracing from $\omega = -\infty$ and proceed to $\omega = +\infty$. Observe the net number of 360° counterclockwise rotations of the $1 + GH(j\omega)$ phasor and call this number (C). (A clockwise rotation is simply a negative counterclockwise rotation.) Next let $G(s)H(s)$ be expressed as a ratio of two polynomials:

$$G(s)H(s) = \frac{N(s)}{D(s)}$$

Find the number of roots of the equations $D(s) = 0$ whose real parts are positive. [This is usually done by observation, since the polynomial $D(s)$ is almost always in factored form. If $D(s)$ is in expanded form, Routh's[4] criterion is a convenient way of finding the number of roots with positive real parts.] Call this number (P). Then let (Z) be the zeros of $1 + G(s)H(s) = [D(s) + N(s)]/D(s)$ with positive real parts. The zeros of $1 + G(s)H(s)$ satisfy the equation $D(s) + N(s) = 0$. If any of the zeros of $1 + G(s)H(s)$ have positive real parts, then these zeros become the poles of $G' = G/(1 + GH)$ and the system is unstable. Hence, for stability, there must be no zeros of $1 + G(s)H(s)$ with positive real parts, and

$$(Z) = (P) - (C) = 0 \tag{21-63}$$

or

$$(C) = (P)$$

for stability.

Returning to the example illustrated in Fig. 21-22, we count no net rotations of $1 + GH(j\omega)$ for case a and two clockwise or (-2) counter-

---

[4]Routh, E. J., *Dynamics of a System of Rigid Bodies*, 3d ed., Macmillan & Co., Ltd., London, 1877.

clockwise rotations for case *b*. Observing the denominator of **G(s)H(s)**, it is apparent that all roots lie in the left half of the complex plane, and thus no poles of **G(s)H(s)** have positive real parts. From Eq. 21-63,

$$(Z) = (0) - (0) = 0$$

for case *a* and the system is stable; for case *b*,

$$(Z) = (0) - (-2) = 2$$

therefore the system is unstable.

The Nyquist plot sometimes requires many repetitive calculations for various values of ω if $G(j\omega)H(j\omega)$ involves a great many terms. Primarily because of this, many designers find the Bode plot more convenient and easier to use than the Nyquist plot. This is particularly true when one must consider the insertion of corrective networks into the system in order to improve the overall system performance.

The Bode plot, or 20 log $|GH(j\omega)|$ and θ versus log ω, is usually drawn by using straight-line approximations instead of tediously calculating each point before plotting the actual curve. This same type of plot was encountered in Chapter 13 when the frequency response of the *R-C*-coupled amplifier was considered. The salient point to remember is that 20 log $1/\omega$ and 20 log ω have a slope of -6 and 6 db/octave, respectively. Although the angle θ is also important, it is not always necessary to consider it explicitly in every analysis, and so it will be ignored here. The two primary reasons for not emphasizing the θ versus log ω plot in this chapter are 1) because this particular curve is a bit more tedious to plot than is the straight-line approximation of 20 log $|GH|$ versus log ω, and 2) because the θ versus log ω plot merely aids in interpreting and refining the results already obtained by the 20 log $|GH|$ versus log ω curve. To illustrate the Bode plot, let us consider the example for $D = 90$

$$|G(j\omega)H(j\omega)| = \frac{90}{\sqrt{4 + \omega^2}\,\sqrt{9 + \omega^2}\,\sqrt{25 + \omega^2}}$$

$$= \frac{3}{[1 + (\omega/2)^2]^{1/2}[1 + (\omega/3)^2]^{1/2}[1 + (\omega/5)^2]^{1/2}}$$

and

$$20 \log |GH(j\omega)| = 20 \log 3 - \left\{ 20 \log \left[1 + \left(\frac{\omega}{2}\right)^2\right]^{1/2} \right.$$

$$\left. + 20 \log \left[1 + \left(\frac{\omega}{3}\right)^2\right]^{1/2} + 20 \log \left[1 + \left(\frac{\omega}{5}\right)^2\right]^{1/2} \right\} \qquad (21\text{-}64)$$

The straight-line approximation to Eq. 21-64 is shown in Fig. 21-23a. The first term, 20 log 3, contributes a constant decibel factor. The other terms contribute a -6-db/octave slope beginning at the corner or breakpoint frequencies of 2, 3, and 5 radians/sec.

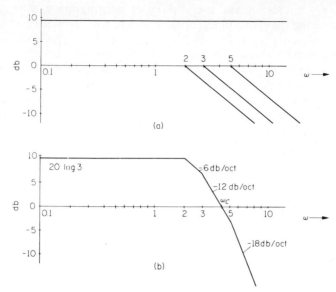

FIG. 21-23. a) Straight-line approximation of Eq. 21-64.
b) Combined straight-line approximation of part a.

Figure 21-23b shows the combination of the individual curves plotted in Fig. 21-23a. The actual curve can be sketched with a fair degree of accuracy by simply following the straight-line approximation and recalling that, at the corner frequencies, each denominator or numerator time constant contributes a -3-db or $+3$-db attenuation, respectively. The Bode Plot for $D = 630$ is shown in Fig. 21-24.

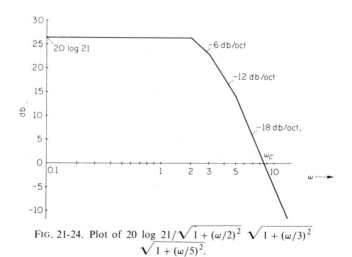

FIG. 21-24. Plot of $20 \log 21/\sqrt{1 + (\omega/2)^2} \ \sqrt{1 + (\omega/3)^2}$
$$\sqrt{1 + (\omega/5)^2}.$$

One very interesting and useful piece of information can be gleaned from comparing these two plots for D = 90 and 630. Notice that the 0-db crossover frequency ω_c in Fig. 21-23b occurs on the -12-db/octave portion of the curve, whereas in Fig. 21-24 the crossover occurs when the curve has a slope of -18 db/octave. From previous results it is known that when D = 90 the system is stable, but when D = 630 the system is unstable. *It can be shown that if the 0-db crossover frequency occurs when the negative slope of the Bode curve is greater than -12 db/octave, the system is unstable.* The θ versus log ω plot can be used as an aid in understanding the reason for this. As with the Nyquist criterion, the Bode criterion determines closed-loop stability from the easier-to-analyze open-loop transfer function.

21-12. Corrective Networks

If a system is known to be closed-loop unstable, it is often possible to make it closed-loop stable by the proper addition of a simple *R-C* corrective network. There are three popular *R-C* networks that are often used—the lead, the lag, and the lag-lead networks—but only the lead network will be discussed in this section. To be effective, a corrective network must be designed to operate at the correct frequency or frequencies. Although design of corrective networks can be done by using the Nyquist or root-locus plots, only the Bode-plot method will be given here.

In designing the corrective network, attention is focused on the 0-db crossover frequency. If the crossover occurs at a negative slope which is greater than -12 db/octave, efforts are made to cause the slope to be -12 or, better, -6 db/octave at the crossover frequency. If a term such as $(1 + \tau s)$ is inserted in the numerator function of $\mathbf{G(s)H(s)}$ for D = 630, the transfer function becomes

$$\mathbf{G(s)H(s)} = \frac{630(1 + \tau s)}{(s + 2)(s + 3)(s + 5)} = \frac{21(1 + \tau s)}{\left(1 + \frac{s}{2}\right)\left(1 + \frac{s}{3}\right)\left(1 + \frac{s}{5}\right)}$$

or

$$\mathbf{G}(j\omega)\mathbf{H}(j\omega) = \frac{21(1 + j\omega\tau)}{\left(1 + j\frac{\omega}{2}\right)\left(1 + j\frac{\omega}{3}\right)\left(1 + j\frac{\omega}{5}\right)}$$

The first question is, "What must be the value of τ?" From Fig. 21-24 the crossover frequency is 8.5 radians/sec, so obviously, $1/\tau$ cannot be greater than this, for it would not cause the slope of the decibel curve at the crossover frequency to change. There are several values that $1/\tau$ could have, so let us choose $1/\tau$ = 7 as a first try. The Bode plot for this follows from

$$G(j\omega)H(j\omega) = \frac{21\left(1 + j\frac{\omega}{7}\right)}{\left(1 + j\frac{\omega}{2}\right)\left(1 + j\frac{\omega}{3}\right)\left(1 + j\frac{\omega}{5}\right)} \qquad (21\text{-}65)$$

and is shown in Fig. 21-25.

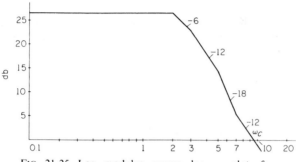

FIG. 21-25. Log modulus versus log ω plot for Eq. 21-65.

This new Bode curve has a crossover frequency of 9.5 and a slope of -12 db/octave at the crossover. Since -12 db/octave tends to be the borderline of stability, it is advisable to introduce a second lead network at this same frequency so that the crossover occurs at a slope of -6 db/octave. Figure 21-26 is the resulting Bode plot. The new crossover fre-

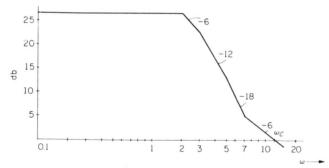

FIG. 21-26. Bode plot of

$$20 \log \left| \frac{21(1 + j\omega/7)^2}{(1 + j\omega/2)(1 + j\omega/3)(1 + j\omega/5)} \right|$$

quency is 12.5 radians/sec at the slope of -6 db/octave, and the system is definitely stable.

Now it is necessary to find the *R-C* circuit that has this particular

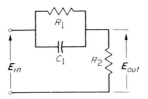

FIG. 21-11c (repeated).

transfer function. By clever hindsight, it is realized that the phase-lead network of Fig. 21-11c has the proper transfer function. Figure 21-11c and Eq. 21-35 (somewhat rearranged) are repeated here for easy reference.

$$\frac{E_{out}}{E_{in}}(s) = \left(\frac{R_2}{R_1 + R_2}\right)\frac{1 + \tau_1 s}{1 + \tau_2 s} \tag{21-66}$$

where

$$\tau_1 = R_1 C_1$$

$$\tau_2 = \frac{R_1 R_2 C_1}{R_1 + R_2}$$

Although Eq. 21-66 does not have exactly the proper form required in the original corrective analysis, the denominator term can be made inconsequential as far as system stability is concerned. Let $\tau_1 = 1/7$ and $\tau_2 = 1/20$ radians/sec. This choice is made by observing that the -6-db/octave crossover occurs at 12.5 radians/sec, and that it makes little difference what the slope of the decibel curve becomes for frequencies greater than this. One choice of values might be

$$R_1 = 1.43 \text{ megohms}$$
$$C_1 = 0.1 \text{ } \mu\text{f}$$
$$R_2 = 0.77 \text{ megohm}$$

In order to produce two similar time constants, we cascade two of these phase-lead networks by means of an isolation stage such as a transistor or vacuum-tube amplifier with a gain of α^2, where $\alpha = (R_1 + R_2)/R_2 = 20/7$. This is to compensate for the $R_2/(R_1 + R_2)$ attenuation factor of the lead network as $\omega \rightarrow 0$. The block diagram of the adjusted servo system, with a factor of $D = 630$, appears in Fig. 21-27.

The open-loop transfer function corresponding to the diagram shown in Fig. 21-27 is

$$G(s)H(s) = \frac{630(20/7)^2(s + 7)^2}{(s + 2)(s + 3)(s + 5)(s + 20)^2}$$

$$= \frac{5180(s + 7)^2}{(s + 2)(s + 3)(s + 5)(s + 20)^2}$$

The Bode plot for this transfer function is shown in Fig. 21-28, and the corresponding Nyquist plot is sketched in Fig. 21-29. Both plots indi-

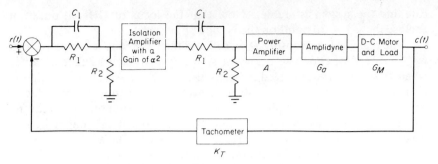

FIG. 21-27. Block diagram of corrected network.

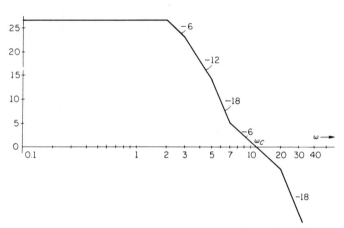

FIG. 21-28. Bode diagram of $\mathbf{GH}(j\omega) = 21(1 + j\omega/7)^2/(1 + j\omega/2)(1 + j\omega/3)(1 + j\omega/5)(1 + j\omega/20)^2$.

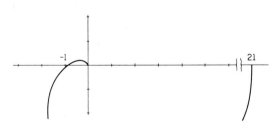

FIG. 21-29. Nyquist plot of $\mathbf{GH}(j\omega) = 21(1 + j\omega/7)^2/(1 + j\omega/2)\ (1 + j\omega/3)\ (1 + j\omega/5)\ (1 + j\omega/20)^2$.

cate that the system is stable. Note that the locus of $\mathbf{GH}(j\omega)$ does not enclose the $-1 + j0$ point on the Nyquist plot.

The reasons for choosing $D = 630$ and then applying a corrective network, instead of choosing a known stable system such as that obtained when $D = 90$, is too involved to be discussed here. This sort of manipulation is necessary for certain systems, since stability is not the only factor to be considered. A system may have a sluggish response, or it may respond too rapidly for any or all gain factors. When this happens, it is necessary to resort to various corrective networks. The analog computer can be a valuable aid in determining the final system performance or in optimizing the performance once a particular corrective network has been chosen.

The foregoing discussion should give some insight as to the methods used in servo analysis. Bode, Nyquist, and root-locus analysis techniques are still popular tools with servo designers, but digital or analog computer simulation is almost always used to complete the design of complex control systems.

Problems

21-1. An amplifier without feedback has a gain of 120. If negative feedback with $\beta = -0.05$ is added, what is the amplifier output if the input is 1.5 volts?

21-2. A negative-feedback amplifier has a gain of 33. For a given output, the input must be 2.6 volts, while for the same output without feedback, the input is 0.3 volt. What is the gain $\mathbf{A}$ before feedback, and what is the corresponding value of β?

21-3. An amplifier without feedback has a signal output of 90 volts with an internally introduced noise output of 1.6 volts. Determine the value of β needed to reduce the noise output to 0.2 volt with the same signal output. The amplifier gain with feedback is -180.

21-4. An R-C-coupled amplifier has a mid-frequency gain of $+330$. The respective values of f_1 and f_2 are 40 and 18,000 Hz for the amplifier without feedback. If negative feedback with $\beta = -0.02$ is added, what are the new gains at 40 and 18,000 Hz?

21-5. Find the new values of f_1 and f_2 for the feedback amplifier in Prob. 21-4.

21-6. The emitter follower circuit as presented in Chapter 7 was analyzed using conventional circuit analysis methods. Identify (or derive) expressions for A and β and confirm that the feedback gain expression (Eq. 21-4) gives the same A' as given in the Chapter 7 development. Use α_{fe} instead of β for the transistor current gain to avoid confusion with the feedback factor β.

21-7. Use conventional circuit analysis methods to derive the expression for the voltage gain for the common emitter amplifier with an un-bypassed emitter-biasing resistor. The equivalent circuit to be analyzed is given in the accompanying figure. Identify A, β, and A' for voltage gain. Show that for $\alpha_{fe} > 10$, $\beta \simeq R_E/R_l$.

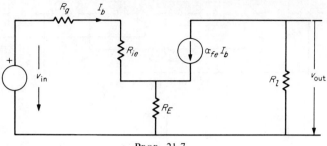

PROB. 21-7

21-8. Show that the block diagram of the accompanying illustration can be reduced to one overall $C/R(s)$ transfer function given by

$$\frac{C}{R}(s) = \frac{G_1 G_2 G_3 G_4}{1 + G_2 H_2 + G_3 G_4 H_4 + G_2 G_3 H_3 + G_2 G_3 G_4 H_2 H_4}$$

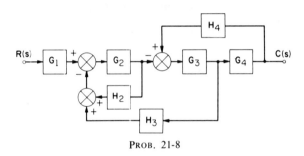

PROB. 21-8

21-9. Determine $C(s)/R(s)$ from the servo block diagrams shown in a, b, c, and d of the accompanying illustration.

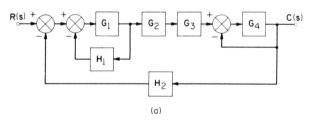

(a)

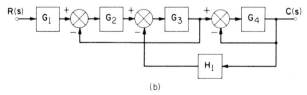

(b)

PROB. 21-9

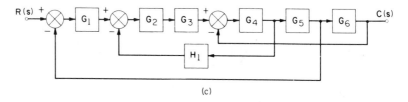

(c)

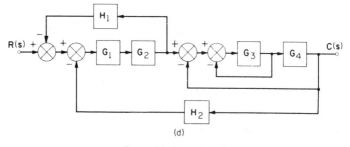

(d)

PROB. 21-9 (*continued*)

21-10. Plot the poles and zeros of the following transfer functions:

a) $\dfrac{500(s^2 + 6s + 50)(s + 2)}{s(s + 3)(s + 7)(s^2 + 2s + 8)}$

b) $\dfrac{20(s + 3)}{s^2(s + j4)(s - j4)}$

c) $\dfrac{150}{(s + 1)(s^2 - s + 10)}$

d) $\dfrac{0.6}{s(1 + 0.1s)(1 + 0.4s)}$

21-11. If the transfer functions in Prob. 21-10 represent the forward loop transfer function $G(s)$ of various servo systems, a) which of the systems are open-loop stable and which are unstable? b) What are the open-loop-system damping factors, damping ratios, undamped natural frequencies ω_n, and damped natural frequencies ω_D?

21-12. Determine the stability of each closed-loop system by calculating and plotting the poles and zeros of the closed-loop transfer function if $H(s) = 1$ and $G(s)H(s)$ equals:

a) $\dfrac{1 + 3s}{s(1 + 0.3s)}$

b) $\dfrac{15(s^2 + 7s + 2)}{s^4 + s^3 + 14s^2 + 52s + 10}$

c) $\dfrac{4(1 + 0.02s)}{0.01s^2 + 0.25s + 1}$

d) $\dfrac{1}{s^3 + 0.1s^2 + 0.15s + 0.375}$

21-13. Write the open-loop transfer function $G(s)H(s)$ for the pole-zero plots shown in parts a and b of the accompanying illustration.

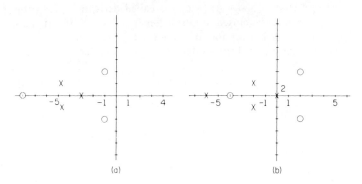

(a)   (b)

PROB. 21-13

21-14. Given

$$G(s)H(s) = \frac{K_1(s + 5)}{s(s + 8)(s + 9)(s + 12)}$$

a) Plot the poles and zeros of $G(s)H(s)$ on the complex s plane. b) Indicate the real-axis portion of the root locus.

21-15. A feedback control system has the following open-loop transfer function:

$$G(s)H(s) = \frac{K_1(s^2 + 5s + 15)}{s^2(s + 10)(s^2 + 4s + 20)}$$

a) Plot the poles and zeros on the complex s plane. b) Determine the real-axis portion of the root locus.

21-16. The closed-loop transfer function of a system is known to be

$$\frac{C}{R}(s) = \frac{85(s + 1)}{(s + 3)(s + 5)(s + 10)}$$

a) Find the output $c(t)$ if $r(t)$ is a unit impulse function. b) Is the system stable or unstable?

21-17. The closed-loop transfer function of a servo system is determined from a root-locus plot as

$$\frac{C}{R}(s) = \frac{140(s^2 + 3s + 16)}{(s + 1)(s^2 + 7s + 24)(s^2 + 4s + 30)}$$

a) What are the closed-loop damping factors and the damping ratios? b) What are the closed-loop undamped and damped natural frequencies of oscillation? c) If $r(t)$ is a unit step function, determine $c(t)$.

21-18. The accompanying illustration shows the complete root-locus plot for the open-loop transfer function

$$G(s)H(s) = \frac{K_1(s + 4)}{s(s + 2)(s + 6)(s + 10)}$$

For $K_1 = 51$ and $H(s) = 5(s + 4)$, which is called derivative plus proportional feedback: a) Determine the closed-loop transfer function. b) If $r(t)$ is a unit step function, find $c(t)$. c) What are the damping factors, damping ratios, and actual frequencies of oscillation? d) Is the system stable or unstable?

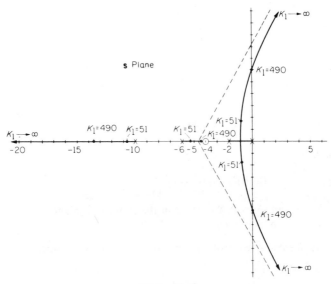

PROB. 21-18

21-19. Repeat Prob. 21-18 for $K_1 = 490$.

21-20. Simulate the following transfer functions, using an analog computer. (Do not worry about time or amplitude scaling.)

a) $\dfrac{C}{R}(s) = \dfrac{25(1 + 0.2s)}{(1 + 0.1s)(1 + 0.05s)}$ b) $\dfrac{C}{R}(s) = \dfrac{2}{(s + 3)(s + 6)(s + 15)}$

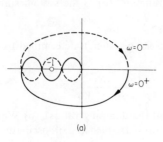

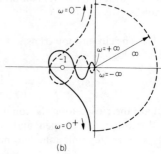

PROB. 21-20

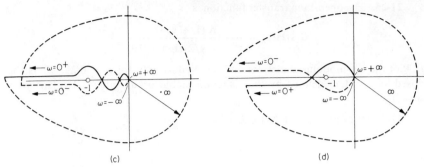

(c) (d)

PROB. 21-20 (*continued*)

21-21. Simulate the complete closed-loop servo system on an analog computer if $H(s) = 1$ for the following open-loop transfer functions:

a) $G(s)H(s) = \dfrac{10(1 + 0.6s)}{(1 + 0.02s)(1 + 0.2s)}$

b) $G(s)H(s) = \dfrac{180}{s(s + 2)(s + 4)(s + 12)}$

c) $G(s)H(s) = \dfrac{90(s + 1)}{s^2(s + 0.5)(s + 3)(s + 7)}$

21-22. Determine whether or not the Nyquist plots in the accompanying figure, represent stable systems. There are no roots with positive real parts in $G(s)H(s)$.

21-23. a) Plot the log-modulus Bode plot for the open-loop transfer function given in Prob. 21-14. b) Does the Bode plot indicate stability or instability?

21-24. a) Determine $G(s)H(s)$ from the Bode plot shown in the accompanying illustration. b) Is the system closed-loop stable? Why?

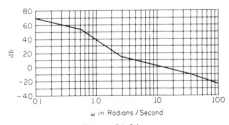

PROB. 21-24

21-25. The open-loop transfer function is

$$G(s)H(s) = \frac{K(1 + s/4)}{s^2(1 + s/2)(1 + s/10)}$$

a) Plot the log-modulus Bode plot and the Nyquist plot for $K = 0.25$. b) Is the closed-loop system defined by this function stable or unstable? Explain your answer.

21-26. Repeat Prob. 21-25 for $K = 4$.

A / *Determinants for*

Algebra

Perhaps the most obvious way to solve a set of linear simultaneous algebraic equations is to use the process of substitution or elimination; however, the use of determinants is a much more orderly and systematic way of effecting the solution. The rules for the determinant reduction of simultaneous equations are rather easy to learn and apply, but no attempt will be made to derive the rules shown here. To illustrate the rules for determinants, consider the following set of linear simultaneous algebraic equations,

$$\left.\begin{array}{l} a_1x_1 + b_1x_2 + c_1x_3 + d_1x_4 = m_1 \\ a_2x_1 + b_2x_2 + c_2x_3 + d_2x_4 = m_2 \\ a_3x_1 + b_3x_2 + c_3x_3 + d_3x_4 = m_3 \\ a_4x_1 + b_4x_2 + c_4x_3 + d_4x_4 = m_4 \end{array}\right\} \qquad \text{(A-1)}$$

where x_1, x_2, x_3, and x_4 are the unknowns, and the a, b, c, and d coefficients are known, as are the forcing functions, which are noted by the symbol m.

The first determinant of interest is the *characteristic* determinant of the set of equations. This determinant (noted by the symbol Δ) is formed by arranging the elements (coefficients) of Eq. A-1 in a grid of rows and columns exactly as shown in Eq. A-1.

$$\Delta = \begin{vmatrix} a_1 & b_1 & c_1 & d_1 \\ a_2 & b_2 & c_2 & d_2 \\ a_3 & b_3 & c_3 & d_3 \\ a_4 & b_4 & c_4 & d_4 \end{vmatrix}$$

For each unknown, such as x_2, to be evaluated, another determinant Δ_{x_2} is formed exactly like Δ, except that the m constants are substituted for the b coefficients of the unknown x_2 which is being evaluated, or

$$\Delta_{x_2} = \begin{vmatrix} a_1 & m_1 & c_1 & d_1 \\ a_2 & m_2 & c_2 & d_2 \\ a_3 & m_3 & c_3 & d_3 \\ a_4 & m_4 & c_4 & d_4 \end{vmatrix}$$

The order of a determinant is equal to the number of rows or columns (*the number of rows always equals the number of columns in a determinant*) *in the array*, and, in this example, the order is 4.

As will be shown, each determinant is simply a number (real or complex) if the coefficients are expressed as numbers, and a polynomial if the coefficients are expressed as algebraic quantities. The value of each unknown, such as x_2, is the ratio of Δ_{x_2} to Δ, or

$$x_2 = \frac{\Delta_{x_2}}{\Delta}$$

Similarly,

$$x_1 = \frac{\Delta_{x_1}}{\Delta} \qquad x_3 = \frac{\Delta_{x_3}}{\Delta} \qquad x_4 = \frac{\Delta_{x_4}}{\Delta}$$

A-1. Evaluation of the Determinant by the Laplace Development

The evaluation of a determinant by the Laplace development proceeds by expanding it into a sum of determinants, of progressively lower order, until the second (or possibly the third) order has been reached. After this, expansion into a sum of simple algebraic terms is made.

First, consider the characteristic determinant Δ. Select any column (usually the one with the most zeros) and set down the first element from this column as a coefficient or multiplier of a new determinant, called a *minor—a first minor for the first element*. This new determinant, or minor, consists of what is left of the original determinant after removing both the column and the row from which the first element was taken. Designate this minor as M_{rc}; that is, the determinant formed by striking out the rth row and the cth column. Now each minor M_{rc} must be multiplied by the factor $(-1)^{r+c}$. This product term is called the *cofactor of the element* in the rth row and the cth column of the original determinant. Expansion of the determinant on any given column is then accomplished by summing the products of the column elements and their respective cofactors. Consider the expansion of Δ on the first column:

$$\Delta = a_1(-1)^{1+1} \begin{vmatrix} a_1 & b_1 & c_1 & d_1 \\ a_2 & b_2 & c_2 & d_2 \\ a_3 & b_3 & c_3 & d_3 \\ a_4 & b_4 & c_4 & d_4 \end{vmatrix} + a_2(-1)^{2+1} \begin{vmatrix} a_1 & b_1 & c_1 & d_1 \\ a_2 & b_2 & c_2 & d_2 \\ a_3 & b_3 & c_3 & d_3 \\ a_4 & b_4 & c_4 & d_4 \end{vmatrix}$$

$$+ a_3(-1)^{3+1} \begin{vmatrix} a_1 & b_1 & c_1 & d_1 \\ a_2 & b_2 & c_2 & d_2 \\ a_3 & b_3 & c_3 & d_3 \\ a_4 & b_4 & c_4 & d_4 \end{vmatrix} + a_4(-1)^{4+1} \begin{vmatrix} a_1 & b_1 & c_1 & d_1 \\ a_2 & b_2 & c_2 & d_2 \\ a_3 & b_3 & c_3 & d_3 \\ a_4 & b_4 & c_4 & d_4 \end{vmatrix}$$

Now, either the 3 × 3 determinants can be expanded by the usual cross-multiplication technique (which will be explained later), or a second expansion can be used to reduce the determinants to a second-order sum by repeating the same process just used. Note how the $(-1)^{r+c}$ factor has been included in the expansion.

$$
\begin{aligned}
\Delta = a_1 b_2 \begin{vmatrix} b_2 & c_2 & d_2 \\ b_3 & c_3 & d_3 \\ b_4 & c_4 & d_4 \end{vmatrix} &- a_1 b_3 \begin{vmatrix} b_2 & c_2 & d_2 \\ b_3 & c_3 & d_3 \\ b_4 & c_4 & d_4 \end{vmatrix} + a_1 b_4 \begin{vmatrix} b_2 & c_2 & d_2 \\ b_3 & c_3 & d_3 \\ b_4 & c_4 & d_4 \end{vmatrix} \\
- a_2 b_1 \begin{vmatrix} b_1 & c_1 & d_1 \\ b_3 & c_3 & d_3 \\ b_4 & c_4 & d_4 \end{vmatrix} &+ a_2 b_3 \begin{vmatrix} b_1 & c_1 & d_1 \\ b_3 & c_3 & d_3 \\ b_4 & c_4 & d_4 \end{vmatrix} - a_2 b_4 \begin{vmatrix} b_1 & c_1 & d_1 \\ b_3 & c_3 & d_3 \\ b_4 & c_4 & d_4 \end{vmatrix} \\
+ a_3 b_1 \begin{vmatrix} b_1 & c_1 & d_1 \\ b_2 & c_2 & d_2 \\ b_4 & c_4 & d_4 \end{vmatrix} &- a_3 b_2 \begin{vmatrix} b_1 & c_1 & d_1 \\ b_2 & c_2 & d_2 \\ b_4 & c_4 & d_4 \end{vmatrix} + a_3 b_4 \begin{vmatrix} b_1 & c_1 & d_1 \\ b_2 & c_2 & d_2 \\ b_4 & c_4 & d_4 \end{vmatrix} \\
- a_4 b_1 \begin{vmatrix} b_1 & c_1 & d_1 \\ b_2 & c_2 & d_2 \\ b_3 & c_3 & d_3 \end{vmatrix} &+ a_4 b_2 \begin{vmatrix} b_1 & c_1 & d_1 \\ b_2 & c_2 & d_2 \\ b_3 & c_3 & d_3 \end{vmatrix} - a_4 b_3 \begin{vmatrix} b_1 & c_1 & d_1 \\ b_2 & c_2 & d_2 \\ b_3 & c_3 & d_3 \end{vmatrix}
\end{aligned}
$$

The final expansion of the 2 × 2 determinants is performed by taking the sum of the products and elements of each diagonal: downward to the right, and attaching a plus sign to the product of the elements; and upward to the right, attaching a minus sign to the product of these elements.

$$\Delta = a_1 b_2 (c_3 d_4 - c_4 d_3) - a_1 b_3 (c_2 d_4 - c_4 d_2) + a_1 b_4 (c_2 d_3 - c_3 d_2) \cdots$$

A-2. Expanding a 3 × 3 Determinant

The expansion of a 3 × 3 determinant can be accomplished without resorting to the 2 × 2 expansion just described. Consider the following steps in evaluating a 3 × 3 determinant:

Step 1. Original determinant:

$$\Delta = \begin{vmatrix} a_1 & b_1 & c_1 \\ a_2 & b_2 & c_2 \\ a_3 & b_3 & c_3 \end{vmatrix}$$

Step 2. Original determinant with columns 1 and 2 repeated:

$$\begin{vmatrix} a_1 & b_1 & c_1 & | & a_1 & b_1 \\ a_2 & b_2 & c_2 & | & a_2 & b_2 \\ a_3 & b_3 & c_3 & | & a_3 & b_3 \end{vmatrix}$$

Step 3.

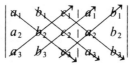

Step 1 shows the initial 3 × 3 determinant Δ. Step 2 shows the initial determinant with columns 1 and 2 placed, as shown, to the right of the initial determinant. Step 3 shows how to multiply the proper diagonal elements, as follows:

$$\Delta = a_1 b_2 c_3 + b_1 c_2 a_3 + c_1 a_2 b_3 - (a_3 b_2 c_1 + b_3 c_2 a_1 + c_3 a_2 b_1)$$

Note that the down-to-the-right products are positive quantities and the up-to-the-right products are negative quantities, as indicated in the diagram of Step 3. The a, b, and c elements can be real numbers, complex numbers, or algebraic quantities.

A-3. Operations on Determinants

Several operations may be performed on determinants without changing their values. Since these operations may expedite the evaluation of the determinants, some of the more useful are presented here.

1. All rows may be interchanged with all columns. Simply exchange column 1 and row 1, column 2 and row 2, and so on, keeping the elements in the same sequence in the new row as in the old column, and vice versa. For example,

$$\begin{vmatrix} a_1 & b_1 & c_1 \\ a_2 & b_2 & c_2 \\ a_3 & b_3 & c_3 \end{vmatrix} = + \begin{vmatrix} a_1 & a_2 & a_3 \\ b_1 & b_2 & b_3 \\ c_1 & c_2 & c_3 \end{vmatrix}$$

2. Any two columns may be interchanged with only a change in the sign of the determinant. For example,

$$\begin{vmatrix} a_1 & b_1 & c_1 \\ a_2 & b_2 & c_2 \\ a_3 & b_3 & c_3 \end{vmatrix} = - \begin{vmatrix} b_1 & a_1 & c_1 \\ b_2 & a_2 & c_2 \\ b_3 & a_3 & c_3 \end{vmatrix} = - \begin{vmatrix} c_1 & b_1 & a_1 \\ c_2 & b_2 & a_2 \\ c_3 & b_3 & a_3 \end{vmatrix}$$

3. All the elements of any column may be increased or decreased by the amount of the corresponding elements in any other column or by any fixed multiple of these respective amounts. For example,

$$\begin{vmatrix} a_1 & b_1 & c_1 \\ a_2 & b_2 & c_2 \\ a_3 & b_3 & c_3 \end{vmatrix} = \begin{vmatrix} (a_1 \pm kb_1) & b_1 & c_1 \\ (a_2 \pm kb_2) & b_2 & c_2 \\ (a_3 \pm kb_3) & b_3 & c_3 \end{vmatrix}$$

4. Any *factor* common to all the elements of any one column may be considered as a *factor of the determinant* and be moved to that position, or vice versa. For example,

$$\begin{vmatrix} a_1 & kb_1 & c_1 \\ a_2 & kb_2 & c_2 \\ a_3 & kb_3 & c_3 \end{vmatrix} = k \begin{vmatrix} a_1 & b_1 & c_1 \\ a_2 & b_2 & c_2 \\ a_3 & b_3 & c_3 \end{vmatrix} = \begin{vmatrix} ka_1 & b_1 & c_1 \\ ka_2 & b_2 & c_2 \\ ka_3 & b_3 & c_3 \end{vmatrix}$$

5. *As a corollary to the preceding rules, all operations applicable to columns are equally valid for rows.*

B / *Standard Resistance Values*

Since it would be far too costly to try to manufacture resistors with *all* possible resistance values, standard values have been adopted by the electrical industry. A list of standard resistor values with ±5 percent tolerance is given in Table B-1. Note how the resistance values are selected so that the 5 percent tolerance values tend to overlap. For instance select any two adjacent resistance values such as 1000 ohms and 910 ohms and note that

$$1000 - 910 = 90$$

is less than

$$1000 \times 0.05 + 910 \times 0.05 = 95.5$$

On the other hand $910 - 820 = 90$ while $0.05 \times 910 + 0.05 \times 820 = 86.5$ do not quite give an overlap in resistance values. This is considered acceptable.

If a table of 10 percent tolerance resistors is established then only one half of the values given in Table B-1 are required to give the necessary overlap. Similar tables can be established for 20 or 1 percent tolerances.

TABLE B-1

STANDARD NOMINAL RESISTANCE VALUES 5% TOLERANCE

Ohms										Megohms				
1.0	3.6	12	43	150	510	1800	5600	18000	56000	0.24	0.39	1.3	3.9	10.0
1.1	3.9	13	47	160	560	2000	6200	20000	62000	0.27	0.43	1.5	4.3	11.0
1.2	4.3	15	51	180	620	2200	6800	22000	68000	0.30	0.47	1.6	4.7	12.0
1.3	4.7	16	56	200	680	2400	7500	24000	75000	0.33	0.51	1.8	5.1	13.0
1.5	5.1	18	62	220	750	2700	8200	27000	82000	0.36	0.56	2.0	5.6	15.0
1.6	5.6	20	68	240	820	3000	9100	30000	91000		0.62	2.2	6.2	16.0
1.8	6.2	22	75	270	910	3300	10000	33000	100000		0.68	2.4	6.8	18.0
2.0	6.8	24	82	300	1000	3600	11000	36000	110000		0.75	2.7	7.5	20.0
2.2	7.5	27	91	330	1100	3900	12000	39000	120000		0.82	3.0	8.2	22.0
2.4	8.2	30	100	360	1200	4300	13000	43000	130000		0.91	3.3	9.1	
2.7	9.1	33	110	390	1300	4700	15000	47000	150000		1.0	3.6		
3.0	10	36	120	430	1500	5100	16000	51000	160000		1.1			
3.3	11	39	130	470	1600				180000		1.2			
									200000					
									220000					

C / *Voltmeters and*

Ammeters

Since voltage and current are the most important of all circuit variables, it is only natural that voltmeters and ammeters are by far the most widely used instruments to meter electrical phenomena. There are three basic electromechanical movements in common usage today. They are: 1) the D'Arsonval movement, which operates on direct current (d-c) only, is the most commonly used movement in both panel and electronic voltmeters and ammeters, 2) the iron vane meter movement, which is rugged and inexpensive, is used in many alternating current (a-c) panel or bench type voltmeters or ammeters, and 3) the electrodynamometer movement, which is a rather expensive and insensitive movement, can be used as a voltmeter, or an ammeter, but has a more important usage as a wattmeter. Although the iron vane movement is perhaps the cheapest of the three afore mentioned meter movements, the D'Arsonval movement has the advantages of being more sensitive more linear and more versatile than the iron vane movement. Although the D'Arsonval movement inherently measures direct currents and voltages, it can easily be converted into an accurate a-c voltmeter or ammeter by using a rectifier. Only the theory and application of the D'Arsonval movement will be discussed in this section. The interested reader can refer to one of many texts on instruments to learn more about the other meter movements.[1]

C-1. The D'Arsonval Meter

The D'Arsonval meter operates on the principle as set forth by Ampere's rule; namely, that a force is exerted on a current-carrying conductor which is situated in a magnetic field. This same law is the underlying principle that explains how all electric motors run.

It is a basic and fundamental law that a current-carrying conductor lying in a magnetic field tends to move at right angles to both the direction of the field and the direction of the current. Ampere's rule states this law mathematically as follows:

$$\mathbf{F}_m = l\mathbf{I} \times \mathbf{B} \qquad \text{Newtons} \qquad \text{(C-1)}$$

[1] See M. B. Stout, *Basic Electrical Measurements*, 2d ed. (New York: Prentice-Hall, 1960).

where $\mathbf{F}_m$ is the force exerted on a current carrying conductor of length l situated in a magnetic field whose flux density is $\mathbf{B}$ webers/sq. meter. As the cross product indicates, the force on the current-carrying conductor is always normal to the direction of motion of the charges (or current $\mathbf{I}$) and thus no work is done by the magnetic field. The end view of a coil of wire placed in a magnetic field between two north and south pole pieces is shown in Fig. C-1. The crossed lines indicate currents going into the coil

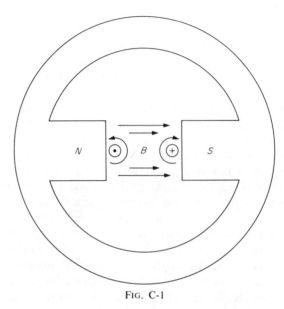

FIG. C-1

and the dots indicate current coming out of the coil. To determine the direction of motion of the current carrying conductors recall the right-hand rule in which the thumb of the right hand is placed in the direction of conventional current flow with the fingers indicating the magnetic flux lines generated by this flow of current. For the current flowing into the coil (into the page) the flux lines due to this current are additive on the top side of the conductors and subtractive on the lower side relative to the existing magnetic field. This means that the vector direction of the magnetic force on these conductors will be downward. By the same token, the magnetic force on the conductors which have the current flowing out of the page will be upward. The net effect is a clockwise rotation of the coil of wire shown in Fig. C-1. In a motor, the coil would be allowed to turn continuously, but in a D'Arsonval meter movement the coil is constrained to move less than 180°.

A simplified diagram of the basic D'Arsonval meter movement is shown in Fig. C-2. A coil of very fine wire is situated between the poles

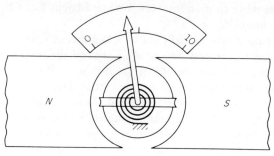

of a permanent magnet by means of two pivots supported on jewel bearings. Although the coil may be wound with or without a bobbin, a bobbin is generally used and is usually made of fiber, plastic, or aluminum. In order to improve the linearity of the meter movement the poles of the magnet are cylindrical in shape; in addition, a cylindrical soft iron core is placed inside of the bobbin. Since the bobbin is constrained to turn for less than 180°, it is easy to support the iron core without unduly constraining the movement of the bobbin. There are two definite advantages for making the pole faces and core cylindrical: 1) the length of the air gap is minimized, and 2) the flux between the pole faces and the core is for all practical purposes radial and because of the radial field the deflections of the coil or bobbin are directly proportional to the current in the coil (refer to Eq. C-1) so that a uniform, linear scale can be achieved. A pair of spiral springs are attached so that one end of the spring is affixed to the shaft of the bobbin and the other end to the main body of the meter movement. Recall that Hooke's law for a spring states that

$$F_s = K\theta \text{ Newtons} \qquad (C-2)$$

where F_s is the force stored in the spring for a given rotational angle θ of the bobbin or coil. When a current is passed through the coil of wire a magnetic force is exerted on the coil which causes rotation. The coil rotates until the restoring force F_s in the spring equals F_m, the magnetic force. If a needle or pointer is then attached to the bobbin of the coil, the deflection angle θ can be read, and of course θ is directly related to the current i passing through the coil. Equating Eqs. C-1 and C-2 results in

$$\theta = \frac{(Bl)}{K} I \text{ radians} \qquad (C-3)$$

The current from the external circuit into which the meter movement is placed is introduced to the coil by attaching lead-in wires to the spiral restraining springs. Although other points could be discussed in regard to the operation of the D'Arsonval movement the preceding discussion should be adequate to give the reader some insight as to how this simple meter movement operates.

The D'Arsonval Movement Used as a Voltmeter. As mentioned in the preceding sections, the coil wire wound around the bobbin is exceedingly fine in order to reduce the weight of the bobbin and thus produce a more sensitive meter movement. Because the wire is fine, only a limited amount of current can be passed through the coil without destroying it. Some of the more sensitive meter movements can cause a full scale deflection of the pointer needle with only 50 or less microamps passing through the moving coil. Actually, even more sensitive meter movements of the D'Arsonval type can be constructed. These are called D'Arsonval galvanometers and can produce a full scale deflection for currents as small as 10^{-9} amperes (1 nanoamp). These movements are quite fragile and cannot be moved about easily. For general usage typical D'Arsonval meter movements produce a full scale meter deflection for a current of from 0.2 to 50 milliamperes. This allows the meter movement to be of relatively rugged construction so that the meter can be carried about as a portable instrument without undue concern about damage caused by mechanical vibrations.

Let us assume that meter movements which produce a full scale deflection with 20 milliamps passing through the coil are available. It is shown in Fig. C-3 how this meter movement can be used to measure various voltages. For instance, suppose it is desired to convert the 20

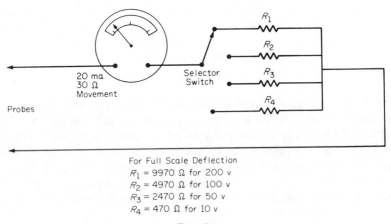

For Full Scale Deflection
$R_1 = 9970 \ \Omega$ for 200 v
$R_2 = 4970 \ \Omega$ for 100 v
$R_3 = 2470 \ \Omega$ for 50 v
$R_4 = 470 \ \Omega$ for 10 v

FIG. C-3

milliamp meter movement into a voltmeter capable of measuring up to 100 volts full scale. Assume that the resistance of the moving coil in the movement is 30 ohms. If 100 volts were applied directly across the coil of the meter movement, a current of $100/30 = 3.3$ amperes would try to flow through the coil. The result would be the almost immediate vaporization of a portion of the coil and the consequent destruction of the movement. One must keep in mind that the meter deflects to full scale when 20 milliamps flows through the coil. Thus, a resistor should be connected in series with the coil to limit the current when large voltages are to be measured. To create a hundred volt full scale voltmeter, simply note that 20 milliamps will flow through the coil if the total resistance between the voltmeter terminals is $100/20 \times 10^{-3} = 5000$ ohms. This means that a series resistor of $5000 - 30 = 4970$ ohms must be placed in series with the coil of the meter movement.

The reader should verify that a series resistor of 470 ohms is needed to convert the voltmeter to 10 volts full scale and 9970 ohms is needed to convert the meter so that it will read 200 volts on full scale deflection.

The D'Arsonval Movement Used as an Ammeter. The very same D'Arsonval movement can be used as either a voltmeter or an ammeter with a simple change in the circuitry. As described above, a set of series resistors is needed to convert the movement into a multirange voltmeter, but to convert the meter to a multirange ammeter, it is necessary to connect a set of parallel resistors with the movement. This is shown in Fig. C-4.

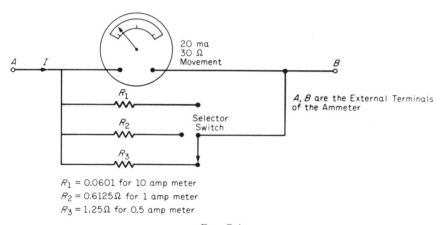

$R_1 = 0.0601$ for 10 amp meter
$R_2 = 0.6125\,\Omega$ for 1 amp meter
$R_3 = 1.25\,\Omega$ for 0.5 amp meter

Fig. C-4

Again keep in mind that the basic meter movement deflects to full scale when 20 milliamps flow through the coil. Suppose it is desired to convert the basic 20 milliammeter movement into an ammeter that will read full scale when 1 ampere flows into the ammeter. Since only 20 milliamperes can be allowed to flow through the coils the remaining 980 milliamperes must be shunted around the coils. This is accomplished by the use of parallel or shunt resistors. To calculate the value of the necessary shunt resistor, first calculate the value of the voltage drop across the coil and shunt resistor when 20 milliamps flow through the coil, or 20 $\times$

$10^{-3} \times 30 = 0.6$ volts. Since the same 0.6 volts also is applied across the shunt resistor, the necessary value of the shunt resistance will be 0.6 volts divided by .98 amps or 0.6125 ohms.

The reader should confirm that to convert the 20 milliamp movement into an ammeter that reads 500 milliamperes full scale or 10 amperes full scale a respective shunt resistor of 1.25 ohms and 0.061 ohm is required.

appendix

D / *Fourier Series Analysis*

The description of a function in terms of a series representation is a powerful tool in many mathematical analyses. The importance and usefulness of infinite series, power series, trigonometric series, and so on as aids in the solution of engineering problems cannot be overemphasized. Here, only the Fourier series, in trigonometric form, will be presented.

If a function $f(t)$ is known to be periodic, to have only a finite number of finite discontinuities, then the function can be represented by a Fourier series as follows:

$$f(t) = A_0 + a_1 \cos \omega t + a_2 \cos 2\omega t + \cdots + a_n \cos n\omega t + \cdots$$
$$+ b_1 \sin \omega t + b_2 \sin 2\omega t + \cdots + b_n \sin n\omega t + \cdots \quad \text{(D-1)}$$

or, written in summation form,

$$f(t) = A_0 + \sum_{n=1}^{\infty} (a_n \cos n\omega t + b_n \sin n\omega t) \quad \text{(D-2)}$$

where ω is called the fundamental frequency, 2ω the second harmonic, and $n\omega$ the nth harmonic.

The constants A_0, a_n, and b_n can be evaluated by either analytical or graphical means. The analytical method is more accurate and is usually preferable if a mathematical description for $f(t)$ is conveniently available. The analytical evaluation of the constants is achieved by using various trigonometric identities and mathematical manipulations.

To evaluate A_0, we integrate Eq. D-1 over a full cyclic period of 2π radians. Since ω is a constant, it is usually convenient to express $f(t)$ as $f(\omega t)$ so as to have uniform limits of integration, or

$$\int_0^{2\pi} f(\omega t)\, d\omega t = \int_0^{2\pi} A_0\, d\omega t + \int_0^{2\pi} a_1 \cos \omega t\, d\omega t$$

$$+ \cdots + \int_0^{2\pi} a_n \cos n\omega t\, d\omega t + \int_0^{2\pi} b_1 \sin \omega t\, d\omega t$$

$$+ \cdots + \int_0^{2\pi} b_n \sin n\omega t\, d\omega t \quad \text{(D-3)}$$

By inspection (or by consulting definite integral tables), it is evident that when all sine and cosine terms are integrated or averaged over a full cycle, the result is zero. The first term on the right-hand side of Eq. D-3 is the only term not equaling zero, so, upon integrating and solving for A_0,

$$A_0 = \frac{1}{2\pi} \int_0^{2\pi} f(\omega t)\, d\omega t \tag{D-4}$$

The evaluation of the a_n constants is facilitated by multiplying Eq. D-1 by $\cos n\omega t$ and then integrating over a full cycle, or

$$\int_0^{2\pi} f(\omega t) \cos n\omega t\, d\omega t = \int_0^{2\pi} A_0 \cos n\omega t\, d\omega t$$

$$+ \int_0^{2\pi} a_1 \cos \omega t \cos n\omega t\, d\omega t + \cdots + \int_0^{2\pi} a_n \cos^2 n\omega t\, d\omega t$$

$$+ \int_0^{2\pi} b_1 \sin \omega t \cos n\omega t\, d\omega t$$

$$+ \cdots + \int_0^{2\pi} b_n \sin n\omega t \cos n\omega t\, d\omega t \tag{D-5}$$

From definite integral tables we can find that

$$\int_0^{2\pi} \cos m\omega t \sin n\omega t\, d\omega t = 0 \qquad \text{for all } m \text{ and } n$$

Also

$$\int_0^{2\pi} \cos m\omega t \cos n\omega t\, d\omega t = \int_0^{2\pi} \sin m\omega t \sin n\omega t\, d\omega t$$

$$= 0 \qquad \text{for } m \neq n$$

$$= \pi \qquad \text{for } m = n$$

Therefore, all terms on the right-hand side of Eq. D-5 vanish except the terms involving $\cos^2 n\omega t$, or

$$\int_0^{2\pi} f(\omega t) \cos n\omega t\, d\omega t = a_n \pi$$

or

$$a_n = \frac{1}{\pi} \int_0^{2\pi} f(\omega t) \cos n\omega t\, d\omega t \tag{D-6}$$

The b_n constants can be similarly evaluated by multiplying both sides of Eq. D-1 by $\sin n\omega t$ and integrating over a full cycle. The result is

$$b_n = \frac{1}{\pi} \int_0^{2\pi} f(\omega t) \sin n\omega t\, d\omega t \tag{D-7}$$

D-1. The Square Wave

As an example of how to calculate the Fourier series coefficients, let us consider the square wave shown in Fig. D-1. Before proceeding with

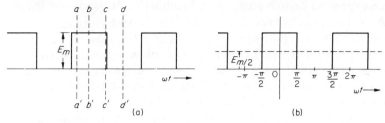

FIG. D-1. A square wave a) with a choice of *y* axes and b) with a particular choice of axes.

the calculations, several observations should be made. Since there often is freedom concerning the choice of a particular set of coordinate axes for the $f(\omega t)$ curve, it is important to make a wise choice. In Fig. D-1a, let us assume that the horizontal axis is fixed but that the *y* axis can be chosen anywhere. Four choices for a *y*-axis location are indicated in Fig. D-1a as line aa', bb', cc', or dd'. From an analytical standpoint, any choice for a *y* axis will produce an acceptable Fourier series representation of the square wave, but a judicious choice will simplify both the calculations for the Fourier series coefficients and the number of terms in the series itself.

Again, before making the actual calculations for the Fourier series coefficients, the following rules pertaining to symmetry should be studied.[1]

1. If a function $f(\omega t)$ has equal area above and below the *x* axis in a complete cycle, $A_0 = 0$.

2. If a function has half-wave symmetry such that $f(\omega t) = -f(\omega t + \pi)$, A_0 and all even harmonics are zero.

3. If a function has even symmetry such that $f(\omega t) = f(-\omega t)$, all sine terms equal zero, though A_0 will not necessarily be zero.

4. If a function has odd symmetry such that $f(\omega t) = -f(-\omega t)$, A_0 and all cosine terms equal zero.

5. If a function has both half-wave symmetry and quarter-wave symmetry $[f(\omega t) = f(\pi - \omega t)]$, the function is describable by a series of either odd cosine or odd sine terms.

6. Always check to see if a simple *x*-axis translation can be made such that there is symmetry as listed for rules 2, 4, and 5. If so, statements, 2, 4, and 5 are still valid except that $A_0 \neq 0$.

To continue with the problem of the square wave and the proper choice of a *y* axis, let us apply the preceding six rules concerning symmetry. The choice of aa' for a *y* axis is not desirable, because no symmetry exists about this axis. The same thing could be said about axis cc'. Both axes bb' and dd' do result in symmetry about the *y* axis and should

[1]For a full discussion of symmetry in relation to Fourier series, see Tang, K. Y., *Alternating Current Circuits*, 3d ed., International Textbook Company, Scranton, Pa., 1960, p. 468.

be considered as desirable choices. Arbitrarily, let us choose the y axis as bb'. (The choice of axes is reflected in Fig. D-1b.)

By inspection, it is evident that all of curve $f(\omega t)$ lies above the x axis, and so $A_0 \neq 0$. However, as indicated in rule 6, a simple x-axis translation (the dashed line) can be made such that the new $A_0 = 0$, and desirable symmetry becomes apparent. Applying each rule for this new x axis, we find that rules 1, 2, and 5 are pertinent. Thus, before we even begin the Fourier series coefficient calculations, a great deal is known about the series, with the result that unnecessary, tedious calculations are avoided.

The actual value of A_0 corresponding to the originally chosen x axis is

$$A_0 = \frac{1}{2\pi} \int_{-\pi/2}^{\pi/2} E_m \, d\omega t$$

$$= \frac{E_m}{2}$$

where $f(\omega t) = 0$ between $-\pi \leq \omega t \leq -\pi/2$ and $\pi/2 \leq \omega t \leq \pi$.

According to rules 3 and 5, only odd cosine terms should appear in the Fourier series. We can use either the original or the translated x axis to calculate the cosine coefficients, because simple x-axis translation does not affect the value of these coefficients. Choosing the original x axis because $f(\omega t)$ is zero over a portion of the cycle, we have

$$a_n = \frac{1}{\pi} \int_{-\pi/2}^{\pi/2} E_m \cos n\omega t \, d\omega t$$

$$= \frac{1}{\pi} \left(\frac{E_m}{n} \sin n\omega t \right) \Big|_{-\pi/2}^{\pi/2}$$

$$= \frac{2E_m}{\pi n} \sin \frac{n\pi}{2}$$

The Fourier series can be written as

$$f(\omega t) = \frac{E_m}{2} + \frac{2E_m}{\pi} \cos \omega t - \frac{2E_m}{3\pi} \cos 3\omega t + \frac{2E_m}{5\pi} \cos 5\omega t \cdots \qquad \text{(D-8)}$$

and, as anticipated, only odd cosine terms appear in the expansion. Equation D-8 can be written in summation form as

$$f(\omega t) = \frac{E_m}{2} + \frac{2E_m}{\pi} \sum_{n=1}^{\infty} \frac{\sin n\pi/2}{n} \cos n\omega t \qquad \text{(D-9)}$$

The Fourier series of two other important wave shapes will now be considered.

D-2. The Half-Wave Rectified Sine Wave

Figure D-2 shows a half-wave rectified sine wave with the y axis chosen to give even symmetry. Thus all sine terms in the Fourier series

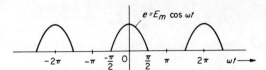

FIG. D-2. A half-wave rectified sine wave.

are zero. A_0 is evaluated as

$$A_0 = \frac{1}{2\pi} \int_{-\pi/2}^{\pi/2} E_m \cos \omega t \, d\omega t$$

$$= \frac{E_m}{\pi}$$

The a_n constants are

$$a_n = \frac{1}{\pi} \int_{-\pi/2}^{\pi/2} E_m \cos \omega t \cos n\omega t \, d\omega t$$

$$= \frac{E_m}{\pi} \left[\frac{\sin (n - 1) \omega t}{2(n - 1)} + \frac{\sin (n + 1) \omega t}{2(n + 1)} \right]\Big|_{-\pi/2}^{\pi/2} \qquad n \neq 1$$

$$= \frac{E_m}{\pi(n + 1)(n - 1)} \left\{ \left[(n + 1) \sin (n - 1)\frac{\pi}{2} \right] \right.$$

$$\left. + \left[(n - 1) \sin (n + 1)\frac{\pi}{2} \right] \right\} \qquad \text{(D-10a)}$$

or

$$a_n = \frac{2E_m}{\pi(n + 1)(n - 1)} \sin (n - 1)\frac{\pi}{2} \qquad \text{(D-10b)}$$

To evaluate a_1, we substitute $n = 1$ into Eq. D-10a, or

$$a_1 = \frac{1}{\pi} \int_{-\pi/2}^{\pi/2} E_m \cos^2 \omega t \, d\omega t$$

$$= \frac{E_m}{2}$$

Therefore, $f(\omega t)$ for a half-wave rectified sine wave, with axes chosen as in Fig. D-2, is

$$f(\omega t) = \frac{E_m}{\pi} + \frac{E_m}{2} \cos \omega t + \frac{2E_m}{\pi} \sum_{n \text{ even}} \frac{\sin [(n - 1)\pi/2] \cos n\omega t}{(n + 1)(n - 1)} \qquad \text{(D-11)}$$

D-3. Full-Wave Rectified Sine Wave

It can be shown that the Fourier series representation of a full-wave rectified sine wave, with coordinate axes chosen as shown in Fig. D-3, is

$$f(\omega t) = \frac{2E_m}{\pi} - \frac{4E_m}{\pi} \sum_{n \text{ even}} \frac{\cos n\omega t}{(n + 1)(n - 1)} \qquad \text{(D-12)}$$

and, as expected, all sine terms are zero because the function $f(\omega t)$ is even.

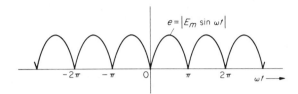

FIG. D-3. A full-wave rectified sine wave.

D-4. Reasons for Obtaining a Fourier Series

The Fourier series of a periodic wave shape should not simply be considered mathematical fiction or manipulation. The frequencies found in the square wave, rectified sine waves, and so forth, actually exist and can be measured by such devices as harmonic wave analyzers. One can also proceed in the opposite direction and construct almost any periodic function by properly combining sine and cosine functions.

The practical importance of the Fourier series is obvious if we consider the vast array of mathematical techniques that have been developed to handle d-c and sinusoidal functions in engineering problems. The filter problem in Chapter 15 illustrates how the performance of an electrical filter can be described in terms of conventional a-c circuit analysis, once the Fourier series representation of a nonsinusoidal function is known.

E / Power Relations in a Transistor Circuit Using Integral Calculus

The power relations in a transistor circuit as presented in Chapter 6 were developed without the need for integrals or differentials. An integral calculus approach will prove useful and informative and is presented here for the convenience of the reader who is familiar with calculus.

Figure E-1 is the basic common-emitter amplifier circuit. This circuit will be used to define the total power delivered to the transistor and to the

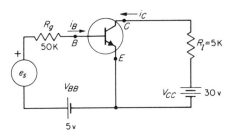

FIG. E-1. The basic common-emitter amplifier configuration used for power calculations.

resistances R_g and R_l, and the power delivered by the sources V_{BB}, V_{CC}, and e_s. The definition of instantaneous power in any portion of a circuit is the product of the appropriate instantaneous voltage and current, or

$$p = vi \qquad \text{(E-1)}$$

To find the total instantaneous power p_t delivered to the transistor the power input to the base p_b and collector p_c must be summed, or

$$p_t = p_b + p_c \qquad \text{(E-2)}$$

where the instantaneous collector power p_c is defined as

$$p_c = v_{CE} i_C \qquad \text{(E-3)}$$

and the instantaneous base power p_b is defined as

$$p_b = v_{BE} i_B \tag{E-4}$$

If $e_s = E_s \sqrt{2} \sin \omega t$, the equations for i_B, v_{BE}, i_C, and v_{CE} are given as follows: The base current i_B is

$$i_B = \frac{V_{BB} + e_s}{R_g}$$

where $R_g \gg R_{in}$; therefore

$$i_B = I_B + \frac{E_s \sqrt{2} \sin \omega t}{R_g}$$

$$= I_B + I_b \sqrt{2} \sin \omega t \tag{E-5}$$

The base-to-emitter voltage v_{BE} is

$$v_{BE} = V_{BB} + e_s - i_B R_g$$

$$= V_{BB} + E_s \sqrt{2} \sin \omega t - I_B R_g - I_b \sqrt{2} R_g \sin \omega t \tag{E-6}$$

The collector current i_C is

$$i_C = I_C + I_c \sqrt{2} \sin \omega t \tag{E-7}$$

The collector-to-emitter voltage v_{CE} is

$$v_{CE} = V_{CC} - i_C R_l = V_{CC} - (I_C + I_c \sqrt{2} \sin \omega t) R_l \tag{E-8}$$

Almost all electrical and electronic equipment is rated in terms of its allowable temperature rise. The temperature rise is usually caused by internal energy absorption, and the total energy is the product of average power and time, or

$$W = Pt \tag{E-9}$$

Thus an electronic device can just as easily be rated in terms of the allowable average power delivered to the device as in terms of the allowable temperature rise. For a repetitive input signal (such as a sinusoidal input) the average power is defined as

$$P = \frac{1}{2\pi} \int_0^{2\pi} p(\omega t) \, d\omega t$$

or

$$P = \frac{1}{2\pi} \int_0^{2\pi} v(\omega t) i(\omega t) \, d\omega t \tag{E-10}$$

The average base power P_B is

$$P_B = \frac{1}{2\pi} \int_0^{2\pi} v_{BE} i_B \, d\omega t$$

$$= \frac{1}{2\pi} \int_0^{2\pi} (V_{BB} + E_s \sqrt{2} \sin \omega t - I_B R_g - I_b \sqrt{2} R_g \sin \omega t)$$

$$(I_B + I_b \sqrt{2} \sin \omega t) \, d\omega t$$

$$= \frac{1}{2\pi} \int_0^{2\pi} [V_{BB}I_B - I_B^2 R_g + (I_B E_s \sqrt{2} - I_B I_b \sqrt{2} R_g + V_{BB} I_b \sqrt{2}$$

$$- I_B R_g I_b \sqrt{2}) \sin \omega t + 2(E_s I_b - I_b^2 R_g) \sin^2 \omega t] \, d\omega t \qquad \text{(E-11)}$$

The integration of Eq. E-11 is straightforward and, if necessary, can be looked up in any standard set of integral tables. After integrating and substituting in limits,

$$P_B = V_{BB}I_B - I_B^2 R_g + E_s I_b - I_b^2 R_g$$
$$= (V_{BB} - I_B R_g)I_B + (E_s - I_b R_g)I_b$$
$$= V_{BE}I_B + V_{be} I_b \qquad \text{(E-12)}$$

The first term on the right-hand side of Eq. E-12 ($V_{BE}I_B$) is the average or quiescent base power, and the second term ($V_{be}I_b$) is the a-c input power. The average collector power P_C is

$$P_C = \frac{1}{2\pi} \int_0^{2\pi} v_{CE} i_C d\omega t$$

$$= \frac{1}{2\pi} \int_0^{2\pi} [V_{CC} - (I_C + I_c \sqrt{2} \sin \omega t)R_l][I_C + I_c \sqrt{2} \sin \omega t] \, d\omega t$$

After cross-multiplying and collecting terms,

$$P_C = \frac{1}{2\pi} \int_0^{2\pi} [V_{CC}I_C - I_C^2 R_l + (-2I_C I_c R_l \sqrt{2} + V_{CC}I_c \sqrt{2}) \sin \omega t$$

$$- 2I_c^2 R_l \sin^2 \omega t] \, d\omega t \qquad \text{(E-13)}$$

and, after integrating and substituting in limits,

$$P_C = V_{CC}I_C - I_C^2 R_l - I_c^2 R_l$$
$$= (V_{CC} - I_C R_l)I_C - I_c^2 R_l$$
$$= V_{CE}I_C - I_c^2 R_l \qquad \text{(E-14)}$$

The first term on the right-hand side of Eq. E-14 ($V_{CE}I_C$) is the average or quiescent power supplied to the collector by the collector supply voltage V_{CC}. The second term ($I_c^2 R_l$) is the a-c power delivered to the load resistance R_l. The total average power P_T delivered to the transistor is

$$P_T = P_B + P_C$$
$$= V_{BE}I_B + V_{be}I_b + V_{CE}I_C - I_c^2 R_l \qquad \text{(E-15)}$$

where the no-signal or quiescent power P_{TQ} delivered to the transistor is

$$P_{TQ} = V_{BE}I_B + V_{CE}I_C \qquad \text{(E-16)}$$

If, in Eq. E-15, $I_c^2 R_l > V_{be}I_b$ (and it usually is), the transistor power with an a-c signal applied is less than with no a-c signal applied. Thus a transistor actually consumes less energy when a-c signals are being applied to the circuit. To obtain maximum transistor utilization, many electronic circuits designed for continuous a-c operation can take advantage of this fact, but it must be remembered that rated transistor dissipations can be exceeded if the transistor is operated under quiescent conditions.

Next let us consider the power delivered by the three energy sources e_s, V_{BB}, and V_{CC}. Again using Eq. E-10, the power delivered by the signal source e_s is

$$P_s = \frac{1}{2\pi} \int_0^{2\pi} e_s i_B \, d\omega t$$

$$= \frac{1}{2\pi} \int_0^{2\pi} E_s \sqrt{2} \sin \omega t (I_B + I_b \sqrt{2} \sin \omega t) \, d\omega t$$

and, after integrating and substituting in limits,

$$P_s = E_s I_b \tag{E-17}$$

The power delivered by the base-biasing source V_{BB} is

$$P_{BB} = \frac{1}{2\pi} \int_0^{2\pi} V_{BB} i_B \, d\omega t$$

$$= \frac{1}{2\pi} \int_0^{2\pi} V_{BB} (I_B + I_b \sqrt{2} \sin \omega t) \, d\omega t$$

and, integrating and substituting in limits,

$$P_{BB} = V_{BB} I_B \tag{E-18}$$

Finally, the power delivered by the collector source V_{CC} is

$$P_{CC} = \frac{1}{2\pi} \int_0^{2\pi} V_{CC} i_C \, d\omega t$$

$$= \frac{1}{2\pi} \int_0^{2\pi} V_{CC} (I_C + I_c \sqrt{2} \sin \omega t) \, d\omega t$$

or

$$P_{CC} = V_{CC} I_C \tag{E-19}$$

It should be noted that with certain biasing methods the collector source V_{CC} also supplies the quiescent base power.

A résumé of the power relations in the transistor circuit of Fig. E-1 follows:

The power delivered to the base-emitter portion of a transistor is

$$P_B = V_{BB} I_B + V_{be} I_b \tag{E-20}$$

The power delivered to the collector-emitter portion of a transistor is

$$P_C = V_{CE} I_C - I_c^2 R_l \tag{E-21}$$

The power delivered by the signal source e_s is

$$P_s = E_s I_b \tag{E-22}$$

The power delivered by the base-biasing source V_{BB} is

$$P_{BB} = V_{BB} I_B \tag{E-23}$$

The power delivered by the collector supply source V_{CC} is

$$P_{CC} = V_{CC} I_C \qquad \text{(E-24)}$$

An example will help to clarify and identify all these power relations.

EXAMPLE E-1. Figure E-2 is similar to the figure used in Example 6-2. The

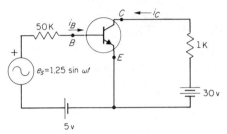

FIG. E-2. The circuit used in Example
E-1.

analysis used in Example 6-2 showed that i_B and i_C had the following values:

$$i_B = 100 + 50 \sin \omega t \qquad \mu a$$
$$i_C = 17 + 9 \sin \omega t \qquad ma$$

Find the average power delivered to the transistor and the average power delivered by the voltage sources

Solution: The power delivered to the base is

$$P_B = V_{BE} I_B + V_{be} I_b$$

where $V_{BE} = 0.66$ volt for $I_B = 100 \ \mu a$ and $V_{CE} = 13$ volts (see Fig. 6-4a) and

$$V_{be} = I_b R_{in} = I_b R_{ie}$$

where R_{ie} is the dynamic input resistance which was found, in an earlier calculation, to equal approximately 200 ohms. Therefore,

$$P_B = 0.66 \times 100 \times 10^{-6} + 200 \left(\frac{50 \times 10^{-6}}{\sqrt{2}} \right)^2$$

$$= 66 \times 10^{-6} + 0.25 \times 10^{-6} \ \text{watt}$$

$$\simeq 66 \ \mu\text{watts}$$

From this analysis it is evident that most of the small base power is delivered by the biasing supply.

The power delivered to the collector is

$$P_C = V_{CE} I_C - I_c^2 R_l$$

where

$$V_{CE} = V_{CC} - I_C R_l$$
$$= 30 - 17 \times 10^{-3} \times 1 \times 10^3 = 13 \ \text{volts}$$

which is the same as the value for V_{CE} read from the graphical characteristics of

Fig. 6-9b. Therefore,

$$P_C = 13 \times 17 \times 10^{-3} - \left(\frac{9}{\sqrt{2}} \times 10^{-3}\right)^2 1000$$

$$= 221 \times 10^{-3} - 40.5 \times 10^{-3} = 180.5 \, \text{mwatts}$$

The total power delivered to the transistor is the sum of the base and collector powers, or

$$P_T = P_B + P_C$$
$$= 0.066 \times 10^{-3} + 180.5 \times 10^{-3} = 180.566 \, \text{mwatts}$$

This illustrates a fact that is true in almost every common-emitter circuit — that the collector input power is much greater than the base input power. The power delivered by the signal source is

$$P_s = E_s I_b$$
$$= \frac{2.5}{\sqrt{2}} \times \frac{50}{\sqrt{2}} \times 10^{-6} = 0.0625 \, \text{mwatt}$$

The power delivered by the base supply is

$$P_{BB} = V_{BB} I_B$$
$$= 5 \times 100 \times 10^{-6} = 0.5 \, \text{mwatt}$$

The power delivered by the collector supply is

$$P_{CC} = V_{CC} I_C$$
$$= 30 \times 17 \times 10^{-3} = 510 \, \text{mwatts}$$

It is now evident that almost all the power delivered to a transistor common-emitter circuit comes from the collector power supply. Although the power requirements of the signal source e_s seem small (6.25×10^{-5} watt), it is important to stress that the signal source must be capable of delivering some power to the circuit. The reader may think that this power-delivering capability is of no importance until it is pointed out that some signals may contain less than a microwatt of power (10^{-6} watt); for example, radio signals that are received from our earth satellites or space vehicles fall in this category.

The reader may now ask why there is a discrepancy between the transistor collector power ($P_C = 180.5$ mwatts) and the collector supply power ($P_{CC} = 510$ mwatts). The difference between these two powers is contained in the load resistance R_l, or

$$P_{R_l} = P_{CC} - P_C$$
$$= 510 - 180 = 330 \, \text{mwatts}$$

As a check, we can calculate P_{R_l} from the basic definition of power, or

$$P_{R_l} = \frac{1}{2\pi} \int_0^{2\pi} (i_C)^2 R_l \, d\omega t$$

$$= \frac{1}{2\pi} \int_0^{2\pi} (I_C + I_c \sqrt{2} \sin \omega t)^2 \, R_l \, d\omega t$$

$$= \frac{1}{2\pi} \int_0^{2\pi} (I_C^2 + 2I_C I_c \sqrt{2} \sin \omega t + 2I_c^2 \sin^2 \omega t) \, R_l \, d\omega t$$

and, after integrating and substituting in limits,

$$P_{R_l} = (I_C^2 + I_c^2) \, R_l \tag{E-25}$$

Then, for Example 6-4,

$$P_R = \left[(17 \times 10^{-3})^2 + \left(\frac{9}{\sqrt{2}} \times 10^{-3} \right)^2 \right] 1000$$

$$= 330 \text{ mwatts} \qquad (check)$$

F | -3db Frequencies With and Without Feedback

The derivation of the relation between f_1 and f'_1, or f_2 and f'_2 (the -3db frequencies without and with feedback) for a R-C coupled amplifier can be obtained by referring to Fig. F-1 (which is the same as Fig. 21-5

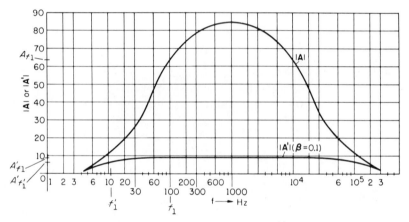

FIG. F-1. Figure 21-5 repeated with the $-3db$ frequencies identified.

with the -3db frequencies identified). In the development to follow it will be helpful to keep in mind that the form of the gain expression, both with and without feedback are identical. (Refer to Eqs. 13-43 and 13-45 along with Eq. 21-4.) First note that the lower -3db frequency f_1 is

$$A_{f_1} = \frac{A_{\text{mid}} \angle 45°}{\sqrt{2}} \qquad (F\text{-}1)$$

and A_{f_1} with feedback becomes A'_{f_1} or

$$A'_{f_1} = \frac{A_{f_1}}{1 - A_{f_1}\beta} = \frac{A'_{\text{mid}}}{1 - jf'_1/f_1} = \frac{A_{\text{mid}}}{1 - A_{\text{mid}}\beta} \qquad (F\text{-}2)$$

$$\frac{}{1 - jf'_1/f_1}$$

where the unprimed and primed quantities refer to the amplifier without and with feedback β respectively. Combining Eqs. F-1, and F-2 obtains

$$\frac{\dfrac{A_{mid}\,\underline{/45^\circ}}{\sqrt{2}}}{1 - \dfrac{A_{mid}\beta\,\underline{/45^\circ}}{\sqrt{2}}} = \frac{A_{mid}}{(1 - A_{mid}\beta)\left(1 - j\dfrac{f_1'}{f_1}\right)} \qquad \text{(F-3)}$$

and since $1\underline{/45^\circ} = (1/\sqrt{2})(1 + j1)$ Eq. F-3 becomes

$$\frac{\dfrac{(1 + j1)}{2}}{1 - \dfrac{A_{mid}\beta}{2}(1 + j1)} = \frac{1}{(1 - A_{mid}\beta)\left(1 - j\dfrac{f_1'}{f_1}\right)} \qquad \text{(F-4)}$$

Using some algebraic manipulations results in

$$\frac{1}{1 - j\dfrac{1}{1 - A_{mid}\beta}} = \frac{1}{1 - j\dfrac{f_1'}{f_1}} \qquad \text{(F-5)}$$

For Eq. F-5 to be satisfied requires that

$$f_1' = f_1/(1 - A_{mid}\beta) \qquad \text{(F-6)}$$

which is the desired relationship.

It is left as an exercise for the reader to show that

$$A_{f_2}' = \frac{A_{f_2}}{1 - A_{f_2}\beta} = \frac{A_{mid}'}{1 - j\dfrac{f_2}{f_2'}} = \frac{A_{mid}/1 - A_{mid}\beta}{1 - j\dfrac{f_2}{f_2'}} \qquad \text{(F-7)}$$

which results in the relationship

$$f_2' = f_2(1 - A_{mid}\beta) \qquad \text{(F-8)}$$

appendix

G / *The Inductance and Capacitance Parameters*

The three fundamental electrical circuit parameters are resistance, *R*, inductance, *L*, and capacitance, *C*. The resistance parameter and its functional relationship in conjunction with Ohm's law were discussed in some depth in Chapters 1 and 2. The two additional circuit parameters of inductance and capacitance were mentioned within the first eleven chapters but their functional "Ohm's law" relationship was not described in detail. The purpose of this appendix is to describe the meaning of the terms inductance and capacitance and to give their usual electrical circuit voltage-current relationships.

G-1. Inductance

Just as Ohm performed experiments with conductors, Michael Faraday and Joseph Henry performed experiments with a coil of conductive material, such as a coil of copper wire. They observed that if the voltage which was applied to the coil terminals was suddenly changed, the current did not immediately follow this change. If the same coil of wire was uncoiled to form a long straight wire, the current then responded instantly to any voltage change at the terminals, just as in the case of Ohm's experiment. What was the nature of this newfound phenomenon that caused a coil of wire to exhibit a voltage-current relation different from that in a straight piece of wire? To explain his observations, Faraday postulated the concept of a magnetic flux. Today, it is readily accepted that a current-carrying conductor generates a magnetic field about the wire and that if the wire is wound into a coil, the magnetic-flux lines generated by one of the turns link all or a portion of the other turns of the coil. This reinforcing of the magnetic-flux lines is what causes a different behavior of the voltage and current in a straight piece of wire as compared to a coil of wire. Actually, even with a straight wire the current does not immediately follow a sudden change in voltage, but Faraday could not observe this, for the time difference between the voltage excitation and the resultant current response was too small for his crude instruments to detect.

From his experiments, Faraday deduced that the current through the coil did not immediately follow the variations in the applied voltage because the magnetic-flux lines induced a voltage in the coil. This induced voltage tended to oppose the applied voltage which, of course, delayed the current variation. Faraday also observed that if a coil of wire was moved through a fixed magnetic field, a voltage was induced in the wire. From these observations, he postulated his now-famous law of magnetic induction, which is called Faraday's law. This law states that an induced voltage due to magnetic induction is equal to the rate of change of flux linkages, or

$$v = \frac{dN\phi}{dt} \tag{G-1}$$

where N is the number of turns in a coil of wire, and ϕ is the magnetic flux linking the turns. For a coil of N turns, Eq. G-1 reduces to

$$v = N\frac{d\phi}{dt} \tag{G-2}$$

Figure G-1 illustrates the proper voltage, current, and flux-linkage relationships which correspond to Eq. G-2. The interpretation is quite simple. If one assumes that a voltage drop v is suddenly applied to the terminals of a coil, as shown in Fig. G-1, then a current i will begin to

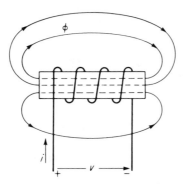

FIG. G-1.   The voltage-current flux-linkage relationships corresponding to $v = N(d\phi/dt)$.

flow, as indicated on the diagram. The right-hand rule can be used to show the proper direction of the magnetic flux ϕ. Since the flux is increasing, $d\phi/dt$ is positive, and Eq. G-2 is satisfied.

Since it is known that current generates a magnetic flux, it is possible to observe that

$$N\phi = Li \tag{G-3}$$

where the flux linkages $N\phi$ are proportional to the current i. Just as Ohm's law experimentally verified that voltage and current in a resistive circuit are related by a constant R, Eq. G-3 states that flux linkage and current are related by a constant L. The constant L is called the *inductance* of the coil of wire; it is a measure of the ability of a coil to generate magnetic flux. The physical coil is called an inductor. The dimensions of L are flux linkage per ampere or weber-turns per ampere. One weber-turn per ampere is commonly called a *henry*, named in honor of Joseph Henry, who also did pioneer work in magnetic phenomena and was a contemporary of Faraday.

Substituting Eq. G-3 into Eq. G-1 results in

$$v = \frac{dLi}{dt}$$

$$= L \frac{di}{dt} \tag{G-4}$$

if L can be assumed to be constant. Figure G-2 shows the symbol for an inductor or the inductance parameter (⎍⎍⎍) and the proper

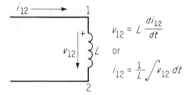

FIG. G-2. The descriptive symbol for the inductance parameter, L, and the usual voltage-current relationships.

voltage-current relationships in order for L to be a positive constant in Eq. G-4.

By equating Eq. G-2 and Eq. G-4, one obtains

$$L = N \frac{d\phi}{di} \quad \text{henry} \tag{G-5}$$

which can also be obtained from Eq. G-3 by taking the derivative of the equation with respect to the current i.

G-2. Capacitance

The discovery that electric energy could be stored between 2 isolated conductors preceded by almost 100 years the formulation of Ohm's law for resistors and Faraday's law for inductors. As will be recalled electric energy is dissipated, but not stored, in a resistor. Energy is stored in the magnetic field of an inductor, and, in order to create the magnetic field, a flow of charges (current) is required. Electric energy is stored in the electric field of a capacitor, and the electric field is created by the deposition of static electric charges on two isolated surfaces.

Experimental evidence shows that the charge q which is deposited on a set of capacitor plates is proportional to the voltage difference v between the plates, or

$$q = Cv \qquad \text{(G-6)}$$

The capacitance C is thus seen to be the measure of the ability of a capacitor to store charge. Taking the time derivative of both sides of Eq. G-6, one obtains

$$\frac{dq}{dt} = \frac{dCv}{dt} \qquad \text{(G-7)}$$

If the capacitance C is considered to be a constant, and if i is substituted for dq/dt, Eq. G-7 becomes

$$i = C\frac{dv}{dt} \qquad \text{(G-8)}$$

Since C can usually be considered a constant for a given capacitor, Eq. G-8 is the basic equation that describes the voltage-current characteristics of a capacitor with the capacitance parameter C. *In honor of Faraday, the unit for capacitance is named the farad.*

Figure G-3 shows the symbol for a capacitor or the capacitance parameter (——‖———) and the proper voltage-current relationships in order for C to be a positive constant in Eq. G-8.

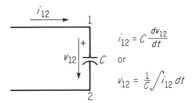

Fig. G-3. The descriptive symbol for the capacitance parameter C, and the usual voltage-current relationships.

Index